Discrete Variable Methods in Ordinary Differential Equations

Peter Henrici
Professor of Mathematics
University of California, Los Angeles

DISCRETE VARIABLE METHODS IN ORDINARY DIFFERENTIAL EQUATIONS

JOHN WILEY & SONS, INC., NEW YORK · LONDON

Library of Congress Catalog Card Number: 61-17359

Printed in the United States of America

to Eleonore

Preface

This book is intended both as a teaching tool on the senior level and as an up-to-date presentation of significant theoretical results. Reflecting my attitude toward teaching numerical analysis, the elementary parts of the book stress basic concepts and questions of intrinsic interest rather than cookbook-like enumeration of specialized techniques. Much of the advanced material in the book, in particular in the field of error propagation, is new.

The book originated in lecture notes for a course entitled "Numerical Methods in Differential Equations" which I have taught repeatedly to mathematicians and mathematically inclined engineers and physicists. The subject matter covered in this course is roughly equivalent to Chapters 1 and 2 and the first half of Chapter 5 of the present book. The prerequisites for this material are a first course in differential equations, some advanced calculus, and a smattering of machine computation. (At the University of California, Los Angeles, the latter can be acquired by taking a one-unit lower division course in coding.) Chapters 3, 4, and 7 require some knowledge of matrices and linear algebra. Some parts of Chapters 5 and 6 presuppose a working knowledge of elementary complex variable theory. Apart from these basic prerequisites the book is self-contained.

The problems found at the end of each chapter are mainly theoretical. It is clear that a student who wishes to comprehend fully the algorithms developed in the book ought to apply them to concrete practical problems. However, the size of a problem which the student can reasonably be expected to work depends largely on the computing equipment at his disposal. For this reason the selection of numerical problems has, with some exceptions, been left to the instructor.

Only a few of the many influences that have shaped my view of the subject can be mentioned here. The chapters on multistep methods owe much of their present form to two important papers by Dahlquist [1956,

1959]. The sections on numerical stability are based on Rutishauser's work [1952], and those on the Runge-Kutta method on Gill's [1951]. For matters of notation, computational know-how, and historical references I have relied heavily on Collatz [1960].

In addition, I am indebted to many friends for help and encouragement. My colleagues T. Ferguson, J. W. Green, P. Hoel, M. R. Hestenes, and C. B. Tompkins have, among others, commented on various parts of early versions of the manuscript and suggested improvements. So did several of my students, notably M. S. Lynn and W. P. Timlake. Thanks are also due to a number of mathematicians at Space Technology Laboratories, Inc., for helpful discussions, and to the Missile and Space Division of the Lockheed Aircraft Corporation for sponsoring the numerical experiments discussed in Chapter 6. Special thanks are due to G. Birkhoff and G. E. Forsythe, who took an early and continuing interest in the manuscript, and to Jim Douglas, Jr., who read the entire manuscript and proposed many substantial improvements. Recognition is due to the Office of Ordnance Research, U.S. Army, and to the Office of Naval Research, U.S. Navy, for partial support of the work on the book. Finally, I wish to thank Ruthanne Clark, Eva Lauer, and Eleonore Henrici for carefully typing the final manuscript as well as several of its preliminary versions, and Beatrice Carson, of John Wiley & Sons, for her splendid editorial work.

P. HENRICI

Los Angeles, Calif.
September 1961

Contents

Introduction

0.1. Definitions. Classification of Problems

Let $-\infty < a \leq b < \infty$. We denote by $[a, b]$ the set of all real numbers x such that $a \leq x \leq b$. If $-\infty \leq a \leq b \leq \infty$, we denote by (a, b) the set of all x such that $a < x < b$. The symbols $(a, b]$ and $[a, b)$ are used in a similar manner. Let p be an integer ≥ 1, let $a < b$, and let the real-valued function $F(x, y_0, y_1, \cdots, y_p)$ be defined for $x \in [a, b]$ and for $(y_0, y_1, \cdots, y_p) \in D$, where D is a domain in the real Euclidean $(p + 1)$-space. The equation

$$F(x, y, y', \cdots, y^{(p)}) = 0 \tag{0-1}$$

is called a *differential equation of order p*. A function $y(x)$ defined and p times differentiable on a subinterval I of $[a, b]$ such that the point $(y(x), y'(x), \cdots, y^{(p)}(x))$ is in D for $x \in I$ and that

$$F(x, y(x), y'(x), \cdots, y^{(p)}(x)) = 0, \qquad x \in I \tag{0-2}$$

is called a *solution* of the differential equation. It is well known from elementary examples that a given differential equation may have many solutions. For instance, if $p = 1$ and $F(x, y_0, y_1) = y_0 - y_1$, then every function $y(x) = Ce^x$ ($C =$ const.) is a solution. In order to pin down a solution of a given differential equation, it is usually necessary to specify some additional properties of it, such as its value and the values of some of its derivatives at specified points. It turns out that for an equation of order p just p conditions are needed in general to fix the solution. An important special case is represented by the conditions

$$\begin{aligned} y(a) &= \eta_0 \\ y'(a) &= \eta_1 \\ &\cdots\cdots\cdots\cdots \\ y^{(p-1)}(a) &= \eta_{p-1} \end{aligned} \tag{0-3}$$

where $\eta_0, \eta_1, \cdots, \eta_{p-1}$ are given constants. The problem of determining $y(x)$ such that (0-2) and (0-3) are satisfied is called an *initial value problem.*

A *boundary value problem* is one in which the function $y(x)$ and/or some of its derivatives are specified at several distinct points.

The simplest initial value problem arises for $p = 1$. It is usually assumed that (0-1) has been solved for the derivative of highest order. The first order initial value problem then appears in the form

$$y' = f(x, y), \qquad y(a) = \eta \tag{0-4}$$

where η is a constant. In order to avoid cumbersome discussions, we shall always assume that the function $f(x, y)$ is defined for $x \in [a, b]$ and for all finite y. Initial value problems of this type are studied in Chapters 1, 2, and 5. Initial value problems for first order systems of differential equations are considered in Chapter 3, and initial value problems for equations of order >1 in Chapters 4 and 6. Chapter 7 deals with boundary value problems for second order equations.

0.2. Necessity of Numerical Methods for Solving Differential Equations

The mathematical existence of a solution of an initial value problem such as (0-4) can be established under fairly general conditions on the function $f(x, y)$. A review of the basic existence theorem is given in Chapter 1, and for systems of equations in Chapter 3. What is wanted in many applications of differential equations is not only a proof of the mathematical existence of a solution—this can frequently be inferred from extramathematical considerations—but also the (approximate) numerical values of the solution for some specified range of values of the independent variable. In this respect the situation is less satisfactory. For some practically important, but nevertheless rather special, classes of differential equations the solutions can be represented in closed form, i.e., as finite combinations of elementary functions such as polynomials, exponential functions, and logarithms, and of indefinite integrals of such functions. Many other differential equations, on the other hand, cannot be solved in this manner. It has been proved that even relatively innocent-looking equations such as

$$y' = x^2 + y^2 \qquad \text{or} \qquad y'' = 6y^2 + x$$

cannot be solved in terms of elementary functions. Indeed, in spite of excellent collections of explicit solutions (see Kamke [1943]), it is probably safe to say that most differential equations cannot be solved explicitly. It also should not be overlooked that even in cases where an explicit solution exists the problem of finding its numerical values may not be entirely trivial. This is, to some extent, even true of the trivial initial value problem $y' = y$, $y(0) = 1$, where, in order to find numerical values of the solution, one has either to calculate or to look up in a table and possibly interpolate values of e^x. Another example is the solution $y(x) = e^{-x^2}\int_0^x e^{t^2}\,dt$ of the equation $y' = 1 - 2xy$; in order to determine values of $y(x)$ one has to calculate an integral that is not expressible in terms of elementary functions and that is not adequately tabulated. In the face of the obvious limitations of explicit solutions, mathematicians have since the early times of analysis tackled problems in differential equations by approximate methods of wider applicability. Two time-honored examples are the method of series expansions and the Picard-Lindelöf iteration method.

The study of series solutions of differential equations has attracted some of the best mathematical minds and has led to important developments in the theory of special analytic functions. The situations in which a differential equation can be expected to have a simple series solution are fairly well understood today. They too form a fairly restricted class. For instance, although the differential equation

$$y'' + \frac{A}{x}y' + \left(\frac{B}{x^2} + C + Dx^2\right)y = 0$$

(A, B, C, D = const.) can be solved elegantly in terms of certain confluent hypergeometric series, no such simple solution exists if the term Dx^2 is replaced by Dx^3. Even if a series solution exists, its convergence may be slow for most values of the independent variable. In a nonlinear differential equation the determination of all but the very first few coefficients of a series solution is usually a difficult undertaking. For equations with jump discontinuities the situation is even worse.

The method of successive approximations is an important tool in the theoretical study of (partial as well as ordinary) differential equations. Its value as an effective means for the numerical calculation of the solution, however, has frequently been exaggerated. The author knows of no instance of a numerical solution by successive approximations of a differential equation that could not have been solved much more easily by some other method.

0.3. Discrete Variable Methods

In this volume we shall deal with those methods for approximating solutions of ordinary differential equations that are based on the principle of *discretization.* These methods have the common feature that no attempt is made to approximate the exact solution $y(x)$ over a continuous range of the independent variable. Approximate values are sought only on a set of discrete points $x_0, x_1, x_2, \cdots$. Frequently, though not always, the points x_n are equidistant. If they are, we write $x_n = a + nh$, $n = 0, 1, 2, \cdots$. The quantity h is then called the *stepsize, stepwidth,* or simply the *step* of the method. Generally speaking, a discrete variable method for solving a differential equation consists in an algorithm which, corresponding to each lattice point x_n, furnishes a number y_n which is to be regarded as an approximation to the value $y(x_n)$ of the exact solution at the point x_n. One of the main questions we shall be concerned with is the size of the quantity $e_n = y_n - y(x_n)$, to be called the *discretization error,* in particular as a function of the step h.

Contrary to some of the methods mentioned earlier, discrete variable methods enjoy the advantage of being almost universally applicable. In the case of the initial value problem (0-4), for example, the only requirement for the applicability of most discrete variable methods is the ability to calculate a good approximation to the value of $f(x, y)$ for a given x and y.* It is true that the function f may have to be evaluated a large number of times in order to keep the discretization error sufficiently small. Once a deterrent to the application of discrete variable methods, this fact causes no concern today, when large numerical calculations, especially if of a repetitive nature, can be performed efficiently and reliably on automatic digital computers.

Among the discrete variable methods for the solution of initial value problems we can distinguish between *one-step methods* and *multistep methods.* In a one-step method the value of y_{n+1} can be found if only y_n is known; no knowledge or remembrance of any of the preceding values $y_{n-1}, y_{n-2}, \cdots$ is required. In a multistep method the calculation of y_{n+1} requires the explicit knowledge not only of y_n but also of a certain number of preceding values $y_{n-1}, y_{n-2}, \cdots$. A method is called a *k-step method* if the values $y_n, y_{n-1}, \cdots, y_{n+1-k}$ are required for the calculation of y_{n+1}.

In one-step methods the starting point does not play a special role. In fact, every lattice point can be considered a new starting point. This makes it easy to change the stepsize h. Multistep methods require a special starting procedure, since at the points $x_0, x_1, \cdots, x_{k-1}$ some of the values y_{n-j} required for the computation are missing, and a special computation may also be necessary at points where the stepsize is changed. All this tends to increase the length and complexity of the required machine codes. On the other hand, multistep methods may be more

* In some cases, e.g., when $f(x, y)$ is given empirically or by a numerical table, this may be a nontrivial requirement.

accurate than one-step methods. Further relative merits of the two types of methods are discussed later. Since the underlying theory is different in the two cases, the two classes of methods have to be treated separately. One-step methods are considered in Chapters 1 through 4, multistep methods in Chapters 5 through 7.

Notes

§0.2. The methods of series expansion and of successive approximation are treated in most books on differential equations. Their use for numerical purposes is discussed, for example, by Levy and Baggott [1934], Milne [1953], Warga [1953], Cernysenko [1958], Franklin [1952].

part I ONE-STEP METHODS FOR INITIAL VALUE PROBLEMS

chapter

1 Euler's Method for a Single Equation of the First Order

Euler's method (see Euler [1913], p. 422; Euler [1914], p. 271) is the simplest not only of all one-step methods but of all methods for the approximate solution of initial value problems. The use of Euler's method in practical numerical problems is not recommended, because its accuracy is very limited. We shall nevertheless study Euler's method in considerable detail, because it exhibits very clearly certain features which are peculiar also to more complicated methods. In particular, the basic phenomena of *discretization error* and *round-off error* are present here and can be studied easily because of the analytic simplicity of the method. A constant step h is assumed throughout the chapter.

1.1. Introduction

1.1-1. Definition of Euler's method. In Euler's method the values y_n are calculated recursively according to the following formulas:

(1-1*a*) $$y_0 = y(a) = \eta$$

(1-1*b*) $$y_{n+1} = y_n + hf(x_n, y_n), \qquad n = 0, 1, 2, \cdots$$

These formulas lend themselves to an obvious geometric interpretation. We look at the differential equation $y' = f(x, y)$ as an equation defining a field of directions in the strip $a \le x \le b$ of the (x, y)-plane. The problem of solving the differential equation is then geometrically equivalent to the problem of determining a curve $y = y(x)$ passing through the given initial point (x_0, y_0) and having its slope at each point agree with the slope prescribed by the direction field. The points (x_n, y_n) defined by (1-1) may be considered as the vertexes of a polygonal graph which passes through the correct initial point and has the property that each link has the direction prescribed by the direction field at its left end point.

Formulas (1-1) also admit of several analytical interpretations.

(i) If we approximate the derivative appearing in the differential equation at the point (x_n, y_n) by a divided forward difference, we obtain

$$\frac{y_{n+1} - y_n}{h} = f(x_n, y_n)$$

Solving for y_{n+1} yields (1-1*b*).

(ii) Integrating the identity $y'(t) = f(t, y(t))$ between the limits x and $x + k$, we obtain

$$y(x + k) - y(x) = \int_x^{x+k} f(t, y(t))\, dt$$

In particular, if $x = x_n$ and $k = h$,

$$y(x_{n+1}) - y(x_n) = \int_{x_n}^{x_{n+1}} f(t, y(t))\, dt$$

Approximating the integral by a crude rule for numerical integration (length of interval times value of integrand at left end point) and identifying $y(x_n)$ with y_n, we again obtain (1-1).

(iii) We finally may assume the possibility of expanding the solution in a *Taylor series* around the point x_n:

$$y(x_n + h) = y(x_n) + hf(x_n, y(x_n)) + \tfrac{1}{2}h^2 y''(x_n) + \cdots$$

Formulas (1-1) are the result of truncating this series after the linear term in h.

Each of these interpretations points the way to a class of generalizations of Euler's method which is discussed in later chapters. It is interesting to note that the generalization indicated by (i) (*numerical differentiation*), which seems to be the most straightforward, has proved to be the least fruitful of the three.

1.1-2. Computational considerations. Not because of the value of the method as a computational procedure, but only in order to facilitate comparison with more complicated procedures to be studied later, we present in Fig. 1.1 a flow diagram, in conventional notation, for the solution of the initial value problem (0-4) by Euler's method on a digital computer.

For the same reason we list in Table 1.1 the first few values in the numerical integration of the initial value problem

$$y' = x - y^2, \qquad y(0) = 0 \tag{1-2}$$

by Euler's method, using the step $h = 0.1$. The same problem is treated in §2.1-2 and §5.1-2 by more advanced methods. Arrows indicate the order in which the values are computed.

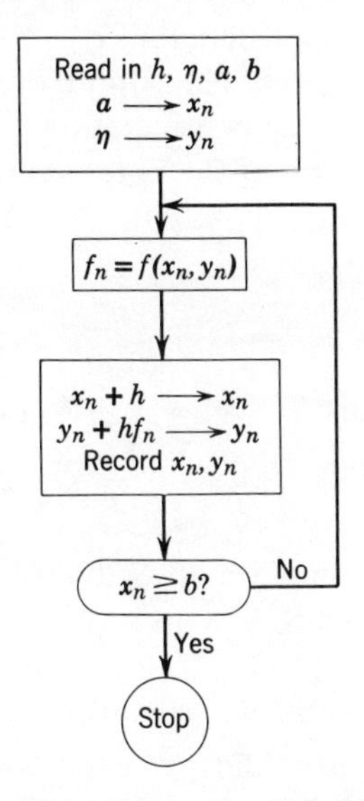

Fig. 1.1. Flow diagram for Euler's method.

In any numerical method for solving differential equations the bulk of the numerical work is likely to be spent in evaluating the function $f(x, y)$ for different values of its arguments. It has therefore become customary

Table 1.1

Numerical Illustration for Euler's Method

x_n	y_n	$f(x_n, y_n)$
0	0	0
0.1	0	0.10000
0.2	0.01000	0.19990
0.3	0.02999	0.29910
0.4	0.05990	

to measure the computational work required for any method by the number of times the function $f(x, y)$ has to be evaluated per integration step. We shall call this number the *specific number of substitutions*. For Euler's method this number is clearly 1.

1.1-3. An analytical example. If $f(x, y)$ is a sufficiently simple function, it may be possible to solve the recurrence relations (1-1) for y_n and to find an explicit expression for y_n as a function of n and h. Such an explicit solution is rarely of practical interest, because it can usually only be found in cases where the differential equation itself can be solved in closed form. However, it can be helpful in the study of theoretical properties of the method under consideration.

Let us find the explicit form of the Euler approximation to the solution of the initial value problem $y' = y$, $y(0) = 1$. Here we have $f(x, y) = y$, hence (1-1b) reduces to

$$y_{n+1} = y_n + hy_n = (1 + h)y_n, \qquad n = 0, 1, \cdots$$

In view of $y_0 = 1$ we thus find $y_1 = 1 + h$, $y_2 = (1 + h)y_1 = (1 + h)^2$, and generally

$$y_n = (1 + h)^n, \qquad n = 1, 2, \cdots$$

Since $n = x/h$, the value approximating the solution at the point $x_n = x$ is thus given by the formula

$$y_n = (1 + h)^{x/h} = [(1 + h)^{1/h}]^x$$

As $h \to 0$ this tends to e^x according to a well-known theorem of calculus (Taylor [1955], p. 74). We have thus shown that by decreasing the mesh size the exact solution of the initial value problem can be approximated arbitrarily well in the special example under consideration.

1.1-4. The error of a numerical method. The error of a numerical step-by-step method for solving a differential equation is due to two sources. The first source is the fact that the numbers y_n furnished by the method with a fixed step h, even if calculated to an infinite number of decimal places, will only rarely agree with the corresponding values $y(x_n)$ of the true solution. This discrepancy must be expected, since a numerical algorithm is by necessity finite in nature, whereas the solutions of even very simple differential equations such as $y' = y$ are transcendental functions and cannot therefore be calculated with a finite number of rational operations. The difference

$$e_n = y_n - y(x_n)$$

where y_n denotes the exact number produced by the algorithm, will be called the *discretization error*. The term *truncation error* is also frequently used.

The second source of error is the fact that in most cases the numbers y_n cannot be calculated with infinite precision because of the limited accuracy of any computing equipment. We shall denote by $\tilde{y}_n$ the values that are actually calculated in place of the y_n. The difference

$$r_n = \tilde{y}_n - y_n$$

will be called the *round-off error*.

For the *total error* $\tilde{y}_n - y(x_n)$ we find by the triangle inequality

$$|\tilde{y}_n - y(x_n)| = |(\tilde{y}_n - y_n) + (y_n - y(x_n))| \leq |e_n| + |r_n|$$

In order to distinguish between $\tilde{y}_n$ and y_n, we shall frequently call y_n the *theoretical approximation* and $\tilde{y}_n$ the *numerical approximation* to the true solution of the differential equation, which always will be denoted by $y(x)$.

Once a method for solving a differential equation is mathematically completely defined [by formulas such as (1-1)], the values of y_n and hence the discretization errors e_n can be considered as well-defined numbers. The round-off errors r_n, on the other hand, are not determined by the mathematical formulation of the method. They depend on the number of digits carried, on the number system employed by the machine, on the location of the decimal point (fixed or floating operations), on the sequence in which the numerical operations are arranged, on the precision of the subroutines which may be used in the evaluation of $f(x, y)$, and on other factors. It is therefore not surprising that the discretization error lends itself more easily to mathematical analysis than the round-off error. Paradoxically, however, it is the complete unpredictability of the latter that makes it possible to arrive at certain completely realistic and practical statements concerning it.

1.1-5. Various types of error estimates. Many algorithms for the solution of numerical problems (not only those for solving differential equations) can be thought of as a numerical operation that depends on a parameter p. The underlying idea is that for a limiting value of the parameter, say for $p \to \infty$, the "true" solution of the problem is obtained. The parameter p may assume continuous or discrete values. In Newton's method for solving an equation of the form $f(x) = 0$ the role of p is played by the index of x in the iteration scheme $x_{p+1} = x_p - f(x_p)/f'(x_p)$. In the problem of solving a differential equation over the fixed interval $[a, b]$, p appears in the form $(b - a)/h$, where h is the step of the adopted method. No numerical method can be considered satisfactory unless something is known about the behavior of the error $e(p)$ of the method as a function of p. There are several possible levels of achievement in the study of the error.

(i) It may simply be known that the method is *convergent*.* By this is meant that

$$e(p) \to 0 \qquad (p \to \infty)$$

(ii) Something may be known about the *speed of convergence*. This knowledge may be expressed by saying that with a certain function $\varphi(p)$, tending to zero as $p \to \infty$, and with an unspecified constant C there holds for all sufficiently large p the estimate $|e(p)| \leq C\,\varphi(p)$. This result is frequently stated in the form

$$e(p) = O(\varphi(p))$$

to be read: $e(p)$ is *of the order of* $\varphi(p)$.

(iii) More precisely, it may be known that

$$|e(p)| \leq \varphi(p) \qquad \text{for all } p \geq p_0$$

where p_0 is a specified number and $\varphi(p)$ is again a specified function tending to zero as $p \to \infty$. Such a result is called a *bound* for the error.

(iv) It may finally be known that

$$\frac{e(p)}{\varphi(p)} \to 1 \qquad (p \to \infty)$$

where again $\varphi(p) \to 0$ as $p \to \infty$. This is called an *asymptotic formula* for the error.

Of the several results described, the error bound is the only one that enables us to pick a value of p such that the error $|e(p)|$ can be guaranteed to be less than a preassigned number. It is possible to obtain error bounds for many algorithms of numerical analysis, including most discrete variable methods in differential equations. From a practical point of view these bounds are frequently of limited usefulness, because the function $\varphi(p)$ turns out to be much larger than the actual error. In such cases it is desirable to supplement the bound by an asymptotic formula or even by a statement concerning the order of the error. There will be ample opportunity to illustrate this point in the following text.

Results of types (ii), (iii), and (iv) imply convergence, and a result of type (i) will be of little practical interest when results of the other types are available. A result of type (i), however, may be of considerable theoretical interest. In many cases it is possible to prove (i) under weaker assumptions that any of the other results. Indeed, sometimes the convergence of a numerical method can be proved *without assuming the*

* It may seem to go without saying that convergence should be required for any useful method. However, some numerically very effective techniques of approximation exist (e.g., asymptotic expansions of definite integrals containing large parameters) that do not satisfy the requirement of convergence.

existence of a solution of the problem. One then may succeed in deducing the existence of a solution from the convergence of the numerical method. Euler's method, as is shown below, is a case in point.

1.2. The Existence of a Solution of the Initial Value Problem

In this section we shall give an elementary and constructive proof of the existence of a unique solution of the basic initial value problem (0-4).

1.2-1. Statement of the theorem. We shall assume that the real function $f(x, y)$ satisfies the following conditions:

(A) $f(x, y)$ is defined and continuous in the strip $a \leq x \leq b$, $-\infty < y < \infty$, where a and b are finite.

(B) There exists a constant L such that for any $x \in [a, b]$ and any two numbers y and y^*

$$|f(x, y) - f(x, y^*)| \leq L\,|y - y^*| \tag{1-3}$$

We shall prove:

THEOREM 1.1. *Let $f(x, y)$ satisfy conditions (A) and (B) above, and let η be a given number. Then there exists exactly one function $y(x)$ with the following three properties:*

(i) *$y(x)$ is continuous and differentiable for $x \in [a, b]$;*
(ii) *$y'(x) = f(x, y(x))$, $x \in [a, b]$;*
(iii) *$y(a) = \eta$.*

In short, the initial value problem (0-4) has a unique solution. The proof, to be given in §§1.2-2 through 1.2-7, consists in showing that a certain sequence of approximate solutions, which are defined by means of Euler's method, converges to a function $y(x)$ with the desired properties.

Condition (B) above is known as a *Lipschitz condition.* It is certainly satisfied if $f(x, y)$ has a continuous derivative with respect to y which is bounded in the strip under consideration. For then we have by the mean value theorem

$$f(x, y) - f(x, y^*) = \frac{\partial f}{\partial y}(x, \bar{y})(y - y^*)$$

where $\bar{y}$ is a value between y and y^*, and (B) follows immediately. However, as is shown by the example $f(x, y) = |y|$, the existence of $\partial f/\partial y$ is not necessary for (B). On the other hand, the example $f(x, y) = |y|^{1/2}$ shows that continuity of f is not sufficient.

It also must be noted that the term $L\,|y - y^*|$ on the right of (B) cannot be replaced by any expression of the form $L\,|y - y^*|^\alpha$ with $\alpha \neq 1$. If $\alpha > 1$, then the solution does not necessarily exist in the whole interval $[a, b]$, as is shown by the problem ($\varepsilon > 0$)

$$y' = |y|^{1+\varepsilon}, \qquad y(0) = 1$$

whose solution $y(x) = (1 - \varepsilon x)^{-1/\varepsilon}$ ceases to exist at $x = \varepsilon^{-1}$. If $\alpha < 1$, the solution is not necessarily unique. This is shown by the initial value problem ($0 < \varepsilon < 1$)

$$y' = |y|^{1-\varepsilon}, \qquad y(0) = 0$$

which has, in addition to the solution $y(x) = 0$, the infinitely many solutions $y(x) = 0$, $0 \leq x \leq c$; $y(x) = [\varepsilon(x - c)]^{1/\varepsilon}$, $x \geq c$, where $c > 0$ is arbitrary.

1.2-2. Definition of the approximate solutions $y_p(x)$. We shall write

$$h_p = \frac{b - a}{2^p} \qquad (p = 0, 1, 2, \cdots) \tag{1-4}$$

and denote by $y_p(x)$ the piecewise linear function whose graph is the polygon with vertexes at the approximate solution points (x_n, y_n) provided by Euler's method, using the step $h = h_p$. Each successive $y_p(x)$ is thus obtained by halving the step in Euler's method (see Fig. 1.2).

In order to describe the functions $y_p(x)$ analytically, we denote† for any $x \in (a, b]$ by $x_{[p]}$ the nearest lattice point to the left of x (but not coinciding with x) in the pth subdivision; in symbols,

$$x_{[p]} = a + nh_p \tag{1-5}$$

where n is the unique integer satisfying

$$a + nh_p < x \leq a + (n + 1)h_p \tag{1-6}$$

To avoid complications, we set $a_{[p]} = a$ for all p. We observe that for $x > a$

$$x_{[p]} < x \leq x_{[p]} + h_p \tag{1-7}$$

and for $p < q$

$$x_{[p]} \leq x_{[q]} \tag{1-8}$$

† This notation is used only in the present section.

With this notation, the functions $y_p(x)$ introduced above may be defined by the relations

$$
\text{(1-9)} \quad \begin{aligned} & y_p(a) = \eta \\ & y_p(x) = y_p(x_{[p]}) + (x - x_{[p]})f(x_{[p]}, y_p(x_{[p]})), \qquad a < x \le b \end{aligned}
$$

We note that the functions $y_p(x)$ are continuous in the interval $[a, b]$.

As a first step towards the proof of Theorem 1.1 we shall prove:

LEMMA 1.1. *The sequence of functions $y_p(x)$ converges for $p \to \infty$, uniformly for $x \in [a, b]$, to a continuous function $y(x)$.*

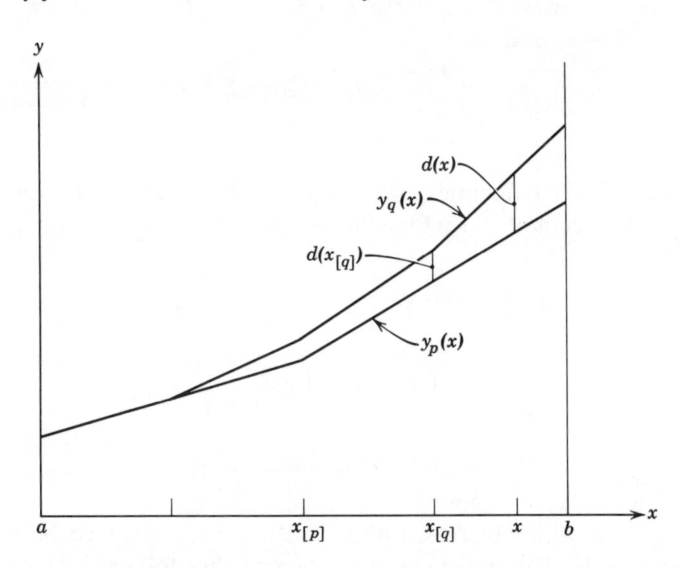

Fig. 1.2. The approximations $y_p(x)$ and $y_q(x)$ for $p = 1$, $q = 2$.

The proof is accomplished by showing that for each x in $[a, b]$ the values $y_p(x)$ $(p = 0, 1, 2, \cdots)$ form a *Cauchy sequence.* More precisely, we shall show that for each $\varepsilon > 0$ there exists an integer $P = P(\varepsilon)$ such that

$$
\text{(1-10)} \qquad |y_p(x) - y_q(x)| < \varepsilon
$$

for all $x \in [a, b]$ whenever $p > P$, $q > P$. By a fundamental principle of analysis (see Taylor [1955], p. 594, Theorem I) this implies the existence of a function $y(x)$ such that $\lim_{p\to\infty} y_p(x) = y(x)$ uniformly for $x \in [a, b]$. By another theorem (see Taylor [1955], p. 596, Theorem II) the function $y(x)$, being the uniform limit of a sequence of continuous functions, is itself continuous.

The proof of (1-10) is given in §1.2-5. It requires some auxiliary results which are derived in §§1.2-3 and 1.2-4.

1.2-3. Solution of a recurrent inequality. Not only in the proof of Lemma 1.1, but also on many later occasions, we shall have to deal with sequences of numbers ξ_n ($n = 0, 1, 2, \cdots$) that are known to satisfy inequalities of the form

$$|\xi_{n+1}| \le A\,|\xi_n| + B \tag{1-11}$$

where A and B are certain nonnegative constants independent of n. It is then desirable to have an estimate for $|\xi_n|$ in terms of $|\xi_0|$ rather than $|\xi_{n-1}|$. Such an estimate is given by the following lemma:

AUXILIARY LEMMA 1.2. *If the numbers* ξ_n *satisfy* (1-11) *for* $n = 0, 1, 2, \cdots, N-1$, *then*

$$|\xi_n| \le A^n\,|\xi_0| + \begin{cases} \dfrac{A^n - 1}{A - 1}B, & A \ne 1, \\ nB, & A = 1, \end{cases} \qquad n = 1, 2, \cdots, N \tag{1-12}$$

For $n = 1$ (1-12) is identical with (1-11) and thus true by hypothesis. Assuming the truth of (1-12) for a value of $n < N$, then by applying (1-11) we find (if $A \ne 1$)

$$\begin{aligned} |\xi_{n+1}| &\le A\left\{A^n\,|\xi_0| + \frac{A^n - 1}{A - 1}B\right\} + B \\ &= A^{n+1}\,|\xi_0| + \left(A\,\frac{A^n - 1}{A - 1} + 1\right)B \\ &= A^{n+1}\,|\xi_0| + \frac{A^{n+1} - 1}{A - 1}B \end{aligned}$$

which is (1-12) with n increased by 1. A similar argument establishes the result if $A = 1$. The statement of the lemma thus follows by induction.

In many applications of Lemma 1.2 A is a quantity of the form $1 + \delta$, where δ is a small positive number. We then may make use of the inequality

$$A = 1 + \delta < e^\delta \qquad (\delta > 0)$$

which follows immediately from the Taylor expansion of e^δ, and write (1-12) in the form

$$|\xi_n| \le e^{n\delta}\,|\xi_0| + \frac{e^{n\delta} - 1}{\delta}B \tag{1-13}$$

which on occasion is more readily amenable to simplifications than (1-12).

We shall apply Lemma 1.2 in order to obtain a bound for the values y_n obtained in the solution of the initial value problem (0-4), using an arbitrary step h. We start by noting that condition (B), if used with $y^* = 0$, implies

$$|f(x, y) - f(x, 0)| \le L\,|y|$$

It follows that $|f(x, y)| \le L\,|y| + c$, where

$$c = \max_{x \in [a,b]} |f(x, 0)| \tag{1-14}$$

From the recurrence relation (1-1*b*) we now readily find

$$\begin{aligned} |y_{n+1}| &\le |y_n| + h\,|f(x_n, y_n)| \\ &\le (1 + hL)\,|y_n| + hc, \qquad n = 0, 1, 2, \cdots \end{aligned}$$

Using $|y_0| = |\eta|$ and applying the result of Lemma 1.2 in the form (1-13), identifying y_n with ξ_n, δ with hL, and B with hc, we obtain

$$|y_n| \le e^{nhL}\,|\eta| + \frac{e^{nhL} - 1}{L}\,c$$

In view of $0 \le nh = x_n - a \le b - a$ we find the estimate

$$|y_n| \le Y, \qquad x_n \in [a, b] \tag{1-15}$$

where the constant

$$Y = e^{(b-a)L}\,|\eta| + \frac{e^{(b-a)L} - 1}{L}\,c \tag{1-16}$$

is independent of h. Since the functions $y_p(x)$ $(p = 0, 1, 2, \cdots)$ are composed of straight line segments joining the points (x_n, y_n) obtained with certain special values of h, we also have

$$|y_p(x)| \le Y, \qquad x \in [a, b] \tag{1-17}$$

with the same definition of Y.

The significance of the last result rests on the fact that it permits us to state that the approximate solutions $y_p(x)$ all stay within a certain *compact* (i.e., bounded and closed) region of the (x, y)-plane.

1.2-4. Modulus of continuity. We denote by R the rectangular region in the (x, y)-plane defined by the inequalities $a \le x \le b$, $|y| \le Y$, where Y is defined by (1-16). By hypothesis (A), $f(x, y)$ is continuous in R. For any $\delta \ge 0$ we now define the quantity

$$\omega(\delta) = \max |f(x, y) - f(x^*, y)| \tag{1-18}$$

where the maximum is taken with respect to all points (x, y) and (x^*, y) in R such that $|x - x^*| \le \delta$. We shall call $\omega(\delta)$ the *modulus of continuity*† of the function $f(x, y)$. It follows immediately from the definition that

(i) *If* $0 \le \delta \le \delta'$, *then*

$$\omega(\delta) \le \omega(\delta') \tag{1-19}$$

† More properly, it should be called the *uniform* modulus of continuity. The term "modulus of continuity" is commonly used only if f is a function of a single variable.

The following simple properties of the function $\omega(\delta)$ will also be required.

(ii) *If* $\alpha > 0$, $\beta > 0$, *then*

$$\omega(\alpha + \beta) \le \omega(\alpha) + \omega(\beta) \tag{1-20}$$

Proof. In all expressions below, the maximum is taken with respect to all points (x, y), (x^{**}, y), and (x^*, y) in R, subject to the restrictions indicated. We have

$$\begin{aligned}
\omega(\alpha + \beta) &= \max_{|x - x^*| \le \alpha + \beta} |f(x, y) - f(x^*, y)| \\
&\le \max_{\substack{|x - x^{**}| \le \alpha \\ |x^{**} - x^*| \le \beta}} \{|f(x, y) - f(x^{**}, y)| + |f(x^{**}, y) - f(x^*, y)|\} \\
&\le \max_{|x - x^{**}| \le \alpha} |f(x, y) - f(x^{**}, y)| \\
&\quad + \max_{|x^{**} - x^*| \le \beta} |f(x^{**}, y) - f(x^*, y)| \\
&= \omega(\alpha) + \omega(\beta)
\end{aligned}$$

proving (1-20).

(iii) $$\lim_{\delta \to 0} \omega(\delta) = 0$$

Proof. If the assertion were untrue, then there would exist a number $\vartheta > 0$ and two sequences of points (x_n, y_n) and (x_n^*, y_n), all lying in R, such that

$$|x_n - x_n^*| < 1/n \tag{1-21}$$

and

$$|f(x_n, y_n) - f(x_n^*, y_n)| \ge \vartheta \tag{1-22}$$

for $n = 1, 2, \cdots$. Since R is compact, the sequence (x_n, y_n) has, by virtue of the Bolzano-Weierstrass theorem (see Taylor [1955], p. 487), a limit point (x, y) in R, and we can extract a subsequence, whose elements we shall denote by (x_{n_k}, y_{n_k}) $(k = 1, 2, \cdots)$, which converges to (x, y). The sequence $(x_{n_k}^*, y_{n_k})$ then also converges to (x, y) by virtue of (1-21). By (1-22), for each index k at least one of the values $f(x_{n_k}, y_{n_k})$, $f(x_{n_k}^*, y_{n_k})$ differs by at least $\vartheta/2$ from $f(x, y)$, which is incompatible with the continuity of f at the point (x, y). This contradiction establishes (iii).

(iv) *The function* $\omega(\delta)$ *is continuous for* $\delta \ge 0$.

Proof. For $\delta = 0$ continuity is an immediate consequence of (iii). In order to prove continuity for $\delta > 0$, let $\varepsilon > 0$ be given. By (iii) there exists $\vartheta > 0$ such that $\omega(\gamma) < \varepsilon$ for $0 \le \gamma < \vartheta$. By (i) and (ii) we then have

$$0 \le \omega(\delta + \gamma) - \omega(\delta) < \varepsilon$$

and also

$$0 \leq \omega(\delta) - \omega(\delta - \gamma) = \omega(\delta - \gamma + \gamma) - \omega(\delta - \gamma) < \varepsilon$$

for $0 \leq \gamma < \vartheta$. Together these relations imply

$$|\omega(\delta + \gamma) - \omega(\delta)| < \varepsilon, \qquad |\gamma| < \vartheta$$

proving continuity at the point δ.

1.2-5. Proof of Lemma 1.1. We now shall estimate the difference

$$d(x) = y_p(x) - y_q(x) \tag{1-23}$$

(see Fig. 1.2) where p and q are two nonnegative integers, $p < q$, for arbitrary $x \in [a, b]$. This estimate is obtained via an estimate for $d(x_{[q]})$. The following lemma is instrumental in both estimates.

LEMMA 1.3. *For* $t \in [a, b]$,

$$|d(t)| \leq [1 + (t - t_{[q]})L]\, |d(t_{[q]})| + (t - t_{[q]})\Omega_p \tag{1-24}$$

where

$$\Omega_p = \omega(h_p) + LMh_p, \qquad M = \max_{(x,y) \in R} |f(x, y)| \tag{1-25}$$

Proof. We note that by (1-9),

$$y_q(t) - y_q(t_{[q]}) = (t - t_{[q]})f(t_{[q]}, y_q(t_{[q]}))$$
$$y_p(t) - y_p(t_{[q]}) = (t - t_{[q]})f(t_{[p]}, y_p(t_{[p]}))$$

Subtracting the second equation from the first, we find

$$d(t) - d(t_{[q]}) = (t - t_{[q]})[f(t_{[q]}, y_q(t_{[q]})) - f(t_{[p]}, y_p(t_{[p]}))]$$

The expression in brackets can be written in the form

$$\begin{aligned} &f(t_{[q]}, y_q(t_{[q]})) - f(t_{[q]}, y_p(t_{[q]})) \\ &+ f(t_{[q]}, y_p(t_{[q]})) - f(t_{[p]}, y_p(t_{[q]})) \\ &+ f(t_{[p]}, y_p(t_{[q]})) - f(t_{[p]}, y_p(t_{[p]})) \end{aligned}$$

Here the differences in each line can be estimated separately. The first difference is by condition (B) bounded by $L\,|d(t_{[q]})|$. The second difference is bounded by $\omega(\delta)$, where $\delta = t_{[q]} - t_{[p]}$ and thus, using (1-7) and (1-19), by $\omega(h_p)$. The absolute value of the third difference does by (B) not exceed

$$L\,|y_p(t_{[q]}) - y_p(t_{[p]})|$$

and this number is by (1-9) bounded by LMh_p. The statement of Lemma 1.3 then follows immediately.

We now apply Lemma 1.3 with $t = t_\nu = a + \nu h_q$, $\nu = 1, 2, \cdots, k$, where $t_k = x_{[q]}$. In this case, $t - t_{[q]} = h_q$, and (1-24) yields

$$|d(t_\nu)| \leq A\,|d(t_{\nu-1})| + B$$

where $A = 1 + h_q L$, $B = h_q \Omega_p$. Applying the auxiliary Lemma 1.2 with $\xi_\nu = d(t_\nu)$, we get in view of $d(a) = 0$

$$|d(x_{[q]})| \leq \Omega_p \frac{e^{(x[q]-a)L} - 1}{L} \tag{1-26}$$

Using Lemma 1.3 with $t = x$, we find

$$|d(x)| \leq [1 + (x - x_{[q]})L]\,|d(x_{[q]})| + (x - x_{[q]})\Omega_p$$

In view of

$$1 + (x - x_{[q]})L \leq e^{(x-x[q])L}$$

the last inequality may be weakened into

$$|d(x)| \leq e^{(x-x[q])L}\,|d(x_{[q]})| + \frac{e^{(x-x[q])L} - 1}{L}\,\Omega_p$$

Combining this with the estimate (1-26) for $d(x_{[q]})$, we obtain

$$|d(x)| \leq \Omega_p \left\{ e^{(x-x[q])L} \frac{e^{(x[q]-a)L} - 1}{L} + \frac{e^{(x-x[q])L} - 1}{L} \right\}$$

or, after reduction,

$$|d(x)| \leq \Omega_p \frac{e^{(x-a)L} - 1}{L} \tag{1-27}$$

This is the desired estimate for $|d(x)|$.

The expression on the right of (1-27) does not depend on q. Since, by proposition (iii) of §1.2-4, $\Omega_p = \omega(h_p) + h_p LM \to 0$ as $p \to \infty$, it can be made arbitrarily small by taking p large enough. The Cauchy criterion (1-10) is thus satisfied, and Lemma 1.1 is proved.

1.2-6. Proof that $y(x)$ is a solution. We now turn to the proof that the limit function $y(x)$ is differentiable and satisfies the given differential equation. Our proof is completely elementary in the sense that no use is made of the concept of the integral.

We shall require certain functions $f_p(x)$ which for $p = 0, 1, 2, \cdots$ are defined as follows:

$$f_p(x) = \begin{cases} f(a, \eta), & x = a \\ f(x_{[p]}, y_p(x_{[p]})), & a < x \leq b \end{cases}$$

The functions $f_p(x)$ are constant in each interval $x_{[p]} < x \leq x_{[p]} + h_p$, and their constant value is the slope of the approximation $y_p(x)$ in that interval. We assert that

$$\lim_{p\to\infty} f_p(t) = f(t, y(t)) \qquad \text{uniformly for } a \leq t \leq b \tag{1-28}$$

In fact,

$$\begin{aligned} f_p(t) - f(t, y(t)) = f(t_{[p]}, y_p(t_{[p]})) - f(t_{[p]}, y_p(t)) \\ + f(t_{[p]}, y_p(t)) - f(t_{[p]}, y(t)) \\ + f(t_{[p]}, y(t)) - f(t, y(t)) \end{aligned}$$

The three differences on the right are bounded by the quantities LMh_p, $[\omega(h_p) + LMh_p][e^{(b-a)L} - 1]$, and $\omega(h_p)$, respectively, which do not depend on t and can be made arbitrarily small by choosing p large enough.

We shall now obtain an estimate for the quantity

$$\frac{y(z) - y(x)}{z - x} - f(x, y(x))$$

where $a \leq x < z \leq b$. We set

$$\Delta_p = z_{[p]} - x_{[p]}, \qquad \nu_p = \Delta_p/h_p$$

We have for $\nu_p \geq 1$, using (1-9),

$$\begin{aligned} y_p(z_{[p]}) - y_p(x_{[p]}) &= \sum_{\nu=1}^{\nu_p} \{y_p(x_{[p]} + \nu h_p) - y_p(x_{[p]} + (\nu - 1)h_p)\} \\ &= h_p \sum_{\nu=0}^{\nu_p-1} f_p(x_{[p]} + \nu h_p) \\ &= h_p \sum_{\nu=0}^{\nu_p-1} \{f_p(x_{[p]} + \nu h_p) - f_p(x_{[p]})\} + \Delta_p f_p(x_{[p]}) \end{aligned}$$

Using the estimate

$$\begin{aligned} |f_p(s) - f_p(t)| &\leq |f(s_{[p]}, y_p(s_{[p]})) - f(t_{[p]}, y_p(s_{[p]}))| \\ &\quad + |f(t_{[p]}, y_p(s_{[p]})) - f(t_{[p]}, y_p(t_{[p]}))| \\ &\leq \omega(|s_{[p]} - t_{[p]}|) + LM\,|s_{[p]} - t_{[p]}| \end{aligned}$$

we find

$$|f_p(x_{[p]} + \nu h_p) - f_p(x_{[p]})| \leq \omega(\Delta_p) + LM\nu h_p$$

Thus, using $\sum_{\nu=0}^{\nu_p-1} \nu \leq \frac{1}{2}\nu_p^2 = \frac{1}{2}h_p^{-2}\Delta_p^2$,

$$\begin{aligned} |y_p(z_{[p]}) - y_p(x_{[p]}) - \Delta_p f_p(x_{[p]})| &\leq h_p \sum_{\nu=0}^{\nu_p-1} [\omega(\Delta_p) + LMh_p\nu] \\ &\leq \Delta_p\omega(\Delta_p) + \tfrac{1}{2}LM\Delta_p^2 \end{aligned}$$

We now let $p \to \infty$, keeping x and z fixed. We have

$$x_{[p]} \to x, \qquad z_{[p]} \to z, \qquad \text{and thus} \qquad \Delta_p \to \Delta = z - x$$

Furthermore,

$$y_p(z_{[p]}) \to y(z), \qquad y_p(x_{[p]}) \to y(x)$$

and, in view of (1-27),

$$f_p(x_{[p]}) = f_p(x) \to f(x, y(x))$$

We also make use of the fact that $\omega(\delta)$ is continuous (see §1.2-4). Consequently,

$$|y(z) - y(x) - \Delta f(x, y(x))| \le \Delta\omega(\Delta) + \tfrac{1}{2}LM\Delta^2$$

or, after dividing by Δ,

$$\left|\frac{y(z) - y(x)}{\Delta} - f(x, y(x))\right| \le \omega(\Delta) + \tfrac{1}{2}LM\Delta$$

Since the expression on the right can be made arbitrarily small by choosing Δ small enough, it has been shown that $y(x)$ is differentiable from the right with the value $f(x, y(x))$. That $y(x)$ is also differentiable from the left and with the same value follows from the continuity of $f(x, y(x))$.

A shorter but less elementary proof of the fact that $y(x)$ satisfies the differential equation can be based on the concept of the definite integral (see §3.1-5).

1.2-7. Proof that $y(x)$ is the only solution. It has been shown above that the sequence of functions $y_p(x)$ converges to some solution $y(x)$ of the initial value problem $y' = f(x, y)$, $y(a) = \eta$. There still might be other solutions of the same problem. In order to complete the proof of Theorem 1.1, we now shall prove that this is impossible by showing that if $z(x)$ is *any* solution of the initial value problem the sequence $\{y_p(x)\}$ necessarily converges to it.

By definition of solution, $z(x)$ is continuous on $[a, b]$. It follows by (A) that $z'(x) = f(x, z(x))$ is also continuous on the same closed interval. We let

$$Z_1 = \max_{x \in [a,b]} |z'(x)| \tag{1-29}$$

We now put

$$d(x) = y_p(x) - z(x)$$

and shall estimate this difference by a method very similar to that used in §1.2-5. For any $t \in [a, b]$ we have

$$y_p(t) - y_p(t_{[p]}) = (t - t_{[p]})f(t_{[p]}, y_p(t_{[p]}))$$

and by the mean value theorem, if τ denotes some number between $t_{[p]}$ and t,

$$z(t) - z(t_{[p]}) = (t - t_{[p]})f(\tau, z(\tau))$$

Subtracting the last relation from the preceding one and interpolating suitable values, we find

$$\begin{aligned} d(t) - d(t_{[p]}) = (t - t_{[p]})\{f(t_{[p]}, y_p(t_{[p]})) &- f(\tau, y_p(t_{[p]})) \\ + f(\tau, y_p(t_{[p]})) &- f(\tau, z(t_{[p]})) \\ + f(\tau, z(t_{[p]})) &- f(\tau, z(\tau))\} \end{aligned}$$

The three differences in braces are bounded by $\omega(h_p)$, $L|d(t_{[p]})|$, and, through another application of the Lipschitz condition and the mean value theorem, by LZ_1h_p. We thus find

$$|d(t)| \le [1 + (t - t_{[p]})L]\,|d(t_{[p]})| + (t - t_{[p]})\Omega_p' \tag{1-30}$$

where

$$\Omega_p' = \omega(h_p) + LZ_1h_p \tag{1-31}$$

From (1-30) there follows the bound

$$|d(x)| \le \Omega_p' \frac{e^{(x-a)L} - 1}{L} \tag{1-32}$$

in exactly the same way as (1-27) followed from (1-24). Since $\Omega_p' \to 0$ as $p \to \infty$, it follows that $y_p(x) \to z(x)$ and hence that $z(x) = y(x)$. This uniqueness result completes the proof of Theorem 1.1.

1.2-8. Some special cases of the existence theorem. Conditions (A) and (B) are satisfied in particular if the function $f(x, y)$ does not depend on y and is a continuous function of x in the interval $[a, b]$. We thus have proved the existence of *the indefinite integral of a continuous function.* The proof offered above is indeed a valid proof even of this elementary theorem, since nowhere in it has the integral concept been used.

Another case in which the conditions are satisfied is presented by the linear differential equation

$$y' = g(x)y + p(x)$$

where $g(x)$ and $p(x)$ are continuous in $[a, b]$. The Lipschitz condition (B) then holds with

$$L = \max_{x \in [a,b]} |g(x)|$$

In situations where condition (B) is violated for large values of $|y - y^*|$, such as for $f(x, y) = x^2 + y^2$, it sometimes nevertheless is possible to assert the existence of a solution by the following mental experiment. One modifies the definition of $f(x, y)$ for large values of y, for instance by setting $f(x, y) = f(x, Y)(y \geq Y)$ and $f(x, y) = f(x, -Y)(y \leq -Y)$, where Y is a suitable (large) number. The existence theorem then becomes applicable. If the solution $y(x)$ satisfies $|y(x)| \leq Y$, it is a solution of the differential equation as originally given.

In the following we shall always assume that the solution $y(x)$ of the given initial value problem exists. If the Lipschitz condition does not hold for unrestricted values of y, we shall assume that it holds at least in a neighborhood of the solution $y(x)$. In this case the theoretical analysis applies only to considerations that take place within that neighborhood.

1.3. The Discretization Error of Euler's Method

1.3-1. Some a priori bounds. By an *a priori bound* of the error of an approximation is meant a bound which can be calculated (in principle) directly from the data of the problem without first determining the approximation itself. The data of the problem consist, in our case, of the function $f(x, y)$ and of η, the prescribed value of $y(a)$.

A first a priori bound can be obtained from the proof of Theorem 1.1. If $e_n = y_n - y(x_n)$ denotes the error at the point $x = x_n = a + nh$ of the approximate value y_n obtained with the step $h = h_p = (b - a)2^{-p}$, then relation (1-32) implies that

$$|e_n| \leq [\omega(h_p) + h_p L Z_1] \frac{e^{(x_n - a)L} - 1}{L}, \qquad x_n \in [a, b]$$

Here Z_1 denotes an upper bound for $|z'(x)|$, where $z(x)$ denotes an arbitrary solution of the initial value problem. In view of the uniqueness now proved, we have $z(x) = y(x)$, the limit of the polygonal functions $y_p(x)$. Each $y_p(x)$ stays in the rectangle R defined in §1.2-4; the same is therefore true for the limit function $y(x)$. Since $y'(x) = f(x, y(x))$, it follows that the constant Z_1 can be replaced by M, where

$$M = \max_{(x,y) \in R} |f(x, y)| \tag{1-33}$$

Furthermore, although (1-32) has been proved only for $h = h_p$, virtually the same proof applies for arbitrary $h > 0$. The estimate thus obtained is valid also for a vanishing Lipschitz constant, provided that the expression $[e^{(x_n - a)L} - 1]L^{-1}$ is replaced by $x_n - a$. In view of the frequent occurrence of this exponential expression and its limiting form as $L \to 0$, it is convenient to define for arbitrary $x \geq 0$ and $L \geq 0$ the function

$$E_L(x) = \begin{cases} \dfrac{e^{Lx} - 1}{L}, & L > 0 \\ x, & L = 0 \end{cases}$$

This function will be referred to as the *Lipschitz function.*

The result thus obtained may now be stated as follows:

THEOREM 1.2. *If* $f(x, y)$ *satisfies the conditions of Theorem* 1.1, *the errors* e_n *of the approximate values* y_n *defined by Euler's method, using an arbitrary step* $h > 0$, *satisfy*

$$|e_n| \leq [\omega(h) + hLM]E_L(x_n - a), \qquad x_n \in [a, b] \tag{1-34}$$

where $\omega(h)$ *and* M *are defined by* (1-18) *and* (1-33), *respectively.*

The bound (1-34) is not easily applicable in practice because of the appearance of the somewhat elusive function $\omega(h)$. A more definite statement can be made if it is known that the exact solution has a continuous second derivative.

THEOREM 1.3. *Let* $f(x, y)$ *satisfy condition* (*B*), *and let the exact solution* $y(x)$ *of the initial value problem be twice continuously differentiable in* $[a, b]$. If

$$N(x) = \tfrac{1}{2} \max_{t \in [a,x]} |y''(t)| \tag{1-35}$$

the error of Euler's method satisfies

$$|e_n| \leq hN(x_n)E_L(x_n - a) \tag{1-36}$$

It is typical for similar theorems on more refined methods that a statement is made only about the approximate values at the lattice points. The simple proof, which we shall give *ab ovo* and without reference to the existence theorem, is also typical for more complicated situations.

By definition

$$y_{m+1} = y_m + hf(x_m, y_m), \qquad m = 0, 1, \cdots, n - 1$$

For the exact solution we have, using a form of Taylor's formula (see Taylor [1955], p. 112, Theorem III),

$$y(x_{m+1}) = y(x_m) + hf(x_m, y(x_m)) + \tfrac{1}{2}h^2y''(\xi)$$

where $x_m < \xi < x_{m+1}$.

Subtracting, we find for the error $e_m = y_m - y(x_m)$

$$e_{m+1} = e_m + h[f(x_m, y_m) - f(x_m, y(x_m))] - \tfrac{1}{2}h^2y''(\xi)$$

Taking absolute values, we get

$$|e_{m+1}| \le |e_m| + h\,|f(x_m, y_m) - f(x_m, y(x_m))| + \tfrac{1}{2}h^2\,|y''(\xi)| \tag{1-37}$$

The second term on the right can be estimated by condition (B). The result is

$$|f(x_m, y_m) - f(x_m, y(x_m))| \le L\,|y_m - y(x_m)| = L\,|e_m|$$

The third term can be estimated for $x_{m+1} \le x_n$ by $N(x_n)$. The resulting bound for $|e_{m+1}|$ can be written in the form

$$|e_{m+1}| \le A\,|e_m| + B, \qquad m = 0, 1, \cdots, n-1 \tag{1-38}$$

where

$$A = 1 + hL, \qquad B = h^2 N(x_n)$$

Here Lemma 1.2 is applicable. Using the fact that $e_0 = 0$ (no initial error), we get

$$|e_n| \le B\,\frac{A^n - 1}{A - 1} \le hN(x_n)E_L(x_n - a)$$

the desired result.

There remains the problem of estimating $N(x)$. Since the solution to be approximated is not known, the definition (1-35) can of course not be used. Assuming, however, that $f(x, y)$ has continuous first partial derivatives in the region R, we can differentiate the identity

$$y'(x) = f(x, y(x))$$

and find

$$\begin{aligned} y''(x) &= f_x(x, y(x)) + f_y(x, y(x))y'(x) \\ &= f_x(x, y(x)) + f_y(x, y(x))f(x, y(x)) \end{aligned}$$

Hence,

$$2N(x) \le \max_{(x,y)\in R} |f_x + f_y f|$$

We conclude this section with a result which is similar in content to Theorem 1.3, but is valid under somewhat relaxed conditions.

THEOREM 1.4. *Let $y(x)$ and $N(x)$ be defined as in Theorem 1.3, and let $f(x, y)$ satisfy the same conditions. Let $\{y_n\}$ be any sequence of numbers satisfying*

$$\begin{aligned} y_0 &= \eta \\ y_{n+1} &= y_n + hf(x_n, y_n) + \theta_n h^2 C, \qquad n = 0, 1, 2, \cdots \end{aligned} \tag{1-39}$$

where $C \ge 0$ is a constant, and the θ_n are numbers which may vary from step to step but always satisfy $|\theta_n| \le 1$. Then

$$|y_n - y(x_n)| \le h(N(x_n) + C)E_L(x_n - a) \tag{1-40}$$

The point of Theorem 1.4 consists in showing that even if the recurrence relations (1-1) are not exactly satisfied the values y_n may still converge to $y(x_n)$, provided that the discrepancy between (1-1) and (1-39) is not too great. For the moment the quantities $\theta_n h^2 C$ may be thought of as round-off errors, although in later applications of Theorem 1.4 their meaning will be different.

The proof of Theorem 1.4 proceeds in the same way as did the proof of Theorem 1.3. In place of (1-37) we now obtain

$$(1\text{-}41) \qquad |e_{m+1}| \le |e_m| + h\,|f(x_m, y_m) - f(x_m, y(x_m))| + \tfrac{1}{2}h^2\,|y''(\xi)| + h^2 C$$

because of the presence of the extra term on the right of (1-39). Using the Lipschitz condition, we can rewrite (1-41) in the form (1-38), where A is the same as before, but B is now $N(x_n) + C$. Completing the proof as above, we obtain (1-40).

1.3-2. Numerical illustration. The estimate (1-36) is very typical for many classical error estimates. The error at a fixed point x, arrived at by a step h which divides $x - a$, is shown to be bounded by a power of h (here the first power), multiplied by an expression which may depend on x but is independent of h. It is by no means certain, however, that in any concrete case the error will be as large as indicated by (1-36).

As an illustration, we consider the numerical solution of the two differential equations

$$(1\text{-}42) \qquad y' = y$$

$$(1\text{-}43) \qquad y' = -y$$

under the initial condition $y(0) = 1$. The true solutions, of course, are $y = e^x$ and $y = e^{-x}$. We shall compare the actual error with the bound given by (1-36). In both equations, $L = 1$. We are fair to the theoretical bound and determine N from the true solutions. Since the second derivatives are e^x and e^{-x}, respectively, we have

$$2N(x) = e^x \text{ for (1-42)} \qquad \text{and} \qquad 2N(x) = 1 \text{ for (1-43)}$$

Hence the error will be bounded by the expressions

$$(1\text{-}44) \qquad \tfrac{1}{2}he^x(e^x - 1)$$

$$(1\text{-}45) \qquad \tfrac{1}{2}h(e^x - 1)$$

respectively. These theoretical error bounds are compared with the actual errors in Tables 1.2a and 1.2b. In both computations sufficiently many digits have been carried that the results are correct to the number of digits given in the tables.

The results show clearly that the bound (1-36) does not give realistic results in the two cases considered. If e^{-5} were to be determined with an error not exceeding 10^{-3} by solving (1-43), the bound (1-36) would require choosing a step h such that

$$\tfrac{1}{2}h(e^5 - 1) \leq 10^{-3}$$

i.e., $h \leq 1/73707$. Actually, $h = 1/64$ is amply sufficient, as Table 1.2b shows.

Table 1.2a

Integration of $y' = y$, $y(0) = 1$ by Euler's Method, Using $h = 2^{-6}$

x_n	1	2	3	4	5
y_n	2.69735	7.27567	19.62499	52.93537	142.7850
e_n	−0.02093	−0.11339	−0.46055	−1.66278	−5.6282
Error bound (1-44)	0.03649	0.36882	2.99487	22.86218	170.9223

Table 1.2b

Integration of $y' = -y$, $y(0) = 1$ by Euler's Method, Using $h = 2^{-6}$

x_n	1	2	3	4	5
y_n	0.364987	0.133215	0.048622	0.017746	0.006477
e_n	−0.002892	−0.002120	−0.001165	−0.000570	−0.000261
Error bound (1-45)	0.013424	0.049914	0.149106	0.418735	1.151666

In deriving the bound (1-45) we took advantage of our knowledge of the exact solution of the given equation. In more realistic situations, where the exact solution is not known, the theoretical bounds must be expected to produce even less satisfactory results. The need for more realistic appraisals of the truncation error is thus clearly indicated.

1.3-3. An a posteriori bound for the error. We now derive a result that does not show how large the error in Euler's method might conceivably be but how large (approximately) it actually is. Similar results are given later for more accurate methods.

Inequality (1-36) may be interpreted as saying that the numbers $h^{-1}|e_n|$ are bounded as $h \to 0$, when $x_n = a + nh$ is fixed. Let us look at the values $h^{-1}e_n$ in the particular case of the problem $y' = -y$, $y(0) = 1$. Table 1.3

gives the values of y_n, e_n, and $h^{-1}e_n$ for $x_n = 1$, computed with the steps $h = 2^{-k}$, $k = 1, 2, \cdots$.

It appears from this table that the numbers $h^{-1}e_n$ are not only bounded but also tend to a definite limit. We shall show that this is indeed so, not only in the present special case, but for the approximate solution of any differential equation by Euler's method. In the present case the limit will be shown to be $-\frac{1}{2}e^{-1} = -0.183940$.

Table 1.3

$k = -\log_2 h$	y_n	e_n	$h^{-1}e_n$
1	0.250000	−0.117879	−0.235758
2	0.316406	−0.051473	−0.205893
3	0.343609	−0.024270	−0.194164
4	0.356074	−0.011805	−0.188885
5	0.362055	−0.005824	−0.186373
6	0.364987	−0.002892	−0.185147
7	0.366438	−0.001441	−0.184540
8	0.367160	−0.000719	−0.184239
.	.		
.	.		
.	.		
∞	0.367879		

We now assume that the function $f(x, y)$ not only satisfies conditions (A) and (B) but also has continuous and bounded first and second derivatives in the region R (see §1.2-4). Under this hypothesis the third derivative of the true solution $y(x)$ exists, and we may write

$$y(x_{n+1}) = y(x_n) + hf(x_n, y(x_n)) + \tfrac{1}{2}h^2y''(x_n) + \tfrac{1}{6}h^3y'''(\xi)$$

where $x_n < \xi < x_{n+1}$. Subtracting this from the corresponding relation satisfied by the approximate values,

$$y_{n+1} = y_n + hf(x_n, y_n)$$

we obtain [where $y_n = y(x_n) + e_n$]

(1-46)

$$e_{n+1} = e_n + h[f(x_n, y(x_n) + e_n) - f(x_n, y(x_n))] - \tfrac{1}{2}h^2y''(x_n) - \tfrac{1}{6}h^3y'''(\xi)$$

By Taylor's formula, the expression in brackets can be written in the form

$$f_y(x_n, y(x_n))e_n + \tfrac{1}{2}f_{yy}(x_n, y^*)e_n^2$$

where y^* is a value between $y(x_n)$ and y_n. We now divide the resulting relation by h and introduce the quantities

$$\bar{e}_n = h^{-1}e_n$$

to be called the *magnified errors*. By our previous result the magnified errors are bounded by the constant

$$C_1 = NE_L(b - a)$$

Equation (1-46) now can be rewritten in the form

$$\bar{e}_{n+1} = \bar{e}_n + h[f_y(x_n, y(x_n))\bar{e}_n - \tfrac{1}{2}y''(x_n)] + h^2 r_n \tag{1-47}$$

where

$$|r_n| \le |f_{yy}(x_n, y(x_n))| \cdot C_1^2 + \tfrac{1}{6}|y'''(\xi)| \le C_2$$

C_2 being another constant. Defining the function

$$g(x) = f_y(x, y(x))$$

we can look at (1-47) as the result of applying Euler's method to the solution of a new differential equation for a function $e(x)$,

$$e'(x) = g(x)e(x) - \tfrac{1}{2}y''(x) \tag{1-48}$$

making at each step an additional error not exceeding h^2C_2. The initial value $\bar{e}_0$ is zero, because $e_0 = 0$. To this equation we can apply Theorem 1-4, with the following result:

THEOREM 1.5. *Let the function $f(x, y)$ have continuous first and second derivatives in the region R (see §1.2-4). Then the error e_n of the Euler approximation to the solution $y(x)$ of $y' = f(x, y)$, $y(a) = \eta$, can be written in the form*

$$e_n = he(x_n) + O(h^2) \tag{1-49}$$

where $e(x)$ is the solution of

$$e'(x) = f_y(x, y(x))e(x) - \tfrac{1}{2}y''(x) \tag{1-50}$$

satisfying $e(a) = 0$. More precisely,

$$\left|\frac{1}{h}e_n - e(x_n)\right| \le hC_3 \cdot E_{L_1}(x_n - a) \tag{1-51}$$

where

$$C_3 = \tfrac{1}{2}\max_{a\le x\le b}|e''(x)| + C_2$$

$$L_1 = \max_{a\le x\le b}|f_y(x, y(x))|$$

The constant C_3 can be bounded in terms of bounds on derivatives of $f(x, y)$ (see Problem 7 at the end of this chapter). The function $e(x)$ defined by (1-50) will be called the *magnified error function* of the problem.

1.3-4. Application. We apply the results of Theorem 1.5 to the two examples considered previously. In the first example, $y' = y$, we have $y(x) = e^x$, and thus

$$f_y(x, y(x)) = 1, \qquad y'' = e^x$$

The equation for $e(x)$ is

$$e'(x) = e(x) - \tfrac{1}{2}e^x, \qquad e(0) = 0$$

and has the solution

$$e(x) = -\tfrac{1}{2}xe^x$$

Thus Theorem 5 yields

$$e_n = -\tfrac{1}{2}x_n e^{x_n}h + O(h^2)$$

In the second example, $y' = -y$, the true solution is $y(x) = e^{-x}$; hence

$$f_y(x, y(x)) = -1, \qquad y'' = e^{-x}$$

The magnified error function now satisfies

$$e'(x) = -e(x) - \tfrac{1}{2}e^{-x}, \qquad e(0) = 0$$

and thus is given by

$$e(x) = -\tfrac{1}{2}xe^{-x} \tag{1-52}$$

Thus we have by Theorem 1.5

$$e_n = -\tfrac{1}{2}x_n e^{-x_n}h + O(h^2) \tag{1-53}$$

In particular, for $x_n = 1$, $n = 1/h$,

$$\lim_{h\to 0} h^{-1}e_n = -\tfrac{1}{2}e^{-1}$$

in agreement with the values of Table 1.3.

In Table 1.4 we compare for $y' = -y$ the actual errors e_n corresponding to $h = 1/64$ with the values $he(x_n)$ as given by (1-52). The function $e(x)$ is seen to reflect very accurately the behavior of the actual error.

In practical problems the true solution is, of course, unknown, and the function $e(x)$ is not as easily determined. We shall see later, however, that even then there exist practical methods for determining $e(x)$ and that formulas analogous to (1-49) can be put to good use.

Table I.4

Comparison of Actual Error and Error Computed by Magnified Error Function

x_n	1	2	3	4	5
e_n	−0.002892	−0.002120	−0.001165	−0.000570	−0.000261
$he(x_n)$	−0.002874	−0.002114	−0.001167	−0.000572	−0.000263

The mere fact that (1-49) holds is helpful even in cases where $e(x)$ is not known. Let us deviate from our usual notation and denote by $y(x, h)$ the approximate solution obtained at the point x with the step h (which necessarily must divide x). Thus

$$y(x, h) = y_{(x-a)/h}$$

Relation (1-49) can then be read thus:

$$y(x, h) = y(x) + he(x) + O(h^2) \tag{1-54}$$

If $y(x, h)$ is calculated for two different values of h, say h and qh, where q is some ratio like $\frac{1}{2}$, then we may eliminate the unknown value $e(x)$ from the two relations (1-54) and

$$y(x, qh) = y(x) + qhe(x) + O((qh)^2)$$

In view of $O(h^2) + O((qh)^2) = O(h^2)$, the result is

$$y(x) = \frac{y(x, qh) - qy(x, h)}{1 - q} + O(h^2) \tag{1-55}$$

The $O(h^2)$ term in (1-55) will of course be zero only in utterly trivial cases (which do not even include our academic examples $y' = \pm y$). Nevertheless, if h is small, the value obtained by omitting $O(h^2)$ in (1-55) will be

Table I.5

Deferred Approach to the Limit

k	$y(1, 2^{-k})$	Improved Values
4	0.356074	
5	0.362055	0.368036
6	0.364987	0.367919
7	0.366438	0.367889
8	0.367160	0.367882
.	.	
.	.	
.	.	
∞	0.367879	

markedly better than either $y(x, h)$ or $y(x, qh)$. We illustrate this point in Table 1.5 by applying (1-55) to the values y_n given in Table 1.3, using always the last two available values ($q = \frac{1}{2}$).

The above extrapolation technique was first used extensively by L. F. Richardson [1927] and was called by him the *deferred approach to the limit*. Less picturesquely we may call it *extrapolation to* $h = 0$ or simply *h-extrapolation*. The idea occurs in many other branches of numerical analysis; it is applicable whenever the error in a numerical process depending upon a parameter has a simple asymptotic behavior.

1.4. The Round-off Error in Euler's Method

1.4-1. The local round-off error. Except in the unlikely case that all parameters entering the computation are rational numbers and rational arithmetic is used throughout, the relations (1-1) defining Euler's method are never satisfied exactly because of round-off errors. The study of round-off error is not simple, because the ordinary laws of arithmetic no longer hold. In order to put such a study on a firm basis, it is necessary to state the underlying assumptions very carefully and explicitly.

Unless stated otherwise, we shall assume that all arithmetic operations are performed in fixed point arithmetic. Then all numbers carried in the computation are integral multiples of a smallest positive number which the machine can handle. This number will be denoted by u and will be called the *basic unit* of the machine.† If x is any number falling within the range of the machine, then x^* will denote a correctly rounded machine representation of x. The words "correctly rounded" mean that

$$|x - x^*| \leq \tfrac{1}{2}u$$

It will always be assumed that the step h and the initial point a are exact machine numbers, i.e., $h^* = h$, $a^* = a$. This implies that the lattice points x_n are calculated exactly. For simplicity we shall also assume that $y_0^* = y_0$, although this assumption is less essential.

Under these assumptions, the values $\tilde{y}_n$ actually calculated are connected by relations of the following form:

$$\text{(1-56)} \qquad \begin{aligned} \tilde{y}_0 &= y_0 \\ \tilde{y}_{n+1} &= \tilde{y}_n + (h\tilde{f}(x_n, \tilde{y}_n))^*, \qquad n = 0, 1, 2, \cdots \end{aligned}$$

Here $\tilde{f}(x_n, \tilde{y}_n)$ denotes an approximate value of $f(x_n, \tilde{y}_n)$. Ideally, $\tilde{f} = f^*$; we must allow for the possibility, however, that the function $f(x, y)$ is not

† For the IBM 704, 709, and 7090 machines, if fixed point arithmetic is used, we have $u = 2^{-35}$.

evaluated with the greatest possible accuracy, for instance because already in this computation there is an accumulation of round-off error, or because inaccurate subroutines are used.

The second equation (1-56) is hard to work with analytically. We therefore replace it by the relation

$$\tilde{y}_{n+1} = \tilde{y}_n + hf(x_n, \tilde{y}_n) + \varepsilon_{n+1} \tag{1-57}$$

The quantity ε_{n+1}, called the *local round-off error*, is defined by this equation. Thus,

$$\varepsilon_{n+1} = (h\tilde{f}(x_n, \tilde{y}_n))^* - hf(x_n, \tilde{y}_n)$$

In order to estimate the size of ε_{n+1}, we write

$$\varepsilon_{n+1} = \pi_{n+1} + \rho_{n+1}$$

where

$$\pi_{n+1} = (h\tilde{f}(x_n, \tilde{y}_n))^* - h\tilde{f}(x_n, \tilde{y}_n)$$

$$\rho_{n+1} = h[\tilde{f}(x_n, \tilde{y}_n) - f(x_n, \tilde{y}_n)]$$

The quantity π_{n+1} may be called the *induced* error. It is induced by rounding the product $h\tilde{f}$ and can be as large as $\frac{1}{2}u$. The quantity ρ_{n+1} will be called the *inherent* error. It is caused by the inaccuracy of the evaluation of the function f. Even if this inaccuracy should amount to several units in the least significant digit, the inherent error would be of the order of hu and thus would under the circumstances considered be much smaller than the induced error.

1.4-2. An a priori bound for the accumulated round-off error. In the remainder of §1.4 we shall be concerned with the problem of estimating the accumulated round-off error $r_n = \tilde{y}_n - y_n$ under the sole assumption that

$$|\varepsilon_n| \le \varepsilon, \qquad n = 1, 2, \cdots \tag{1-58}$$

where ε is a constant. If only induced error is present, then ε may be taken as $\frac{1}{2}u$. This bound is likely to be correct also in the presence of inherent error, because the effect of any error in f is greatly weakened by the subsequent multiplication by the small number h. We now can perform our analysis in a similar manner to that used in the proof of Theorem 1.3. Subtracting from (1-57) the corresponding relation satisfied by the exact solution y_n of the difference equation and putting $r_n = \tilde{y}_n - y_n$, we find

$$r_0 = 0$$

$$r_{n+1} = r_n + h[f(x_n, \tilde{y}_n) - f(x_n, y_n)] + \varepsilon_{n+1}, \qquad n = 0, 1, 2, \cdots$$

Using the Lipschitz condition and (1-58), it follows that

$$|r_{n+1}| \leq (1 + hL)\,|r_n| + \varepsilon, \qquad n = 0, 1, \cdots$$

This recursive system can be solved by Lemma 1.2. We formulate the result in the following theorem:

THEOREM 1.6. *If $f(x, y)$ satisfies condition (B) and if the local round-off errors satisfy* (1-46), *the accumulated round-off error r_n of Euler's method, started with the exact value, obeys*

$$|r_n| \leq \frac{\varepsilon}{h} E_L(x_n - a), \qquad n = 0, 1, \cdots \tag{1-59}$$

If ε is one-half of the basic unit, (1-59) expresses the fact that after n steps the accumulated round-off error is at most $(2h)^{-1}E_L(x_n - a)$ times the basic unit. For $L = 0$ this expression reduces to $(2h)^{-1}(x_n - a) = \frac{1}{2}n$. This result admits of a simple interpretation: $L = 0$ implies that $f(x, y)$ is independent of y. The solution of

$$y' = f(x), \qquad y(a) = \eta$$

by Euler's method amounts to forming the sum $\sum_{m=0}^{n-1} hf(x_m)$, and the round-off error is obviously bounded by $n/2$ times the basic unit.

For the initial value problem $y' = y$, $y(0) = 1$ we have $L = 1$, and the above result yields

$$|r_n| \leq \frac{u}{2h}(e^{x_n} - 1)$$

In particular, for $h = 10^{-3}$, $n = 1000$ we find

$$|r_{1000}| \leq \tfrac{1}{2}u \cdot 10^3(e - 1) = 859u$$

The same result is also obtained for $y' = -y$, since in this case, too, we have $L = 1$.

1.4-3. The dependence of the accumulated round-off error on the local round-off errors. As in the case of the discretization error, more realistic estimates can be obtained by assuming that the exact solution $y(x)$ is (at least approximately) known. As a preparation we shall derive a formula which exhibits the dependence of r_n on the local round-off errors. In order not to tackle too many difficulties at once, in the present chapter we consider only the case of a linear differential equation†

$$y' = g(x)y + p(x) \tag{1-60}$$

† General differential equations are studied in Chapter 2.

Here $g(x)$ and $p(x)$ are assumed to be continuous in the interval $[a, b]$. The theoretical approximation is in this instance given by

$$y_{n+1} = y_n + h[g(x_n)y_n + p(x_n)] \tag{1-61}$$

The numerical approximation $\tilde{y}_n$ satisfies

$$\tilde{y}_{n+1} = \tilde{y}_n + h[g(x_n)\tilde{y}_n + p(x_n)] + \varepsilon_{n+1} \tag{1-62}$$

Subtracting (1-61) from (1-62), we find a relation for the accumulated round-off error r_n,

$$r_{n+1} = r_n + hg(x_n)r_n + \varepsilon_{n+1} \tag{1-63}$$

We may consider the last relation as a difference equation for the variable r_n, to be solved under the initial condition $r_0 = 0$. We seek a solution of this equation in the form

$$r_n = \sum_{m=1}^{n} d_{n,m}\varepsilon_m \tag{1-64}$$

which exhibits the dependence of r_n on all preceding local round-off errors ε_m. In order to determine the constants $d_{n,m}$ we substitute into (1-63):

$$\sum_{m=1}^{n+1} d_{n+1,m}\varepsilon_m = \sum_{m=1}^{n} d_{n,m}\varepsilon_m + hg(x_n)\sum_{m=1}^{n} d_{n,m}\varepsilon_m + \varepsilon_{n+1}$$

This relation must be satisfied no matter what the individual round-off errors ε_m are. Comparing the coefficients of ε_m $(m = 1, 2, \cdots, n)$ on each side, we find

$$d_{n+1,m} = d_{n,m} + hg(x_n)d_{n,m} \quad (n = 0, 1, \cdots;\ m = 1, 2, \cdots, n) \tag{1-65a}$$

Comparing the coefficients of ε_{n+1}, there results

$$d_{n+1,n+1} = 1 \qquad (n = 0, 1, \cdots) \tag{1-65b}$$

Conversely, if the coefficients $d_{n,m}$ are defined by (1-65) and r_n by (1-64), then r_n is easily seen to satisfy (1-63).

Let us apply this result to the equation $y' = y$. In this case $g(x) \equiv 1$, and the solution of (1-65) is found to be

$$d_{n,m} = (1 + h)^{n-m} \qquad (n \geq m)$$

Thus,

$$r_n = \sum_{m=1}^{n} (1 + h)^{n-m}\varepsilon_m$$

In general, the system (1-65) will not be solvable explicitly. Even then we can obtain an explicit representation for r_n, at the expense of some accuracy. Keeping in (1-65a) the index m fixed and concentrating on the dependence on n, we recognize that this equation may be looked at as the result of applying Euler's method to the differential equation

$$d'_m(x) = g(x)d_m(x) \tag{1-66a}$$

for an unknown function $d_m(x)$. Relation (1-65b) furnishes the necessary initial condition

$$d_m(x_m) = 1 \tag{1-66b}$$

This initial value problem can be solved explicitly. Setting

$$G(x) = \int_a^x g(t)\, dt$$

the solution is given by

$$d_m(x) = e^{G(x)-G(x_m)} \tag{1-67}$$

By Theorem 1.3 the quantities $d_{n,m}$ differ from $d_m(x_n)$ by less than Ch, where C is a constant which does not depend on h, n, or m. Thus,

$$|d_{n,m} - \exp\,[G(x_n) - G(x_m)]| \le Ch, \qquad a \le x_m \le x_n \le b$$

Replacing $d_{n,m}$ in (1-64) by $d_m(x_n)$ and estimating the error committed, we find

$$r_n = \sum_{m=1}^{n} \exp\,[G(x_n) - G(x_m)]\varepsilon_m + \theta hC \sum_{m=1}^{n} |\varepsilon_m| \tag{1-68}$$

where $-1 \le \theta \le 1$. If we disregard the correction term, then this equation shows that the influence of local error ε_m on the accumulated error r_n at a later point is magnified if and only if $G(x_n) - G(x_m) > 0$, i.e., if

$$\int_{x_m}^{x_n} g(t)\, dt > 0$$

1.4-4. An improved bound. Equation (1-64) is useful for obtaining an improved bound for $|r_n|$ under the assumption that

$$|\varepsilon_m| \le \varepsilon \qquad (m = 1, 2, \cdots) \tag{1-69}$$

Rewriting (1-65a) in the form

$$d_{n+1,m} = (1 + hg(x_n))d_{n,m}$$

we see at once that in view of $d_{m,m} = 1$ all coefficients $d_{n,m}$ are positive if $h < h_0$, where

$$h_0 = \begin{cases} \infty, & \text{if } g(x) \ge 0,\ x \in [a, b] \\ \min\limits_{x\in[a,b]} (-g(x))^{-1}, & \text{otherwise} \end{cases}$$

Thus, if (1-69) holds and if $h < h_0$,

$$|r_n| \leq \varepsilon \sum_{m=1}^{n} d_{n,m}$$

In order to express the last sum in closed form, we write

$$m_n = h \sum_{m=1}^{n} d_{n,m} \tag{1-70}$$

Using (1-65), we find

$$\begin{aligned} m_{n+1} - m_n &= h\left(\sum_{m=1}^{n+1} d_{n+1,m} - \sum_{m=1}^{n} d_{n,m}\right) \\ &= h\left(d_{n+1,n+1} + \sum_{m=1}^{n} (d_{n+1,m} - d_{n,m})\right) \\ &= h\left(1 + \sum_{m=1}^{n} hg(x_n)d_{n,m}\right) \\ &= h[1 + g(x_n)m_n] \end{aligned}$$

Thus,

$$m_{n+1} = m_n + h[g(x_n)m_n + 1]$$

Obviously, $m_0 = 0$. Thus, by Theorem 1.3,

$$m_n = m(x_n) + \theta_n Ch \qquad (|\theta_n| \leq 1) \tag{1-71}$$

where the function $m(x)$ is the solution of the initial value problem

$$\begin{aligned} m'(x) &= g(x)m(x) + 1 \\ m(a) &= 0 \end{aligned} \tag{1-72}$$

and C is a constant which depends only on the given differential equation. We have thus proved:

THEOREM 1.7. *If the local round-off errors in the solution of* (1-60) *by Euler's method satisfy* (1-69), *and if* $h < h_0$, *then the accumulated error satisfies*

$$|r_n| \leq \frac{\varepsilon}{h} \{m(x_n) + O(h)\} \tag{1-73}$$

where $m(x)$ *is defined by* (1-72).

For the problem $y' = y$, $y(0) = 1$ we find $m(x) = e^x - 1$, and thus, assuming $\varepsilon = u/2$,

$$|r_n| \leq \frac{u}{2h} [(e^{x_n} - 1) + O(h)]$$

If the $O(h)$ term is neglected, this yields for $h = 10^{-3}$, $x_n = 1$

$$|r_{1000}| \leq \tfrac{1}{2}u \cdot 10^3(e - 1) = 859u$$

and there is no improvement over the a priori bound of Theorem 1.6.

If, on the other hand, the initial value problem to be solved is $y' = -y$, $y(0) = 1$, we have $m(x) = 1 - e^{-x}$ and hence, with the same data as above,

$$|r_{1000}| \leq \tfrac{1}{2}u \cdot 10^3(1 - e^{-1}) = 326u$$

which is an improvement over the previous result. The difference could be made even more pronounced by considering a larger value of x_n.

1.5. Random Variables

The bounds for the accumulated round-off error given above have been derived under the pessimistic hypothesis that all round-off errors have constant sign and maximum value and thus reinforce each other systematically. Such a regular distribution of round-off errors, although theoretically possible, is not likely ever to arise in practice, and the bound (1-73), although theoretically the "best possible," must be expected to vastly overestimate the actual error. Thus it is desirable to have on hand other appraisals for the round-off errors which indicate the "average" or "normal" growth of the round-off error. Such appraisals can be obtained by considering the local round-off errors as *random variables*.

It is not assumed that the reader has any knowledge of random variables. Before discussing their application to round-off errors, we shall present some elementary facts concerning them.*

1.5-1. Definition of a random variable; distributions. A *random variable*, denoted until further notice by the symbol ξ, is the numerical result of performing a random experiment which can be repeated a large number of times under uniform conditions. One of the classical examples is the throwing of a die. We shall assume that ξ has associated with it a *probability distribution*. For our purposes the following definition of this concept is sufficient. Let an arbitrary interval I be given, and let the experiment be performed N times, where N is large. Let N_I be the number of times where ξ, the outcome of the experiment, is a number belonging to I. If

$$\lim_{N\to\infty} \frac{N_I}{N} = P(\xi \in I) \tag{1-74}$$

exists, then it is called the *probability* that ξ belongs to I. If the limit exists for all intervals I, the variable ξ is said to possess a probability distribution.

* Only the barest outline of the theory can be presented here. For a more thorough treatment the reader is referred to standard texts such as (in increasing order of difficulty) Hoel [1954], Feller [1957], Cramér [1946], and Loève [1955].

For intervals I of the form $a < x \leq b$ the probability $P(\xi \in I)$ can be expressed in terms of a function of a single variable. For, if we define

$$F(x) = P(\xi \leq x) \tag{1-75}$$

then clearly

$$\begin{aligned} P(a < \xi \leq b) &= P(\xi \leq b) - P(\xi \leq a) \\ &= F(b) - F(a) \end{aligned}$$

The function $F(x)$ is called the distribution function of the variable ξ. By definition of probability, $F(x)$ is a nondecreasing function whose domain is the real line and whose range lies in the interval $[0, 1]$.

If the distribution function $F(x)$ is continuous and piecewise continuously differentiable, the derivative $p(x) = F'(x)$ is called the *frequency function* or *probability density* of the random variable. Indeed, since

$$\lim_{\Delta x \to 0} \frac{F(x + \Delta x) - F(x)}{\Delta x} = p(x)$$

the probability that ξ lies between x and $x + \Delta x$ is approximately $p(x)\,\Delta x$. Moreover, if $\lim_{x \to -\infty} F(x) = 0$, then

$$P(\xi \leq x) = F(x) = \int_{-\infty}^{x} \frac{dF(t)}{dx}\,dt = \int_{-\infty}^{x} p(t)\,dt \tag{1-76a}$$

In many applications (such as playing dice) the random variable ξ assumes only the (finitely or infinitely many) discrete values x_ν ($\nu = 0, 1, 2, \cdots$)* Let p_ν be the probability that ξ assumes the value x_ν. Then the distribution function is constant between any two neighboring points x_ν and has a jump $\Delta F_\nu = p_\nu$ at the point x_ν. The probability that $\xi \leq x$ is then given by

$$P(\xi \leq x) = \sum_{x_\nu \leq x} \Delta F_\nu = \sum_{x_\nu \leq x} p_\nu \tag{1-76b}$$

If the notation of the Stieltjes integral is adopted, then both (1-76*a*) and (1-76*b*) can be combined into the single formula

$$P(\xi \leq x) = \int_{-\infty}^{x} dF(x) \tag{1-76}$$

Since in this book only distribution functions of the two types just discussed will be considered, it suffices for our purposes to define (1-76) by either (1-76*a*) or (1-76*b*), whichever is applicable.

* These values are not related to the lattice points x_n considered above.

Examples. (a) Assume that ξ is capable of assuming only values in the interval $[a, b]$, and that it assumes all values in this interval with equal probability. The distribution function is then given by

$$F(x) = \begin{cases} 0, & x \leq a \\ \dfrac{x - a}{b - a}, & a \leq x \leq b \\ 1, & b \leq x \end{cases}$$

The probability density is zero outside $[a, b]$ and has the constant value $p = 1/(b - a)$ inside $[a, b]$. Such a distribution is called *rectangular* (see Fig. 1.3).

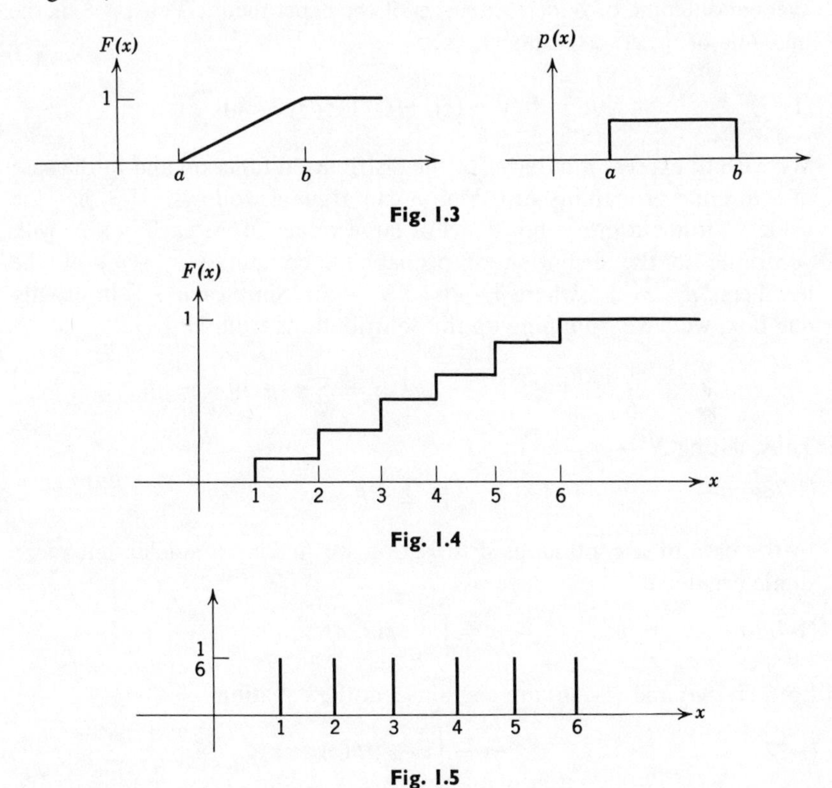

Fig. 1.3

Fig. 1.4

Fig. 1.5

(b) A simple example of a discrete distribution is the number of spots on the face of a thrown die. Here ξ can only have one of the six values $x_\nu = \nu$, $\nu = 1, 2, \cdots, 6$. For an ideal die, $p_\nu = \frac{1}{6}$. The distribution function is graphically represented in Fig. 1.4. The probability density $p(x)$ does not exist in this case. However, the probabilities p_ν can be represented graphically by attaching a vertical line segment of length p_ν to the point x_ν (see Fig. 1.5).

1.5-2. Mean and variance of a random variable. There are a great variety of possible distribution functions; every nondecreasing function $F(x)$ which is continuous on the right and satisfies $F(-\infty) = 0$ and $F(\infty) = 1$ defines a distribution function of a certain random variable. For many practical purposes, however, a distribution function is sufficiently characterized by as few as two numerical parameters, called the *mean* and the *variance* of the distribution.

The mean μ of the random variable ξ is the limit as $N \to \infty$ of the average outcome of N performances of the experiment. Thus, if ξ_i is the outcome of the ith experiment, then

$$\mu = \lim_{N\to\infty} \frac{1}{N}(\xi_1 + \xi_2 + \cdots + \xi_N) \tag{1-77}$$

We wish to express μ in terms of the distribution function and in the case of a discrete probability distribution can argue as follows. If ξ_i has the value x_ν, it is put into a box B_ν. For large values of N, each box B_ν will, according to the definition of probability, contain $N(p_\nu + \delta_\nu)$ of the numbers $\xi_1, \cdots, \xi_N$, where $\delta_\nu \to 0$ as $N \to \infty$. Since each ξ_i is in exactly one box, we have, summing up the contributions from all boxes,

$$\frac{1}{N}(\xi_1 + \xi_2 + \cdots + \xi_N) = \sum_\nu x_\nu(p_\nu + \delta_\nu)$$

Thus, letting $N \to \infty$,

$$\mu = \sum_\nu x_\nu p_\nu \tag{1-78a}$$

In the case of a continuous distribution we find by a similar but more subtle argument

$$\mu = \int_{-\infty}^{\infty} xp(x)\,dx \tag{1-78b}$$

Both (1-78a) and (1-78b) are contained in the equation

$$\mu = \int_{-\infty}^{\infty} x\,dF(x) \tag{1-78}$$

which, after the heuristic explanation given above, may be accepted as the definition of the mean value μ. Frequently we write

$$\mu = E(\xi)$$

and call $E(\xi)$ the *expected value* of the random variable ξ. As to the above examples, we easily find $E(\xi) = \frac{1}{2}(b + a)$ for (a) and $E(\xi) = \frac{7}{2}$ for (b).

If ξ is any random variable, and if $f(x)$ is any function of x, then $f(\xi)$ is again a random variable. If its expected value exists, we can obtain it by multiplying $f(x)$ with the probability that $\xi = x$ and summing with respect to all values of x. Thus

$$E(f(\xi)) = \sum_{\nu} f(x_\nu)p_\nu \tag{1-79a}$$

in the discrete case, and

$$E(f(\xi)) = \int_{-\infty}^{\infty} f(x)p(x)\,dx \tag{1-79b}$$

in the continuous case. Both cases are contained in the equation

$$E(f(\xi)) = \int_{-\infty}^{\infty} f(x)\,dF(x) \tag{1-79}$$

As an application of these rules we note that, if a and b are arbitrary constants,

$$E(a\xi) = \int_{-\infty}^{\infty} ax\,dF(x) = aE(\xi) \tag{1-80}$$

and

$$E(\xi + b) = \int_{-\infty}^{\infty} (x + b)\,dF(x) = E(\xi) + b \tag{1-81}$$

Equation (1-78a) admits of a simple mechanical interpretation. If a particle of mass p_ν is attached to the point $x = x_\nu$ of the x-axis, then, since $\sum_\nu p_\nu = 1$, the common center of gravity of all particles is evidently located at the point $x = \mu$. A similar interpretation is also possible in the case of a continuous distribution. Thus, the expected value may be looked at as a rough *measure of location* of a random variable. It does not say anything, however, about the spread or the dispersion of ξ. The identically vanishing random variable has mean zero, and so does a random variable which assumes each of the values $\pm 10^6$ with probability $\frac{1}{2}$. For many purposes it is convenient to introduce as a *measure of dispersion* a nonnegative quantity called the *variance* of ξ and defined by

$$\operatorname{var}(\xi) = E((\xi - \mu)^2) \tag{1-82}$$

The variance thus is the expected value of the square of the deviation of ξ from its mean μ. Using the above formulas with $f(\xi) = (\xi - \mu)^2$, we find the explicit expressions

$$\operatorname{var}(\xi) = \sum_{\nu} (x_\nu - \mu)^2 p_\nu \tag{1-83a}$$

in the discrete case, and

$$\operatorname{var}(\xi) = \int_{-\infty}^{\infty} (x - \mu)^2 p(x)\,dx \tag{1-83b}$$

in the continuous case. Both formulas are special cases of

$$\operatorname{var}(\xi) = \int_{-\infty}^{\infty} (x - \mu)^2 \, dF(x) \tag{1-83}$$

For the special distributions considered above we find in case (a)

$$\operatorname{var}(\xi) = \frac{1}{b - a} \int_a^b \left(x - \frac{b + a}{2} \right)^2 dx = \frac{(b - a)^2}{12}$$

and in case (b)

$$\begin{aligned} \operatorname{var}(\xi) &= \tfrac{1}{6}[(-\tfrac{5}{2})^2 + (-\tfrac{3}{2})^2 + (-\tfrac{1}{2})^2 + (\tfrac{1}{2})^2 + (\tfrac{3}{2})^2 + (\tfrac{5}{2})^2] \\ &= \tfrac{35}{12} \end{aligned}$$

We also note the following relations, which follow immediately from the definition of the variance and from (1-80) and (1-81): If a and b are any two constants, then

$$\operatorname{var}(a\xi) = a^2 \operatorname{var}(\xi) \tag{1-84}$$

$$\operatorname{var}(\xi + b) = \operatorname{var}(\xi) \tag{1-85}$$

Mechanically the variance corresponds to the moment of inertia around the center of gravity of the system of particles considered above.

If $\operatorname{var}(\xi) = \sigma^2$, then the nonnegative quantity σ has the same dimension as ξ and μ. It is called the *standard deviation* of the random variable ξ and may at the moment be considered a measure for the "average" deviation of a random variable from its expected value. The deeper significance of σ will become apparent in connection with sums of random variables.

1.5-3. Functions of several random variables. Let ξ and η be two random variables arising from two experiments that are executed simultaneously, and let their distribution functions be $F(x)$ and $G(y)$, respectively. We also introduce the joint distribution function $F(x, y)$ of ξ and η, defined as the probability that simultaneously $\xi \leq x$ and $\eta \leq y$:

$$F(x, y) = P(\xi \leq x \;\&\; \eta \leq y)$$

If $f(x, y)$ is any numerical-valued function of x and y, then $f(\xi, \eta)$ is again a random variable. We obtain its expected value by multiplying $f(x, y)$ by the probability that $\xi = x$, $\eta = y$, and summing with respect to all x and y. The result can be written in the general form

$$E(f(\xi, \eta)) = \int_{-\infty}^{\infty} \int_{-\infty}^{\infty} f(x, y) \, dF(x, y) \tag{1-86}$$

We wish to evaluate this integral in the special cases $f(\xi, \eta) = \xi + \eta$ and $f(\xi, \eta) = \xi\eta$.

We note that for $f(\xi, \eta) = \xi$ the integral (1-86) must yield the expected value of the variable ξ, the other variable η simply being disregarded. Thus

$$\int_{-\infty}^{\infty} \int_{-\infty}^{\infty} x\, dF(x, y) = E(\xi)$$

and similarly

$$\int_{-\infty}^{\infty} \int_{-\infty}^{\infty} y\, dF(x, y) = E(\eta)$$

But from this there follows immediately the basic result

$$E(\xi + \eta) = \int_{-\infty}^{\infty} \int_{-\infty}^{\infty} (x + y)\, dF(x, y) = E(\xi) + E(\eta) \tag{1-87}$$

which may be stated thus: *The expected value of the sum equals the sum of the expected values.*

No such simple rule exists in general for the product of two random variables. Fortunately, a simple result holds if the two random variables are *independent.* Two variables ξ and η are called independent if for any two intervals I and J the probability that simultaneously $\xi \in I$ and $\eta \in J$ equals the product of the individual probabilities that $\xi \in I$ and $\eta \in J$. In practical applications two random variables are considered independent if the two experiments defining ξ and η do not interfere with each other. In the language of probabilities, the condition of independence reads

$$P(\xi \in I \,\&\, \eta \in J) = P(\xi \in I) P(\eta \in J)$$

If this is applied to the intervals $I = [-\infty, x]$, $J = [-\infty, y]$, we find the fundamental relation

$$F(x, y) = F(x) G(y)$$

for the joint distribution function of two independent random variables. The last identity makes it possible to evaluate (1-86) for $f(\xi, \eta) = \xi\eta$; we find

$$E(\xi\eta) = \int_{-\infty}^{\infty} \int_{-\infty}^{\infty} xy\, dF(x, y) = \int_{-\infty}^{\infty} x\, dF(x) \int_{-\infty}^{\infty} y\, dG(y) = E(\xi)E(\eta) \tag{1-88}$$

In words: *The expected value of the product of two independent random variables equals the product of the expected values.*

We are now in a position to calculate the variance of a sum of two independent random variables. By definition,

$$\text{var}\,(\xi + \eta) = E((\xi + \eta - E(\xi + \eta))^2)$$

Introducing the new random variables

$$\xi' = \xi - E(\xi), \qquad \eta' = \eta - E(\eta)$$

we have

$$E(\xi') = 0, \qquad E(\xi'^2) = \text{var}\,(\xi)$$
$$E(\eta') = 0, \qquad E(\eta'^2) = \text{var}\,(\eta)$$

and hence, using both (1-87) and (1-88),

$$\begin{aligned}\text{var}\,(\xi + \eta) &= E((\xi' + \eta')^2)\\ &= E(\xi'^2 + 2\xi'\eta' + \eta'^2)\\ &= E(\xi'^2) + 2E(\xi')E(\eta') + E(\eta'^2)\end{aligned}$$

and finally

$$\text{var}\,(\xi + \eta) = \text{var}\,(\xi) + \text{var}\,(\eta) \tag{1-89}$$

It follows that *the variance of a sum of two independent random variables is equal to the sum of the variances.*

If the variables are not independent, the term $E(\xi'\eta')$ does not in general vanish. It is called the *covariance* of the two variables ξ' and η'.

The fundamental relations (1-87), (1-88), and (1-89) are by induction easily seen to hold for any finite number of random variables. Thus, if $\xi_1, \xi_2, \cdots, \xi_n$ are n random variables arising from simultaneously performed experiments, then

$$E(\xi_1 + \xi_2 + \cdots + \xi_n) = E(\xi_1) + E(\xi_2) + \cdots + E(\xi_n) \tag{1-90}$$

If, in addition, any two of the variables ξ_i are mutually independent, then also

$$E(\xi_1\xi_2 \cdots \xi_n) = E(\xi_1)E(\xi_2) \cdots E(\xi_n) \tag{1-91}$$

and

$$\text{var}\,(\xi_1 + \xi_2 + \cdots + \xi_n) = \text{var}\,(\xi_1) + \text{var}\,(\xi_2) + \cdots + \text{var}\,(\xi_n) \tag{1-92}$$

1.5-4. Sums of large numbers of random variables. In most of the later applications we shall be concerned with random variables of the form

$$r_n = d_{n1}\xi_1 + d_{n2}\xi_2 + \cdots + d_{nn}\xi_n \tag{1-93}$$

where the d_{nm} ($n = 1, 2, \cdots;\ m = 1, 2, \cdots, n$) are constants and the ξ_m are independent random variables having means and variances

$$E(\xi_m) = \mu_m, \qquad \text{var}\,(\xi_m) = \sigma_m^2, \qquad m = 1, 2, \cdots$$

Using the relations of the preceding two sections it is easy to calculate the mean and the variance of r_n. In view of (1-90) and (1-80) we find

$$E(r_n) = d_{n1}\mu_1 + d_{n2}\mu_2 + \cdots + d_{nn}\mu_n \tag{1-94}$$

Similarly (1-92) and (1-84) yield

$$\text{var}(r_n) = d_{n1}^2\sigma_1^2 + d_{n2}^2\sigma_2^2 + \cdots + d_{nn}^2\sigma_n^2 \tag{1-95}$$

The values $E(r_n)$ and var (r_n) convey a certain amount of information about the location and the dispersion of the random variable r_n, but it must be realized that in general they do not fully determine the distribution function corresponding to r_n. The explicit determination of the distribution function of a random variable of the form (1-93), even if the distributions of the ξ_m are known, is a very difficult problem, except in certain simple cases. Under these circumstances it is all the more gratifying that a very simple situation exists if n, the number of terms in (1-93), approaches infinity. It turns out that in this case under certain weak conditions the distribution function of r_n (after suitable normalization) approaches a simple function which is totally independent of the distributions of the variables ξ_m appearing in (1-93). This is the content of the so-called *central limit theorems* of probability theory. We state below a particular central limit theorem in a form which makes it immediately applicable to our problems.*

THEOREM 1.8 (LIMIT THEOREM). *Let ξ_m $(m = 1, 2, \cdots)$ be independent random variables with means μ_m and variances σ_m^2 such that the normalized variables $(\xi_m - \mu_m)/\sigma_m$ are uniformly bounded. Let the random variables r_n be defined by* (1-93), *and assume that*

$$\lim_{n\to\infty} \frac{d_{nm}\sigma_m}{[\text{var}(r_n)]^{1/2}} = 0 \qquad (m = 1, 2, \cdots) \tag{1-96}$$

uniformly in m. If $F_n(x)$ $(n = 1, 2, \cdots)$ are the distribution functions of the normalized random variables

$$\bar{r}_n = \frac{r_n - E(r_n)}{[\text{var}(r_n)]^{1/2}} \tag{1-97}$$

then

$$\lim_{n\to\infty} F_n(x) = \Phi(x) \tag{1-98}$$

where

$$\Phi(x) = \frac{1}{(2\pi)^{1/2}} \int_{-\infty}^{x} e^{-t^2/2}\, dt \tag{1-99}$$

A proof of this theorem is outside the scope of this book.

* This formulation has been adapted from Loève [1955], p. 295. It should be noted that the classical forms of the limit theorem are not immediately applicable to our case, because here the coefficients d_{nm} depend on n as well as m.

The function $\Phi(x)$ is known as the *normal distribution function*, and its derivative

$$\varphi(x) = \frac{1}{(2\pi)^{1/2}} e^{-x^2/2} \tag{1-100}$$

as the *normal frequency function*. A random variable whose distribution is the normal distribution function is called *normally distributed*. The limit theorem says, in effect, that under condition (1-96) the variables r_n, if suitably normalized, are for large values of n approximately normally distributed. More specifically, it enables us to state that for large n and for given values of x and y, $x < y$, the probability that

$$xs_n \le r_n - m_n \le ys_n, \qquad \text{where } s_n = (\text{var}\,(r_n))^{1/2}, \qquad m_n = E(r_n)$$

is approximately $\Phi(y) - \Phi(x)$. In particular, if $x = -y$, the probability that $|r_n - m_n| > ys_n$ is approximately

$$\pi(y) = 1 - (\Phi(y) - \Phi(-y))$$

In the table below we give values of the function $\pi(x)$ for $x = 1, 2, \cdots, 5$.

x	1	2	3	4	5
$\pi(x)$	0.31731	0.04500	0.00270	0.00006	0.00000

It follows that for a normally distributed variable only about 32% of the deviations from the mean exceed the standard deviation, and only about .27% exceed three times the standard deviation.

Further values of the normal distribution function may be taken from extensive existing tables (National Bureau of Standards [1942]).

1.6. Probabilistic Theory of Round-off Errors

Resuming the discussion of round-off error begun in §1.4, we now introduce the assumption that the local round-off errors are *random variables*.

This assumption calls for immediate criticism. Strictly speaking, the solution of a given initial value problem by a well-defined method, using digital computing equipment which is assumed to be in proper working order, is a fully determined event. If the calculation were repeated several times under the same conditions, exactly the same round-off errors would result each time. Using a sufficient amount of labor, it would be possible to figure out in advance what the round-off error at each step was going to be. The same objection can be raised, however, against that classical experiment in probability theory, the throwing of a die. This, too, is a strictly determined event, determined by the initial conditions, at least as long as the laws of Newtonian mechanics are assumed to hold. By integrating the equations of motion it is theoretically possible to determine in advance the outcome of throwing a die as a function of the initial conditions. If a die does not always show the same number of spots, this must be due to a change in the initial conditions. We have come to accept the outcome of the throwing of a die as a random event because its dependence on the initial conditions is very complex. In like manner, we may consider the local round-off errors as random events, because they, too, depend in a very complicated manner on the initial values assigned to the solution. In both cases, the solution of a differential equation and the throwing of a die, the *sample space*, i.e., the totality of all experiments which determine the random variable, is obtained by changing the initial conditions.

1.6-1. The distribution of the local round-off error. We recall the conventions and definitions introduced in §1.4-1, in particular the decomposition of the local round-off error into inherent and induced error. As mentioned there, the major contribution to the local round-off error will usually arise from the induced error. We shall for the time being neglect the inherent error altogether and set $\varepsilon_m = \pi_m$. The absence of the inherent error can be realized experimentally, e.g., by making $f(x, y) = cy$, c integral.

We now turn to the distribution of the induced error π_m. By definition, π_m is the error arising in rounding the product hf^*. Since both h and f^* are integral multiples of u, hf^* is an integral multiple of u^2. If both h and f^* are positive and if, as customary, a positive number exactly midway between the two nearest machine numbers is rounded up, then the round-off error of the product (in the binary system) has one of the $1/u = 2^N$ values

$$-\tfrac{1}{2}u + u^2,\ -\tfrac{1}{2}u + 2u^2,\ \cdots,\ \tfrac{1}{2}u - u^2,\ \tfrac{1}{2}u \tag{1-101}$$

We are interested in the distribution of the round-off error in all products of the form hf^*, where h is a constant and f^* is arbitrary. It would seem natural to assume that all errors (1-101) are equally probable. However, a simple example shows that this need not be true. If $h = \frac{1}{2}$, the only possible errors in hf^* in the binary system are

$$0 \quad \text{(when the last digit of } f^* \text{ is 0)}$$

and

$$\tfrac{1}{2}u \quad \text{(when the last digit of } f^* \text{ is 1)}$$

If the two values of the last digit of f^* are considered equally probable, then both round-off errors have equal probability $\frac{1}{2}$. Thus, among the 2^N values (1-101) only two occur in this case. We conclude that the actual distribution of round-off error in products hf^* with fixed h and random f^* depends on h.

The general result on the distribution of round-off error in the product of a fixed and of a random binary number is as follows:

THEOREM 1.9. *If $u = 2^{-N}$, if h has the constant value*

$$h = h_1 2^{-1} + h_2 2^{-2} + \cdots + h_k 2^{-k}$$

where $k \leq N$, $h_i = 0, 1$ $(i = 1, \cdots, k-1)$, $h_k = 1$, if z is a random number of the form

$$z = z_1 2^{-1} + z_2 2^{-2} + \cdots + z_N 2^{-N}$$

where $z_i = 0, 1$ $(i = 1, \cdots, N)$ and where for each $i \geq N - k + 1$ these two values of z_i are considered equally probable, then the round-off error in the product hz assumes with probability 2^{-k} each of the 2^k values

$$-\tfrac{1}{2}u + uv, \; -\tfrac{1}{2}u + 2uv, \cdots, \tfrac{1}{2}u - uv, \; \tfrac{1}{2}u$$

where $v = 2^{-k}$.

Proof. For any x we denote, as customary, by $[x]$ the greatest integer not exceeding x (frequently called the *integral part of* x). We shall denote by $\{x\}$ the number

$$\{x\} = x - [x]$$

to be called the *fractional part of* x. By definition,

$$0 \leq \{x\} < 1$$

and, for any integer m,

$$\{x + m\} = \{x\} \tag{1-102}$$

The round-off error r of the product hz depends on the quantity

$$q = \{2^N hz\}$$

If $0 \leq q < \frac{1}{2}$, then $r = -qu$; if $\frac{1}{2} \leq q < 1$, then $r = (1 - q)u$. The theorem will be proved if we can show that for each integer ν, $0 \leq \nu < 2^k$, and for preassigned values $z_1, \cdots, z_{N-k}$ there exists exactly one set of values $z_{N-k+1}, \cdots, z_N$ such that

$$q = \nu v \tag{1-103}$$

By straightforward multiplication we find, using $h_k = 1$,

$$\begin{aligned} q = \{ & z_N 2^{-k} + (z_{N-1} + h_{k-1} z_N) 2^{-k+1} \\ & + (z_{N-2} + h_{k-1} z_{N-1} + h_{k-2} z_N) 2^{-k+2} + \cdots \\ & + (z_{N-k+1} + h_{k-1} z_{N-k+2} + \cdots + h_1 z_N) 2^{-1} \} \end{aligned} \tag{1-104}$$

We now let

$$vv = q_1 2^{-1} + q_2 2^{-2} + \cdots + q_k 2^{-k} \tag{1-105}$$

where the q_i are either 0 or 1, and define $z_N, z_{N-1}, \cdots, z_{N-k+1}$ recursively by the condition that they be either 0 or 1 and satisfy the relations

$$\begin{aligned} z_N &= q_k \\ z_{N-1} + h_{k-1} z_N &= q_{k-1} + 2m_{k-1} \\ z_{N-2} + h_{k-1} z_{N-1} + h_{k-2} z_N &= q_{k-2} + 2m_{k-2} - m_{k-1} \\ &\cdots\cdots\cdots \\ z_{N-k+1} + h_{k-1} z_{N-k+2} + \cdots + h_1 z_N &= q_1 + 2m_1 - m_2 \end{aligned}$$

where the m_i $(i = 1, \cdots, k-1)$ are suitably chosen integers. Inserting these values z_i in (1-104), we find

$$\begin{aligned} q &= \{q_k 2^{-k} + (2m_{k-1} + q_{k-1})2^{-k+1} + (2m_{k-2} + q_{k-2} - m_{k-1})2^{-k+2} + \cdots \\ &\qquad + (2m_1 + q_1 - m_2)2^{-1}\} \\ &= \{q_k 2^{-k} + q_{k-1} 2^{-k+1} + \cdots + q_1 2^{-1} + m_1\} \\ &= vv \end{aligned}$$

by (1-105) and (1-102). Every one of the 2^k equations (1-103) is thus satisfied by one of the 2^k sets of values $(z_{N-k+1}, \cdots, z_N)$. Since no set can satisfy two equations, the theorem is proved.

Theorem 1.9 shows that the distribution of π_m depends on the last nonzero digit of h. The distribution function is given by

$$F_{u,v}(x) = \begin{cases} 0, & x < -\frac{1}{2}u + uv \\ vv, & x_\nu \le x < x_{\nu+1} \\ 1, & x \ge \frac{1}{2}u \end{cases} \tag{1-106}$$

where

$$x_\nu = -\tfrac{1}{2}u + \nu uv \qquad (\nu = 1, \cdots, 2^k)$$

It will be noted that the graph of the function $F_{u,v}(x)$ is symmetric with respect to the point $(\frac{1}{2}uv, \frac{1}{2})$ and *not* $(0, \frac{1}{2})$. This is due to the fact that, under the circumstances described, rounding up is slightly more frequent than rounding down. The probability diagram corresponding to $F_{u,v}(x)$ shows the constant probability $p_\nu = v$ attached to the points $x = x_\nu$ $(\nu = 1, \cdots, 2^k)$.

It remains to determine the expected value and the variance of the random variable π_m. Both quantities are most easily calculated by observing that the points x_ν and $x_{v^{-1}-\nu+1}$ are symmetrically located with respect to the point $x = \frac{1}{2}uv$. We thus obtain

$$\mu = E(\pi_m) = v \sum_{\nu=1}^{v^{-1}} x_\nu = v \sum_{\nu=1}^{\frac{1}{2}v^{-1}} (x_\nu + x_{v^{-1}-\nu+1}) = \tfrac{1}{2}uv \tag{1-107}$$

Similarly, putting $\lambda = \nu - \frac{1}{2}v^{-1}$, we have

$$\sigma^2 = \text{var}\,(\pi_m) = v \sum_{\nu=1}^{v^{-1}} (x_\nu - \tfrac{1}{2}uv)^2 = 2v \sum_{\lambda=1}^{\frac{1}{2}v^{-1}} (\lambda - \tfrac{1}{2})^2 u^2 v^2$$

Using the formula

$$1^2 + 3^2 + \cdots + (2n-1)^2 = \tfrac{1}{3}n(2n-1)(2n+1)$$

we easily find

$$\sigma^2 = \tfrac{1}{12}u^2(1 - v^2) \tag{1-108}$$

For small values of v the graph of the function $F_{u,v}(x)$ approaches that of the function

$$F_u(x) = \begin{cases} 0, & x \le -\frac{1}{2}u \\ \frac{1}{2} + x/u, & -\frac{1}{2}u \le x \le \frac{1}{2}u \\ 1, & x \ge \frac{1}{2}u \end{cases} \tag{1-109}$$

corresponding to a continuous and uniform distribution of the values π_m in the interval $[-\frac{1}{2}u, \frac{1}{2}u]$. The expected value and the variance of this distribution are easily found to be

$$\mu = 0, \qquad \sigma^2 = \tfrac{1}{12}u^2 \tag{1-110}$$

and these indeed are the values approached by (1-107) and (1-108) when $v \to 0$. For most practical purposes it is sufficient to approximate the distribution of π_m by $F_u(x)$.

1.6-2. The distribution of the accumulated round-off error. We now apply to relation (1-64) the general results on random variables outlined in §§1.5-3 and 1.5-4, assuming that $\varepsilon_m = \pi_m$ and π_m has one of the distributions $F_{u,v}(x)$ or $F_u(x)$.

From (1-93) we immediately get

$$E(r_n) = \sum_{m=1}^{n} d_{n,m} E(\varepsilon_m)$$

It would not be realistic at this point to assume that $E(\varepsilon_m) = \mu$ since in most computing machines the absolute value is rounded, not the number itself. This means that for negative products hg^* we have $E(\varepsilon_m) = -\mu$. Thus, all we can say in general is that

$$|E(\varepsilon_m)| \le \mu$$

If h is sufficiently small and all $d_{n,m}$ are positive, it follows that

$$|E(r_n)| \le \mu \sum_{m=1}^{n} d_{nm} = \frac{\mu}{h} m_n \tag{1-111}$$

where m_n is defined by (1-70). Using (1-71) we find

$$|E(r_n)| \leq \frac{\mu}{h}\{m(x_n) + O(h)\} \tag{1-112}$$

where $m(x)$ is the function defined by (1-72). Comparing this result with Theorem 1.7, we find that $|E(r_n)|$ grows at most like μ/ε times max $|r_n|$. For the distribution $F_{u,v}(x)$ we have $\mu/\varepsilon = v$.

If we wish to determine var (r_n) in a similar manner, we must make the important hypothesis that *the random variables ε_m are independent.* It does not seem easy either to prove this hypothesis or to reduce it to a simpler one, and examples have been produced (Huskey [1949]) which tend to show that under extreme conditions the ε_m may to a certain extent be dependent. For these reasons we shall consider our assumption as a working hypothesis which has to be (and will be) verified by numerical experimentation.

If the ε_m are independent, and if var $(\varepsilon_m) = \sigma^2$, we find by applying (1-94) that

$$\text{var}\,(r_n) = \frac{\sigma^2}{h} v_n \tag{1-113}$$

where

$$v_n = h \sum_{m=1}^{n} d_{nm}^2 \tag{1-114}$$

We try to evaluate the sums v_n, called *reduced variances*, by establishing a difference equation for them. We have

$$\begin{aligned} v_{n+1} - v_n &= h\left\{\sum_{m=1}^{n+1} d_{n+1,m}^2 - \sum_{m=1}^{n} d_{n,m}^2\right\} \\ &= h\left\{d_{n+1,n+1}^2 + \sum_{m=1}^{n} (d_{n+1,m}^2 - d_{n,m}^2)\right\} \\ &= h\left\{1 + \sum_{m=1}^{n} (d_{n+1,m} - d_{n,m})(d_{n+1,m} + d_{n,m})\right\} \end{aligned}$$

Using (1-65) repeatedly, we deduce that

$$\begin{aligned} v_{n+1} - v_n &= h\left\{1 + g(x_n)h \sum_{m=1}^{n} d_{n,m}[2d_{n,m} + hg(x_n)d_{n,m}]\right\} \\ &= h\{1 + 2g(x_n)v_n\} + h^2 g(x_n)^2 v_n \end{aligned}$$

From (1-67) we know that the quantities v_n are bounded. The values $g(x_n)^2$ are also bounded, because $g(x_n)$ is continuous. We thus can write

$$v_{n+1} - v_n = h\{1 + 2g(x_n)v_n\} + O(h^2)$$

This is a difference equation to which Theorem 1.4 can be applied. Since $v_0 = 0$, it follows that

$$v_n = v(x_n) + O(h) \tag{1-115}$$

where the function $v(x)$, called the *reduced variance function*, is defined by

$$\begin{aligned} v(a) &= 0 \\ v'(x) &= 2g(x)v(x) + 1 \end{aligned} \tag{1-116}$$

We have thus found that

$$\operatorname{var}(r_n) = \frac{\sigma^2}{h}\{v(x_n) + O(h)\} \tag{1-117}$$

This result at once permits the qualitative conclusion that the standard deviation of r_n, which was seen to be typical for the "average" round-off error, is of the order of $h^{-1/2}$, which is by a factor $h^{1/2}$ better than the largest theoretically possible round-off error given by (1-73).

It remains to determine the distribution of r_n. Since the variables π_m are bounded, the conclusion of the central limit theorem (Theorem 1.8) is applicable if we can show that the quantities

$$D_{n,m} = \frac{d_{n,m}}{(d_{n,1}^2 + d_{n,2}^2 + \cdots + d_{n,n}^2)^{1/2}}$$

tend to zero, uniformly with respect to m, as $n \to \infty$, $nh = x > a$. Defining

$$G(x) = \int_a^x g(t)\,dt$$

the explicit solution of (1-116) can be written in the form

$$v(x) = e^{2G(x)} \int_a^x e^{-2G(t)}\,dt$$

It follows that $v(x) > 0$. By (1-67),

$$D_{n,m} = h^{1/2}\frac{d_{n,m}}{v_n^{1/2}} = h^{1/2}\left\{\frac{e^{G(x_n)-G(x_m)}}{[v(x_n)]^{1/2}} + O(h)\right\}$$

and the required result is obvious.

It is now easy to see that a hypothesis such as (1-96) is necessary for the validity of the limit theorem. Suppose that the values $d_{n,m}$ are given by

$$d_{n,m} = \begin{cases} 1, & n = m \\ 0, & n > m \end{cases}$$

This corresponds to an infinitely rapid decay of the influence of a local round-off error on the accumulated round-off error. Clearly, the distribution of r_n will always be the same as the distribution of ε_n, and thus will not in general be normal.

In deriving the results of this section, no use has been made of any special properties of the distribution of ε_m. We may therefore sum up the results in a general way as follows:

THEOREM 1.10. *If the local round-off errors ε_m in the solution of $y' = g(x)y + p(x)$ by Euler's method are bounded, independent random variables with $|E(\varepsilon_m)| \le \mu$ and $\operatorname{var}(\varepsilon_m) = \sigma^2$, then the accumulated round-off error r_n is a random variable satisfying*

$$|E(r_n)| \le \frac{\mu}{h}\{m(x_n) + O(h)\}$$

(1-118)

$$\operatorname{var}(r_n) = \frac{\sigma^2}{h}\{v(x_n) + O(h)\}$$

where $m(x)$ and $v(x)$ are defined by (1-72) *and* (1-116) *respectively. Moreover, the distribution of the normalized variable $[r_n - E(r_n)]/[\operatorname{var}(r_n)]^{1/2}$ approaches the normal distribution as $h \to 0$, $nh = x - a$.*

1.6-3. Numerical example. We shall apply the results obtained above to the differential equations $y' = \pm y$. These equations are linear; moreover, since $g(x) = \pm 1$, there is no inherent error. We thus expect the results of §1.6-2 to hold.

For $y' = y$ we have already determined

$$m(x) = e^x - 1$$

Integrating (1-116) we easily find

$$v(x) = \tfrac{1}{2}(e^{2x} - 1)$$

We thus expect, approximately,

$$E(r_n) = \frac{\mu}{h}(e^{x_n} - 1)$$

$$\operatorname{var}(r_n) = \frac{\sigma^2}{2h}(e^{2x_n} - 1)$$

In particular, for $n = 10^3$, $h = 10^{-3}$, $x_n = 1$ we find, using the values of μ and σ^2 given by the continuous distribution $F_u(x)$,

$$E(r_{1000}) = 0$$

$$\operatorname{var}(r_{1000}) = \frac{u^2 \cdot 10^3}{24}(e^2 - 1) = 266.2u^2$$

According to the central limit theorem, the distribution of r_{1000} should be approximately normal.

The following experiment was performed with the intention of testing these results. The equation $y' = y$ was solved by Euler's method for the 500 different initial conditions

$$y_0 = y_{0,q} = \tfrac{1}{8} + q\Delta$$

where $q = 0, 1, \cdots, 499$, and

$$\Delta = \tfrac{1}{3} \cdot 2^{-13} \qquad \text{(rounded)}$$

working with $u = 2^{-28}$. The numerical end values $\tilde{y}_{n,q}$ were compared with the theoretical values

$$y_{n,q} = y_{0,q}(1 + h)^{h^{-1}}$$

calculated from the formula

$$(1 + h)^{h^{-1}} = e^{h^{-1}\log(1+h)} = e\{1 - \tfrac{1}{2}h + \tfrac{11}{24}h^2 - \tfrac{21}{48}h^3 + \cdots\}$$

The round-off errors

$$r_{n,q} = \tilde{y}_{n,q} - y_{n,q}$$

were recorded and experimental values of the expectation and the variances were calculated from the formulas

$$E(r_n)_e = \frac{1}{500}\sum_{q=0}^{499} r_{n,q}$$

$$\text{var}\,(r_n)_e = \frac{1}{500}\sum_{q=0}^{499} (r_{n,q} - E(r_n)_e)^2$$

The following numerical values were obtained for $n = 1000$:

$$E(r_n)_e = -0.2u, \qquad \text{var}\,(r_n)_e = 266.7u^2$$

The distribution of the values $r_{n,q}$ is shown in Fig. 1.6. The height of each rectangle is proportional to the number of values $r_{1000,q}$ in the interval making up its base. Also shown, in the same scale, is the normal distribution.

The excellent agreement between the predicted and the experimental values is evident, and the actual distribution obviously resembles the normal distribution.* We conclude that in this case the statistical theory of round-off is vindicated by the facts.

Similar results were obtained for the differential equation $y' = -y$. For this equation one finds

$$v(x) = \tfrac{1}{2}(1 - e^{-2x})$$

* Refined methods for testing statistical hypotheses are available, but they lie outside the scope of this book.

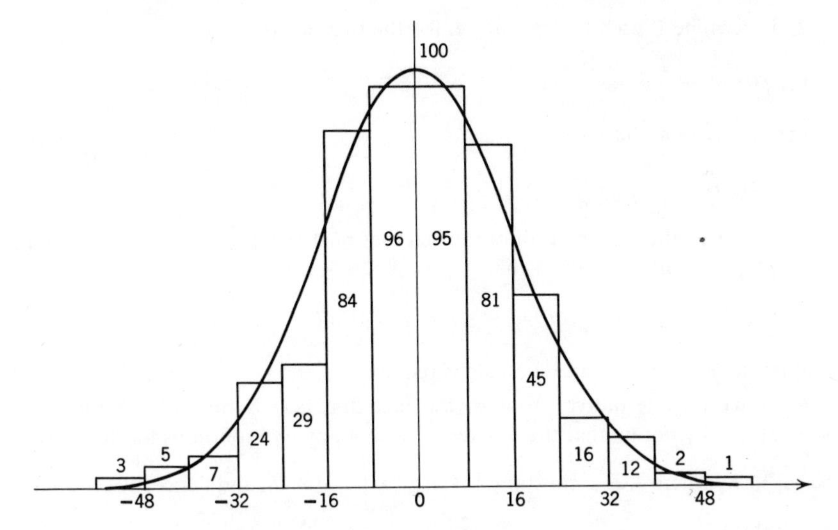

Fig. 1.6 Distribution of round-off errors.

and hence, using the same data as above,

$$E(r_{1000}) = 0$$

$$\text{var}\,(r_{1000}) = \frac{u^2 \cdot 10^3}{24}(1 - e^{-2}) = 36.0u^2$$

A numerical experiment, performed under the same conditions as the one described above, yielded the values

$$E(r_{1000})_e = -0.14u$$

$$\text{var}\,(r_{1000})_e = 39.4u^2$$

again in satisfactory agreement with the theory.

1.7. Problems for Solution

An asterisk denotes problems of above average difficulty.

Section 1.2

1. Show that Euler's method fails to approximate the solution $y = (\frac{2}{3}x)^{3/2}$ of the initial value problem $y' = y^{1/3}$, $y(0) = 0$. Explain.

2. Determine Lipschitz constants L for the functions

(a) $f(x, y) = \dfrac{2}{x} y \quad (x \geq 1)$

(b) $f(x, y) = \arctan y$

(c) $f(x, y) = \dfrac{(x^3 - 2)^{27}}{17x^2 + 4}$

3*. Give an alternative proof of the fact that $y(x) = \lim_{p\to\infty} y_p(x)$ is a solution of $y' = f(x, y)$ by using the fact that

$$y_p(x) = y_0 + \int_a^x f_p(t)\, dt$$

and letting $p \to \infty$. (Indicate all steps required.)

4*. Give an alternative proof of the fact that $y(x)$ is the *only* solution of $y' = f(x, y)$ by noting that the difference $\Delta(x)$ of any two solutions has to satisfy $|\Delta(x)| \leq K$, where K is a constant, $|\Delta(x)| \leq L \int_a^x |\Delta(t)|\, dt$, and hence, by repeated integration,

$$|\Delta(x)| \leq K \frac{L^n}{n!} (x - a)^n$$

for every positive integer n, which is possible only if $\Delta(x) = 0$.

Section 1.3

5. Determine analytically the Euler approximation to the initial value problem $y' = 2x^{-1}y$, $y(1) = 1$. Find also the exact solution of the problem and determine the magnified error function. Verify relation (1–49). [*Hint:* Obtain a simple recurrence relation for the quantities y_n.]

6. How large (approximately) is the discretization error of the approximation to the solution of

$$y' = \left(\frac{1}{x} + 1\right) y, \qquad y(1) = e$$

obtained by Euler's method?

7. Obtain a bound for the constant C_3 appearing in Theorem 1.5 in terms of bounds on the function $f(x, y)$ and its derivatives.

8*. Consider the linear difference equation for a function $y(x, h)$,

$$\begin{aligned} y(x + h, h) - y(x, h) &= h[f(x)y(x, h) + g(x)] \\ y(a, h) &= y_0 \end{aligned} \tag{1-119}$$

where h plays the role of a parameter. Show that (1-119) is formally satisfied by the infinite series (not necessarily convergent)

$$y(x, h) = \sum_{p=0}^{\infty} h^p \eta_p(x) \tag{1-120}$$

where

$$\eta_0' = f(x)\eta_0 + g(x), \qquad \eta_0(a) = y_0$$

and

$$\begin{aligned}
\eta_1' &= f(x)\eta_1 - \tfrac{1}{2}\eta_0'' \\
\eta_2' &= f(x)\eta_2 - \tfrac{1}{2}\eta_1'' - \tfrac{1}{6}\eta_0''' \\
&\cdots\cdots\cdots\cdots\cdots\cdots \\
\eta_k' &= f(x)\eta_k - \tfrac{1}{2}\eta_{k-1}'' - \cdots - \frac{1}{(k+1)!}\eta_0^{(k+1)} \\
&\cdots\cdots\cdots\cdots\cdots\cdots
\end{aligned}$$

$$\eta_k(a) = 0, \qquad k = 1, 2, \cdots$$

9*. Show that the series (1-120) converges for

(a) $f(x) = 1, \quad g(x) = 0, \quad a = 0$

(b) $f(x) = -1/x, \quad g(x) = 0, \quad a = 1$

Determine the radii of convergence and calculate the first few functions $\eta_p(x)$.

10. The initial value problem

$$y' = -2y + \tfrac{1}{2}x^2 - \tfrac{1}{3}x^3 - \tfrac{1}{6}x^4, \qquad y(0) = \tfrac{1}{8} \tag{1-121}$$

has the exact solution

$$y(x) = \tfrac{1}{8} - \tfrac{1}{4}x + \tfrac{1}{4}x^2 - \tfrac{1}{12}x^4$$

Using Euler's method with the steps $h = 2^{-p}$ ($p = 1, 2, \cdots, 8$) the following approximations $y(1, 2^{-p})$ for $y(1) = \frac{1}{24}$ were obtained:

p	$y(1, 2^{-p})$
1	0.036458333
2	0.040486653
3	0.041414746
4	0.041608960
5	0.041652882
6	0.041663299
7	0.041665834
8	0.041666459

Verify that in this case, judging from the numerical values, not only $\lim_{h\to 0} h^{-1}e(1, h)$, but also $\lim_{h\to 0} h^{-2}e(1, h)$ appears to exist, contrary to what can be expected in general from Theorem 1.5. Verify also that extrapolation to the limit does not improve convergence in this case. Explain by determining the magnified error function.

11*. By using the results of Problem 8, show that the error $e(1, h)$ of the approximation considered in Problem 10 satisfies

$$\lim_{h\to 0} h^{-2}e(1, h) = \frac{5e^{-2} - 1}{24} = 0.0135\cdots$$

12. Determine analytically the values y_n obtained by solving the initial value problem $y' = x - x^3$, $y(0) = 0$ by Euler's method. Calculate the error $e_n = y_n - y(x_n)$ and compare it with the approximation furnished by the magnified error function. Is there a value of x (besides 0) for which the error is particularly small when h is small?

[Use the formulas

$$1 + 2 + \cdots + n = \frac{n(n+1)}{2}$$

$$1^3 + 2^3 + \cdots + n^3 = \left[\frac{n(n+1)}{2}\right]^2 \quad]$$

13*. It is known that

$$\lim_{n\to\infty}\left(1 + \frac{x}{n}\right)^n = e^x$$

If K is a constant and $|\theta_n| \le 1$ $(n = 1, 2, \cdots)$, prove also that

$$\lim_{n\to\infty}\left(1 + \frac{x}{n} + \theta_n\frac{K}{n^2}\right)^n = e^x$$

[Use Theorem 1.4.]

Section 1.4

14. Determine the maximum permissible local round-off error ε if the accumulated round-off error in the numerical integration of the problem $y' = (-1/x)y, y(1) = 1$, using $h = 10^{-2}$ should after 1000 steps not exceed 10^{-4}

(a) by using Theorem 1.6;
(b) by using Theorem 1.7.

Section 1.6

15. Determine the distribution function $F(x)$ of the round-off error in the product hf if both h and f are arbitrary N-digit binary numbers which assume all values with equal probability. Determine the mean and the variance for this distribution.

16. Determine the distribution of the round-off errors of the product hf under the hypothesis of Theorem 1.9, but assuming that all products hf are rounded to the nearest N-digit number not exceeding hf (rounding by chopping). Also find μ and σ^2 for this distribution.

17. Perform a small numerical experiment on the distribution of accumulated round-off error in solving $y' = y$ by Euler's method, using decimal computation and always rounding *down*. Compare the results with the values predicted by Theorem 1.9 in conjunction with the results of Problem 16. [Working, in the notation of §1.6-3, with $y_0 = 1$, $\Delta = 0.10101$, $q = 0, \cdots, 9$, $u = 10^{-5}$, $h = 0.125$ yields $E(r_8)_e = -5.6u$, var $(r_8)_e = 1.36u^2$.]

18. Give an alternative proof of (1-117) by introducing into (1-114) the approximate values $d_m(x_n)$ defined by (1-67) and considering the sum as an approximation to a definite integral.

19*. If the functions $m(x)$ and $v(x)$ are defined by (1-72) and (1-116) respectively, show that for $x > a$

$$[v(x)]^{1/2} \geq \frac{1}{(x-a)^{1/2}} m(x)$$

As an application prove the inequality

$$\left(\frac{e^{2x}-1}{2}\right)^{1/2} \geq \frac{1}{x^{1/2}} (e^x - 1)$$

[*Hint*: Solve the differential equations in terms of definite integrals and apply the Schwarz inequality.]

Notes

§1.1. Euler's method is also known as the Euler-Cauchy polygon method; Cauchy [1840] has the first precise statement on convergence.

§1.2. The existence theorem given here and its proof differ from Peano's existence theorem (see, e.g., Kamke [1947], p. 59; Coddington and Levinson [1955], p. 6), which is also based on the polygon method, in the fact that the Lipschitz condition is used from the beginning. This makes it possible to avoid an appeal to the (nonconstructive) Ascoli theorem and to prove convergence of the whole sequence $\{y_p(x)\}$ rather than of a subsequence.

§1.3. For a slightly different proof of Theorem 1.3 see Collatz [1960], p. 58.

§1.6. The statistical model of round-off error propagation in the step-by-step integration of differential equations seems to have been introduced by Brouwer [1937]. It was subsequently refined by Rademacher [1948], Papoulis [1952], Mikulaschkova [1957]. Objections against the model (in particular against the hypothesis of independence) are raised by Huskey [1949], Forsythe [1959], van Wijngaarden [1953]. Forsythe [1959] proposes to replace the "natural" rounding error by an artificially generated random error. It would be interesting to test this procedure in the way indicated in §1.6-3.

chapter

2 General One-Step Methods for a Single Equation of the First Order

It has been shown in Chapter 1 that arbitrarily high accuracy can be achieved by Euler's method by choosing the step h sufficiently small, provided that the arithmetic is carried out in such a way that round-off is negligible. The discretization error is roughly proportional to h, which means that the accuracy that can be achieved over a given interval with N evaluations of the function $f(x, y)$ is proportional to $1/N$. It will be shown in this chapter that there are one-step methods which are much more effective than Euler's method in the sense that the accuracy attainable with N evaluations is proportional to $1/N^p$ with $p > 1$.

Any algorithm for solving a differential equation in which the approximation y_{n+1} to the solution at the point x_{n+1} can be calculated if only x_n, y_n, and h are known is called a *one-step method*. It is convenient to write the functional dependence of y_{n+1} on the quantities x_n, y_n, h in the form

$$y_{n+1} - y_n = h\Phi(x_n, y_n;\ h) \tag{2-1}$$

The function Φ will be called the *increment function*. It depends on the given differential equation. For instance, in Euler's method,

$$\Phi(x, y;\ h) = f(x, y) \tag{2-2}$$

In this case Φ happens to be independent of h.

The quantity $h\Phi(x_n, y_n;\ h)$ represents the increment of the approximate solution. In order to facilitate comparison with the increment of the exact solution, we introduce a special notation for the latter. If the function $f = f(x, y)$ satisfies the conditions of Theorem 1.1, and if (x, y) is an arbitrary point of the region $a \le x \le h$, $-\infty < y < \infty$, where the function f is defined, then for $t \in [x, b]$ there exists a solution of the initial value problem

$$z' = f(t, z), \qquad z(x) = y \tag{2-3}$$

By a trivial modification of the same theorem (replace x by $-x$) the solution $z(t)$ can be continued backward through the interval $[a, x]$. The function $z(t)$ thus defined can be described geometrically as *the* solution of the differential equation $z' = f(t, z)$ passing through the point (x, y). For arbitrary h such that $x + h \in [a, b]$ we now define the function Δ, to be called the *exact relative increment*, by

$$\Delta(x, y; h) = \begin{cases} \dfrac{z(x + h) - y}{h}, & h \neq 0 \\ f(x, y), & h = 0 \end{cases} \tag{2-4}$$

By virtue of the fact that $z(t)$ is a solution of (2-3), Δ as a function of h is continuous at $h = 0$. The quantity $h\Delta(x, y; h)$ represents the increment of the exact solution of the given differential equation at the point (x, y); Δ itself for $h \neq 0$ represents a difference quotient.

2.1. Special One-Step Methods

2.1-1 Direct use of Taylor's expansion. If the function $y(x)$ is a solution of $y' = f(x, y)$, where we assume $f(x, y)$ infinitely differentiable, then the higher derivatives of $y(x)$ are completely determined by the function $f(x, y)$. For instance,

$$\begin{aligned} y'' &= \frac{d}{dx} f(x, y) = f_x(x, y) + f_y(x, y) \cdot y' \\ &= f_x(x, y) + f_y(x, y) f(x, y) \end{aligned} \tag{2-5}$$

and y''', y^{IV}, $\cdots$ are obtained in a similar manner. To simplify the notation, we write

$$y'' = f'(x, y), \qquad y''' = f''(x, y)$$

and generally,

$$y^{(p+1)} = f^{(p)}(x, y) \qquad (p = 1, 2, \cdots)$$

The primes denote *total* derivatives with respect to x.

With this notation, the increment of the solution passing through the point (x, y) is given by

$$z(x + h) - z(x) = hf(x, y) + \frac{h^2}{2} f'(x, y) + \cdots + \frac{h^p}{p!} f^{(p-1)}(x, y) + O(h^{p+1})$$

For the exact relative increment we thus find the expansion

$$(2\text{-}6)\quad \Delta(x, y;\ h) = f(x, y) + \frac{h}{2} f'(x, y) + \cdots + \frac{h^{p-1}}{p!} f^{(p-1)}(x, y) + O(h^p)$$

Here p is an arbitrary positive integer.

It seems natural to choose the increment function $\Phi(x, y;\ h)$ in such a manner that it agrees with Δ as closely as possible. One obvious way of achieving this goal is to put

$$(2\text{-}7)\quad \Phi(x, y;\ h) = f(x, y) + \frac{h}{2} f'(x, y) + \cdots + \frac{h^{p-1}}{p!} f^{(p-1)}(x, y)$$

The method defined by (2-7) will be called the *Taylor expansion method of order p*. Euler's method is recovered by putting $p = 1$. The accuracy of the method can be increased arbitrarily by increasing p.

The computational disadvantages of the Taylor expansion method are obvious. Instead of dealing with the one function f, one now has to operate simultaneously with the p functions $f, f', \cdots, f^{(p-1)}$. In the case of automatic computation this increases the space required for storing the code. Moreover, in all but the simplest equations (mostly those which can be integrated in closed form), the complexity of the analytical expressions for the functions $f^{(k)}$ increases as k increases. (The reader is asked to verify this statement for an innocent-looking differential equation such as $y' = x^2 + y^2$.) The practical use of this method is therefore in general not recommended.

2.1-2. Indirect use of Taylor's expansion (methods of Runge-Kutta type). The basic idea here is to produce a formula for Φ which agrees with Δ with an error of $O(h^p)$, where $p > 1$, without making it necessary to compute any derivatives of f. We illustrate the idea by the following simple example. Let us tentatively put

$$\Phi(x, y;\ h) = a_1 f(x, y) + a_2 f(x + p_1 h, y + p_2 h f(x, y))$$

where a_1, a_2, p_1, p_2 are constants to be determined. This quantity can be formed by calculating f twice for two different sets of arguments. Expanding in powers of h, we get

$$\Phi(x, y;\ h) = (a_1 + a_2) f(x, y) + h a_2 [p_1 f_x(x, y) + p_2 f_y(x, y) f(x, y)] + O(h^2)$$

We now impose the condition that this agrees as well as possible with the quantity $\Delta(x, y;\ h)$, which in view of (2-5) may in expanded form be written as follows:

$$\Delta(x, y;\ h) = f(x, y) + \tfrac{1}{2} h [f_x(x, y) + f_y(x, y) f(x, y)] + O(h^2)$$

Comparing the constant terms, we find

$$a_1 + a_2 = 1 \tag{2-8a}$$

Agreement in the linear terms in h requires

$$a_2 p_1 = \tfrac{1}{2}, \qquad a_2 p_2 = \tfrac{1}{2} \tag{2-8b}$$

It can be shown that for the quadratic term no agreement can be obtained without imposing conditions on $f(x, y)$. Solving the equations (2-8), we find the family of solutions

$$a_1 = 1 - \alpha, \qquad a_2 = \alpha, \qquad p_1 = p_2 = \frac{1}{2\alpha}$$

where α is a free parameter, $\alpha \neq 0$. By virtue of its derivation the increment function

$$\Phi(x, y; h) = (1 - \alpha)f(x, y) + \alpha f\left(x + \frac{h}{2\alpha}, y + \frac{h}{2\alpha} f(x, y)\right) \tag{2-9}$$

deviates for $\alpha \neq 0$ from $\Delta(x, y; h)$ by only $O(h^2)$, in spite of the fact that it does not involve $f'(x, y)$. Two particular cases of (2-9) are well known:

(i) $\alpha = \frac{1}{2}$. Here (2-1) takes the form

$$y_{n+1} = y_n + \tfrac{1}{2}h[f(x_n, y_n) + f(x_n + h, y_n + hf(x_n, y_n))] \tag{2-10a}$$

This method is sometimes called the "*improved Euler method*" (Collatz [1960], p. 54); it is also known as the *Heun* method (Heun [1900]; Rademacher [1948]).

(ii) $\alpha = 1$. There results the formula

$$y_{n+1} = y_n + hf(x_n + \tfrac{1}{2}h, y_n + \tfrac{1}{2}hf(x_n, y_n)) \tag{2-10b}$$

which is called the *improved polygon method* or the *modified Euler method* (Collatz [1960], p. 53).

Both methods (2-10) admit of simple graphical interpretations which we leave to the reader. Both methods agree in that they require two evaluations of the function $f(x, y)$ per integration step. If $f(x, y)$ does not actually depend on y and the solution of the differential equation reduces to the evaluation of an integral, (2-10a) reduces to the trapezoidal rule and (2-10b) to the mid-point rule of numerical integration.

The general method defined by (2-9) may also be called a *simplified Runge-Kutta* method. The function (2-9) may be looked at as a weighted average of values of the function f taken at different points, with the peculiarity that the arguments in the second term depend on the value of the first term. The *classical Runge-Kutta method* (Runge [1895]; Kutta [1901])—probably the best known of all one-step methods—is similar in spirit. Here Φ is a weighted average of *four* evaluations of the function f, with the arguments in each evaluation depending on the preceding evaluation. Specifically,

(2-11*a*) $$\Phi(x, y; h) = \tfrac{1}{6}[k_1 + 2k_2 + 2k_3 + k_4]$$

where

(2-11*b*) $$\begin{cases} k_1 = f(x, y) \\ k_2 = f(x + \tfrac{1}{2}h, y + \tfrac{1}{2}hk_1) \\ k_3 = f(x + \tfrac{1}{2}h, y + \tfrac{1}{2}hk_2) \\ k_4 = f(x + h, y + hk_3) \end{cases}$$

A one-step method is called *of order p* if its increment function approximates the exact relative increment Δ with an error of the order of h^p. The integer p is called the *exact order* of the method if this statement is not true for any larger integer. Thus, the Taylor expansion method defined by (2-7) is of the exact order p, except when $f^{(k)}(x, y) = 0$ for $k \geq p$.

The simplified Runge-Kutta method (2-9) by construction is of order 2. It is shown in Chapter 3 that the method defined by (2-11) is of order 4, and that this in general is the exact order of the method. The latter remark may already be inferred from the following consideration: If $f(x, y)$ does not depend on y, then $k_2 = k_3$, and the Runge-Kutta method reduces to Simpson's rule for numerical integration, which is known to be of order 4 in the sense here defined.*

Computational considerations. The flow diagram for the solution of an initial value problem by any one-step method looks like the one for Euler's method (Fig. 1.1), with the exception that in more complicated cases the place of the computation of $f(x_n, y_n)$ is now taken by the evaluation of the more complicated function $\Phi(x_n, y_n; h)$. We shall discuss the evaluation of Φ for the classical Runge-Kutta method (2-11). We may write (2-11) in the form

(2-12*a*) $$\Phi(x, y;\ h) = \sum_{\nu=1}^{4} a_\nu k_\nu$$

where $k_\nu = f(X_\nu, Y_\nu)$ with $X_1 = x$, $Y_1 = y$, and

(2-12*b*) $$X_\nu = x + hp_\nu, \qquad Y_\nu = y + hp_\nu k_{\nu-1}, \qquad \nu = 2, 3, 4$$

* Runge's early investigation of the method was motivated by an attempt to extend Simpson's rule to the integration of differential equations.

The constants a_ν and p_ν have the values

ν	1	2	3	4
a_ν	$\frac{1}{6}$	$\frac{1}{3}$	$\frac{1}{3}$	$\frac{1}{6}$
p_ν	0	$\frac{1}{2}$	$\frac{1}{2}$	1

and are best stored once for all. Accordingly, the evaluation of Φ may itself be made into a cyclic process. Figure 2.1 gives the complete flow

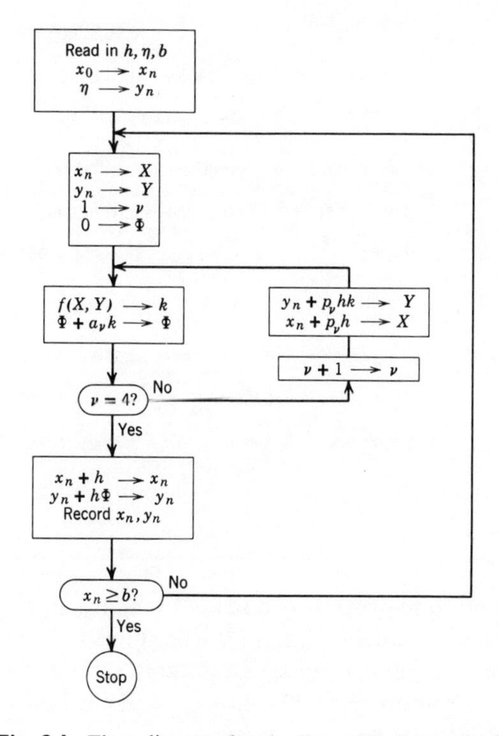

Fig. 2.1. Flow diagram for the Runge-Kutta method.

diagram for the method. (A special device for reducing the round-off error in the Runge-Kutta and similar methods is discussed in §2.3-5.)

As a numerical illustration we show in Table 2.1 the first few steps of the solution of the initial value problem

$$y' = x - y^2, \qquad y(0) = 0$$

(already considered in §1.1-2) by the Runge-Kutta method, using the step $h = 0.1$. Arrows indicate the order in which the values are calculated.

For manual computation the lack of convenient checks must be listed as a disadvantage of the method.

Table 2.1

Numerical Example for Runge-Kutta Method

x_n	y_n		X_ν, Y_ν, $k_\nu = f(X_\nu, Y_\nu)$		
		$\nu = 1$	2	3	4
0	**0**	0	0.05000 00000	0.05000 00000	0.10000 00000
		0	0	0.00250 00000	0.00499 93750
		0	0.05000 00000	0.04999 37500	0.09997 50062
			$\Phi =$ **0.04999 37510**		
0.1	**0.00499 93751**	0.10000 00000	0.15000 00000	0.15000 00000	0.20000 00000
		0.00499 93751	0.00999 81254	0.01249 43770	0.01998 37642
		0.09997 50062	0.14990 00375	0.14984 38905	0.19960 06492
			$\Phi =$ **0.14984 39185**		
0.2	**0.01998 37670**	0.20000 00000	0.25000 00000	0.25000 00000	0.30000 00000
		0.01998 37670	0.02996 37995	0.03243 88755	0.04487 52950
		0.19960 06491	0.24910 21707	0.24891 52805	0.29798 62034
			$\Phi =$ **0.24893 69591**		
0.3	**0.04487 74629**				

2.1-3. Further one-step methods. In addition to the methods described above, a great many others have been proposed in the literature. Reasons of space prevent a detailed description; some additional methods are discussed in Problems 3–9, 12–14, and 18 at the end of the chapter.

2.2. The Discretization Error of the General One-Step Method

2.2-1. Convergence and consistency. Let $\Phi(x, y; h)$ be an increment function of a one-step method for approximating solutions of a given differential equation $y' = f(x, y)$. We shall always assume that the function $f(x, y)$ satisfies the conditions of Theorem 1.1, so that the initial value problem

$$y' = f(x, y), \qquad y(a) = \eta \tag{2-13}$$

has a solution $y(x)$ ($x \in [a, b]$) for arbitrary η. We also shall assume that the increment function $\Phi(x, y; h)$ is defined for $x \in [a, b]$, $y \in (-\infty, \infty)$, and at any given point (x, y) for all sufficiently small $h \geq 0$. Starting with $y_0 = \eta$, it is then possible to calculate the values y_n recursively from (2-1) as long as $x_n \in [a, b]$.

The method defined by the increment function $\Phi(x, y; h)$ is called *convergent* if, for arbitrary η and arbitrary $x \in [a, b]$,

$$\lim_{\substack{h \to 0 \\ x_n = x}} y_n = y(x) \tag{2-14}$$

Euler's method, for instance, is a convergent method by virtue of Theorem 1.2.

A simple necessary and sufficient condition for convergence of a general one-step method is given by the following theorem:

THEOREM 2.1. *Let the function $\Phi(x, y; h)$ be continuous (jointly as a function of its three arguments) in the region defined by $x \in [a, b]$, $y \in (-\infty, \infty)$, $0 \leq h \leq h_0$, where $h_0 > 0$, and let there exist a constant L such that*

$$|\Phi(x, y^*; h) - \Phi(x, y; h)| \leq L|y^* - y| \tag{2-15}$$

for all $(x, y; h)$ and $(x, y^*; h)$ in the region just defined. Then the relation

$$\Phi(x, y; 0) = f(x, y) \tag{2-16}$$

is a necessary and sufficient condition for the convergence of the method defined by the increment function Φ.

Condition (2-16) is called the condition of *consistency*. Roughly speaking, the theorem states that consistency is necessary and sufficient for convergence. The reader will readily convince himself that all the special methods considered in §2.1 are consistent.

Proof. Let $\Phi(x, y; 0) = g(x, y)$. The function $g(x, y)$ satisfies the conditions of Theorem 1.1. It follows that for any η the initial value problem

$$z' = g(x, z), \qquad z(a) = \eta \tag{2-17}$$

has a unique solution $z(x)$. From the proof of Theorem 1.1 it follows moreover that the point $(x, z(x)) \in R$ for $x \in [a, b]$, where R denotes the compact region defined in §1.2-4.

We shall show that the numbers z_n defined by $z_0 = \eta$,

$$z_{n+1} = z_n + h\Phi(x_n, z_n; h), \qquad n = 0, 1, 2, \cdots; \; x_n \in [a, b] \tag{2-18}$$

converge to $z(x)$. By subtracting from (2-18) the trivial relation $z(x_{n+1}) = z(x_n) + h\Delta(x_n, z(x_n); h)$ we find for the error $e_n = z_n - z(x_n)$ the relation

$$e_{n+1} = e_n + h[\Phi(x_n, z_n; h) - \Delta(x_n, z(x_n); h)]$$

By the mean value theorem,

$$\Delta(x_n, z(x_n); h) = \frac{z(x_{n+1}) - z(x_n)}{h} = g(x_n + \theta h, z(x_n + \theta h))$$

where $0 < \theta < 1$. The expression in brackets may thus be written

$$\begin{aligned} &\Phi(x_n, z_n; h) - \Phi(x_n, z(x_n); h) \\ &+ \Phi(x_n, z(x_n); h) - \Phi(x_n, z(x_n); 0) \\ &+ g(x_n, z(x_n)) - g(x_n + \theta h, z(x_n + \theta h)) \end{aligned}$$

The function $\Phi(x, y; h)$, being continuous, is uniformly continuous on the compact set $x \in [a, b]$, $y = z(x)$, $0 \le h \le h_0$. Hence we may conclude as in §1.2-4 that the quantity

$$\zeta(h) = \max_{x \in [a, b]} |\Phi(x, z(x); h) - \Phi(x, z(x); 0)|$$

tends to zero as $h \to 0$. Similarly

$$\chi(h) = \max_{\substack{x \in [a, b] \\ 0 \le k \le h}} |g(x, z(x)) - g(x + k, z(x + k))|$$

tends to zero as $h \to 0$. Using (2-15), we thus obtain the estimate

$$|e_{n+1}| \le |e_n| + h[L\,|e_n| + \zeta(h) + \chi(h)]$$

This is of the form (1-11), with $A = 1 + hL$, $B = h(\zeta(h) + \chi(h))$. It follows that

$$|e_n| \le [\zeta(h) + \chi(h)]E_L(x_n - a)$$

Since the expression on the right tends to 0 as $h \to 0$, it follows that

$$\lim_{\substack{h \to 0 \\ x_n = x}} z_n = z(x), \qquad x \in [a, b]$$

This already establishes the sufficiency of the consistency condition (2-16) for convergence, for if $g(x, y) = f(x, y)$ then $z(x) = y(x)$. In order to show the necessity, assume that the method defined by $\Phi(x, y; h)$ is convergent, but $g(x, y) \ne f(x, y)$ at some point (x, y). It follows from Theorem 1.1 that there exists an η such that the solution $y(t)$ of the initial value problem passes through the point (x, y). The values z_n defined by $z_0 = \eta$ and (2-18) on the one hand tend to $y(t)$ and on the other, by what was proved above, to the solution $z(t)$ of the initial value problem (2-17).

If $z(x) \neq y(x)$, we have an immediate contradiction; if $z(x) = y(x)$, then

$$z'(x) = g(x, y) \neq f(x, y) = y'(x)$$

so that again $z(t) \not\equiv y(t)$. Thus the assumption that $g(x, y) \not\equiv f(x, y)$ leads to a contradiction.

2.2-2. An a priori bound. Although Theorem 2.1 guarantees that a consistent method is convergent, it is hardly useful for purposes of estimating the discretization error. In particular, it fails to indicate the *order* of the (accumulated) discretization error. The following result fills this gap. We denote by $y(x)$ the solution of the initial value problem (2-13) and let Δ be defined by (2-4). We then have

THEOREM 2.2. *Let $\Phi(x, y; h)$ satisfy the conditions of Theorem* 2.1, *and let there exist constants $N \geq 0$ and $p \geq 0$ such that*

$$|\Phi(x, y(x); h) - \Delta(x, y(x); h)| \leq Nh^p, \qquad x \in [a, b];\ h \leq h_0 \tag{2-19}$$

Then, for $x_n \in [a, b]$ and arbitrary $h \leq h_0$,

$$|y_n - y(x_n)| \leq h^p N E_L(x_n - a) \tag{2-20}$$

where E_L denotes theLipschitz function.

The quantity $\Phi - \Delta$ appearing in (2-19) is called the *relative local discretization* (*or truncation*) *error* of the method.* The theorem roughly states that if the relative local discretization error is $O(h^p)$, then the (absolute) accumulated discretization error is of the same order.

Proof. We subtract from (2-1) the identity $y(x_{n+1}) = y(x_n) + h\Delta(x_n, y(x_n); h)$ and, setting $e_n = y_n - y(x_n)$, write the result in the form

$$\begin{aligned} e_{n+1} = e_n + h[&\Phi(x_n, y_n; h) - \Phi(x_n, y(x_n); h) \\ &+ \Phi(x_n, y(x_n); h) - \Delta(x_n, y(x_n); h)] \end{aligned} \tag{2-21}$$

Estimating the expressions on the right by (2-15) and (2-19), we find

$$|e_{n+1}| \leq |e_n| + hL\,|e_n| + h^{p+1}N$$

This inequality is of the form $|e_n| \leq A\,|e_n| + B$, where $A = 1 + hL$, $B = h^{p+1}N$. The desired result follows by applying Lemma 1.2.

The following result is related to Theorem 2.2 in the same manner as Theorem 1.4 is related to Theorem 1.3.

THEOREM 2.3. *Let the function $\Phi(x, y; h)$ satisfy the conditions of Theorem 2.2 and let $\{y_n\}$ be any sequence of numbers satisfying*

$$\begin{aligned} y_0 &= \eta \\ y_{n+1} &= y_n + h[\Phi(x_n, y_n; h) + h^q\theta_n K], \\ &\qquad\qquad n = 0, 1, 2, \cdots;\ x_n \in [a, b] \end{aligned} \tag{2-22}$$

* Relative with respect to the stepsize h.

where $K \geq 0$ and $q \geq 0$ are constants, and where the θ_n are arbitrary numbers such that $|\theta_n| \leq 1$. Then, for $x_n \in [a, b]$ and $h \leq h_0$,

$$|y_n - y(x_n)| \leq h^r N_1 E_L(x_n - a) \tag{2-23}$$

where $r = \min(p, q)$ and $N_1 = Nh_0^{p-r} + Kh_0^{q-r}$.

This theorem guarantees the convergence of the values y_n to the exact solution even if the basic relation (2-1) is satisfied only approximately, provided the error at each step is $O(h^{1+q})$ with $q > 0$.

The proof consists in a straightforward modification of the proof of Theorem 2.2, the constant N being replaced throughout by N_1.

2.2-3. Application to special methods: expressions for L. For concrete applications of the error bound of Theorem 2.2 it is necessary to know explicit values for the constants L and N appearing in (2-15) and (2-19) respectively. In this and the following paragraph we shall calculate such values for the special methods discussed in §2.1.

Assuming that the function $f(x, y)$ is continuous and has continuous derivatives of sufficiently high order, we put

$$L_k = \sup_{\substack{x \in [a, b] \\ y \in (-\infty, \infty)}} \left| \frac{\partial}{\partial y} f^{(k)}(x, y) \right|, \qquad k = 0, 1, \cdots$$

and assume that these quantities are finite. The constant L_0 then may serve as a Lipschitz constant for the function $f(x, y)$ itself.

For the method (2-7) based on Taylor's expansion we find by application of the mean value theorem, if $h \leq h_0$,

$$L \leq L_0 + \frac{h_0}{2} L_1 + \cdots + \frac{h_0^{p-1}}{p!} L_{p-1}$$

For the simplified Runge-Kutta method (2-11) we find, combining the inequalities

$$|f(x, y^*) - f(x, y)| \leq L_0 |y^* - y|$$

and

$$\left| f\left(x + \frac{h}{2\alpha}, y^* + \frac{h}{2\alpha} f(x, y^*)\right) - f\left(x + \frac{h}{2\alpha}, y + \frac{h}{2\alpha} f(x, y)\right) \right|$$

$$\leq L_0 \left| y^* - y + \frac{h}{2\alpha} [f(x, y^*) - f(x, y)] \right|$$

$$\leq L_0 \left(1 + \frac{h}{2|\alpha|} L_0\right) \left| y^* - y \right|$$

for the constant L the bound

$$L \leq |1 - \alpha| L_0 + |\alpha| \left(1 + \frac{h_0 L_0}{2|\alpha|}\right) L_0$$

If $0 < \alpha \leq 1$, this yields

$$L \leq \left(1 + \frac{h_0 L_0}{2}\right) L_0 \tag{2-24}$$

A similar, if more roundabout, calculation yields for the Lipschitz constant of the classical Runge-Kutta method (Carr [1958])

$$L \leq \left(1 + \frac{h_0 L_0}{2} + \frac{h_0^2 L_0^2}{6} + \frac{h_0^3 L_0^3}{24}\right) L_0 \leq \frac{e^{h_0 L_0} - 1}{h_0} \tag{2-25}$$

2.2-4. Application to special methods: expressions for N. To estimate N from (2-19) is, of course, not possible if the exact solution $y(x)$ is not known. From the proof of Theorem 1.1 we know, however, that the exact solution $y(x)$ stays within the compact region R defined in §1.2-4. Thus it is permissible to replace N by an upper bound for

$$h^{-p} |\Phi(x, y;\ h) - \Delta(x, y;\ h)|$$

for $h \leq h_0$ and $(x, y) \in R$. If the method defined by Φ is of order p, and if both Φ and Δ are p times continuously differentiable, then

$$\frac{\partial^k}{\partial h^k} \{\Phi(x, y;\ h) - \Delta(x, y;\ h)\}_{h=0} = 0$$

for $k = 0, 1, \cdots, p - 1$, and we obtain from Taylor's formula†

$$\Phi(x, y;\ h) - \Delta(x, y;\ h) = h^p \left\{ \frac{1}{p!} \frac{\partial^p \Phi}{\partial h^p} (x, y;\ h^*) - \frac{1}{(p+1)!} f^{(p)} (x + h^*;\ z(x + h^*)) \right\}$$

where $0 < h^* < h$. Thus (2-16) is satisfied with

$$N = \max_{\substack{(x,y) \in R \\ h \leq h_0}} \left| \frac{1}{p!} \frac{\partial^p \Phi}{\partial h^p} (x, y;\ h) - \frac{1}{(p+1)!} f^{(p)}(x + h, z(x + h)) \right| \tag{2-26}$$

The application of this result to the method of explicit Taylor expansion is trivial, since

$$\frac{\partial^p \Phi}{\partial h^p} (x, y;\ h) = 0$$

† $z(t)$ denotes the solution of $z' = f(t, z)$ passing through (x, y).

We simply find

$$N = \frac{1}{(p+1)!} \max_{(x,y)\in R} |f^{(p)}(x, y)|$$

For the simplified Runge-Kutta method (2-9) we have $p = 2$. Since

$$\Phi(x, y; h) = (1 - \alpha)f(x, y) + \alpha f(x', y')$$

where

$$x' = x + \frac{h}{2\alpha}, \qquad y' = y + \frac{h}{2\alpha} f(x, y)$$

we find

$$\frac{\partial \Phi}{\partial h}(x, y; h) = \tfrac{1}{2}\{f_x(x', y') + f_y(x', y')f(x, y)\}$$

and

$$\frac{\partial^2 \Phi}{\partial h^2}(x, y; h) = \frac{1}{4\alpha}\{f_{xx}(x', y') + 2f_{xy}(x', y')f(x, y) + f_{yy}(x', y')[f(x, y)]^2\} \tag{2-27}$$

If we determine constants M and K such that $M \geq 1$,

$$|f(x, y)| \leq M, \qquad \left|\frac{\partial^{i+k} f}{\partial x^i\, \partial y^k}\right| \leq \frac{K}{M^{i+k-1}} \qquad (i + k \leq 2) \tag{2-28}$$

for $(x, y) \in R$, expression (2-27) can be estimated as follows:

$$\left|\frac{\partial^2 \Phi}{\partial h^2}(x, y; h)\right| \leq \frac{1}{|\alpha|} KM$$

From

$$f'' = f_{xx} + 2f_{xy}f + f_{yy}f^2 + f_y(f_x + f_y f)$$

we readily find

$$|f''| \leq 4KM + 2K^2M$$

Thus

$$N \leq KM\left(\frac{1}{2|\alpha|} + \frac{2}{3} + \frac{1}{3}K\right) \tag{2-29}$$

for the simplified Runge-Kutta method.

For the classical Runge-Kutta method Bieberbach ([1944], p. 55) found by analogous calculations the simple bound

$$N \leq 6KM(1 + K + K^2 + K^3 + K^4) \tag{2-30}$$

Here it must be assumed that the second bound in (2-28) holds for $i + k \leq 4$. Better bounds for N for both the simplified and the classical Runge-Kutta method can be obtained by a refined analysis (see §3.3-4).

2.2-5. The principal error function. If the method defined by $\Phi(x, y; h)$ has exact order p, and if both $\Phi(x, y; h)$ and $\Delta(x, y; h)$ have continuous derivatives of order $p + 1$ with respect to h which depend continuously also on (x, y), then we may write

$$\Phi(x, y; h) - \Delta(x, y; h) = h^p \varphi(x, y) + O(h^{p+1}) \tag{2-31}$$

where the function $\varphi(x, y)$ is continuous and does not vanish identically. The function $\varphi(x, y)$ will be called the *principal error function* of the method.

The quantity $h[\Phi(x, y; h) - \Delta(x, y; h)]$ represents the error of the method after one step, begun at the point (x, y). This quantity is called the (absolute) *local discretization error* or the *local truncation error* of the method. The order p of the method may be regarded as a first crude measure for the accuracy of the method, inasmuch as it states that the local discretization error is comparable to h^{p+1}, but not to any higher power of h. The principal error function enables us to refine this statement as follows: For sufficiently small values of h the local discretization error is approximately represented by $h^{p+1}\varphi(x, y)$.

We now determine the principal error function for the special methods considered in §2.1.

For Euler's method we have

$$\Phi(x, y; h) - \Delta(x, y; h) = f(x, y) - [f(x, y) + \tfrac{1}{2}hf'(x, y) + O(h^2)]$$

If $f'(x, y) \not\equiv 0$, the exact order of the method is 1, and the principal error function is given by

$$\varphi(x, y) = -\tfrac{1}{2}f'(x, y) \tag{2-32}$$

Similarly, we find for the Taylor expansion method of order p

$$\varphi(x, y) = -\frac{1}{(p + 1)!}f^{(p)}(x, y) \tag{2-33}$$

For the methods of Runge-Kutta type the calculations are less trivial. For the simplified Runge-Kutta method we have to push the expansion of the right-hand side of (2-9) to the terms involving h^2. Applying Taylor's formula for functions of two variables, we find

$$\Phi(x, y; h) = f + \frac{h}{2}(f_x + f_y f) + \frac{h^2}{8\alpha}(f_{xx} + 2f_{xy}f + f_{yy}f^2) + O(h^3)$$

The variables (x, y) are to be substituted in the function f and its derivatives. On the other hand, writing out the derivatives f' and f'' in (2-6), we have

$$\Delta(x, y;\ h) = f + \frac{h}{2}(f_x + f_y f) + \frac{h^2}{6}(f_{xx} + 2f_{xy}f + f_{yy}f^2 + f_x f_y + f_y^2 f) + O(h^3)$$

Comparing the last two relations, there follows

$$\varphi(x, y) = \left(\frac{1}{8\alpha} - \frac{1}{6}\right)(f_{xx} + 2f_{xy}f + f_{yy}f^2) - \frac{1}{6}(f_x f_y + f_y^2 f) \tag{2-34}$$

This may be written more compactly as follows:

$$\varphi(x, y) = \left(\frac{1}{8\alpha} - \frac{1}{6}\right)f'' - \frac{1}{8\alpha}f_y f' \tag{2-35}$$

For the classical Runge-Kutta method we shall obtain in §3.3-5 as a by-product of more general considerations the principal error function

$$\varphi(x, y) = \frac{1}{2880}f^{\text{IV}} - \frac{1}{576}f_y f''' + \frac{1}{288}(f_y^2 - f_{xy} - f_{yy}f)f'' + \frac{1}{192}(2f_{xy}f_y + 3f_{yy}f_y f - 2f_y^3 + f_{yy}f_x)f' \tag{2-36}$$

Again the arguments are understood to be (x, y) everywhere.

Example. In Table 2.2 we list the principal error function corresponding to the equation $y' = y$ for some of the methods which have been discussed. It turns out that for this particular equation $\varphi(x, y) = Ky$, where the value of the constant K depends on the method.

Table 2.2

Method	Order	K
Taylor expansion (2-5)	p	$-\dfrac{1}{(p+1)!}$
Simplified Runge-Kutta (2-10)	2	$-\frac{1}{6}$
Classical Runge-Kutta (2-11)	4	$-\frac{1}{120}$

2.2-6. Asymptotic formula for the error. It would be easy to demonstrate by means of numerical examples that the error estimates derived in §2.2-2 generally overestimate the actual error by a large percentage, especially if the number of steps is large. As in Chapter 1, we shall therefore supplement these results by formulas which *approximate* rather than *bound* the actual error.

We shall assume that the order of the method defined by $\Phi(x, y;\ h)$ is at least 1, and that all partial derivatives of $f(x, y)$ and $\Phi(x, y;\ h)$ of order $\leq (p + 1)$ exist and are continuous and bounded in the region $x \in [a, b]$, $y \in (-\infty, \infty)$, $h \leq h_0$. The term denoted by $O(h^{p+1})$ in (2-31) is then of the form $\theta K h^{p+1}$, where K is a constant and $|\theta| \leq 1$. We once again compare the approximate values y_n defined by (2-1) with the corresponding exact values $y(x_n)$. Since the hypotheses of Theorem 2.2 are satisfied, we know that there exists a constant K_1 such that

$$|y_n - y(x_n)| \leq K_1 h^p \tag{2-37}$$

for $x_n \in [a, b]$ and $h \leq h_0$. With this information we return to relation (2-21). Using $y_n = y(x_n) + e_n$ and expanding in powers of e_n, we get

$$\begin{aligned} &\Phi(x_n, y_n;\ h) - \Phi(x_n, y(x_n);\ h) \\ &= \Phi_y(x_n, y(x_n);\ h)e_n + \tfrac{1}{2}\Phi_{yy}(x_n, y^*;\ h)e_n^2 \\ &= [\Phi_y(x_n, y(x_n);\ 0) + h\Phi_{yh}(x_n, y(x_n);\ h^+)]e_n + \tfrac{1}{2}\Phi_{yy}(x_n, y^*;\ h)e_n^2 \end{aligned} \tag{2-38}$$

Here y^* is a value between $y(x_n)$ and y_n; $0 < h^+ < h$. Since $p \geq 1$, it follows from (2-19) that $\Phi(x, y;\ 0) = f(x, y)$. We shall consistently write

$$g(x) = f_y(x, y(x))$$

Hence,

$$\Phi_y(x_n, y(x_n);\ 0) = g(x_n)$$

Furthermore, $|e_n| \leq K_1 h^p$ and the second derivatives of Φ are bounded. Consequently we may write, using some new constant K_3,

$$\Phi(x_n, y_n;\ h) - \Phi(x_n, y(x_n);\ h) = g(x_n)e_n + \theta K_3 h^{p+1}$$

Here and in the following text $\theta, \theta', \cdots$ denote unspecified numbers of modulus less than 1 which may change from equation to equation.

Substituting in (2-21) and using (2-31), we get

$$e_{n+1} = e_n + h\{g(x_n)e_n + \theta K_3 h^{p+1} + \varphi(x_n, y(x_n))h^p + \theta' K h^{p+1}\} \tag{2-39}$$

We now introduce new numbers $\bar{e}_n$, to be called *magnified errors*, by the relation

$$e_n = \bar{e}_n h^p$$

With these, relation (2-39) assumes the form

$$\bar{e}_{n+1} = \bar{e}_n + h\{g(x_n)\bar{e}_n + \varphi(x_n, y(x_n))\} + \theta'' K_4 h^2 \tag{2-40}$$

where $K_4 = K + K_3$. Relation (2-40) is one to which Theorem 1.4 can be applied. It follows that there exists a constant K_5 such that

$$\bar{e}_n = e(x_n) + \theta K_5 h$$

where the function $e(x)$ is the solution of the initial value problem

$$\begin{aligned} e'(x) &= g(x)e(x) + \varphi(x, y(x)) \\ e(a) &= 0 \end{aligned} \tag{2-41}$$

The function $e(x)$ will be called the *magnified error function.*

We have proved:

THEOREM 2.4. *Let p be the exact order of the method defined by $\Phi(x, y; h)$ and let $\Phi(x, y; h)$ and $f(x, y)$ satisfy the conditions stated at the beginning of* §2.2-6. *Then the discretization error of the approximate solution defined by* (2-1) *satisfies*

$$e_n = h^p e(x_n) + O(h^{p+1}) \tag{2-42}$$

where $e(x)$ denotes the magnified error function defined by (2-41).

For $f(x, y) = y$ we found $\varphi(x, y) = Ky$ for all special methods which have been considered. For the initial value problem $y' = y$, $y(0) = 1$, the initial value problem for $e(x)$ thus reduces to

$$e'(x) = e(x) + Ke^x, \qquad e(0) = 0$$

Integrating, we find

$$e(x) = Kxe^x$$

Thus, for the methods listed in Table 2.2,

$$e_n = Kh^p x_n e^{x_n} + O(h^{p+1})$$

Although the absolute error grows exponentially with x, the relative error grows only linearly.

2.2-7. Application of the asymptotic formula. The most immediate application of Theorem 2.4 is a justification for general one-step methods of Richardson's extrapolation to $h = 0$, discussed for Euler's method in §1.3-4. For greater explicitness we denote by $y(x, h)$ the approximate value obtained at the point x with the step h (it is assumed that $x - a$ is a multiple of h). Equation (2-42) may then be restated as follows:

$$y(x, h) = y(x) + h^p e(x) + O(h^{p+1}) \tag{2-43}$$

Solving the differential equation with two different values of h, say h and qh, where q is neither 0 nor 1, and writing down the corresponding equations (2-43), we may regard the result as a system of two linear equations with the two unknowns $y(x)$ and $e(x)$. Solving for $y(x)$, we find

$$y(x) = \frac{y(x, qh) - q^p y(x, h)}{1 - q^p} + O(h^{p+1}) \tag{2-44}$$

This relation shows that the "extrapolated" value [represented by the fraction in (2-44)] approximates the exact solution with an error whose order exceeds the order of the method by 1. It is evident from (2-44) that the value of q should not be close to 1. In practice $q = 2$ is frequently used. Also, the foregoing analysis assumes that there is no round-off. If the calculated values $y(x, h)$ are contaminated by round-off errors which are comparable to the discretization error, the extrapolated values behave erratically.

Another way to use Theorem 2.4 in order to obtain an approximation to e_n is the following: Integrate the differential equation (2-41) for $e(x)$ numerically along with the given differential equation and calculate e_n by (2-42), neglecting the correction term. If this program is carried out, it seems necessary to know:

(a) the principal error function $\varphi(x, y)$ of the method;
(b) the derivative $f_y(x, y)$;
(c) the true solution of the problem, since it figures as an argument in both f_y and φ.

The last requirement, paradoxical as it may seem at first sight, turns out to be the most trivial of the three. If in place of the true solution $y(x)$ the approximate values y_n are substituted in (2-41), the right-hand side of (2-41), and thus the solution $e(x)$, is falsified by an amount $O(h^p)$, which may be absorbed into the correction term $O(h^{p+1})$ in (2-42).

The above program is rather ambitious for practical purposes, especially since it also involves the calculation of the functions $f_y(x, y)$ and $\varphi(x, y)$. For this reason, most of the existing routines for solving initial value problems by one-step methods shun the integration of the magnified error equation. If any attempt is made at appraising the discretization error, this is usually confined to an approximate evaluation of the local discretization error. If the local error exceeds a preset threshold value, the integration step is diminished; if it is significantly smaller than the permissible threshold, the step is increased. It is clear that this procedure amounts at best to a qualitative control of the error; no reliable quantitative appraisal of the accumulated discretization error is possible in this manner.

For the approximate calculation of the local error Gorn and Moore [1953] devised a simple algorithm which will now be explained.

Let $z(t)$ again denote the solution of the equation $z' = f(t, z)$ passing through the point (x, y). Starting from (x, y), we can obtain an approximate value for $z(x + h)$ with the method defined by $\Phi(x, y; h)$ either by performing one step of length h or two steps of length $\frac{1}{2}h$. We denote the increments by δ_1 and δ_2, respectively, and shall compare them both to the exact increment $h\Delta(x, y;\, h)$.

For the one-step increment we find, if the method is of order p,

$$\begin{aligned}\delta_1 &= h\Phi(x, y;\, h) \\ &= h\Delta(x, y;\, h) + h^{p+1}\varphi(x, y) + O(h^{p+2})\end{aligned} \tag{2-45}$$

The two-step increment is given by

$$\delta_2 = \tfrac{1}{2}h\Phi(x, y;\, \tfrac{1}{2}h) + \tfrac{1}{2}h\Phi(x + \tfrac{1}{2}h, y + \tfrac{1}{2}h\Phi(x, y;\, \tfrac{1}{2}h);\, \tfrac{1}{2}h)$$

In view of

$$\tfrac{1}{2}h\Phi(x, y;\, \tfrac{1}{2}h) = \tfrac{1}{2}h\Delta(x, y;\, \tfrac{1}{2}h) + (\tfrac{1}{2}h)^{p+1}\varphi(x, y) + O(h^{p+2})$$

we have

$$\begin{aligned}\delta_2 = {} & \tfrac{1}{2}h\Delta(x, y;\, \tfrac{1}{2}h) + (\tfrac{1}{2}h)^{p+1}\varphi(x, y) + O(h^{p+2}) \\ & + \tfrac{1}{2}h\Phi(x + \tfrac{1}{2}h, z(x + \tfrac{1}{2}h);\, \tfrac{1}{2}h) \\ & + \tfrac{1}{2}h\Phi_y(x + \tfrac{1}{2}h, z^*;\, \tfrac{1}{2}h)[(\tfrac{1}{2}h)^{p+1}\varphi(x, y) + O(h^{p+2})]\end{aligned} \tag{2-46}$$

where z^* is a value between $z(x + \frac{1}{2}h)$ and $y + \frac{1}{2}h\Phi(x, y;\, \frac{1}{2}h)$. The third line in the expression for δ_2 is $O(h^{p+2})$. The second line may be replaced by

$$\tfrac{1}{2}h\Delta(x + \tfrac{1}{2}h, z(x + \tfrac{1}{2}h);\, \tfrac{1}{2}h) + (\tfrac{1}{2}h)^{p+1}\varphi(x + \tfrac{1}{2}h, z(x + \tfrac{1}{2}h)) + O(h^{p+2})$$

In view of

$$\tfrac{1}{2}h\Delta(x, y;\, \tfrac{1}{2}h) + \tfrac{1}{2}h\Delta(x + \tfrac{1}{2}h, z(x + \tfrac{1}{2}h);\, \tfrac{1}{2}h) = h\Delta(x, y;\, h)$$

and

$$\varphi(x + \tfrac{1}{2}h, z(x + \tfrac{1}{2}h)) = \varphi(x, y) + O(h)$$

it follows that

$$\delta_2 = \Delta(x, y;\, h) + 2(\tfrac{1}{2}h)^{p+1}\varphi(x, y) + O(h^{p+2}) \tag{2-47}$$

We define the quantity

$$\delta = \delta(x, y;\, h) = \delta_2 - \delta_1 \tag{2-48}$$

The value of δ can be computed explicitly for given x, y, and h by calculating the difference of the results of performing one step of length h and two steps of length $\frac{1}{2}h$. Comparing (2-45) and (2-47), we find

$$\delta = -(1 - 2^{-p})h^{p+1}\varphi(x, y) + O(h^{p+2}) \tag{2-49}$$

Solving for $\varphi(x, y)$ we obtain the desired approximate expression for the principal error function. The result is summarized in:

THEOREM 2.5. *Under the conditions of Theorem* 2.4 *the principal error function satisfies*

$$\varphi(x, y) = -\frac{1}{1 - 2^{-p}} h^{-p-1}\delta(x, y;\, h) + O(h) \tag{2-50}$$

where $\delta(x, y;\, h)$ *is the explicitly calculable quantity defined by* (2-48).

The algorithm implied in the definition of δ accomplishes in the small what Richardson's extrapolation to the limit accomplishes in the large. It may therefore be called *local extrapolation to the limit.*

2.2-8. Variable stepsize. For certain differential equations, especially those arising from physical problems, it can be predicted by qualitative arguments that the solution will behave very smoothly in certain ranges of the independent variable and less smoothly in others. An example is the orbit of an interplanetary missile, which is smooth except in the neighborhood of celestial objects. In such cases it would be unwise to carry out the integration with one and the same value of h over the whole range of the independent variable. We shall now modify the statements of Theorems 2.2 and 2.3 to deal with this situation.

We assume that the stepsize actually employed is $\vartheta(x)h$, where h is a constant, basic stepsize and $\vartheta(x)$ is a piecewise continuous function of x such that $0 < \vartheta(x) \le 1$, $x \in [a, b]$. We redefine the lattice points x_n by

$$x_0 = a, \qquad x_{n+1} = x_n + \vartheta(x_n)h, \qquad n = 0, 1, 2, \cdots \tag{2-51}$$

The assumptions made on $\vartheta(x)$ imply that $\vartheta(x)$ is bounded away from zero, i.e., that there exists a constant $c > 0$ such that $\vartheta(x) \ge c$, $a \le x \le b$. This in turn implies that $x_n \ge a + nch$; thus, for each $h > 0$, the integration over the interval $[a, b]$ requires only a finite number of steps.

The approximate values y_n are now generated by the formula

$$y_{n+1} = y_n + \vartheta(x_n)h\Phi(x_n, y_n;\, \vartheta(x_n)h) \tag{2-52}$$

We have the following analog of Theorem 2.2:

THEOREM 2.6. *Let the function* $\Phi(x, y;\, h)$ *in lieu of* (2-19) *satisfy the condition*

$$|\Phi(x, y(x);\, h\vartheta(x)) - \Delta(x, y(x);\, h\vartheta(x))| \le Nh^p \tag{2-53}$$

Then the values y_n *defined by* (2-52) *still satisfy*

$$|y_n - y(x_n)| \le NE_L(x_n - a)h^p, \qquad x_n \in [a, b] \tag{2-54}$$

The proof is completely analogous to that of Theorem 2.2 and therefore omitted.

The analog of Theorem 2.4 is given by

THEOREM 2.7. *Let $\Phi(x, y; h)$ and $f(x, y)$ satisfy the conditions of Theorem 2.3. The error e_n of the approximate solution y_n defined by* (2-52) *then satisfies*

$$e_n = e(x_n)h^p + O(h^{p+1}), \qquad x_n \in [a, b] \tag{2-55}$$

where the function $e(x)$ is defined by

$$\begin{aligned} e(a) &= 0 \\ e'(x) &= g(x)e(x) + [\vartheta(x)]^p \varphi(x, y(x)) \end{aligned} \tag{2-56}$$

Again the proof involves only trivial modifications of the proof of Theorem 2.4 and is therefore left to the reader.

2.2-9. Linear equations; numerical example. Some of the analysis of the preceding sections is simplified if the differential equation to be integrated is of the form

$$y' = g(x)y \tag{2-57}$$

i.e., linear and homogeneous.* If the increment function $\Phi(x, y; h)$ preserves this linearity (which is true for all methods discussed in this chapter), then the principal error function is also linear in y, and we may set

$$\varphi(x, y) = \psi(x)y \tag{2-58}$$

The magnified error function $e(x)$ then satisfies

$$e'(x) = g(x)e(x) + [\vartheta(x)]^p \psi(x)y(x) \tag{2-59}$$

Excluding the trivial case $y(x) \equiv 0$, we define the *relative magnified error function* $r(x)$ by

$$r(x) = \frac{e(x)}{y(x)} \tag{2-60}$$

Inserting $e(x) = r(x)y(x)$ in (2-59), we obtain

$$r'(x)y(x) + r(x)y'(x) = g(x)r(x)y(x) + [\vartheta(x)]^p \psi(x)y(x)$$

In view of (2-57), this simplifies to

$$r'(x) = [\vartheta(x)]^p \psi(x)$$

* Although such differential equations can of course be integrated by quadrature, direct integration by one of the above methods can be preferable for numerical purposes, because this avoids calculating a value of the exponential function.

and we obtain immediately

$$r(x) = \int_a^x \vartheta(t)^p \psi(t)\, dt \tag{2-61}$$

Thus it is possible to calculate the relative magnified error by a simple quadrature, without knowledge of the true solution.

As an example we shall consider the integration of the equation $y' = -16xy$ under the initial condition*

$$y(-0.75) = 2^{-3/2}\pi^{-1/2}e^{-9/2} = 0.00221\ 59242$$

using the Runge-Kutta method with a constant step h. The exact solution is

$$y(x) = 2^{-3/2}\pi^{-1/2}e^{-8x^2}$$

Putting $c = -16$ for compactness, we have

$$\begin{aligned} f(x, y) &= cxy \\ f'(x, y) &= (c + c^2x^2)y \\ f''(x, y) &= (3c^2x + c^3x^3)y \\ f'''(x, y) &= (3c^2 + 6c^3x^2 + c^4x^4)y \\ f^{IV}(x, y) &= (15c^3x + 10c^4x^3 + c^5x^5)y \end{aligned}$$

By virtue of (2-36) we find

$$\begin{aligned} \varphi(x, y) &= \frac{1}{2880}(15c^3x + 10c^4x^3 + c^5x^5)y - \frac{1}{576}cx(3c^2 + 6c^3x^2 + c^4x^4)y \\ &\quad + \frac{1}{288}(c^2x^2 - c)(3c^2x + c^3x^3)y + \frac{1}{192}(2c^2x - 2c^3x^3)(c + c^2x^2)y \\ &= -\frac{c^5x^5}{120}y \end{aligned}$$

The principal error function thus turns out to be a linear function of y, as expected. From (2-61) we now easily get

$$r(x) = -\frac{c^5}{720}(x^6 - a^6)$$

where $a = -0.75$. This being an even function of x, the relative error will be approximately the same at points which lie symmetric about the origin. Since the exact solution is also symmetric, the same is true for the absolute error. Since $r(\pm a) = 0$, $y(-a)$ is obtained with higher than fourth order accuracy.

* Chosen because tables of the function $(2\pi)^{-1/2}e^{-x^2/2}$ are available (National Bureau of Standards [1942]).

In Table 2.3 we give the approximate values y_n corresponding to* $x = -0.50(.25)0.75$, calculated with the steps $h = 2^{-p}$ for $p = 4(1)9$. The last values in each column coincide with the exact solution to the number of digits given.

Table 2.3

x	−0.50	−0.25	0	0.25	0.50	0.75
$p = 4$	0.02693857	0.12069994	0.19899916	0.12069933	0.02694366	0.00222524
5	0.02699083	0.12096227	0.19943300	0.12096226	0.02699099	0.00221620
6	0.02699515	0.12098372	0.19946843	0.12098372	0.02699516	0.00221593
7	0.02699546	0.12098525	0.19947096	0.12098525	0.02699546	0.00221592
8	0.02699548	0.12098535	0.19947113	0.12098535	0.02699548	0.00221592
9	0.02699548	0.12098536	0.19947114	0.12098536	0.02699548	0.00221592

In Table 2.4 the actual errors are compared with the approximate errors

$$e_n = h^4 r(x_n) y(x_n)$$

predicted by the theory.

The prediction of zero error for $x = 0.75$ is vindicated to some extent by the more rapid convergence of the numerical values in the last column of Table 2.3.

Table 2.4

	$x = 0$		$x = 0.25$		$x = 0.50$	
	Actual	Predicted	Actual	Predicted	Actual	Predicted
$p = 4$	−0.00047198	−0.00079376	−0.00028604	−0.00047785	−0.00005183	−0.00009740
5	0.00003814	0.00004961	0.00002311	0.00002987	0.00000449	0.00000608
6	0.00000271	0.00000310	0.00000164	0.00000187	0.00000033	0.00000038
7	0.00000018	0.00000019	0.00000011	0.00000012	0.00000002	0.00000002
8	0.00000001	0.00000001	0.00000001	0.00000001	0.00000000	0.00000000
9	0.00000000	0.00000000	0.00000000	0.00000000	0.00000000	0.00000000

Table 2.5

x	−0.50	−0.25	0	0.25	0.50	0.75
Euler $h = 2^{-11}$	−0.00013981	−0.00078297	−0.00116196	−0.00062636	−0.00017476	−0.00002582
Modified Euler $h = 2^{-10}$	−0.00000085	−0.00000428	−0.00000669	−0.00000428	−0.00000085	−0.00000000

For the sake of comparison, we give in Table 2.5 the errors of the approximate values of the solution of the same initial value problem which were obtained by the Euler method (1-1) and the modified Euler method (2-10*b*). We used the steps $h = 2^{-11}$ for the Euler method and $h = 2^{-10}$

* If $c - a$ is an integral multiple of b, the symbol $a(b)c$ denotes the set of numbers $x = a + nb$, where $n = 0, 1, 2, \cdots, (c - a)b^{-1}$.

for the modified Euler method. These steps require the same number of evaluations of the function $f(x, y)$ as the Runge-Kutta method with the step $h = 2^{-9}$, for which $|e_n| < 10^{-8}$ for all considered values of x.

2.3. The Round-off Error in the General One-Step Method

2.3-1. A priori bounds. We reiterate the assumptions made in §1.4-1 and shall continue to use the notation introduced there. If fixed point operations are assumed, the numerical approximation $\tilde{y}_n$ in place of (2-1) satisfies the relation

(2-62) $$\tilde{y}_{n+1} = \tilde{y}_n + (h\tilde{\Phi}(x_n, \tilde{y}_n;\ h))^*$$

where $\tilde{\Phi}$ denotes a rounded and possibly otherwise approximated value of Φ. We rewrite (2-62) in the form

(2-63) $$\tilde{y}_{n+1} = \tilde{y}_n + h\Phi(x_n, \tilde{y}_n;\ h) + \varepsilon_{n+1}$$

where

(2-64) $$\varepsilon_{n+1} = (h\tilde{\Phi}(x_n, \tilde{y}_n;\ h))^* - h\Phi(x_n, \tilde{y}_n;\ h)$$

is the *local round-off error*. As before, ε_{n+1} can be regarded as the sum of the *inherent error*

(2-65) $$\rho_{n+1} = h[\tilde{\Phi}(x_n, \tilde{y}_n;\ h) - \Phi(x_n, \tilde{y}_n;\ h)]$$

and the *induced error*

(2-66) $$\pi_{n+1} = (h\tilde{\Phi}(x_n, \tilde{y}_n;\ h))^* - h\tilde{\Phi}(x_n, \tilde{y}_n;\ h)$$

In this paragraph we shall establish a bound for the accumulated round-off error $r_n = \tilde{y}_n - y_n$ under the sole assumption that

(2-67) $$|\varepsilon_n| \le \varepsilon, \qquad n = 1, 2, \cdots$$

where ε is a constant. Subtracting from (2-63) the corresponding recurrence relation (2-1) for the theoretical approximation, we find

(2-68) $$r_{n+1} = r_n + h[\Phi(x_n, \tilde{y}_n;\ h) - \Phi(x_n, y_n;\ h)] + \varepsilon_{n+1}$$

If L denotes the Lipschitz constant of the method, we get in view of (2-15) and (2-67) the recursive inequality

$$|r_{n+1}| \le (1 + hL)\,|r_n| + \varepsilon$$

Solving this recurrence relation by use of Lemma 1.1, we obtain

THEOREM 2.8. *If the increment function* $\Phi(x, y; h)$ *satisfies condition* (2-15), *and if the local round-off errors are bounded by* (2-67), *then the accumulated round-off error satisfies*

$$|r_n| \leq \frac{\varepsilon}{h} E_L(x_n - a), \qquad a \leq x_n \leq b \tag{2-69}$$

Values of the constant L for various special methods have been given in §2.2-3.

The above result is interesting in that the estimate (2-69) does not depend on the constants N and p which are typical for the discretization error of the method. This indicates that the propagation of round-off error is not related to the discretization error of a method.

2.3-2. Dependence of the accumulated round-off error on the local round-off errors. Although theoretically valuable, the estimate (2-69) is not likely to excite much practical interest, because it is based on the unrealistic assumption that all local round-off errors have their maximum value and reinforce each other systematically. Also, at each step the Lipschitz estimate (2-15) has been used, which may greatly overestimate the actual discrepancy between the two quantities involved.

As a basis for more realistic appraisals of the round-off error we now shall examine its dependence on the local round-off errors. Our aim is a relation similar to (1-64), our starting point relation (2-68). Reviving the old assumption that Φ has continuous and bounded derivatives of order $p + 1$, we have (since $p \geq 1$)

$$\begin{aligned} \Phi(x_n, \tilde{y}_n; h) - \Phi(x_n, y_n; h) &= \Phi_y(x_n, y(x_n); h) r_n \\ &+ \Phi_{yy}(x_n, y^+; h) \frac{(\tilde{y}_n - y(x_n))^2}{2} - \Phi_{yy}(x_n, y^{++}; h) \frac{(y_n - y(x_n))^2}{2} \end{aligned} \tag{2-70}$$

where y^+ and y^{++} are certain intermediate values. Expanding Φ_y in powers of h and using

$$\Phi(x, y; 0) = f(x, y),$$

we find, writing $g(x) = f_y(x, y(x))$,

$$\Phi_y(x_n, y(x_n); h) = g(x_n) + h\Phi_{yh}(x_n, y(x_n); h^+)$$

where $0 < h^+ < h$.

In order to arrive at significant results, it is now necessary to let the bound ε for the local round-off error depend on h in such a manner that

$$\varepsilon \leq K h^{q+1}, \qquad h < h_0, \tag{2-71}$$

where q and K are independent of h and $q \geq 1$. We also shall assume that

$$Nh^{p+1} \leq \varepsilon \tag{2-72}$$

where N is the constant occurring in the bound (2-19) for the local discretization error. Equations (2-71) and (2-72) are compatible for all sufficiently small values of h only if $q \leq p$.

The above hypotheses are appropriate when the local round-off error dominates the local discretization error. This situation arises automatically if the method is so accurate that the local discretization error is small compared to the basic unit u.

According to the Theorems 2.2, 2.3, and 2.8 the above hypotheses guarantee that

$$|\tilde{y}_n - y_n| = |r_n| \leq \frac{\varepsilon}{h} E = h^q KE$$

$$|\tilde{y}_n - y(x_n)| = |r_n + e_n| \leq \frac{2\varepsilon}{h} E = 2h^q KE, \qquad x_n \in [a, b]$$

where $E = E_L(b - a)$. Denoting by M an upper bound for the second derivatives of Φ, we thus conclude from (2-70) that

$$\Phi(x_n, \tilde{y}_n; h) - \Phi(x_n, y_n; h) = g(x_n)r_n + \theta_{n+1}\varepsilon K_1$$

where $|\theta_n| \leq 1$, $n = 0, 1, \cdots$, and

$$K_1 = ME(1 + \tfrac{5}{2}h_0^{q-1}KE)$$

It follows that (2-68) can be replaced by the relation

$$r_{n+1} = r_n + hg(x_n)r_n + \theta_{n+1}h\varepsilon K_1 + \varepsilon_{n+1}$$

This is identical with (1-63), except that the place of the undetermined quantities ε_{n+1} is now taken by the equally undetermined quantities $\theta_{n+1}h\varepsilon K_1 + \varepsilon_{n+1}$. As in §1.4-3, it follows that

$$r_n = r_n^{(1)} + r_n^{(2)} \tag{2-73}$$

where

$$r_n^{(1)} = \sum_{m=1}^{n} d_{n,m}\varepsilon_m \tag{2-73a}$$

$$r_n^{(2)} = h\varepsilon K_1 \sum_{m=1}^{n} d_{n,m}\theta_m \tag{2-73b}$$

The coefficients $d_{n,m}$ are defined by the relations (1-65) [with $g(x) = f_y(x, y(x))$]. As shown in §1.4-3, they satisfy

$$d_{n,m} = d_m(x_n) + O(h)$$

where

$$d_m(x) = e^{G(x)-G(x_m)}, \qquad G(x) = \int_a^x g(t)\,dt$$

The component $r_n^{(1)}$ of r_n is called the *primary component* of the accumulated round-off error. It is the error which would occur in the numerical integration of the linear equation $y' = g(x)y$ by Euler's method if the local errors were the same. If the function $m(x)$ is defined by (1-72), then we have

$$h \sum_{m=1}^{n} d_{m,n} = m(x_n) + O(h), \qquad x_n \in [a, b]$$

For sufficiently small values of h, $d_{m,n} > 0$. If (2-67) holds, it follows that

(2-74) $$|r_n^{(1)}| \le \frac{\varepsilon}{h}\{m(x_n) + O(h)\}$$

The component $r_n^{(2)}$ is called the *secondary component* of the accumulated round-off error. We owe its presence to the nonlinearity of the given differential equation and the complexity of the method defined by Φ. However, it follows by the above reasoning that

(2-75) $$|r_n^{(2)}| \le \varepsilon K_1\{m(x_n) + O(h)\}$$

The results (2-74) and (2-75) show that, under the hypothesis that the local round-off error dominates the local discretization error, the accumulated primary round-off error dominates the accumulated secondary round-off error as $h \to 0$. The propagation of primary round-off is the same for all one-step methods.

2.3-3. An a posteriori bound for variable local round-off error. The assumption (2-67) is primarily appropriate for fixed decimal (or binary) point arithmetic, where all numbers entering the computation are calculated with the same precision. With a view towards relaxing these rigid assumptions, we now replace (2-67) by the following more general condition:

(2-76) $$|\varepsilon_n| \le \varepsilon p(x_n)$$

Here $p(x)$ is a known nonnegative, piecewise smooth function of x. The last mentioned property means that the interval $[a, b]$ can be split into a finite number of subintervals in each of which the function $p(x)$ is continuous and has a continuous derivative with finite limits at the two end points. The quantity ε continues to be subjected to conditions

(2-71) and (2-72). An application of condition (2-76) is made in §2.3-5. Another application, to floating point arithmetic, is discussed in §3.4-8.

If an estimate for $r_n^{(1)}$ is desired, we now have to estimate the sum

$$m_n = h \sum_{m=1}^{n} d_{n,m} p_m$$

where $p_m = p(x_m)$. We seek an approximate closed expression for m_n by establishing a difference equation. If $p(x)$ is differentiable at $x = x_n$, then we have by (1-65)

$$\begin{aligned} m_{n+1} - m_n &= h\left\{\sum_{m=1}^{n+1} d_{n+1,m} p_m - \sum_{m=1}^{n} d_{n,m} p_m\right\} \\ &= h\left\{d_{n+1,n+1} p_{n+1} + \sum_{m=1}^{n} (d_{n+1,m} - d_{n,m}) p_m\right\} \\ &= h\left\{p(x_{n+1}) + g(x_n) \sum_{m=1}^{n} d_{n,m} p_m\right\} \\ &= h\{p(x_n) + g(x_n) m_n\} + O(h^2) \end{aligned}$$

The last relation may be looked at as an attempt to solve by Euler's method the differential equation

$$m'(x) = g(x)m(x) + p(x) \tag{2-77a}$$

By assumption, the interval $[a, b]$ may be subdivided into a finite number of subintervals in each of which $p(x)$ is continuous. By Theorem 1.4, m_n thus approximates an appropriate solution of (2-77*a*) in each of these subintervals. It follows that

$$m_n = m(x_n) + O(h) \tag{2-78}$$

where $m(x)$ is the continuous solution of (2-77*a*) satisfying

$$m(a) = 0 \tag{2-77b}$$

Observing that $r^{(2)} = O(\varepsilon)$, we obtain

THEOREM 2.9. *If the local round-off errors are bounded by* (2-76), *where* $p(x)$ *is a piecewise smooth function and* ε *satisfies conditions* (2-71) *and* (2-72), *then the accumulated round-off error satisfies*

$$|r_n| \le \frac{\varepsilon}{h}\{m(x_n) + O(h)\}, \qquad x_n \in [a, b] \tag{2-79}$$

where $m(x)$ *is defined by* (2-77).

2.3-4. Statistical estimates. In view of the numerical results presented in §1.6 we must expect that the bound (2-79) vastly overestimates the actual round-off error in practically all except very short integrations.

Realistic statements about the size of the round-off error that must actually be expected in numerical integrations can be found by the statistical methods introduced in §§1.5 and 1.6. We reintroduce at this point the hypothesis that the local round-off errors may be treated as random variables. For the experimental verification of this hypothesis we shall, as before, generate many solutions of one and the same differential equation by varying the initial conditions.

In making this hypothesis, it is not necessary to assume that the *whole* local round-off error is a random variable. If the local error has a component that is not random, this component may be split off and estimated separately by the methods of the preceding sections. For simplicity of notation, however, we shall denote the random part of the local error again by ε_m.

As in Chapter 1, we shall assume that the variables ε_m have a known distribution. Using the results of §1.4, we shall then determine the approximate distribution of the portion $r_n^{(1)}$ of the accumulated round-off error which depends directly on the ε_m. (No statistical statement is possible in general about $r_n^{(2)}$ because there is no reason why the quantities θ_m should be random.) To allow for greater generality, however, we now shall allow that the distribution of the local round-off error varies from step to step. Specifically, we shall assume that

$$\begin{aligned} |E(\varepsilon_m)| &\le \mu p(x_m) \\ \operatorname{var}(\varepsilon_m) &= \sigma^2 q(x_m) \end{aligned} \tag{2-80}$$

where $p(x)$ and $q(x)$ are nonnegative, piecewise smooth functions in $[a, b]$ which are assumed to be known. The nonnegative quantities μ and σ are assumed to be independent of x or m, although they may depend on h in view of (2-67) and (2-71).

If

$$r_n^{(1)} = \sum_{m=1}^{n} d_{n,m}\varepsilon_m$$

we have by (1-94)

$$|E(r_n^{(1)})| \le \mu \sum_{m=1}^{n} d_{nm}p_m = \frac{\mu}{h} m_n$$

or, in view of (2-78),

$$|E(r_n^{(1)})| \le \frac{\mu}{h}\{m(x_n) + O(h)\},$$

If the ε_m are independent, we find for the variance of the accumulated error, using (1-95)

$$\operatorname{var}(r_n^{(1)}) = \sigma^2 \sum_{m=1}^{n} d_{nm}^2 q_m$$

Thus it becomes necessary to find an approximation for the sum

$$v_n = h \sum_{n=1}^{n} d_{n,m}^2 q_m$$

In view of (1-67), the quantities v_n are bounded for $x_n \in [a, b]$.

If $q(x)$ is continuous at the point $x = x_n$, we find, using once again the difference equations (1-65) satisfied by $d_{n,m}$,

$$\begin{aligned} v_{n+1} - v_n &= h\left\{\sum_{m=1}^{n+1} d_{n+1,m}^2 q_m - \sum_{m=1}^{n} d_{n,m}^2 q_m\right\} \\ &= h\{q_{n+1} + \sum_{m=1}^{n} (d_{n+1,m}^2 - d_{n,m}^2) q_m\} \\ &= h\{q_{n+1} + \sum_{m=1}^{n} (d_{n+1,m} - d_{n,m})(d_{n+1,m} + d_{n,m}) q_m\} \\ &= h\{q_{n+1} + hg(x_n) \sum_{m=1}^{n} d_{n,m}^2 (2 + hg(x_n)) q_m\} \end{aligned}$$

and thus, using the boundedness of $q(x)$ and v_n,

$$v_{n+1} - v_n = h\{2g(x_n)v_n + q_{n+1}\} + O(h^2)$$

By Theorem 1.4 it follows that in each of the intervals where $q(x)$ is continuous,

$$v_n = v(x_n) + O(h) \tag{2-81}$$

where the function $v(x)$ is a solution of the differential equation

$$v'(x) = 2g(x)v(x) + q(x) \tag{2-82a}$$

Piecing the solutions continuously together at the points of discontinuity of $q(x)$, it follows that (2-81) holds in the whole interval $[a, b]$, provided that $v(x)$ denotes the continuous and piecewise differentiable solution of (2-82*a*) satisfying

$$v(a) = 0 \tag{2-82b}$$

We summarize the results of this section in the following theorem.

THEOREM 2.10. *Let the local round-off errors* ε_n *be independent random variables which satisfy the conditions of Theorem* 2.9 *and let their expected values and variances satisfy* (2-80), *where* $p(x)$ *and* $q(x)$ *are piecewise smooth functions. Then the primary component* $r_n^{(1)}$ *of the accumulated round-off error is a random variable such that*

$$|E(r_n^{(1)})| \le \frac{\mu}{h}\{m(x_n) + O(h)\} \tag{2-83}$$

$$\text{var}\,(r_n^{(1)}) = \frac{\sigma^2}{h}\{v(x_n) + O(h)\}, \qquad x_n \in [a, b] \tag{2-84}$$

where $m(x)$ *and* $v(x)$ *are defined by* (2-77) *and* (2-82) *respectively.*

This result confirms the earlier observation that the round-off error of a one-step method is in no way connected with the truncation error of the method.

2.3-5. Partial double precision. One way of diminishing the effects of round-off consists in performing all computations in double precision. From a theoretical point of view this simply amounts to working with a basic unit u^2 in place of u. Another less drastic step in the direction of minimizing round-off is the use of a mode of operation called *partial double precision*. This is a simple computational device which, at very little cost, totally eliminates the induced error (which was seen to be the major source of local round-off error) at the expense of increasing somewhat the inherent error. This is achieved by carrying the values $\tilde{y}_n$ in double precision. All other quantities are calculated in single precision. As before, it is assumed that the step h and the abscissas x_n are exact single precision numbers.

Denoting as before by z^* the (single precision) rounded value of a number z, the algorithm of partial double precision is given by

$$\begin{aligned} \tilde{y}_0 &= y_0 \\ \tilde{y}_{n+1} &= \tilde{y}_n + h\tilde{\Phi}(x_n, \tilde{y}_n^*; h), \qquad n = 0, 1, 2, \cdots \end{aligned} \tag{2-85}$$

Here the $\tilde{y}_n$ are double precision numbers. The two important features of the algorithm are:

(i) the product $h\tilde{\Phi}$ is left unrounded, and is in its entirety added to $\tilde{y}_n$;

(ii) the function $\tilde{\Phi}$ is evaluated with the more significant portion of $\tilde{y}_n$ only. Thus the time required for computing $\tilde{\Phi}$ is not increased in comparison with ordinary single precision operation.

The local round-off error ε_{n+1} was defined by

$$\tilde{y}_{n+1} = \tilde{y}_n + h\Phi(x_n, \tilde{y}_n; h) + \varepsilon_{n+1}$$

In the present case we thus have

$$\varepsilon_{n+1} = h\tilde{\Phi}(x_n, \tilde{y}_n^*; h) - h\Phi(x_n, \tilde{y}_n; h)$$

In order to estimate ε_{n+1}, we write

$$\begin{aligned} \varepsilon_{n+1} = h[\tilde{\Phi}(x_n, \tilde{y}_n^*; h) - \Phi(x_n, \tilde{y}_n^*; h)] \\ + h[\Phi(x_n, \tilde{y}_n^*; h) - \Phi(x_n, \tilde{y}_n; h)] \end{aligned} \tag{2-86}$$

If the function Φ is evaluated with an error not exceeding a fixed quantity $\delta > 0$, we thus have

$$|\varepsilon_{n+1}| \leq h\delta + \tfrac{1}{2}hLu$$

or less pessimistically,

$$|\varepsilon_{n+1}| \leq h\delta + \tfrac{1}{2}hu[g(x_n) + O(h)] \tag{2-87}$$

If the $O(h)$ term is neglected, the last relation is of the form (2-76), where $\varepsilon = h\delta$, $p(x) = 1 + \frac{1}{2}u\delta^{-1}g(x)$.

Relation (2-86) can also be made the basis of a statistical treatment. If the accumulated error is small enough, we have

$$\Phi(x_n, \tilde{y}_n^*; h) - \Phi(x_n, \tilde{y}; h) = g(x_n)(\tilde{y}_n^* - \tilde{y}_n) + O(h)$$

where $g(x) = f_y(x, y(x))$. It is reasonable to assume that the quantity $\tilde{y}_n^* - \tilde{y}_n$ behaves like a random variable with the rectangular distribution defined by the function $F_u(x)$ (see §1.6-1). If $\tilde{\Phi} - \Phi$ is also considered a random variable, and if it is assumed that

$$|E(\tilde{\Phi} - \Phi)| \leq mu, \qquad \operatorname{var}(\tilde{\Phi} - \Phi) = s^2u^2$$

where m and s^2 are constants,† then it follows that ε_{n+1} is a random variable such that

$$|E(\varepsilon_{n+1})| \leq hmu \tag{2-88}$$

and, if the variables $\tilde{\Phi} - \Phi$ and $\tilde{y}^* - \tilde{y}$ are independent,

$$\operatorname{var}(\varepsilon_{n+1}) = h^2(s^2 + \tfrac{1}{12}g^2(x_n))u^2 + O(h^4u^2) \tag{2-89}$$

It is thus seen that for partial double precision the variance of the local round-off error depends on x, even though a fixed binary point is employed. We note that the variance of the local error is now of the order of h^2u^2. This represents an improvement over ordinary single precision by a factor h^2. The standard deviation of the accumulated round-off error is thus diminished by a factor of the order of h.

2.3-6. Variable step. It remains to discuss the propagation of round-off error in the case where consecutive lattice points are calculated by the formula

$$x_{n+1} = x_n + \vartheta(x_n)h \tag{2-90}$$

where $\vartheta(x)$ is a positive, piecewise continuous function which is independent of h or n. As before, we shall assume that the numbers x_n are calculated exactly. This requirement practically restricts $\vartheta(x)$ to the class of piecewise constant functions; this, however, is the situation which is most frequently encountered in practice when the stepsize is varied.

† If $\tilde{\Phi} = \Phi^*$, we may assume that $m = 0$, $s^2 = \frac{1}{12}$.

We obtain a recurrence relation for the accumulated round-off error by subtracting from the exact relation

$$y_{n+1} = y_n + \vartheta(x_n)h\Phi(x_n, y_n;\ \vartheta(x_n)h)$$

the corresponding relation satisfied by the numerical values,

$$\tilde{y}_{n+1} = \tilde{y}_n + \vartheta(x_n)h\Phi(x_n, \tilde{y}_n;\ \vartheta(x_n)h) + \varepsilon_{n+1} \tag{2-91}$$

where ε_{n+1} denotes the local round-off error. Linearizing as in §2.3-2 and making the same assumptions on ε_{n+1} we find

$$r_{n+1} = r_n + h\vartheta(x_n)g(x_n)r_n + \eta_{n+1} \tag{2-92}$$

where

$$\eta_{n+1} = \varepsilon_{n+1} + \theta_{n+1}h\varepsilon K_1$$

The further analysis is similar to (but not identical with) that of §1.4-3. Assuming that

$$r_n = \sum_{m=1}^{n} d_{n,m}\eta_m \tag{2-93}$$

for arbitrary values of η_m, we find upon substituting in (2-92) that the coefficients $d_{n,m}$ must satisfy

$$\begin{aligned} &d_{n+1,n+1} = 1, \qquad n = 0, 1, 2, \cdots \\ &d_{n+1,m} = d_{n,m} + h\vartheta(x_n)g(x_n)d_{n,m}, \quad m = 1, 2, \cdots, \quad n = m, m+1, \cdots \end{aligned} \tag{2-94}$$

By Theorem 2.6 it follows that

$$d_{n,m} = d_m(x_n) + O(h) \tag{2-95}$$

where

$$d'_m(x) = g(x)d_m(x), \qquad d_m(x_m) = 1 \tag{2-96}$$

Both for the a posteriori bound and for the statistical work we have to evaluate the sum

$$m_n = h\sum_{m=1}^{n} d_{n,m}p_m \tag{2-97}$$

where $p_m = p(x_m)$, $p(x)$ being defined as in §2.3-4. Using the relations (2-94), one easily finds the formula

$$m_{n+1} = m_n + h[\vartheta(x_n)g(x_n)m_n + p(x_n)] + O(h^2)$$

which may also be written as follows:

$$m_{n+1} = m_n + \vartheta(x_n)h[g(x_n)m_n + \vartheta(x_n)^{-1}p(x_n)] + O(h^2)$$

By virtue of Theorem 2.6 it follows that

$$m_n = m(x_n) + O(h) \tag{2-98}$$

where

$$m'(x) = g(x)m(x) + \vartheta(x)^{-1}p(x), \qquad m(a) = 0 \tag{2-99}$$

For the approximate determination of the variance of the accumulated error it is necessary to calculate

$$v_n = h \sum_{m=1}^{n} d_{n,m}^2 q_m \tag{2-100}$$

where $q_m = q(x_m)$. Proceeding as in §2.3-4, we find that

$$v_{n+1} = v_n + h[2\vartheta(x_n)g(x_n)v_n + q(x_n)] + O(h^2)$$

Rewriting this relation in the form

$$v_{n+1} = v_n + \vartheta(x_n)h[2g(x_n)v_n + \vartheta(x_n)^{-1}g(x_n)] + O(h^2)$$

and applying Theorem 2.6 it follows that

$$v_n = v(x_n) + O(h) \tag{2-101}$$

where

$$v'(x) = 2g(x)v(x) + \vartheta(x)^{-1}q(x), \qquad v(a) = 0 \tag{2-102}$$

Once the sums (2-97) and (2-100) have been calculated, the discussion proceeds as in §§2.3-3 and 2.3-4. We thus obtain the following final theorem, which embodies the most comprehensive result of its type in this section.

THEOREM 2.11. *Let the points* (x_n, y_n) *be determined by* (2-52), *and let the local round-off errors satisfy* (2-76), *where* ε *is such that* (2-71) *and* (2-72) *hold. The accumulated round-off error then satisfies*

$$|r_n| \le \frac{\varepsilon}{h}\{m(x_n) + O(h)\}, \qquad x_n \in [a, b] \tag{2-103}$$

where $m(x)$ *is defined by* (2-99). *If, furthermore, the local round-off errors are mutually independent random variables satisfying* (2-80), *then the primary component* $r_n^{(1)}$ *of the accumulated round-off error is a random variable with*

$$|E(r_n^{(1)})| \le \frac{\mu}{h}\{m(x_n) + O(h)\} \tag{2-104}$$

$$\operatorname{var}(r_n^{(1)}) = \frac{\sigma^2}{h}\{v(x_n) + O(h)\}, \quad x_n \in [a, b] \tag{2-105}$$

where $m(x)$ *is the same as above, and* $v(x)$ *is given by* (2-102).

The presence of the term $\vartheta(x)^{-1}$ in the definitions of $m(x)$ and $v(x)$ suggests that the regions where the step is small, and hence $\vartheta(x)^{-1}$ is large, contribute comparatively more to the accumulated round-off error than the regions where a large step is employed.

2.3-7. Numerical illustration. The results of the preceding section depend essentially on the hypothesis that the local round-off errors can be treated as random variables, and the results concerning the variance make it necessary to assume that these variables are independent. Rather than attempting to prove these hypotheses, we shall present some numerical evidence in support of them.

We choose as our example the numerical integration of the initial value problems

$$y' = -16xy$$

$$y(-0.75) = y_{0,q} = y_{0,0}(1 + q\Delta), \qquad q = 0, 1, \cdots, Q-1$$

where

$$y_{0,0} = 0.0022159242 \cdot 2^{-14}$$
$$\Delta = \tfrac{1}{3} \cdot 2^{-8}$$
$$Q = 500$$

by (I) the modified Euler method; (II) the Runge-Kutta method.

The numerical and theoretical approximations at the point x_n corresponding to the initial value $y_{0,q}$ are denoted by $\tilde{y}_{n,q}$ and $y_{n,q}$, respectively.

We first assume h to be fixed, $h = 2^{-6}$, and consider the round-off error at different values of x. Working with a fixed binary point ($u = 2^{-36}$), it is natural to assume that the local round-off errors have the fixed rectangular distribution defined by the function $F_u(x)$ (see §1.6-1). This implies that

$$E(\varepsilon_m) = 0, \qquad \text{var}\,(\varepsilon_m) = \tfrac{1}{12}u^2$$

It is thus expected that

$$E(r_n) = 0, \qquad \text{var}\,(r_n) = \frac{2^6}{12}\,u^2 v(x_n)$$

for both methods under consideration, where

$$v'(x) = -32xv(x) + 1, \qquad v(-0.75) = 0$$

Values of $v(x)$ (found by numerical integration) are given in Table 2.6.

Table 2.6

x	-0.50	-0.25	0	0.25	0.50	0.75
$v(x)$	6.5	132.7	361.1	133.0	6.7	0.1

Table 2.7

x_n		−0.50	−0.25	0	0.25	0.50	0.75
$u^{-1}E(r_n)$	predicted	0.00	0.00	0.00	0.00	0.00	0.00
	method I	0.71	3.37	5.83	3.48	0.68	0.04
	method II	1.05	5.02	7.78	5.02	1.03	0.05
$u^{-2}\operatorname{var}(r_n)$	predicted	34.6	707.7	1925.8	709.3	35.7	0.5
	method I	42.0	844.1	2304.3	848.8	43.1	0.5
	method II	41.6	842.2	2282.6	841.6	41.6	0.5

In Table 2.7 (see also Fig. 2.2) these predicted values are compared with the values of the mean and the variance calculated numerically by the formulas

$$E(r_n) = \frac{1}{Q}\sum_{q=0}^{Q-1}(\tilde{y}_{n,q} - y_{n,q}), \qquad \operatorname{var}(r_n) = \frac{1}{Q}\sum_{q=0}^{Q-1}(\tilde{y}_{n,q} - E(r_n))^2$$

The "exact" values $y_{n,q}$ were also obtained by numerical computation, but working with higher precision. It is remarkable how the numerical solutions, after being rather widely dispersed at $x = 0$, come into focus again at $x = 0.75$, exactly as predicted by the theory.

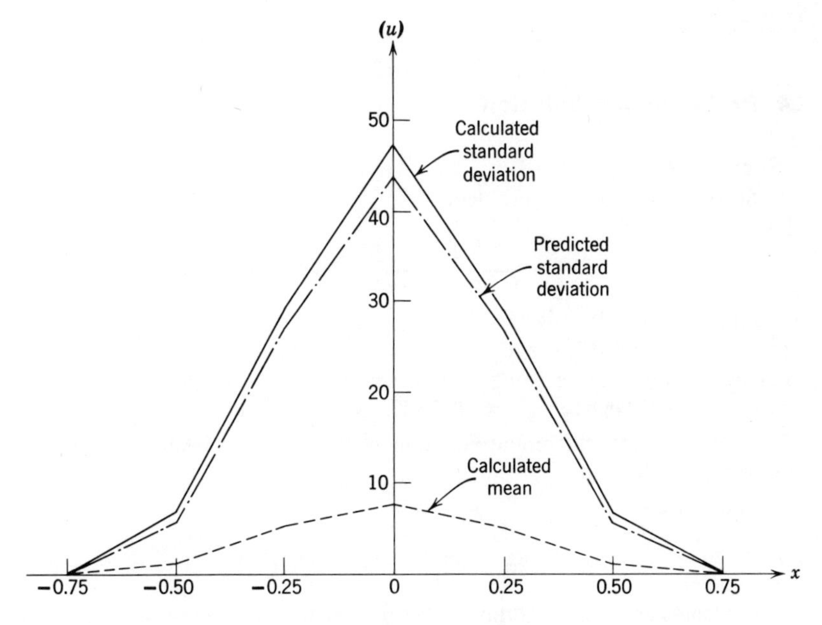

Fig. 2.2. Propagation of round-off error in the Runge-Kutta method.

Instead of keeping h fixed and letting x vary, we may also study the dispersion of the numerical values $y_{n,q}$ as a function of h for a fixed value of x_n. The predicted variance at $x_n = 0$ is given by

$$u^{-2} \operatorname{var}(r_n) \approx \frac{1}{h} \cdot \frac{1}{12} \cdot 361.1$$

The experimental values in Table 2.8 have been calculated by the modified Euler method. The agreement is not very good for large values of h, but it becomes better with decreasing h. This may be explained by

Table 2.8

	h	2^{-4}	2^{-5}	2^{-6}	2^{-7}	2^{-8}
$u^{-2} \operatorname{var}(r_n)$	predicted	481.4	962.9	1925.9	3851.7	7703.5
	experimental	848.9	1380.7	2304.6	4556.0	7801.0

the fact that in the above analysis the local round-off error has been assumed to consist only of induced error. This is asymptotically true as $h \to 0$; for large values of h, however, the inherent error cannot entirely be neglected.

2.4. Problems for Solution

Section 2.1

1. Solve the initial value problem

$$y' = \frac{1}{1 - 0.2 \cos y}, \qquad y(0) = 0$$

in the interval $[0, 2\pi]$:

(a) by the modified Euler method, using $h = \pi/5$;

(b) by the fourth order Runge-Kutta method, using $h = 2\pi/5$.

[(a) $y_{10} = 6.27246$, (b) $y_5 = 6.28176$.]

2. Prove that the differential equation $y' = -2ax$ is solved exactly by the modified Euler method.

3. The relation

$$y_{n+1} - y_n = \tfrac{1}{2}h[f(x_n, y_n) + f(x_{n+1}, y_{n+1})] \tag{2-106}$$

defines a one-step method, although because of the occurrence of y_{n+1} on the right-hand side the increment function is defined only implicitly in this case.

Assuming that (2-106) is solved for y_{n+1} and that the solution has an expansion of the form $y_{n+1} = y_n + c_1 h + c_2 h^2 + \cdots$, determine the order of the method.

4. Show that the implicit one-step method defined by

$$y_{n+1} = y_n + \tfrac{1}{6}h[4f(x_n, y_n) + 2f(x_{n+1}, y_{n+1}) + hf'(x_n, y_n)]$$

is of order 3.

5. Let $g(x, y) = f'(x, y) = f_x + f_y f$. Prove that the one-step method defined by the increment function

$$\Phi(x, y;\ h) = f(x, y) + \tfrac{1}{2}hg(x + \tfrac{1}{3}h, y + \tfrac{1}{3}hf(x, y))$$

has order 3.

6. Consider the one-step method defined by

$$\Phi(x, y;\ h) = f(x + \tfrac{1}{2}h, y + \tfrac{1}{2}h\bar{\Phi}(x, y;\ \tfrac{1}{2}h)) \tag{2-107}$$

where $\bar{\Phi}$ denotes some other increment function. Using the fact that for a twice continuously differentiable function $z(t)$

$$h^{-1}\int_x^{x+h} z(t)\, dt = z(x + \tfrac{1}{2}h) + \tfrac{1}{24}h^2 z''(\xi)$$

where $x < \xi < x + h$, show that the order of the method defined by (2-107) is at least 2 provided that the order of the method defined by $\bar{\Phi}$ is positive.

7. If $z(t)$ is three times continuously differentiable, then

$$h^{-1}\int_x^{x+h} z(t)\, dt = \tfrac{1}{4}z(x) + \tfrac{3}{4}z(x + \tfrac{2}{3}h) + \tfrac{1}{216}h^3 z'''(\xi)$$

where $x < \xi < x + h$. If the order of the method defined by the increment function $\bar{\Phi}$ is at least 2, show that the order of the method defined by

$$\Phi(x, y;\ h) = \tfrac{1}{4}f(x, y) + \tfrac{3}{4}f(x + \tfrac{2}{3}h, y + \tfrac{2}{3}h\bar{\Phi}(x, y;\ \tfrac{2}{3}h)) \tag{2-108}$$

is at least 3.

8*. *One-step methods based on quadrature.* Let the constants w_k, $\theta_k \in [0, 1]$ $(k = 1, 2, \cdots, \nu)$, $C \neq 0$ and the positive integer μ be such that for any sufficiently differentiable function $z(t)$ a number $\xi \in (x, x + h)$ can be selected such that the following *quadrature formula* holds:

$$h^{-1}\int_x^{x+h} z(t)\, dt = \sum_{k=1}^{\nu} w_k z(x + \theta_k h) + Ch^{\mu} z^{(\mu)}(\xi) \tag{2-109}$$

Let the functions $\Phi_k(x, y;\ h)$ $(k = 1, 2, \cdots, \nu)$ be arbitrary increment functions.

Show that the order of the one-step method defined by

$$\Phi(x, y;\, h) = \sum_{k=1}^{\nu} w_k f(x + \theta_k h, y + \theta_k h \bar{\Phi}_k(x, y;\, \theta_k h)) \tag{2-110}$$

is at least min $(\mu, \bar{p} + 1)$, where $\bar{p}$ denotes the smallest of the orders of the methods defined by the functions $\bar{\Phi}_k$.

9*. Let the general Runge-Kutta method be defined as an explicit one-step method in which the increment function is defined solely in terms of the function $f(x, y)$. Using the fact that there exist formulas of the form (2-109) for arbitrary μ, prove that there exist Runge-Kutta methods of arbitrary order. [*Hint:* Assuming the existence of a Runge-Kutta method of order n, generate one of order $n + 1$ by combining it with a quadrature formula with $\mu = n + 1$.]

10*. Using the fact (Hildebrand [1956], Chapter VIII) that there exist quadrature formulas with $\nu = \frac{1}{2}\mu$ for all even μ and with $\nu = \frac{1}{2}(\mu + 1)$, $\theta_1 = 0$ for all odd μ, show that the number of substitutions required for a Runge-Kutta method of order $2n$ does not exceed $(n!)^2$.

Section 2.2

11. Determine the magnified error function for the numerical solution of the initial value problem

$$y' = -2xy, \qquad y(-2) = e^{-4}$$

by the Heun method (2-10*a*). Are there any points where the error is smaller than $O(h^2)$?

12. Determine the principal error function of the method (2-106) and express it in terms of derivatives of the exact solution.

13. Denoting by $\bar{\varphi}(x, y)$ the principal error function of the method defined by $\bar{\Phi}(x, y;\, h)$, derive a formula for the principal error function of the method given by (2-107). [Distinguish the cases where the order $\bar{p}$ of $\bar{\Phi}$ satisfies $\bar{p} = 1$ and $\bar{p} > 1$.]

14*. Show that if $\bar{p} \geq \mu$ for the method of Problem 8, the principal error function is given by

$$\varphi(x, y) = -Cf^{(\mu)}(x, y)$$

15. Show that for the problem $y' = ay$, $y(0) = 1$ the modified Euler method is identical with the Taylor method of order 2, and the classical Runge-Kutta method with the Taylor method of order 4. Find closed expressions for the principal error functions.

16. In an experiment involving the integration of a differential equation over a fixed interval by the same method using different values of h, the discretization error assumed (approximately) the following values as a function of h:

h	e_n
$h = 2^{-4}$	$e_n = 523 \cdot 10^{-8}$
2^{-5}	$65 \cdot 10^{-8}$
2^{-6}	$8 \cdot 10^{-8}$
2^{-7}	$1 \cdot 10^{-8}$

What can you say about the order of the method?

17. Using the step sizes $h = 2^{-p} \cdot \frac{1}{9}$ $(p = 0, 1, 2, \cdots)$, the following values were obtained for $x_n = \frac{8}{9}$ in the numerical integration of the problem $y' = -3y$, $y(0) = \frac{1}{8}$ by the (ordinary) Euler method:

p	y_n
0	0.004877305
1	0.006760986
2	0.007721044
3	0.008202987
4	0.008444179
5	0.008564801
6	0.008625115
7	0.008655273
8	0.008670352
9	0.008677891
10	0.008681659

(Exact result: $y(\frac{8}{9}) = \frac{1}{8}e^{-8/3} \approx 0.008685431$.) Verify that the values obtained by extrapolation to the limit $h = 0$ have an error which appears to be $O(h^3)$ rather than $O(h^2)$, as ordinarily may be expected. Explain either by expanding y_n in powers of h or by using the results of Problem 8 of Chapter 1.

18. Heun [1900] proposed a Runge-Kutta method with the increment function

$$\Phi(x, y;\, h) = \tfrac{1}{4}(k_1 + 3k_3)$$

where

$$k_{\nu+1} = f(x + \tfrac{1}{3}\nu h,\, y + \tfrac{1}{3}\nu h k_\nu), \qquad \nu = 0, 1, 2$$

Prove that this is a third order method and determine the principal error function.

19. The initial value problem

$$y' = -y \operatorname{sign} x, \qquad y(-0.5) = 0.1$$

can be solved by the Runge-Kutta method in two ways: Writing

$$\Phi(x, y;\, h) = \sum_{\nu=1}^{4} a_\nu f(X_\nu, Y_\nu)$$

where the values of a_ν, X_ν, Y_ν are defined by (2-12*b*), we may either put

(i) $$f(X_\nu, Y_\nu) = \begin{cases} +Y_\nu, & \text{if } X_\nu < 0 \\ -Y_\nu, & \text{if } X_\nu \geq 0 \end{cases}$$

or

(ii) $$f(X_\nu, Y_\nu) = \begin{cases} +Y_\nu, & \text{if } x < 0 \\ -Y_\nu, & \text{if } x \geq 0 \end{cases}$$

The following numerical values result at $x_n = 0.5$ for the steps $h = 2^{-p}$, $p = 2, 3, \cdots$:

p	y_n (i)	y_n (ii)
2	0.091659898	0.10000683
3	0.095838792	0.10000021
4	0.097916628	0.10000000
5	0.098958330	0.10000000
6	0.099479166	
7	0.099739583	
8	0.099934895	
9	0.099967447	
10	0.099983723	

Verify that in case (i) the error is of the order of h and give an explanation. [In case (ii) the error, owing to special circumstances, happens to be of the order of h^5, although ordinarily it could only be expected to be of the order of h^4.]

20*. *Minimizing the error for a given number of steps.* If the initial value problem (1-1) is solved by a one-step method of order k and principal error function $\varphi(x, y)$, using a variable step defined by the step modifier $\vartheta(x)$, then by solving (2-56) the magnified error is easily seen to be given by

$$e(x) = u(x) \int_a^x u(t)^{-1} \varphi(t, y(t)) [\vartheta(t)]^k \, dt \tag{2-111}$$

where

$$u(x) = \exp \int_a^x f_y(t, y(t)) \, dt$$

If the solution is wanted only for the fixed value $x = b$, one may consider the problem of minimizing (2-111), considered as a function of $\vartheta(t)$, under the side condition that

$$\int_a^b \frac{1}{\vartheta(t)} \, dt = 1$$

which for small values of h approximately expresses the fact that the number of steps is constant. [The function $\vartheta(t)^{-1}$ is the "density" of the lattice points.] Assuming that $\varphi(t, y(t)) > 0$, show that the solution† of this minimum problem is given by

$$\vartheta(t) = \frac{A}{g(t)}$$

† The solution of this problem is due to D. D. Morrison.

where

$$g(t) = [u(t)^{-1}\varphi(t, y(t))]^{1/(k+1)}$$

$$A = \int_a^b g(t)\, dt$$

As an application, show that the problem $y' = y$, $y(0) = 1$ is best solved with a *uniform* step for any b.

[*Hint:* Use Hölder's inequality

$$\int_a^b \alpha\beta\, dt \le \left(\int_a^b \alpha^p\, dt\right)^{1/p} \left(\int_a^b \beta^q\right)^{1/q}\Bigg]$$

Section 2.3

21. Under the usual statistical assumptions on the local round-off error, what is the probability (approximately) that the round-off error in the numerical solution of the problem discussed in §2.3-7 by a one-step method, using single precision arithmetic with the step $h = 2^{-8}$, will exceed:

(i) $175u$ at $x = 0$, (ii) $2.8u$ at $x = 0.75$?

22. *Round-off in floating operations.* If $y' = g(x)y$, and if $p = y$, $q = y^2$, prove that the functions $m(x)$ and $v(x)$ defined by (2-77) and (2-82) respectively can also be defined as the solutions of the second order equations

$$m'' = 2gm' + (g' - g^2)m$$

and

$$v'' = 4gv' + (2g' - 4g^2)v$$

satisfying the initial conditions

$$m(a) = 0, \qquad m'(a) = y(a)$$

and

$$v(a) = 0, \qquad v'(a) = [y(a)]^2$$

Also show that under the same conditions the functions

$$\mu(x) = \frac{m(x)}{y(x)}, \qquad \omega(x) = \frac{v(x)}{y^2(x)}$$

(relative measures of the round-off error) satisfy

$$\mu(x) = \omega(x) = x - a$$

As an application, compare the propagation of round-off in the solution of $y' = -y$, $y(0) = 1$ by a one-step method for floating and fixed point operations.

23. *Inherent round-off error in the Runge-Kutta method.* Let the increment function for the classical Runge-Kutta method be schematically written in the form

$$\Phi = \tfrac{1}{6}k_1 + \tfrac{1}{3}k_2 + \tfrac{1}{3}k_3 + \tfrac{1}{6}k_4$$

A numerical value of Φ can be calculated either by

$$\text{(A)} \qquad \tilde{\Phi} = [(\tfrac{1}{6})^*(k_1^* + 2k_2^* + 2k_3^* + k_4^*)]^*$$

or by

$$\text{(B)} \qquad \tilde{\Phi} = [(\tfrac{1}{6})^*k_1^*]^* + [(\tfrac{1}{3})^*k_2^*]^* + [(\tfrac{1}{3})^*k_3^*]^* + [(\tfrac{1}{6})^*k_4^*]^*$$

We assume that the quantities $k_\nu^* - k_\nu$ and the round-off errors in products with nonterminating binary fractions are random variables with mean 0 and variance $\sigma^2 = \frac{1}{12}u^2$. Show that in case (A)

$$E(\tilde{\Phi} - \Phi) = [(\tfrac{1}{6})^* - \tfrac{1}{6}]\Phi, \qquad \operatorname{var}(\tilde{\Phi} - \Phi) = \tfrac{23}{216}u^2$$

and in case (B), provided that $(\frac{1}{6})^* + (\frac{1}{3})^* + (\frac{1}{3})^* + (\frac{1}{6})^* = 1$,

$$E(\tilde{\Phi} - \Phi) = 0, \qquad \operatorname{var}(\tilde{\Phi} - \Phi) = \tfrac{77}{216}u^2$$

(The result is important for partial double precision.)

Notes

§2.1-2. The classical references on the Runge-Kutta method are Runge [1895], Heun [1900], Kutta [1901]. Modified versions have been proposed by Gill [1951], Albrecht [1955], Conte and Reeves [1956], Blum [1957], Collatz [1960], p. 64. Combinations of the Runge-Kutta idea and of the use of higher derivatives are advocated by Zurmühl [1948] and Fehlberg [1958]. Runge-Kutta methods of order exceeding 4 are given by Nyström [1925] and Huta [1956, 1957]. The programming of the Runge-Kutta method on a digital computer is discussed by Murray [1950], Gill [1951], Blum [1957], Martin [1958], Romanelli [1960]. A number of papers deal with one-step methods based on integration formulas. Quadrature formulas involving higher derivatives are used by Milne [1949] and, in a very similar manner, by Lotkin [1952] (see Problems 3 and 4). All these formulas are applications of a basic formula due to Obrechkoff [1942]. The idea of using Gaussian or similar quadrature formulas for the numerical integration of ordinary differential equations appears in Hammer and Hollingsworth [1955], Morrison and Stoller [1958], Korganoff [1958] (see Problems 6, 7, 8, 13, 14). A considerable body of literature has been published in the U.S.S.R. on a method due to Caplygin, which at the same time yields bounds for the error; see Babkin [1948], Luzin [1951], Azbelev [1952, 1953, 1955], Voronovskaya [1955]. Weissinger [1953] has a method for implicit differential equations. For still other methods see Vlasow and Čarnyi [1950], Bukovics [1950], Bückner [1952].

§§2.2-2, 2.2-3, 2.2-4. Bounds for the local discretization error of the classical Runge-Kutta method (for single equations) are given (without proof) by Bieberbach [1944], p. 54, and Lotkin [1951]. The propagated error is studied by

Bukovics [1953, 1954], Ceschino [1954], Carr [1958]. Milne [1950] gives a numerical example.

§§2.2-6, 2.2-7, 2.2-8. For a heuristic treatment of the asymptotic behavior of the discretization error in one-step and multistep methods, see Lotkin [1954]. Methods for estimating the local discretization error are given by Gorn and Moore [1953], Morel [1956], Kuntzmann [1959a]. Variable step is treated by Garfinkel [1954], Ceschino [1956].

§2.3-5. The method of partial double precision is proposed by Young [1955]. Other ways to "control" round-off error are given by Gill [1951] and Blum [1957].

chapter

3 General One-step Methods for Systems of Equations of the First Order

3.1. Theoretical Introduction

3.1-1. Definitions. A system of ordinary differential equations of the first order is a system of equations of the form

$$
\begin{aligned}
y^{1\prime} &= f^1(x, y^1, y^2, \cdots, y^s) \\
y^{2\prime} &= f^2(x, y^1, y^2, \cdots, y^s) \\
&\cdots\cdots\cdots\cdots\cdots\cdots \\
y^{s\prime} &= f^s(x, y^1, y^2, \cdots, y^s)
\end{aligned}
\tag{3-1}
$$

where the functions $f^1, f^2, \cdots, f^s$ are given functions of their $s + 1$ arguments. (The superscripts in this chapter always denote indices, never powers.) A set of functions $y^1(x), y^2(x), \cdots, y^s(x)$ which are defined and differentiable in an interval $[a, b]$ and satisfy identically in x the relation

$$y^{i\prime}(x) = f^i(x, y^1(x), y^2(x), \cdots, y^s(x)), \qquad i = 1, 2, \cdots, s \tag{3-2}$$

is called a *solution* of the system. A problem which arises frequently in practice and which will be studied in the present chapter is to find a solution of the system (3-1) which satisfies given initial conditions of the form

$$y^i(a) = \eta^i, \qquad i = 1, 2, \cdots, s \tag{3-3}$$

where the numbers η^i are preassigned constants. This problem will be called the *initial value problem*. Other conditions, more complicated than (3-3), also arise in practice.

Systems of ordinary differential equations arise in several ways.

(i) Theoretically, every ordinary differential equation of order higher than 1 can be reduced to a system of first order equations. A differential equation of order m is given by

$$y^{(m)} = f(x, y, y', y'', \cdots, y^{(m-1)}) \tag{3-4}$$

where f is a given function of its $m + 1$ arguments. (Superscripts in parentheses denote derivatives, as usual.) The reduction to a system is accomplished by setting

$$y = y^1, \qquad y' = y^2, \cdots, y^{(m-1)} = y^m \tag{3-5}$$

If the functions $y^1, y^2, \cdots, y^m$ satisfy the system

$$\begin{aligned} y^{1\prime} &= y^2 \\ y^{2\prime} &= y^3 \\ &\cdots\cdots\cdots\cdots \\ y^{m\prime} &= f(x, y^1, y^2, \cdots, y^m) \end{aligned} \tag{3-6}$$

which is a special case of (3-1), then the function $y(x) = y^1(x)$ evidently satisfies (3-4). As an example we consider the second order equation

$$y'' = -y$$

This is reduced to a system by setting

$$y = y^1, \qquad y' = y^2$$

The system then takes the form

$$\begin{aligned} y^{1\prime} &= y^2 \\ y^{2\prime} &= -y^1 \end{aligned} \tag{3-7}$$

Some authorities (Milne [1953], p. 82; Gill [1951], p. 96) recommend this reduction of equations of higher order to a system of equations of the first order also for numerical purposes; others (Collatz [1960], p. 117) take the opposite position, arguing that reduction to a first order system increases both the error and the necessary number of operations. Methods for the direct integration of equations of higher order are presented and analyzed in Chapters 4 and 6. However, the theoretical and experimental results to be presented in these chapters indicate that the truncation error is generally not increased and the round-off error is frequently substantially decreased when an equation of higher order is first reduced to a first order system and then solved by an equivalent method for such systems.

(ii) Systems of ordinary differential equations also arise in a natural way from many physical problems. Classical examples are electric circuits with more than one loop and mechanical problems with several degrees of freedom. More specific examples are the equations of motion of a gyroscope, the fundamental equations of exterior ballistics, and the equations governing the flight of a rocket.

(iii) Recently, systems of ordinary differential equations have also been used to obtain approximations to solutions of partial differential equations (Anonymous [1957], p. 73). We illustrate the method by applying it to the problem of finding a function $u(x, t)$ satisfying the partial differential equation

$$\frac{\partial u}{\partial t} = \frac{\partial^2 u}{\partial x^2} \qquad (0 \leq x \leq L,\ t \geq 0) \tag{3-8}$$

and the side conditions

$$u(x, 0) = f(x) \qquad (0 \leq x \leq L)$$

$$u(0, t) = g(t), \qquad u(L, t) = h(t) \qquad (t \geq 0)$$

where $f(x)$, $g(t)$, and $h(t)$ are given functions.

The usual numerical approach to this problem consists in discretizing the differential equation (3-8) in both the x- and the t-direction. The method under discussion differs from this in that the discretization is performed only with respect to one of the two variables. With respect to the other variable, the differential character of the problem is retained. If we choose to discretize in the x-direction, then we approximate $\partial^2 u/\partial x^2$ for each $t \geq 0$ and for $x = x_n = nk$ ($k = L/N$, $n = 1, 2, \cdots, N - 1$) by

$$\frac{1}{k^2}(u(x_{n+1}, t) - 2u(x_n, t) + u(x_{n-1}, t))$$

Setting

$$u^n(t) = u(x_n, t)$$

and retaining the differential character in the t-direction, we thus replace (3-8) by the system of ordinary differential equations

$$\frac{du^n}{dt} = k^{-2}(u^{n+1} - 2u^n + u^{n-1}) \qquad (n = 1, 2, \cdots, N - 1)$$

where $u^0 = g(t)$, $u^N = h(t)$. The initial conditions are

$$u^n(0) = f(x_n) \qquad (n = 1, \cdots, N - 1)$$

The method is easily adapted to more complicated partial differential equation problems; for error estimates, see Rothe [1930], Budak [1956], Douglas [1956], Franklin [1959].

3.1-2. Vector notation. The subsequent analysis is greatly simplified, both conceptually and formally, by considering the quantities y^i ($i =$ 1, 2. $\cdots$, s) as components of the *vector*

$$\mathbf{y} = \begin{pmatrix} y_1 \\ y_2 \\ \cdot \\ \cdot \\ \cdot \\ y_s \end{pmatrix}$$

Consequently, we shall also write $f^i(x, y^1, \cdots, y^s) = f^i(x, \mathbf{y})$. Combining the s functions $f^i(x, \mathbf{y})$ into another vector

$$\mathbf{f}(x, \mathbf{y}) = \begin{pmatrix} f^1(x, \mathbf{y}) \\ f^2(x, \mathbf{y}) \\ \cdot \\ \cdot \\ \cdot \\ f^s(x, \mathbf{y}) \end{pmatrix}$$

we can write the system (3-1) in the much more compact form

$$\mathbf{y}' = \mathbf{f}(x, \mathbf{y}) \tag{3-9a}$$

If the vector $\boldsymbol{\eta}$ is defined by

$$\boldsymbol{\eta} = \begin{pmatrix} \eta^1 \\ \eta^2 \\ \cdot \\ \cdot \\ \cdot \\ \eta^s \end{pmatrix}$$

then the initial condition (3-3) is given by

$$\mathbf{y}(a) = \boldsymbol{\eta} \tag{3-9b}$$

We shall need a vector analog for the absolute value of a real or complex number. If $\mathbf{v}$ is a vector with real or complex components v^i ($i =$ 1, 2, $\cdots$, s), we shall in this chapter use the analog defined by

$$\|\mathbf{v}\| = |v^1| + |v^2| + \cdots + |v^s| \tag{3-10}$$

We note that for any real or complex number λ and any two vectors $\mathbf{v}$ and $\mathbf{w}$ the following relations hold:

$$\|\mathbf{v}\| \geq 0; \qquad \|\mathbf{v}\| = 0 \qquad \text{if and only if } \mathbf{v} = \mathbf{0} \tag{3-11a}$$

$$\|\lambda \mathbf{v}\| = |\lambda| \, \|\mathbf{v}\| \tag{3-11b}$$

$$\|\mathbf{v} + \mathbf{w}\| \leq \|\mathbf{v}\| + \|\mathbf{w}\| \qquad \text{(triangle inequality)} \tag{3-11c}$$

Here, the product of a vector with a scalar and the sum of two vectors are defined in the natural manner.

Relations (3-11) characterize the number $\|\mathbf{v}\|$ as a *norm* of the vector $\mathbf{v}$. Other definitions of the norm might be used (and one indeed will be in Chapter 7), but the above is the most convenient for the purpose on hand.

If $\mathbf{v}_n$ $(n = 0, 1, 2, \cdots)$ denotes a sequence of vectors, then the notation

$$\lim_{n\to\infty} \mathbf{v}_n = \mathbf{w} \qquad \text{means that} \qquad \lim_{n\to\infty} \|\mathbf{v}_n - \mathbf{w}\| = 0$$

or, equivalently, that the sequences v_n^i have the limits w^i $(i = 1, 2, \cdots, s)$.

If each component v^i of a vector $\mathbf{v}$ depends on a variable t, $\mathbf{v}$ should, strictly speaking, be called a *vector-valued function* of t. However, if no misunderstanding is possible, the word "vector-valued" will frequently be omitted. By

$$\lim_{t\to t_0} \mathbf{v}(t) = \mathbf{z} \qquad \text{is meant that} \qquad \lim_{t\to t_0} \|\mathbf{v}(t) - \mathbf{z}\| = 0$$

A function $\mathbf{v}(t)$ is called continuous at $t = t_0$ if

$$\lim_{t\to t_0} \mathbf{v}(t) = \mathbf{v}(t_0)$$

By $\mathbf{v}'(t)$ and $\int_a^b \mathbf{v}(t)\,dt$ are meant the vectors whose components are, respectively, the derivatives and integrals of the components of $\mathbf{v}(t)$. If $\varphi(t) \to 0$ for $t \to t_0$, then

$$\mathbf{v}(t) = O(\varphi(t)), \qquad t \to t_0$$

means that

$$\|\mathbf{v}(t)\| = O(\varphi(t)), \qquad t \to t_0$$

For vectors $\mathbf{v}$ depending on several variables the concepts of limit, continuity, and order are defined analogously.

3.1-3. The existence of a solution of the initial value problem. We assume that the vector-valued function $\mathbf{f}(x, \mathbf{y})$ of the scalar variable x and the vector $\mathbf{y} = (y^1, \cdots, y^s)$ satisfies the following two hypotheses:

(A) $\mathbf{f}(x, \mathbf{y})$ is defined and continuous in the region $a \le x \le b$, $-\infty < y^i < \infty$, $i = 1, 2, \cdots, s$.

(B) There exists a constant L, the Lipschitz constant, such that for arbitrary $x \in [a, b]$ and any two vectors $\mathbf{y}$ and $\mathbf{y}^*$

$$\|\mathbf{f}(x, \mathbf{y}) - \mathbf{f}(x, \mathbf{y}^*)\| \le L\,\|\mathbf{y} - \mathbf{y}^*\|$$

We shall prove the following analog of Theorem 1.1:

THEOREM 3.1. *Let the function* $\mathbf{f}(x, \mathbf{y})$ *satisfy conditions (A) and (B) above, and let* $\boldsymbol{\eta}$ *be a given vector. Then there exists exactly one function* $\mathbf{y}(x)$ *with the following three properties:*

(i) $\mathbf{y}(x)$ *is continuous and continuously differentiable for* $x \in [a, b]$;
(ii) $\mathbf{y}'(x) = \mathbf{f}(x, \mathbf{y}(x)), \qquad x \in [a, b]$;
(iii) $\mathbf{y}(a) = \boldsymbol{\eta}$.

In short, the initial value problem (3-9) has a unique solution.

The proof of Theorem 3.1 very much resembles the proof of Theorem 1.1. We shall begin by constructing a sequence of functions $\mathbf{y}_p(x)$ ($p = 0, 1, 2, \cdots$), which will be shown to converge to a continuous function $\mathbf{y}(x)$. This construction is quite analogous to the construction carried out in §§1.2-2 through 1.2-5 and will be described in outline only. We then shall show that $\mathbf{y}(x)$ is a solution of the initial value problem (3-9). These statements could be proved by the completely elementary techniques employed in §§1.2-6 and 1.2-7; for reasons of variety, however, we shall prove them here by different and less elementary methods.

3.1-4. The convergence of the approximate solutions $\mathbf{y}_p(x)$.

Let h_p and $x_{[p]}$ be defined as in §1.2-2. For $p = 0, 1, 2, \cdots$ we define the functions $\mathbf{y}_p(x)$ by

$$\begin{aligned} &\mathbf{y}_p(a) = \boldsymbol{\eta} \\ &\mathbf{y}_p(x) = \mathbf{y}_p(x_{[p]}) + (x - x_{[p]})\mathbf{f}(x_{[p]}, \mathbf{y}_p(x_{[p]})) \end{aligned} \tag{3-12}$$

These functions are trivially continuous in each of the intervals $(x_{[p]}, x_{[p]} + h_p)$, and, by construction, also at the points $x_{[p]}$ themselves. We shall prove the following analog of Lemma 1.1:

LEMMA 3.1. *The sequence of vector-valued functions* $\mathbf{y}_p(x)$ *converges for* $p \to \infty$, *uniformly for* $x \in [a, b]$, *to a continuous function* $\mathbf{y}(x)$.

Proof. Using condition (B) with $\mathbf{y}^* = \mathbf{0}$, we find

$$\|\mathbf{f}(x, \mathbf{y})\| \le L\, \|\mathbf{y}\| + c \tag{3-13}$$

where

$$c = \max_{x \in [a,b]} \|\mathbf{f}(x, 0)\|$$

From (3-12) we get

$$\mathbf{y}_p(x_{[p]} + h_p) = \mathbf{y}_p(x_{[p]}) + h_p \mathbf{f}(x_{[p]}, \mathbf{y}_p(x_{[p]})), \qquad x \in (a, b - h_p]$$

and thus, using (3-13),

$$\|\mathbf{y}_p(x_{[p]} + h_p)\| \le (1 + h_p L)\, \|\mathbf{y}_p(x_{[p]})\| + h_p c$$

Applying Lemma 1.2 with $\xi_n = \|\mathbf{y}(a + nh_p)\|$, $A = (1 + h_pL)$, $B = h_pc$, it follows that

$$\|\mathbf{y}_p(x_{[p]})\| \leq e^{(x[p]-a)L} \|\boldsymbol{\eta}\| + E_L(x_{[p]} - a)c$$

where E_L denotes the Lipschitz function. Since the functions $\mathbf{y}_p(x)$ are linear between any two adjacent points x_p, we have

$$\|\mathbf{y}_p(x)\| \leq Y, \qquad x \in [a, b] \tag{3-14}$$

where

$$Y = e^{(b-a)L} \|\boldsymbol{\eta}\| + E_L(b - a)c \tag{3-15}$$

We denote by R the compact region in $(x, \mathbf{y})$-space defined by $x \in [a, b]$, $\|\mathbf{y}\| \leq Y$. Inequality (3-14) then states that the functions $\mathbf{y}_p(x)$ stay within R for $x \in [a, b]$ and $p = 0, 1, 2, \cdots$. In view of the compactness of R, a function which is continuous in R has a finite maximum in R. We put

$$M = \max_{(x,\mathbf{y})\in R} \|\mathbf{f}(x, \mathbf{y})\| \tag{3-16}$$

Moreover, for any $\delta \geq 0$ we set

$$\omega(\delta) = \max \|\mathbf{f}(x, \mathbf{y}) - \mathbf{f}(x^*, \mathbf{y})\| \tag{3-17}$$

where the maximum is taken with respect to all points $(x, \mathbf{y})$ and $(x^*, \mathbf{y})$ in R such that $|x - x^*| \leq \delta$. As in §1.2-4 we have

$$\lim_{\delta\to 0} \omega(\delta) = 0 \tag{3-18}$$

With a view towards applying the Cauchy criterion, we put

$$\mathbf{d}(x) = \mathbf{y}_p(x) - \mathbf{y}_q(x)$$

where p and q are any two nonnegative integers, $p < q$. The following is the exact multidimensional analog of Lemma 1.3:

LEMMA 3.2. *For* $t \in [a, b]$,

$$\|\mathbf{d}(t)\| \leq [1 + (t - t_{[q]})L] \|\mathbf{d}(t_{[q]})\| + (t - t_{[q]})\Omega_p \tag{3-19}$$

where

$$\Omega_p = \omega(h_p) + LMh_p$$

Proof. As in the proof of Lemma 1.3 we may write

$$\begin{aligned}\mathbf{d}(t) - \mathbf{d}(t_{[q]}) = (t - t_{[q]})\{&\mathbf{f}(t_{[q]}, \mathbf{y}_q(t_{[q]})) - \mathbf{f}(t_{[q]}, \mathbf{y}_p(t_{[q]})) \\ &+ \mathbf{f}(t_{[q]}, \mathbf{y}_p(t_{[q]})) - \mathbf{f}(t_{[p]}, \mathbf{y}_p(t_{[q]})) \\ &+ \mathbf{f}(t_{[p]}, \mathbf{y}_p(t_{[q]})) - \mathbf{f}(t_{[p]}, \mathbf{y}_p(t_{[p]}))\}\end{aligned}$$

The differences of values of the function $\mathbf{f}$ in the first two lines on the right are estimable by $L \|\mathbf{d}(t_{[q]})\|$ and $\omega(h_p)$, respectively. Using (B), we have for the norm of the last difference the estimate

$$L \|\mathbf{y}_p(t_{[q]}) - \mathbf{y}_p(t_{[p]})\|$$

or, in view of (3-12), by virtue of $t_{[q]} - t_{[p]} \le h_p$, LMh_p. The statement of Lemma 3.2 now is evident.

The inequality

$$\|\mathbf{d}(x)\| \le \Omega_p \frac{e^{(x-a)L} - 1}{L} \tag{3-20}$$

now follows exactly as (1-27) followed from (1-25). Since the expression on the right of (3-20) does not depend on q and tends to zero as $p \to \infty$ by virtue of (3-18), uniformly for $x \in [a, b]$, the statement of Lemma 3.1 now also follows.

3.1-5. Completion of the proof of the existence theorem. In order to show that the limit function $\mathbf{y}(x)$ satisfies the given differential equation, we define for $p = 0, 1, 2, \cdots$ vector-valued functions $\mathbf{f}_p(x)$ (depending on the single scalar variable x) by setting

$$\mathbf{f}_p(x) = \begin{cases} \mathbf{f}(a, \boldsymbol{\eta}), & x = a \\ \mathbf{f}(x_{[p]}, \mathbf{y}_p(x_{[p]})), & a < x \le b \end{cases} \tag{3-21}$$

The functions $\mathbf{f}_p(x)$ are constant in each of the intervals $(x_{[p]}, x_{[p]} + h_p]$, and the constant value of each component is the slope of the corresponding component of $\mathbf{y}_p(x)$ in that interval. Thus

$$\mathbf{y}_p(x) - \boldsymbol{\eta} = \int_a^x \mathbf{f}_p(t)\, dt \tag{3-22}$$

We shall prove below that

$$\lim_{p \to \infty} \mathbf{f}_p(t) = \mathbf{f}(t, \mathbf{y}(t)) \tag{3-23}$$

uniformly for $t \in [a, b]$. Assuming this fact, we may let $p \to \infty$ in (3-22) and obtain

$$\mathbf{y}(x) - \boldsymbol{\eta} = \int_a^x \mathbf{f}(t, \mathbf{y}(t))\, dt \tag{3-24}$$

Condition (A) implies that $\mathbf{f}(t, \mathbf{y}(t))$ is continuous; hence the derivative of the integral in (3-24) exists and equals $\mathbf{f}(x, \mathbf{y}(x))$. But then the derivative of the function on the left exists also and has the same value. Thus,

$$\mathbf{y}'(x) = \mathbf{f}(x, \mathbf{y}(x)), \qquad x \in [a, b]$$

as desired.

It remains to establish (3-22). We have

$$\begin{aligned} \|\mathbf{f}_p(t) - \mathbf{f}(t, \mathbf{y}(t))\| &\le \|\mathbf{f}(t_{[p]}, \mathbf{y}_p(t_{[p]})) - \mathbf{f}(t, \mathbf{y}_p(t_{[p]}))\| + \|\mathbf{f}(t, \mathbf{y}_p(t_{[p]})) \\ &\quad - \mathbf{f}(t, \mathbf{y}_p(t))\| + \|\mathbf{f}(t, \mathbf{y}_p(t)) - \mathbf{f}(t, \mathbf{y}(t))\| \\ &\le \omega(h_p) + LMh_p + L\,\|\mathbf{y}_p(t) - \mathbf{y}(t)\| \end{aligned}$$

The last expression can be made arbitrarily small by choosing p large enough, uniformly in t. This establishes (3-23).

In order to show that $\mathbf{y}(x)$ is the *only* solution of the initial value problem, we note that *any* solution of the initial value problem is by definition continuous, and hence bounded, in the interval $[a, b]$. If $\mathbf{y}(x)$ and $\mathbf{z}(x)$ are any two solutions, then their difference

$$\boldsymbol{\delta}(x) = \mathbf{y}(x) - \mathbf{z}(x)$$

is bounded also:

$$\|\boldsymbol{\delta}(x)\| \leq K, \qquad x \in [a, b] \tag{3-25}$$

From

$$\mathbf{y}(x) - \boldsymbol{\eta} = \int_a^x \mathbf{f}(t, \mathbf{y}(t))\, dt$$

$$\mathbf{z}(x) - \boldsymbol{\eta} = \int_a^x \mathbf{f}(t, \mathbf{z}(t))\, dt$$

it follows by subtraction that

$$\boldsymbol{\delta}(x) = \int_a^x [\mathbf{f}(t, \mathbf{y}(t)) - \mathbf{f}(t, \mathbf{z}(t))]\, dt$$

and hence, by the Lipschitz condition,

$$\|\boldsymbol{\delta}(x)\| \leq L \int_a^x \|\boldsymbol{\delta}(t)\|\, dt \tag{3-26}$$

We shall now prove that

$$\|\boldsymbol{\delta}(x)\| \leq K \frac{L^k(x - a)^k}{k!}, \qquad x \in [a, b] \tag{3-27}$$

for $k = 0, 1, 2, \cdots$. By (3-25), this inequality is true for $k = 0$. Assuming its truth for a positive integer k, we find, using (3-26), that

$$\|\boldsymbol{\delta}(x)\| \leq L \int_a^x \frac{KL^k(t - a)^k}{k!}\, dt = K \frac{L^{k+1}(x - a)^{k+1}}{(k + 1)!}$$

establishing (3-27) with k increased by 1. It follows that (3-27) is true for all nonnegative integers k. Since

$$\frac{L^k(b - a)^k}{k!} \to 0, \qquad k \to \infty$$

this is only possible if $\|\boldsymbol{\delta}(x)\| = 0$, $a \leq x \leq b$. It follows that $\mathbf{y}(x) \equiv \mathbf{z}(x)$. This completes the proof of Theorem 3.1.

An important special case in which the existence theorem is applicable is the system of *linear* differential equations

$$\mathbf{y}' = \mathbf{G}(x)\mathbf{y} + \mathbf{q}(x) \tag{3-28}$$

where

$$\mathbf{G}(x) = \begin{pmatrix} g^{11}(x) & g^{12}(x) \cdots g^{1s}(x) \\ g^{21}(x) & g^{22}(x) \cdots g^{2s}(x) \\ \cdots & \\ g^{s1}(x) & g^{s2}(x) \cdots g^{ss}(x) \end{pmatrix}$$

is a $s \times s$ matrix whose elements are continuous functions of x for $x \in [a, b]$, and where $\mathbf{q}(x)$ is likewise continuous. Condition (B) is then satisfied with

$$L = \max_{x \epsilon a,b} \|\mathbf{G}(x)\|$$

where

$$\|\mathbf{G}(x)\| = \max_{1 \le j \le s} \sum_{i=1}^{s} |g^{ij}(x)|$$

3.2. Special One-Step Methods for Systems

3.2-1. Definitions; notational simplification. Let $x \in [a, b]$, and let $\mathbf{y}$ be an arbitrary vector. Denoting by $\mathbf{z}(t)$ the solution of the system of differential equations $\mathbf{z}' = \mathbf{f}(t, \mathbf{z})$ satisfying $\mathbf{z}(x) = \mathbf{y}$, we set

$$\mathbf{\Delta}(x, \mathbf{y};\, h) = \begin{cases} \dfrac{\mathbf{z}(x+h) - \mathbf{z}(x)}{h}, & h \neq 0 \\ \mathbf{f}(x, \mathbf{y}), & h = 0 \end{cases} \tag{3-29}$$

and call $\mathbf{\Delta}$ the *exact relative increment* of the solution of the given differential equation. A *one-step method* for the solution of the initial value problem (3-9) is defined by the formulas

$$\begin{aligned} \mathbf{y}_0 &= \boldsymbol{\eta} \\ \mathbf{y}_{n+1} &= \mathbf{y}_n + h\mathbf{\Phi}(x, \mathbf{y}_n;\, h), \qquad n = 0, 1, \cdots \end{aligned} \tag{3-30}$$

Here the function $\mathbf{\Phi}(x, \mathbf{y};\, h)$, called the *increment function*, is chosen so as to approximate $\mathbf{\Delta}(x, \mathbf{y};\, h)$ as well as possible. If p is the largest integer for which

$$\mathbf{\Phi}(x, \mathbf{y};\, h) - \mathbf{\Delta}(x, \mathbf{y};\, h) = O(h^p) \tag{3-31}$$

then p is called the (exact) *order* of the method defined by (3-30).

Before exhibiting some realizations of the increment function, we shall introduce a notational device which will simplify the subsequent analysis.

In order to dispense with the special role played by the independent variable x we augment the system (3-1) by the differential equation

$$y^{0\prime} = 1 \tag{3-32}$$

to be satisfied by a new function $y^0(x)$, subject to the initial condition

$$y^0(a) = a \tag{3-33}$$

Obviously (3-32) and (3-33) imply $y^0(x) = x$. The system (3-1) can therefore be replaced by the equivalent system for $s + 1$ functions,

$$y^{i\prime} = f^i(y^0, y^1, \cdots, y^s) \qquad (i = 0, \cdots, s) \tag{3-34}$$

where it is understood that

$$f^0(y^0, y^1, \cdots, y^s) = 1 \tag{3-35}$$

The new system (3-34) has the advantage that the variables entering into the functions f^i may all be considered as dependent variables.

For simplicity, we shall continue to denote the dependent variables by $y^1, y^2, \cdots, y^s$, whether x is one of them or not. Consequently we shall write the initial value problem (3-9) in the form

$$\mathbf{y}' = \mathbf{f}(\mathbf{y}), \qquad \mathbf{y}(a) = \boldsymbol{\eta} \tag{3-36}$$

where $\mathbf{y}$, $\mathbf{f}$, and $\boldsymbol{\eta}$ are again vectors with s components. If $\mathbf{f}(\mathbf{y})$ does not depend explicitly on x, neither does the function $\boldsymbol{\Delta}$, nor, as we may assume, the increment function $\boldsymbol{\Phi}$. Thus we shall write

$$\boldsymbol{\Delta} = \boldsymbol{\Delta}(\mathbf{y};\ h), \qquad \boldsymbol{\Phi} = \boldsymbol{\Phi}(\mathbf{y};\ h)$$

We shall now construct some special increment functions $\boldsymbol{\Phi}$.

3.2-2. Use of Taylor's expansion; tensor notation. If the function $\mathbf{y}(x)$ is a solution of (3-36), and if the components of $\mathbf{f}$ are sufficiently differentiable, then the higher derivatives of $\mathbf{y}(x)$ can be expressed in terms of the function $\mathbf{f}$ and its derivatives. For instance,

$$\begin{aligned}\mathbf{y}''(x) &= \frac{d}{dx}\,\mathbf{f}(\mathbf{y}(x)) = \sum_{j=1}^{s} \frac{\partial \mathbf{f}}{\partial y^j}\frac{dy^j}{dx} \\ &= \sum_{j=1}^{s} \frac{\partial \mathbf{f}}{\partial y^j} f^j\end{aligned}$$

Generally we shall write for $k = 1, 2, \cdots$, presuming differentiability,

$$\mathbf{y}^{(k+1)}(x) = \frac{d^k}{dx^k}\,\mathbf{f}(\mathbf{y}(x)) = \mathbf{f}^{(k)}(\mathbf{y}(x))$$

The functions $\mathbf{f}^{(k)}(\mathbf{y})$ are well-determined functions of the vector $\mathbf{y}$. We thus have

$$\boldsymbol{\Delta}(\mathbf{y};\ h) = \mathbf{f}(\mathbf{y}) + \frac{h}{2}\,\mathbf{f}'(\mathbf{y}) + \frac{h^2}{3!}\,\mathbf{f}''(\mathbf{y}) + \cdots \tag{3-37}$$

As in the case of a single equation, a method for approximate integration can be based on a truncated Taylor series. The increment function for the Taylor expansion method of order p is given by

$$\mathbf{\Phi}(\mathbf{y};\, h) = \mathbf{f}(\mathbf{y}) + \frac{h}{2}\mathbf{f}'(\mathbf{y}) + \cdots + \frac{h^{p-1}}{p!}\mathbf{f}^{(p-1)}(\mathbf{y}) \tag{3-38}$$

For $p = 1$ this reduces to Euler's method.

The method defined by (3-38) is of no great practical interest, since the evaluation of the derivatives $\mathbf{f}^{(k)}$ is in general much too involved.* However, the method is of theoretical importance in view of the fact that the methods of Runge-Kutta type, to be discussed below, are based on the idea of approximating (3-38) by expressions which do not involve any functions other than $\mathbf{f}(\mathbf{y})$. For this reason we shall now examine in greater detail the structure of the derivatives $\mathbf{f}^{(k)}(\mathbf{y})$.

In order to simplify the notation, we write for $i, j, k = 1, \cdots, s$

$$\frac{\partial f^i}{\partial y^j} = f^i_j, \qquad \frac{\partial^2 f^i}{\partial y^j\, \partial y^k} = f^i_{jk} \tag{3-39}$$

and use a similar notation also for higher derivatives. By $\mathbf{f}_j$, $\mathbf{f}_{jk}$, $\cdots$ we mean the vectors with the components f^i_j, f^i_{jk}, $\cdots$ $(i = 1, \cdots, s)$. We also adopt the convention that if an index occurs both as a subscript and as a superscript the product should be summed with respect to this index from 1 to s. Thus, for instance,

$$f^i_j f^j = \sum_{j=1}^{s} \frac{\partial f^i}{\partial y^j} f^j \tag{3-40}$$

It is easily verified that sums of products of the form (3-40) can be differentiated like ordinary products. Thus,

$$\begin{aligned} \frac{d}{dx}(f^i_j f^j) &= \left(\frac{d}{dx} f^i_j\right) f^j + f^i_j \left(\frac{d}{dx} f^j\right) \\ &= f^i_{jk} f^j f^k + f^i_j f^j_k f^k \end{aligned}$$

We introduce the abbreviations

$$\begin{aligned} A^i &= f^i & E^i &= f^i_{jkm} f^j f^k f^m \\ B^i &= f^i_j f^j & F^i &= f^i_{jk} f^j_m f^k f^m \\ C^i &= f^i_{jk} f^j f^k & G^i &= f^i_j f^j_{km} f^m f^k \\ D^i &= f^i_j f^j_k f^k & H^i &= f^i_j f^j_k f^k_m f^m \end{aligned} \tag{3-41}$$

* There may be exceptions to this in the case of special differential equations; see, e.g., Miller [1955], Cheney [1960].

where all indices range from 1 to s and the argument in all functions is understood to be $\mathbf{y}$, and we denote by $\mathbf{A}, \mathbf{B}, \cdots$ the vectors with the components $A^i, B^i, \cdots \quad (i = 1, 2, \cdots, s)$.

With these notational devices the first few derivatives $\mathbf{f}^{(k)}$ can be expressed in a formally (and only formally) compact manner as follows:

$$\begin{aligned} \mathbf{f}(\mathbf{y}) &= \mathbf{A} \\ \mathbf{f}'(\mathbf{y}) &= \mathbf{B} \\ \mathbf{f}''(\mathbf{y}) &= \mathbf{C} + \mathbf{D} \\ \mathbf{f}'''(\mathbf{y}) &= \mathbf{E} + 3\mathbf{F} + \mathbf{G} + \mathbf{H} \end{aligned} \tag{3-42}$$

We shall have to expand expressions of the form $\mathbf{f}(\mathbf{y} + h\mathbf{a})$ in powers of h, where $\mathbf{y}$ and

$$\mathbf{a} = \begin{pmatrix} a^1 \\ a^2 \\ \cdot \\ \cdot \\ \cdot \\ a^s \end{pmatrix}$$

are fixed vectors. The result of applying Taylor's theorem for functions of several variables (Taylor [1955], p. 227) can be written in the form

$$f^i(\mathbf{y} + h\mathbf{a}) = f^i + hf^i_j a^j + \tfrac{1}{2}h^2 f^i_{jk} a^j a^k + \tfrac{1}{6}h^3 f^i_{jkl} a^j a^k a^l + O(h^4) \tag{3-43}$$

3.2-3. Methods of Runge-Kutta type. We are now prepared to discuss methods which are only indirectly based on Taylor's expansion. As in §2.1-2, the idea consists in combining values of the function $\mathbf{f}$ (and not of its derivatives!), which are taken at different points, in such a way that the resulting function $\mathbf{\Phi}(\mathbf{y};\ h)$ agrees as well as possible with (3-37) or, in our new notation, with

$$\mathbf{A} + \tfrac{1}{2}h\mathbf{B} + \tfrac{1}{6}h^2(\mathbf{C} + \mathbf{D}) + \tfrac{1}{24}h^3(\mathbf{E} + 3\mathbf{F} + \mathbf{G} + \mathbf{H}) + \cdots \tag{3-44}$$

We explain the analytical technique by discussing the "simplified" Runge-Kutta methods, working with two substitutions. We tentatively put

$$\mathbf{\Phi}(\mathbf{y};\ h) = a_1\mathbf{f}(\mathbf{y}) + a_2\mathbf{f}(\mathbf{y} + ph\mathbf{f}(\mathbf{y}))$$

where the constants a_1, a_2, and p are to be determined. Using (3-43), we have

$$\begin{aligned} \mathbf{f}(\mathbf{y} + ph\mathbf{f}(\mathbf{y})) &= \mathbf{f}(\mathbf{y}) + hp\mathbf{f}_j(\mathbf{y})f^j(\mathbf{y}) + \tfrac{1}{2}(hp)^2\mathbf{f}_{jk}(\mathbf{y})f^j(\mathbf{y})f^k(\mathbf{y}) + O(h^3) \\ &= \mathbf{A} + hp\mathbf{B} + \tfrac{1}{2}(hp)^2\mathbf{C} + O(h^3) \end{aligned}$$

Thus,

(3-45) $$\mathbf{\Phi}(\mathbf{y};\, h) = (a_1 + a_2)\mathbf{A} + a_2 hp\mathbf{B} + \tfrac{1}{2}a_2(hp)^2\mathbf{C} + O(h^3)$$

Equating the constant and the linear term in h with the corresponding terms in (3-44), we obtain the conditions

(3-46) $$\begin{aligned} a_1 + a_2 &= 1 \\ a_2 p &= \tfrac{1}{2} \end{aligned}$$

It is not possible to obtain agreement in the quadratic term, since the vector $\mathbf{D}$ is missing from (3-45). The general solution of (3-46) is

$$a_1 = 1 - \alpha, \qquad a_2 = \alpha, \qquad p = \frac{1}{2\alpha}$$

where $\alpha \neq 0$. Thus the desired increment function is given by

(3-47) $$\mathbf{\Phi}(\mathbf{y};\, h) = (1 - \alpha)\mathbf{f}(\mathbf{y}) + \alpha\mathbf{f}\left(\mathbf{y} + \frac{h}{2\alpha}\mathbf{f}(\mathbf{y})\right), \qquad \alpha \neq 0$$

It deviates from (3-44) by $O(h^2)$. Two evaluations of the vector function $\mathbf{f}(\mathbf{y})$ are required for the computation of $\mathbf{\Phi}$.

We now try to see what we can do with four evaluations. We define the four vectors

$$\begin{aligned} \mathbf{k}_1 &= \mathbf{f}(\mathbf{y}) \\ \mathbf{k}_2 &= \mathbf{f}(\mathbf{y} + hp_1\mathbf{k}_1) \\ \mathbf{k}_3 &= \mathbf{f}(\mathbf{y} + h(p_2 - p_3)\mathbf{k}_1 + hp_3\mathbf{k}_2) \\ \mathbf{k}_4 &= \mathbf{f}(\mathbf{y} + h(p_4 - p_5 - p_6)\mathbf{k}_1 + hp_5\mathbf{k}_2 + hp_6\mathbf{k}_3) \end{aligned}$$

and try to determine the constants $p_1, \cdots, p_6$ as well as $a_1, \cdots, a_4$ such that the expression

(3-48) $$\mathbf{\Phi}(\mathbf{y};\, h) = a_1\mathbf{k}_1 + a_2\mathbf{k}_2 + a_3\mathbf{k}_3 + a_4\mathbf{k}_4$$

agrees with (3-44) as well as possible. Expanding in powers of h by repeated use of (3-43) and neglecting terms which are $O(h^4)$, we find

$$\begin{aligned} \mathbf{k}_1 &= \mathbf{A} \\ \mathbf{k}_2 &= \mathbf{A} + hp_1\mathbf{B} + \tfrac{1}{2}h^2p_1^2\mathbf{C} + \tfrac{1}{6}h^3p_1^3\mathbf{E} \\ \mathbf{k}_3 &= \mathbf{A} + hp_2\mathbf{B} + \tfrac{1}{2}h^2(p_2^2\mathbf{C} + 2p_1p_3\mathbf{D}) + \tfrac{1}{6}h^3(p_2^3\mathbf{E} + 6p_1p_2p_3\mathbf{F} + 3p_1^2p_3\mathbf{G}) \\ \mathbf{k}_4 &= \mathbf{A} + hp_4\mathbf{B} + \tfrac{1}{2}h^2[p_4^2\mathbf{C} + 2(p_1p_5 + p_2p_6)\mathbf{D}] \\ &\quad + \tfrac{1}{6}h^3[p_4^3\mathbf{E} + 6p_4(p_1p_5 + p_2p_6)\mathbf{F} + 3(p_1^2p_5 + p_2^2p_6)\mathbf{G} + 6p_1p_3p_6\mathbf{H}] \end{aligned}$$

Substituting in (3-48) and equating the coefficients of the vectors **A**, $\cdots$, **H** with the corresponding coefficients in (3-44), we get for the ten parameters $a_1, \cdots, a_4, p_1, \cdots, p_6$ the following eight equations:

$$
\begin{aligned}
&a_1 + a_2 + a_3 + a_4 = 1\\
&a_2 p_1 + a_3 p_2 + a_4 p_4 = \tfrac{1}{2}\\
&a_2 p_1^2 + a_3 p_2^2 + a_4 p_4^2 = \tfrac{1}{3}\\
&a_3 p_1 p_3 + a_4(p_1 p_5 + p_2 p_6) = \tfrac{1}{6}\\
&a_2 p_1^3 + a_3 p_2^3 + a_4 p_4^3 = \tfrac{1}{4}\\
&a_3 p_1 p_2 p_3 + a_4 p_4(p_1 p_5 + p_2 p_6) = \tfrac{1}{8}\\
&a_3 p_1^2 p_3 + a_4(p_1^2 p_5 + p_2^2 p_6) = \tfrac{1}{12}\\
&a_4 p_1 p_3 p_6 = \tfrac{1}{24}
\end{aligned}
\tag{3-49}
$$

[For $a_3 = a_4 = 0$ the first two equations reduce to (3-46), as they should.] The general solution of the above system of nonlinear equations is difficult. Kutta [1901] indicated a one-parameter family of solutions, as follows:

(3-50)

a_1	a_2	a_3	a_4
$\frac{1}{6}$	$\frac{2-t}{3}$	$\frac{t}{3}$	$\frac{1}{6}$

p_1	p_2	p_3	p_4	p_5	p_6
$\frac{1}{2}$	$\frac{1}{2}$	$\frac{1}{2t}$	1	$1-t$	t

Until very recently the solution corresponding to $t = 1$ was almost universally used. This, of course, is the classical Runge-Kutta case, where

$$a_1 = a_4 = \tfrac{1}{6}, \qquad a_2 = a_3 = \tfrac{1}{3}$$

and the formulas for the vectors $\mathbf{k}_j$ ($j = 1, 2, 3, 4$) have the simple form

$$
\begin{aligned}
\mathbf{k}_1 &= \mathbf{f}(\mathbf{y})\\
\mathbf{k}_2 &= \mathbf{f}(\mathbf{y} + \tfrac{1}{2}h\mathbf{k}_1)\\
\mathbf{k}_3 &= \mathbf{f}(\mathbf{y} + \tfrac{1}{2}h\mathbf{k}_2)\\
\mathbf{k}_4 &= \mathbf{f}(\mathbf{y} + h\mathbf{k}_3)
\end{aligned}
\tag{3-51}
$$

The Runge-Kutta formulas given in §2.1-2 for the integration of a single differential equation are a special case of this result.

Gill [1951] proposed to use Kutta's solution (3-50) with $t = 1 - 2^{-1/2}$. This choice was dictated by the rather limited storage facilities of the machines available at that time. It was shown by Gill that with this special choice of t four vectors of s components, which have to be stored simultaneously in order to carry out the Runge-Kutta process, become linearly dependent, so that actually only three vectors have to be stored. On the other hand, Gill's choice of t results in an increase in complexity (p_5 is no longer 0) and in a loss of symmetry in the coefficients. Moreover, it has been shown by Blum [1957] that the same saving in storage can also be achieved with the classical Runge-Kutta values, if necessary.

For computational considerations concerning the Runge-Kutta method we refer to §2.1-2.

3.2-4. Methods based on quadrature. The one-step quadrature methods discussed in Problem 8 of Chapter 2 carry over without change to systems of equations. We confine ourselves to a listing of the formulas and postpone a discussion of the error to §3.3.

Let the basic quadrature formula (for a scalar function) be given by (2-109). As in the case of a single equation, we need a method which predicts the values of $\mathbf{y}(x)$ at the points $x_n + p_k h$. We assume that this prediction is by means of methods defined by the increment functions $\bar{\mathbf{\Phi}}_{(k)}(\mathbf{y};\ p_k h)$. (The subscript k indicates that a different predictor formula may be used for each abscissa.) The quadrature method is then defined by

$$\mathbf{\Phi}(\mathbf{y};\ h) = \sum_{k=1}^{\nu} w_k \mathbf{f}(\mathbf{y} + p_k h \bar{\mathbf{\Phi}}_{(k)}(\mathbf{y};\ p_k h)) \tag{3-52}$$

3.3. The Discretization Error of One-Step Methods

Many of the proofs of the theorems in this section are entirely analogous to the proofs of the analogous theorems in §2.2 and are therefore either omitted or presented in abbreviated form.

3.3-1. Convergence and consistency. We shall always assume that the vector function $\mathbf{f}(x, \mathbf{y})$ satisfies the conditions of the existence theorem 3.1. The initial value problem

$$\mathbf{y}' = \mathbf{f}(x, \mathbf{y}), \qquad \mathbf{y}(a) = \boldsymbol{\eta} \tag{3-53}$$

then has a solution $\mathbf{y}(x)$ in the interval $[a, b]$ for an arbitrary initial vector $\boldsymbol{\eta}$. We also shall assume that the increment function $\mathbf{\Phi}(x, \mathbf{y};\ h)$ is defined for $x \in [a, b]$, $y^i \in (-\infty, \infty)$, and, at any given point $(x, \mathbf{y})$, for all $h \geq 0$ such that $x + h \in [a, b]$. It is then possible to calculate, for any $h \geq 0$, vectors $\mathbf{y}_n$ from (3-30) as long as $x_n \in [a, b]$.

The method defined by the increment function $\mathbf{\Phi}(x, \mathbf{y};\ h)$ is said to be *convergent* if, for arbitrary $\boldsymbol{\eta}$ and arbitrary $x \in [a, b]$,

$$\lim_{\substack{h \to 0 \\ x_n = x}} \mathbf{y}_n = \mathbf{y}(x) \tag{3-54}$$

A method is said to be *consistent* with the differential equation (3-53) if

$$\mathbf{\Phi}(x, \mathbf{y};\ 0) = \mathbf{f}(x, \mathbf{y}) \tag{3-55}$$

identically in x and $\mathbf{y}$.

Exactly as in §2.2-1 one can prove the following:

THEOREM 3.2. *Let the function $\mathbf{\Phi}(x, \mathbf{y};\ h)$ be continuous (jointly as a function of its $s + 2$ arguments) in the region described above, and let there exist a constant L such that*

$$\|\mathbf{\Phi}(x, \mathbf{y}^*;\ h) - \mathbf{\Phi}(x, \mathbf{y};\ h)\| \le L\,\|\mathbf{y}^* - \mathbf{y}\| \tag{3-56}$$

for all points $(x, \mathbf{y}^*;\ h)$ and $(x, \mathbf{y};\ h)$ in that region such that $h \le h_0$, where $h_0 > 0$ is some fixed constant. Then a necessary and sufficient condition for the convergence of the method defined by $\mathbf{\Phi}$ is that the method be consistent.

3.3-2. An a priori bound. Let $\mathbf{y}(x)$ denote the solution of the initial value problem, and let $\mathbf{\Delta}(x, \mathbf{y};\ h)$ be defined by (3-29). The following theorem combines, for a system of differential equations, the features of Theorems 2.2 and 2.3.

THEOREM 3.3. *Let $\mathbf{\Phi}(x, \mathbf{y};\ h)$ satisfy the conditions of Theorem 3.2, and let there exist constants $N \ge 0$, $p \ge 0$, and $h_0 > 0$ such that*

$$\|\mathbf{\Phi}(x, \mathbf{y}(x);\ h) - \mathbf{\Delta}(x. \mathbf{y}(x);\ h)\| \le Nh^p, \qquad x \in [a, b],\ h \le h_0 \tag{3-57}$$

Let $\{\mathbf{y}_n\}$ be any sequence of vectors satisfying

$$\begin{aligned} &\mathbf{y}_0 = \boldsymbol{\eta} \\ &\mathbf{y}_{n+1} = \mathbf{y}_n + h[\mathbf{\Phi}(x_n, \mathbf{y}_n;\ h) + h^q K \boldsymbol{\theta}_n], \quad n = 0, 1, 2, \cdots;\ x_n \in [a, b] \end{aligned} \tag{3-58}$$

where $K \ge 0$ and $q \ge 0$ are constants, and where the vectors $\boldsymbol{\theta}_n$ satisfy $\|\boldsymbol{\theta}_n\| \le 1$. Then, for $x_n \in [a, b]$ and $h \le h_0$,

$$\|\mathbf{y}_n - \mathbf{y}(x_n)\| \le h^r N_1 E_L(x_n - a) \tag{3-59}$$

where $r = \min(p, q)$ and $N_1 = Nh_0^{p-r} + Kh_0^{q-r}$.

This theorem provides us with a bound for the error $\mathbf{e}_n = \mathbf{y}_n - \mathbf{y}(x_n)$ even if the relations (3-30) are only approximately satisfied.

Proof. Subtracting from (3-58) the trivial relation

$$\mathbf{y}(x_{n+1}) = \mathbf{y}(x_n) + h\mathbf{\Delta}(x_n, \mathbf{y}(x_n);\ h)$$

we get

$$\mathbf{e}_{n+1} = \mathbf{e}_n + h[\mathbf{\Phi}(x_n, \mathbf{y}_n;\ h) - \mathbf{\Delta}(x_n, \mathbf{y}(x_n);\ h) + h^q K \boldsymbol{\theta}_n] \tag{3-60}$$

By applications of the triangle inequality we have, using (3-56) and (3-57),

$$\begin{aligned} &\|\mathbf{\Phi}(x_n, \mathbf{y}_n;\ h) - \mathbf{\Delta}(x_n, \mathbf{y}(x_n);\ h) + h^q K \boldsymbol{\theta}_n\| \\ &\qquad = \|\mathbf{\Phi}(x_n, \mathbf{y}_n;\ h) - \mathbf{\Phi}(x_n, \mathbf{y}(x_n);\ h) + \mathbf{\Phi}(x_n, \mathbf{y}(x_n);\ h) \\ &\qquad\qquad - \mathbf{\Delta}(x_n, \mathbf{y}(x_n);\ h) + h^q K \boldsymbol{\theta}_n\| \le L\,\|\mathbf{e}_n\| + h^p N + h^q K \end{aligned}$$

Hence from (3-60)

$$\|\mathbf{e}_{n+1}\| \le (1 + hL)\,\|\mathbf{e}_n\| + h^{p+1}N + h^{q+1}K$$

and the result (3-59) follows from Lemma 1.2, setting

$$\xi_n = \|\mathbf{e}_n\|,\ A = 1 + hL,\ B = h^{p+1}N + h^{q+1}K, \qquad \xi_0 = 0.$$

It is not hard to show that the bound (3-59) also holds for *irregularly spaced points* x_n defined by (2-51), provided that the hypothesis (3-57) is replaced by

$$\|\mathbf{\Phi}(x, \mathbf{y}(x);\ h\vartheta(x)) - \mathbf{\Delta}(x, \mathbf{y}(x);\ h\vartheta(x))\| \le h^p N \tag{3-61}$$

3.3-3. Application to special methods: Values for L. We shall assume in this and the next section that the function $\mathbf{f}$ does not depend explicitly on x. As was seen in §3.2-1, this assumption can be made without loss of generality.

We assume that the functions $\mathbf{f}_j^{(p)}(\mathbf{y})$ $(j = 1, \cdots, s)$ are bounded for $y^i \in (-\infty, \infty)$ $(i = 1, \cdots, s)$ and put

$$L_p = \sup_{\substack{j=1,\cdots,s \\ -\infty < y^i < \infty}} \|\mathbf{f}_j^{(p)}(\mathbf{y})\|, \qquad p = 0, 1, 2, \cdots$$

For the method (3-38) based on Taylor's expansion we find by the mean value theorem

$$\Phi^i(\mathbf{y}^*;\ h) - \Phi^i(\mathbf{y};\ h)$$

$$= \left[f_j^i(\mathbf{y}^{[0]}) + \frac{h}{2} f_j^{(1)i}(\mathbf{y}^{[1]}) + \cdots + \frac{h^{p-1}}{p!} f_j^{(p-1)i}(\mathbf{y}^{[p-1]})\right](y^{*j} - y^j)$$

where $\mathbf{y}^{[0]}, \mathbf{y}^{[1]}, \cdots$ denote certain vectors whose components lie between those of $\mathbf{y}^*$ and those of $\mathbf{y}$. It follows that (3-56) holds with

$$L = L_0 + \frac{h}{2} L_1 + \cdots + \frac{h^{p-1}}{p!} L_{p-1} \tag{3-62}$$

In order to determine the constant L for the simplified Runge-Kutta method, we make use of the relation

$$\|\mathbf{f}(\mathbf{y}^*) - \mathbf{f}(\mathbf{y})\| \le L_0\|\mathbf{y}^* - \mathbf{y}\|$$

valid for any two vectors $\mathbf{y}^*$ and $\mathbf{y}$. It follows that

$$\left\|\mathbf{f}\left(\mathbf{y}^* + \frac{h}{2\alpha}\mathbf{f}(\mathbf{y}^*)\right) - \mathbf{f}\left(\mathbf{y} + \frac{h}{2\alpha}\mathbf{f}(\mathbf{y})\right)\right\| \le L_0\bigg[\|\mathbf{y}^* - \mathbf{y}\|$$

$$+ \frac{h}{2|\alpha|}\|\mathbf{f}(\mathbf{y}^*) - \mathbf{f}(\mathbf{y})\|\bigg] \le L_0\left(1 + \frac{h}{2|\alpha|}L_0\right)\|\mathbf{y}^* - \mathbf{y}\|$$

It follows that (3-56) holds with

$$L = L_0\left(|1 - \alpha| + |\alpha| + \frac{h_0 L_0}{2}\right) \tag{3-63}$$

For the classical Runge-Kutta method the analysis is entirely similar. Indicating the dependence of the vectors $\mathbf{k}_i$ on $\mathbf{y}$ by writing† $\mathbf{k}_i = \mathbf{k}_i(\mathbf{y})$, $i = 1, 2, 3, 4$, we have

$$\|\mathbf{k}_1(\mathbf{y}^*) - \mathbf{k}_1(\mathbf{y})\| \le L_0\|\mathbf{y}^* - \mathbf{y}\|$$

and generally for $i = 2, 3, 4$

$$\begin{aligned}\|\mathbf{k}_i(\mathbf{y}^*) - \mathbf{k}_i(\mathbf{y})\| &\le \|\mathbf{f}(\mathbf{y}^* + hp_i\mathbf{k}_{i-1}(\mathbf{y}^*)) - \mathbf{f}(\mathbf{y} + hp_i\mathbf{k}_{i-1}(\mathbf{y}))\| \\ &\le L_0[\|\mathbf{y}^* - \mathbf{y}\| + hp_i\,\|\mathbf{k}_{i-1}(\mathbf{y}^*) - \mathbf{k}_{i-1}(\mathbf{y})\|]\end{aligned}$$

where $p_2 = p_3 = \frac{1}{2}$, $p_4 = 1$. It follows that for $h \le h_0$

$$\|\mathbf{k}_2(\mathbf{y}^*) - \mathbf{k}_2(\mathbf{y})\| \le L_0\left(1 + \frac{h_0L_0}{2}\right)\|\mathbf{y}^* - \mathbf{y}\|$$

$$\|\mathbf{k}_3(\mathbf{y}^*) - \mathbf{k}_3(\mathbf{y})\| \le L_0\left[1 + \frac{h_0L_0}{2}\left(1 + \frac{h_0L_0}{2}\right)\right]\|\mathbf{y}^* - \mathbf{y}\|$$

$$\|\mathbf{k}_4(\mathbf{y}^*) - \mathbf{k}_4(\mathbf{y})\| \le L_0\left\{1 + h_0L_0\left[1 + \frac{h_0L_0}{2}\left(1 + \frac{h_0L_0}{2}\right)\right]\right\}\|\mathbf{y}^* - \mathbf{y}\|$$

Combining these results, we find

$$\begin{aligned}&\|\boldsymbol{\Phi}(\mathbf{y}^*; h) - \boldsymbol{\Phi}(\mathbf{y}; h)\| \\ &\quad\le \frac{1}{6}\{\|\mathbf{k}_1(\mathbf{y}^*) - \mathbf{k}_1(\mathbf{y})\| + 2\,\|\mathbf{k}_2(\mathbf{y}^*) - \mathbf{k}_2(\mathbf{y})\| + 2\,\|\mathbf{k}_3(\mathbf{y}^*) - \mathbf{k}_3(\mathbf{y})\| \\ &\quad+ \|\mathbf{k}_4(\mathbf{y}^*) - \mathbf{k}_4(\mathbf{y})\|\} \le L_0\left[1 + \frac{h_0L_0}{2} + \frac{(h_0L_0)^2}{6} + \frac{(h_0L_0)^3}{24}\right]\|\mathbf{y}^* - \mathbf{y}\|\end{aligned} \tag{3-64}$$

† Here subscripts do not denote derivatives.

The last result implies a value of the constant L in (3-56). This bound differs by only $O(h_0^4)$ from the slightly larger bound

$$L = \frac{e^{h_0 L_0} - 1}{h_0} \tag{3-65}$$

For the methods based on quadrature we have, if $\bar{L}$ denotes a common Lipschitz constant for the predictor methods $\bar{\mathbf{\Phi}}_{(k)}$,

$$\|\mathbf{\Phi}(\mathbf{y}^*;\, h) - \mathbf{\Phi}(\mathbf{y};\, h)\| \le \sum_{k=1}^{\nu} |w_k|\ \|\mathbf{f}(\mathbf{y}^* + p_k h \bar{\mathbf{\Phi}}_{(k)}(\mathbf{y}^*;\, p_k h))$$
$$- \mathbf{f}(\mathbf{y} + p_k h \bar{\mathbf{\Phi}}_{(k)}(\mathbf{y};\, p_k h))\| \le L_0 \sum_{k=1}^{\nu} |w_k|\,(1 + p_k h \bar{L})\ \|\mathbf{y}^* - \mathbf{y}\|$$

If $w_k \ge 0$ $(k = 1, \cdots, \nu)$, we have $\Sigma\, |w_k|\, p_k = \frac{1}{2}$ and hence

$$L = L_0\left(1 + \frac{h_0}{2}\bar{L}\right) \tag{3-66}$$

3.3-4. Application to special methods: values of N. It is known from the proof of Theorem 3.1 that the exact solution $\mathbf{y}(x)$ of the initial value problem (3-53) satisfies

$$\|\mathbf{y}(x)\| \le Y$$

where Y is defined by (3-15). If the method defined by $\mathbf{\Phi}$ is of order p, and if $\mathbf{\Phi}$ and $\mathbf{f}$ are p times continuously differentiable, it follows from

$$\Phi^i(x, \mathbf{y};\, h) - \Delta^i(x, \mathbf{y};\, h)$$
$$= h^p\left\{\frac{1}{p!}\frac{\partial^p \Phi^i}{\partial h^p}(x, \mathbf{y};\, h^+) - \frac{1}{(p+1)!} f^{(p)i}(x + h^+, \mathbf{z}(x + h^+))\right\}$$

where $\mathbf{z}(t)$ denotes the solution passing through $(x, \mathbf{y})$ and $0 < h^+ < h$, that (3-57) holds with

$$N = \frac{1}{(p+1)!} M_p + \frac{1}{p!} \max_{\substack{\|\mathbf{y}\| \le Y \\ h \le h_0}} \left\|\frac{\partial^p \mathbf{\Phi}}{\partial h^p}(\mathbf{y};\, h)\right\| \tag{3-67}$$

where

$$M_p = \max_{\|\mathbf{y}\| \le Y} \|\mathbf{f}^{(p)}(\mathbf{y})\|$$

For the Taylor expansion method (3-38), $\partial^p \mathbf{\Phi}/\partial h^p = \mathbf{0}$; hence

$$N = \frac{1}{(p+1)!} M_p \tag{3-68}$$

In order to write bounds for N for the methods of Runge-Kutta type, we set

$$D_p = \max_{\substack{1 \le j_1, \cdots, j_p \le s \\ \|\mathbf{y}\| \le Y}} \|\mathbf{f}_{j_1 j_2 \cdots j_p}(\mathbf{y})\|$$

For the simplified Runge-Kutta method (3-47) we have, using (3-43),

$$\frac{\partial^2 \mathbf{\Phi}}{\partial h^2}(\mathbf{y}; h) = \frac{1}{4\alpha}\mathbf{f}_{jk}\left(\mathbf{y} + \frac{h}{2\alpha}\mathbf{f}(\mathbf{y})\right) f^j(\mathbf{y}) f^k(\mathbf{y})$$

and hence

$$\left\| \frac{\partial^2 \mathbf{\Phi}}{\partial h^2}(\mathbf{y};\ h) \right\| \le \frac{1}{4|\alpha|} D_2 M_0^2$$

Thus

$$N = \frac{1}{6} M_2 + \frac{1}{8|\alpha|} D_2 M_0^2 \tag{3-69}$$

for this method.

The classical Runge-Kutta method (3-48) was constructed so that (3-57) is valid for $p = 4$. The task of finding a bound for $\|\partial^4\mathbf{\Phi}/\partial h^4\|$ is simplified by the following auxiliary consideration. Let t be a constant, and let $\mathbf{k} = \mathbf{k}(h)$ be some vector whose components depend on h. We shall calculate the first four derivatives of the function

$$\mathbf{g}(h) = \mathbf{f}(\mathbf{y} + th\mathbf{k}(h)) \tag{3-70}$$

Writing

$$\frac{d\mathbf{g}}{dh} = \dot{\mathbf{g}}, \qquad \frac{d^2\mathbf{g}}{dh^2} = \mathbf{g}^{[2]}, \ldots$$

and using a similar notation for other functions depending on h, we find

$$\dot{\mathbf{g}}(h) = t(hk^j)^{\cdot}\mathbf{f}_j \tag{3-71a}$$

$$\mathbf{g}^{[2]}(h) = t^2(hk^j)^{\cdot}(hk^m)^{\cdot}\mathbf{f}_{jm} + t(hk^j)^{[2]}\mathbf{f}_j \tag{3-71b}$$

$$\begin{aligned} \mathbf{g}^{[3]}(h) &= t^3(hk^j)^{\cdot}(hk^m)^{\cdot}(hk^n)^{\cdot}\mathbf{f}_{jmn} \\ &\quad + 3t^2(hk^j)^{[2]}(hk^m)^{\cdot}\mathbf{f}_{jm} + t(hk^j)^{[3]}\mathbf{f}_j \end{aligned} \tag{3-71c}$$

$$\begin{aligned} \mathbf{g}^{[4]}(h) &= t^4(hk^j)^{\cdot}(hk^m)^{\cdot}(hk^n)^{\cdot}(hk^r)^{\cdot}\mathbf{f}_{jmnr} \\ &\quad + 6t^3(hk^j)^{[2]}(hk^m)^{\cdot}(hk^n)^{\cdot}\mathbf{f}_{jmn} \\ &\quad + 4t^2(hk^j)^{[3]}(hk^m)^{\cdot}\mathbf{f}_{jm} \\ &\quad + 3t^2(hk^j)^{[2]}(hk^m)^{[2]}\mathbf{f}_{jm} \\ &\quad + t(hk^j)^{[4]}\mathbf{f}_j \end{aligned} \tag{3-71d}$$

The argument in the functions $\mathbf{f}$ and their derivatives is $\mathbf{y} + th\mathbf{k}$.

The relations (3-51) defining the classical Runge-Kutta method may be written

$$\mathbf{k}_i = \mathbf{f}(\mathbf{y} + p_i h\mathbf{k}_{i-1}), \qquad i = 1, 2, 3, 4 \tag{3-72}$$

where $p_1 = 0$, $p_2 = p_3 = \frac{1}{2}$, $p_4 = 1$. (The vector $\mathbf{k}_0$ need not be defined.) These relations are precisely of the form (3-70), and the formulas (3-71) enable us to express the derivatives of the $\mathbf{k}_i$.

First of all

$$\dot{\mathbf{k}}_1 = \mathbf{k}_1^{[2]} = \mathbf{k}_1^{[3]} = \mathbf{k}_1^{[4]} = \mathbf{0}$$

since $\mathbf{f}(\mathbf{y})$ is independent of h. Consequently, from (3-71),

$$\begin{aligned} \dot{\mathbf{k}}_2 &= \tfrac{1}{2}k_1^j\mathbf{f}_j \\ \mathbf{k}_2^{[2]} &= \tfrac{1}{4}k_1^jk_1^m\mathbf{f}_{jm} \\ \mathbf{k}_2^{[3]} &= \tfrac{1}{8}k_1^jk_1^mk_1^n\mathbf{f}_{jmn} \\ \mathbf{k}_2^{[4]} &= \tfrac{1}{16}k_1^jk_1^mk_1^nk_1^r\mathbf{f}_{jmnr} \end{aligned} \tag{3-73}$$

It is our aim to find bounds for the derivatives of the $\mathbf{k}_i$. In order to write these bounds in the simplest possible manner, we define a new constant K such that

$$D_p \leq \frac{K}{M^{p-1}}, \; p = 0, 1, \cdots, 4 \tag{3-74}$$

where $M = M_0$. It follows that $K \geq 1$. In view of $\|\mathbf{k}_1\| \leq M$ we get from (3-73)

$$\|\mathbf{k}_2^{[\nu]}\| \leq 2^{-\nu}KM, \qquad \nu = 1, 2, 3, 4$$

It follows that

$$\begin{aligned} \|(h\mathbf{k}_2)^{\cdot}\| &\leq \|\mathbf{k}_2\| + \|h\dot{\mathbf{k}}_2\| \\ &\leq M[1 + \tfrac{1}{2}(hK)] \\ &= Me^{\theta} \end{aligned}$$

where $\theta = \frac{1}{2}(hK)$, and for $\nu = 2, 3, 4$,

$$\begin{aligned} \|(h\mathbf{k}_2)^{[\nu]}\| &\leq \|\nu\mathbf{k}_2^{[\nu-1]}\| + \|h\mathbf{k}_2^{[\nu]}\| \\ &\leq 2^{-\nu+1}\nu KMe^{\theta} \end{aligned}$$

Here we have replaced factors of the form $1 + ah$ by e^{ah}, because this makes it possible to estimate products of such factors closely by upper bounds as follows:

$$\prod (1 + a_ih) \leq e^{h\Sigma a_i}$$

From (3-71), letting $\mathbf{g}(h) = \mathbf{k}_3$, $\mathbf{k}(h) = \mathbf{k}_2$, we now readily find

$$\begin{aligned} \|\dot{\mathbf{k}}_3\| &\leq \tfrac{1}{2}KMe^{\theta} \\ \|\mathbf{k}_3^{[2]}\| &\leq \tfrac{1}{4}KM(1 + 2K)e^{2\theta} \\ \|\mathbf{k}_3^{[3]}\| &\leq \tfrac{1}{8}KM(1 + 9K)e^{3\theta} \\ \|\mathbf{k}_3^{[4]}\| &\leq \tfrac{1}{16}KM(1 + 28K + 12K^2)e^{4\theta} \end{aligned}$$

From this we get

$$\begin{aligned} \|(h\mathbf{k}_3)^{\cdot}\| &\leq \|\mathbf{k}_3\| + \|h\dot{\mathbf{k}}_3\| \\ &\leq M[1 + \tfrac{1}{2}(hK)e^{\theta}] \\ &\leq Me^{2\theta'} \end{aligned}$$

where

$$\theta' = \tfrac{1}{4}(hK)e^{\theta} = \tfrac{1}{4}(hK)e^{\frac{1}{2}(hK)} \tag{3-75}$$

In a similar manner we find, using the fact that $\theta \leq 2\theta'$,

$$\begin{aligned} \|(h\mathbf{k}_3)^{[2]}\| &\leq KMe^{5\theta'} \\ \|(h\mathbf{k}_3)^{[3]}\| &\leq \tfrac{3}{4}KM(1 + 2K)e^{8\theta'} \\ \|(h\mathbf{k}_3)^{[4]}\| &\leq \tfrac{1}{2}KM(1 + 9K)e^{9\theta'} \end{aligned}$$

Using (3-71d) with $\mathbf{g}(h) = \mathbf{k}_4$, $\mathbf{k}(h) = \mathbf{k}_3$, we now get by the same method

$$\|\mathbf{k}_4^{[4]}\| \leq \tfrac{1}{2}MK(2 + 19K + 27K^2)e^{9\theta'}$$

By virtue of

$$\frac{\partial^4 \mathbf{\Phi}}{\partial h^4}(\mathbf{y};\, h) = \tfrac{1}{3}\mathbf{k}_2^{[4]} + \tfrac{1}{3}\mathbf{k}_3^{[4]} + \tfrac{1}{6}\mathbf{k}_4^{[4]}$$

we now obtain, collecting the above results,

$$\left\| \frac{\partial^4 \mathbf{\Phi}}{\partial h^4}(\mathbf{y};\, h) \right\| \leq \tfrac{1}{24}MK(5 + 52K + 60K^2)e^{9\theta'} \tag{3-76}$$

where M and K are defined by (3-74) and θ' by (3-75). From formula (3-84), given below, it follows that M_4 can in terms of K and M be bounded as follows:

$$M_4 = \max_{\|\mathbf{y}\| \leq Y} \|\mathbf{f}^{(4)}(\mathbf{y})\| \leq MK(1 + 11K + 7K^2 + K^3) \tag{3-77}$$

We now can apply (3-67). By virtue of $\theta' = O(h)$, $e^{9\theta'} = 1 + O(h)$, and not much is lost if the two bounds (3-76) and (3-77) are combined as follows:

$$N = KM\,\frac{49 + 524K + 468K^2 + 24K^3}{2880}\,e^{9\theta'} \tag{3-78}$$

This is the desired bound for the constant N for the Runge-Kutta method.

In order to estimate N for the methods based on quadrature, we denote by $\bar{N}$ the largest and by $\bar{p}$ the smallest of the constants N and p corresponding to the prediction methods defined by the increment functions $\bar{\mathbf{\Phi}}$ in (3-52). Denoting by $\mathbf{z}(t)$ the solution of $\mathbf{z}' = \mathbf{f}(t, \mathbf{z})$ passing through $(x, \mathbf{y})$, we then can estimate $h\mathbf{\Phi}(x, \mathbf{y};\, h) - h\mathbf{\Delta}(x, \mathbf{y};\, h)$ as follows:

$$\begin{aligned} &\left\| h\sum_{k=1}^{\nu} w_k \mathbf{f}(x + p_k h,\, \mathbf{y} + p_k h\bar{\mathbf{\Phi}}_{(k)}(x, \mathbf{y};\, p_k h)) - \mathbf{z}(x + h) + \mathbf{y} \right\| \\ &\qquad \leq \left\| h\sum_{k=1}^{\nu} w_k \mathbf{f}(x + p_k h,\, \mathbf{z}(x + p_k h)) - \mathbf{z}(x + h) + \mathbf{y} \right\| \\ &\qquad\quad + Lh\sum_{k=1}^{\nu} |w_k|\, \|p_k h\bar{\mathbf{\Phi}}_{(k)}(x, \mathbf{y};\, p_k h) - [\mathbf{z}(x + p_k h) - \mathbf{y}]\| \\ &\qquad \leq \left\| h\sum_{k=1}^{\nu} w_k \mathbf{z}'(x + p_k h) - [\mathbf{z}(x + h) - \mathbf{y}] \right\| + L\bar{N}h\sum_{k=1}^{\nu} |w_k|\, (p_k h)^{\bar{p}+1} \end{aligned}$$

The first term is bounded by $|C|\,h^{\mu+1}M_\mu$. If $w_k \geq 0$ and $C > 0$, we have

$$\sum_{k=1}^{\nu} |w_k|\, p_k^{\bar{p}+1} \leq \int_0^1 t^{\bar{p}+1}\, dt = \frac{1}{\bar{p}+2}$$

hence the second term is estimated by

$$h^{\bar{p}+2} L\bar{N} \frac{1}{\bar{p}+2}$$

It follows that $p = \min(\bar{p}+1, \mu)$, and (3-57) holds with

$$N = \frac{L\bar{N}}{\bar{p}+2} h_0^{\bar{p}+1-p} + CM_\mu h_0^{\mu-\bar{p}} \tag{3-79}$$

Thus the quadrature formula is used to full advantage only if $\bar{p} + 1 \geq \mu$.

3.3-5. The principal error function. If N denotes the constant defined by (3-57) and estimated in the preceding section, then the quantity $h^{p+1}N$ represents a bound for the local discretization error. It is our experience that such bounds overestimate the actual error by a considerable margin, and that a truer indication of the actual size of the error is given by an asymptotic formula for the error. If the method defined by $\mathbf{\Phi}(x, \mathbf{y};\ h)$ has exact order p, and if both $\mathbf{\Phi}(x, \mathbf{y};\ h)$ and $\mathbf{\Delta}(x, \mathbf{y};\ h)$ have continuous derivatives of order $p + 1$, we may write

$$\mathbf{\Phi}(x, \mathbf{y};\ h) - \mathbf{\Delta}(x, \mathbf{y};\ h) = h^p \boldsymbol{\varphi}(x, \mathbf{y}) + O(h^{p+1}) \tag{3-80}$$

where the vector $\boldsymbol{\varphi}$ depends continuously on x and $\mathbf{y}$ and does not vanish identically.

As in the one-dimensional case, the function $\boldsymbol{\varphi}(x, \mathbf{y})$ is called the *principal error function* of the method. The quantities $h^p\boldsymbol{\varphi}(x, \mathbf{y})$ and $h^{p+1}\boldsymbol{\varphi}(x, \mathbf{y})$ are approximations for the relative and absolute local truncation error of the method at the point $(x, \mathbf{y})$.

Using the formula, following from (3-80),

$$\boldsymbol{\varphi}(x, \mathbf{y}) = \frac{1}{p!} \frac{\partial^p \mathbf{\Phi}}{\partial h^p}(x, \mathbf{y};\ 0) - \frac{1}{(p+1)!} \mathbf{f}^{(p)}(x, \mathbf{y}) \tag{3-81}$$

we now shall determine $\boldsymbol{\varphi}$ for some of the special methods which were considered previously. We shall again assume, with no loss of generality, that $\mathbf{f}$, $\mathbf{\Phi}$, $\mathbf{\Delta}$, and hence $\boldsymbol{\varphi}$ do not depend explicitly on x.

For the Taylor expansion method we have immediately

$$\boldsymbol{\varphi}(\mathbf{y}) = -\frac{1}{(p+1)!} \mathbf{f}^{(p)}(\mathbf{y}) \tag{3-82}$$

because $\mathbf{\Phi}$ is a polynomial of degree $p - 1$ in h.

In the case of the simplified Runge-Kutta method defined by (3-47) we find

$$\frac{1}{2}\frac{\partial^2 \mathbf{\Phi}}{\partial h^2}(\mathbf{y};\ 0) = \frac{1}{8\alpha}\,\mathbf{f}_{jk}f^k f^j$$

The argument in the functions **f** and their derivatives is **y**. From this we have to subtract

$$\tfrac{1}{6}\mathbf{f}''(\mathbf{y}) = \tfrac{1}{6}\mathbf{f}_{jk}f^j f^k + \tfrac{1}{6}\mathbf{f}_j f^j_k f^k$$

Since $\mathbf{f}' = \mathbf{f}_k f^k$, the result may be written in the following form:

$$\boldsymbol{\varphi}(\mathbf{y}) = \left(\frac{1}{8\alpha} - \frac{1}{6}\right)\mathbf{f}'' - \frac{1}{8\alpha}\,\mathbf{f}_j f'^j \tag{3-83}$$

It appears that the principal error vector assumes a particularly simple form if we choose $\alpha = \frac{3}{4}$.

The classical Runge-Kutta method can be dealt with similarly. We shall only quote the result of the rather lengthy calculations. Continuing the line of the vectors **A**, **B**, $\cdots$, **H** defined in §3.2-2, we put

$$\begin{aligned}
\mathbf{I} &= \mathbf{f}_{jklm}f^j f^k f^l f^m & \mathbf{N} &= \mathbf{f}_m f^m_{jl} f^l_k f^j f^k\\
\mathbf{J} &= \mathbf{f}_{jkm}f^m_l f^j f^k f^l & \mathbf{P} &= \mathbf{f}_l f^l_m f^m_{jk} f^j f^k\\
\mathbf{K} &= \mathbf{f}_{jm}f^m_{kl} f^j f^k f^l & \mathbf{Q} &= \mathbf{f}_m f^m_l f^l_k f^k_j f^j\\
\mathbf{L} &= \mathbf{f}_{jm}f^m_l f^l_k f^j f^k & \mathbf{R} &= \mathbf{f}_{lm} f^m_k f^l_j f^j f^k\\
\mathbf{M} &= \mathbf{f}_j f^j_{klm} f^k f^l f^m
\end{aligned}$$

the argument **y** being understood everywhere. It can be shown that

$$\frac{1}{4!}\frac{\partial^4 \mathbf{\Phi}}{\partial h^4}(\mathbf{y};\ 0) = \tfrac{1}{576} \times\{5\mathbf{I} + 30\mathbf{J} + 18\mathbf{K} + 24\mathbf{L} + 4\mathbf{M} + 12\mathbf{N} + 6\mathbf{P} + 18\mathbf{R}\}$$

On the other hand,

$$\frac{1}{5!}\mathbf{f}^{\text{iv}}(\mathbf{y}) = \tfrac{1}{120}\{\mathbf{I} + 6\mathbf{J} + 4\mathbf{K} + \mathbf{M} + 3\mathbf{N} + \mathbf{P} + \mathbf{Q} + 3\mathbf{R}\} \tag{3-84}$$

It follows that

$$\boldsymbol{\varphi}(\mathbf{y}) = \tfrac{1}{2880} \times\{\mathbf{I} + 6\mathbf{J} - 6\mathbf{K} + 24\mathbf{L} - 4\mathbf{M} - 12\mathbf{N} + 6\mathbf{P} - 24\mathbf{Q} + 18\mathbf{R}\} \tag{3-85}$$

One may try to simplify this expression by expressing as many as possible of the accumulated derivatives of **f** in terms of components of $\mathbf{f}^{(k)}$, the derivatives of the true solution. The result is

$$\begin{aligned}\boldsymbol{\varphi}(\mathbf{y}) = {} & \tfrac{1}{2880}\mathbf{f}^{\text{iv}} - \tfrac{1}{576}\mathbf{f}_j f'''^j + \tfrac{1}{288}[\mathbf{f}_j f^j_k - \mathbf{f}_{jk}f^j]f''^k\\
& + \tfrac{1}{192}[2\mathbf{f}_{jk}f^j f^k_l - 2\mathbf{f}_{jl}f^j_k f^k_l + \mathbf{f}_{jl}f^j_k f^k]f'^l\end{aligned} \tag{3-86}$$

For $s = 2$, $y^1 = x$, $y^2 = y$, $f^1 = 1$, $f^2 = f(x, y)$, the second component of the vector (3-86) yields (2-36).

In order to discuss the principal error function for the methods based on quadrature, let $\mathscr{A}(\mathbf{y}) = (a^{ij}(\mathbf{y}))$ denote the $s \times s$ matrix whose elements are given by

$$a^{ij}(\mathbf{y}) = f^i_j(\mathbf{y}), \qquad i, j = 1, \cdots, s \tag{3-87}$$

If $\mathbf{k}$ is any vector, we then can write

$$\mathbf{f}(\mathbf{y} + h\mathbf{k}) - \mathbf{f}(\mathbf{y}) = h\mathscr{A}(\mathbf{y})\mathbf{k} + O(h^2)$$

Now let $\mathbf{\Phi}(\mathbf{y};\ h)$ be defined by (3-52), where the increment function $\bar{\mathbf{\Phi}}(\mathbf{y};\ h)$ used for predicting has order $\bar{p} = \mu - 1$ and has associated with it the principal error function $\overline{\boldsymbol{\varphi}}(\mathbf{y})$. We then have

$$\begin{aligned}
\mathbf{\Phi}(\mathbf{y};\ h) &= \sum_{k=1}^{\nu} w_k \mathbf{f}(\mathbf{y} + p_k h \bar{\mathbf{\Phi}}(\mathbf{y};\ p_k h)) \\
&= \sum_{k=1}^{\nu} w_k [\mathbf{f}(\mathbf{y} + p_k h \mathbf{\Delta}(\mathbf{y};\ p_k h)) + (p_k h)^\mu \mathscr{A}(\mathbf{y})\overline{\boldsymbol{\varphi}}(\mathbf{y}) + O(h^{\mu+1})] \\
&= \mathbf{\Delta}(\mathbf{y};\ h) - h^\mu C\mathbf{f}^{(\mu)}(\mathbf{y}) + h^\mu \left(\frac{1}{\mu + 1} - C\mu!\right) \\
&\qquad \times \mathscr{A}(\mathbf{y})\overline{\boldsymbol{\varphi}}(\mathbf{y}) + O(h^{\mu+1})
\end{aligned}$$

We conclude that

$$\boldsymbol{\varphi}(\mathbf{y};\ h) = -C\mathbf{f}^{(\mu)}(\mathbf{y}) + \left(\frac{1}{\mu + 1} - C\mu!\right)\mathscr{A}(\mathbf{y})\overline{\boldsymbol{\varphi}}(\mathbf{y}) \tag{3-88}$$

If $\bar{p} + 1 > \mu$, we have simply

$$\boldsymbol{\varphi}(\mathbf{y}) = -C\mathbf{f}^{(\mu)}(\mathbf{y})$$

In this case the principal error function is proportional to a derivative of the true solution.

Most of the formulas for $\boldsymbol{\varphi}(\mathbf{y})$ given above are too unwieldy for practical computation. What is therefore needed is a numerical method to approximate $\boldsymbol{\varphi}(\mathbf{y})$. In the one-dimensional case such a method was provided by the device of local extrapolation to the limit which formed the content of Theorem 2.5. An entirely analogous device also holds in the case of a system of differential equations. Let $\boldsymbol{\delta}_1$ and $\boldsymbol{\delta}_2$ denote the increments of the numerical solution over an interval of length h, calculated by performing, with the method defined by $\mathbf{\Phi}(\mathbf{y};\ h)$, one step of length h and two steps of length $\frac{1}{2}h$, respectively. In symbols:

$$\begin{aligned}
\boldsymbol{\delta}_1 &= h\mathbf{\Phi}(\mathbf{y};\ h) \\
\boldsymbol{\delta}_2 &= \tfrac{1}{2}h[\mathbf{\Phi}(\mathbf{y};\ \tfrac{1}{2}h) + \mathbf{\Phi}(\mathbf{y} + \tfrac{1}{2}h\mathbf{\Phi}(\mathbf{y});\ \tfrac{1}{2}h)]
\end{aligned}$$

It can then be proved exactly as in §2.2-7 that

$$\boldsymbol{\varphi}(\mathbf{y}) = -\frac{h^{-p-1}}{1 - 2^{-p}}(\boldsymbol{\delta}_2 - \boldsymbol{\delta}_1) + O(h) \tag{3-89}$$

3.3-6. Asymptotic formula for the discretization error. We again allow the functions $\mathbf{f}(x, \mathbf{y})$ and $\boldsymbol{\Phi}(x, \mathbf{y};\ h)$ to depend explicitly on x, but continue to denote by $\mathbf{f}_i$ the vector $\partial\mathbf{f}/\partial y^i$. (The subscript h, however, is reserved for denoting the derivative with respect to h.) It is assumed that the method defined by $\boldsymbol{\Phi}(x, \mathbf{y};\ h)$ has exact order p, that $\boldsymbol{\Phi}(x, \mathbf{y};\ h)$ satisfies the conditions of Theorem 3.3 and, in addition, that it has continuous derivatives of order $p + 2$ with respect to all arguments for $x \in [a, b]$, $y^i \in (-\infty, \infty)$ $(i = 1, \cdots, s)$, and $h \leq h_0$. The principal error function $\boldsymbol{\varphi}(x, \mathbf{y})$ then exists and is continuous and has continuous derivatives. We shall also assume that the second derivatives of $\boldsymbol{\Phi}$ with respect to h and the y^i are bounded:

$$\begin{aligned} &\|\boldsymbol{\Phi}_{hi}(x, \mathbf{y};\ h)\| \leq K_1, \\ &\|\boldsymbol{\Phi}_{ij}(x, \mathbf{y};\ h)\| \leq K_1, \qquad i, j = 1, \cdots, s, \\ &\qquad x \in [a, b], \quad y^i \in (-\infty, \infty) \quad (i = 1, \cdots, s), \quad h \leq h_0 \end{aligned} \tag{3-90}$$

If $\mathbf{y}(x)$ denotes the solution of the initial value problem (3-9), and if the vectors $\mathbf{y}_n$ are defined by (3-30), then according to Theorem 3.3 the errors $\mathbf{e}_n = \mathbf{y}_n - \mathbf{y}(x_n)$ satisfy

$$\|\mathbf{e}_n\| \leq K_2 h^p \tag{3-91}$$

where $K_2 = NE_L(b - a)$.

We return once again to relation (3-60) for the discretization error, which we now write in the form

$$\begin{aligned} \mathbf{e}_{n+1} = \mathbf{e}_n + h[&\boldsymbol{\Phi}(x_n, \mathbf{y}_n;\ h) - \boldsymbol{\Phi}(x_n, \mathbf{y}(x_n);\ h) \\ &+ \boldsymbol{\Phi}(x_n, \mathbf{y}(x_n);\ h) - \boldsymbol{\Delta}(x_n, \mathbf{y}(x_n);\ h)] \end{aligned} \tag{3-92}$$

We have, using Taylor's formula with remainder,

$$\begin{aligned} \boldsymbol{\Phi}(x_n, \mathbf{y}_n;\ h) - \boldsymbol{\Phi}(x_n, \mathbf{y}(x_n);\ h) &= \boldsymbol{\Phi}(x_n, \mathbf{y}(x_n) + \mathbf{e}_n, h) - \boldsymbol{\Phi}(x_n, \mathbf{y}(x_n);\ h) \\ &= \boldsymbol{\Phi}_i(x_n, \mathbf{y}(x_n);\ h)e_n^i + \tfrac{1}{2}\boldsymbol{\theta}_n K_1 K_2^2 h^{2p} \end{aligned}$$

where $\|\boldsymbol{\theta}_n\| \leq 1$. If $\mathscr{A}(x, \mathbf{y}) = (a^{ij})$ denotes the matrix with components

$$a^{ij} = f_j^i(x, \mathbf{y})$$

and, if we define the matrix $\mathbf{G}(x)$ by

$$\mathbf{G}(x) = \mathscr{A}(x, \mathbf{y}(x))$$

then, by expanding in powers of h and using (3-90),

$$\mathbf{\Phi}_i(x_n, \mathbf{y}(x_n); h) = \mathbf{G}(x_n)\mathbf{e}_n + h^{p+1}\boldsymbol{\theta}'_n K_1 K_2$$

where again $\|\boldsymbol{\theta}'_n\| \leq 1$. Finally, letting

$$K_3 = \frac{1}{(p+1)!} \sup_{\substack{x \in [a,b] \\ h \leq h_0}} \left\| \frac{\partial^{p+1}}{\partial h^{p+1}} [\mathbf{\Phi}(x, \mathbf{y}(x); h) - \mathbf{\Delta}(x, \mathbf{y}(x); h)] \right\|$$

we have, again by Taylor's formula,

$$\mathbf{\Phi}(x, \mathbf{y}(x); h) - \mathbf{\Delta}(x, \mathbf{y}(x); h) = h^p \boldsymbol{\varphi}(x, \mathbf{y}(x)) + h^{p+1}\boldsymbol{\theta} K_3$$

where $\boldsymbol{\theta}$ depends on x and h, but $\|\boldsymbol{\theta}\| \leq 1$. Defining the magnified errors $\bar{\mathbf{e}}_n$ by

$$\bar{\mathbf{e}}_n = h^{-p}\mathbf{e}_n, \qquad n = 0, 1, 2, \cdots$$

we thus can write (3-92) in the form

$$\bar{\mathbf{e}}_{n+1} = \mathbf{e}_n + h[\mathbf{G}(x_n)\bar{\mathbf{e}}_n + \boldsymbol{\varphi}(x_n, \mathbf{y}(x_n))] + \boldsymbol{\theta}''_n h^2[K_1K_2 + h_0^{p-1}K_1K_2^2 + K_3]$$

This relation can be thought of as defining [up to $O(h^2)$] an Euler approximation to the solution of the following initial value problem for a function $\mathbf{e}(x)$:

$$\begin{aligned} &\mathbf{e}(a) = \mathbf{0} \\ &\mathbf{e}' = \mathbf{G}(x)\mathbf{e} + \boldsymbol{\varphi}(x, \mathbf{y}(x)) \end{aligned} \tag{3-93}$$

By the hypotheses stated at the beginning of this section, the solution $\mathbf{e}(x)$ of this problem exists and has a continuous second derivative for $x \in [a, b]$. We thus can apply Theorem 3.3, replacing $\mathbf{y}(x)$ by $\mathbf{e}(x)$ and $\mathbf{f}(x, \mathbf{y}) = \mathbf{\Phi}(x, \mathbf{y}; h)$ by $\mathbf{G}(x)\mathbf{e} + \boldsymbol{\varphi}(x, \mathbf{y}(x))$ and setting

$$\begin{aligned} p &= 1 \\ L_1 &= \max_{x \in [a,b]} \left\{ \max_{1 \leq j \leq s} \sum_{i=1}^{s} |g^{ij}(x)| \right\} \\ N_1 &= \frac{1}{2} \max_{x \in [a,b]} \|\mathbf{e}''(x)\| \\ K_4 &= K_1K_2(1 + h^{p-1}K_2) + K_3 \end{aligned} \tag{3-94}$$

It follows that

$$\|\bar{\mathbf{e}}_n - \mathbf{e}(x_n)\| \leq hK_5, \qquad x_n \in [a, b]$$

where

$$K_5 = (N_1 + K_4)E_{L_1}(b - a) \tag{3-95}$$

We have proved the following result:

THEOREM 3.4. *Under the hypotheses stated at the beginning of this section, the discretization error in the solution of the initial value problem*

(3-9) *by the one-step method defined by* $\mathbf{\Phi}(x, \mathbf{y};\ h)$ *satisfies*

$$\mathbf{e}_n = h^p\mathbf{e}(x_n) + h^{p+1}\boldsymbol{\theta}_n K_5 \tag{3-96}$$

where $\mathbf{e}(x)$ *is the solution of* (3-93), $\|\boldsymbol{\theta}_n\| \le 1$, *and* K_5 *is given by* (3-95).

Theoretically this result can serve to improve a given numerical solution by integrating the system (3-93) and approximately calculating the errors $\mathbf{e}_n$ from (3-96), neglecting the term involving $\boldsymbol{\theta}_n$. More practically the theorem justifies the extrapolation to the limit for systems. Defining $\mathbf{y}(x, h)$ as in §2.2-7, we have for $q \ne 1$

$$\mathbf{y}(x) = \frac{\mathbf{y}(x, qh) - q^p\mathbf{y}(x, h)}{1 - q^p} + \boldsymbol{\theta}\,\frac{1 + q}{1 - q^p}\,q^r h^{p+1} K_5 \tag{3-97}$$

where $\|\boldsymbol{\theta}\| \le 1$. The value of $\mathbf{y}(x)$ obtained by putting $\boldsymbol{\theta} = \mathbf{0}$ is in error by only $O(h^{p+1})$.

Theorem 3.4 is easily extended to the case of variable stepsize. If the points x_n are defined by (2-51), and if (3-57) is replaced by (3-61), then (3-96) still holds provided that $\mathbf{e}(x)$ is now defined by

$$\begin{aligned} \mathbf{e}(a) &= \mathbf{0} \\ \mathbf{e}' &= \mathbf{G}(x)\mathbf{e} + [\vartheta(x)]^p\boldsymbol{\varphi}(x, \mathbf{y}(x)) \end{aligned} \tag{3-98}$$

3.3-7. Example. For convenience, we shall write (in this section only)

$$\begin{aligned} y^1 &= y, & e^1 &= \eta, & \varphi^1 &= \varphi \\ y^2 &= z, & e^2 &= \zeta, & \varphi^2 &= \psi \end{aligned}$$

We shall discuss the solution of the nonlinear initial value problem

$$\begin{aligned} y' &= -z, & y(0) &= a^{-1} \\ z' &= -2y^3, & z(0) &= a^{-2} \end{aligned} \tag{3-99}$$

where $a > 0$, by the one-step method (3-52), where the underlying quadrature formula is given by*

$$h^{-1}\int_x^{x+h} f(t)\,dt = \tfrac{1}{4}f(x) + \tfrac{3}{4}f(x + \tfrac{2}{3}h) + \tfrac{1}{216}h^3 f'''(\xi) \tag{3-100}$$

and where $\overline{\mathbf{\Phi}}_1$ is the increment function of the modified Euler method [(3-47) with $\alpha = 1$]. The exact solution of (3-99) is $y(x) = \xi^{-1}$, $z(x) = \xi^{-2}$, where $\xi = x + a$. It follows that

$$\mathscr{A}(\mathbf{y}(x)) = \begin{pmatrix} 0 & -1 \\ -6\xi^{-2} & 0 \end{pmatrix}$$

* This is a special case of a quadrature formula due to Radau; see Hildebrand [1956], p. 338.

According to (3-83), the principal error vector of the modified Euler method has the components

$$\bar{\varphi}(x) = \xi^{-4}, \qquad \bar{\psi}(x) = \tfrac{5}{2}\xi^{-5}$$

Since $\bar{p} = 2$, $\mu = 3$, we have $\bar{p} + 1 = \mu$ and hence $p = 3$. From (3-85) we determine the components of the principal error vector as follows:

$$\varphi(x) = -\tfrac{2}{3}\xi^{-5}, \qquad \psi(x) = -\tfrac{17}{9}\xi^{-6}$$

The components $\eta(x)$ and $\zeta(x)$ of the magnified error vector thus satisfy

$$\begin{aligned} \eta' &= -\zeta - \tfrac{2}{3}\xi^{-5}, & \eta(0) &= 0 \\ \zeta' &= -6\xi^{-2}\eta - \tfrac{17}{9}\xi^{-6}, & \zeta(0) &= 0 \end{aligned} \tag{3-101}$$

Eliminating ζ from the first equation by differentiating and substituting for ζ' from the second, we can integrate the system and find the solution

$$\begin{aligned} \eta(x) &= \frac{1}{126a^4}\left\{2\left(\frac{\xi}{a}\right)^3 - 49\left(\frac{\xi}{a}\right)^{-2} + 47\left(\frac{\xi}{a}\right)^{-4}\right\} \\ \zeta(x) &= \frac{1}{63a^5}\left\{-3\left(\frac{\xi}{a}\right)^2 - 49\left(\frac{\xi}{a}\right)^{-3} + 52\left(\frac{\xi}{a}\right)^{-5}\right\} \end{aligned} \tag{3-102}$$

In Table 3.1 we give the values of the numerical solution, the actual errors, and the errors predicted by (3-96) for $x_n = 6$, using the steps $h = 2^{-p}$ with $p = 1(1)7$.

Table 3.1

p	y_n	$y_n - y(6)$	$h^3\eta(6)$	z_n	$z_n - z(6)$	$h^3\zeta(6)$
1	0.1324308	0.0074308	0.0077580	0.0126438	−0.0029812	−0.0030205
2	.1259683	.0009683	.0009698	.0152445	− .0003805	− .0003776
3	.1251219	.0001219	.0001212	.0155774	− .0000476	− .0000472
4	.1250152	.0000152	.0000152	.0156191	− .0000059	− .0000059
5	.1250019	.0000019	.0000019	.0156243	− .0000007	− .0000007
6	.1250002	.0000002	.0000002	.0156249	− .0000001	− .0000001
7	.1250000	.0000000	.0000000	.0156250	− .0000000	− .0000000

3.4. The Round-off Error in the Integration of Systems by One-Step Methods

3.4-1. An a priori bound for the accumulated round-off error. If fixed point operations are used, the numerical approximation $\tilde{\mathbf{y}}_n$ satisfies in place of (3-30) the relation

$$\tilde{\mathbf{y}}_{n+1} = \tilde{\mathbf{y}}_n + [h\tilde{\mathbf{\Phi}}(x_n, \tilde{\mathbf{y}}_n; h)]^*, \qquad n = 0, 1, 2, \cdots \tag{3-103}$$

where $\tilde{\mathbf{\Phi}}(x_n, \tilde{\mathbf{y}}_n; h)$ denotes a numerical approximation to $\mathbf{\Phi}(x_n, \tilde{\mathbf{y}}_n; h)$ (note that the arguments in both functions are the same), and where the asterisk denotes a rounded value. For analytical investigations it is more convenient to assume that the numerical values $\tilde{\mathbf{y}}_n$ are linked by the relation

$$\tilde{\mathbf{y}}_{n+1} = \tilde{\mathbf{y}}_n + h\mathbf{\Phi}(x_n, \tilde{\mathbf{y}}_n; h) + \boldsymbol{\epsilon}_{n+1} \tag{3-104}$$

where $\boldsymbol{\epsilon}_{n+1}$ is called the vector of the *local round-off errors.* In the situation described above, $\boldsymbol{\epsilon}_{n+1}$ is given by

$$\boldsymbol{\epsilon}_{n+1} = [h\tilde{\mathbf{\Phi}}(x_n, \tilde{\mathbf{y}}_n; h)]^* - h\mathbf{\Phi}(x_n, \tilde{\mathbf{y}}_n; h) \tag{3-105}$$

and is thus easily estimated. Equation (3-104) has the advantage, however, that it is not limited in its application to the rather special case of fixed-point, single precision arithmetical operations (see §3.4-8).

The purpose of this section is to derive a bound for the accumulated round-off error

$$\mathbf{r}_n = \tilde{\mathbf{y}}_n - \mathbf{y}_n$$

under the sole assumption that

$$\|\boldsymbol{\epsilon}_n\| \leq \varepsilon \tag{3-106}$$

where ε is a constant. (Frequently one may assume that $\varepsilon = \frac{1}{2}su$.) Subtracting (3-30) from (3-104), it follows that

$$\mathbf{r}_{n+1} = \mathbf{r}_n + h\{\mathbf{\Phi}(x_n, \tilde{\mathbf{y}}_n; h) - \mathbf{\Phi}(x_n, \mathbf{y}_n; h)\} + \boldsymbol{\epsilon}_{n+1} \tag{3-107}$$

If L denotes the Lipschitz constant of the method defined by $\mathbf{\Phi}$, it follows, using (3-106), that

$$\|\mathbf{r}_{n+1}\| \leq \|\mathbf{r}_n\| + hL\|\mathbf{r}_n\| + \varepsilon$$

Applying Lemma 1.2 and using the fact that $\mathbf{r}_0 = 0$, we deduce from this inequality in a by now familiar manner the following theorem:

THEOREM 3.5. *If the method defined by the increment function* $\mathbf{\Phi}(x, \mathbf{y}; h)$ *satisfies* (3-56), *and if the local round-off errors are bounded according to* (3-106), *then the accumulated round-off error is bounded as follows:*

$$\|\mathbf{r}_n\| \leq \frac{\varepsilon}{h} E_L(x_n - a), \quad x_n \in [a, b] \tag{3-108}$$

Underlying this result is the hypothesis that the round-off errors not only at each lattice point x_n but also in each of the s simultaneous differential equations systematically reinforce each other. The resulting estimate is therefore very pessimistic, and the need for a more realistic appraisal of the effect of the local round-off errors is thus even greater than in the case of a single equation.

3.4-2. Dependence of the accumulated round-off error on the local errors: a special case. We begin by considering the solution of the linear system

$$\mathbf{y}' = \mathbf{G}(x)\mathbf{y} \tag{3-109}$$

by Euler's method. The elements of the $s \times s$ matrix $\mathbf{G}(x)$ are supposed to be given continuous functions of x. The increment function for Euler's method is in this case given by

$$\mathbf{\Phi}(x, \mathbf{y}; h) = \mathbf{G}(x)\mathbf{y}$$

Relation (3-107) thus assumes the simple form

$$\mathbf{r}_{n+1} = \mathbf{r}_n + h\mathbf{G}(x_n)\mathbf{r}_n + \boldsymbol{\epsilon}_{n+1} \tag{3-110}$$

We aim at representing $\mathbf{r}_n$ in terms of the vectors $\boldsymbol{\epsilon}_m$ $(m \leq n)$, as follows:

$$\mathbf{r}_n = \sum_{m=1}^{n} \mathbf{D}_{n,m}\boldsymbol{\epsilon}_m \tag{3-111}$$

Here the $\mathbf{D}_{n,m}$ denote matrices yet to be determined. Introducing (3-111) into (3-110), we find

$$\sum_{m=1}^{n+1} \mathbf{D}_{n+1,m}\boldsymbol{\epsilon}_m = \sum_{m=1}^{n} (\mathbf{I} + h\mathbf{G}(x_n))\mathbf{D}_{n,m}\boldsymbol{\epsilon}_m + \boldsymbol{\epsilon}_{n+1} \tag{3-112}$$

where $\mathbf{I}$ denotes the unit matrix. This identity must hold for all determinations of the vectors $\boldsymbol{\epsilon}_m$. In particular it must hold when only one $\boldsymbol{\epsilon}_m$ is different from zero, and for all possible choices of that sole nonzero $\boldsymbol{\epsilon}_m$. It follows that the multipliers of each $\boldsymbol{\epsilon}_m$ on either side of (3-112) must be identical. This leads to the conditions

(3-113)

$$\mathbf{D}_{n+1,n+1} = \mathbf{I}, \qquad n = 0, 1, 2, \cdots$$
$$\mathbf{D}_{n+1,m} = \mathbf{D}_{n,m} + h\mathbf{G}(x_n)\mathbf{D}_{n,m}, \qquad m = 1, 2, 3, \cdots, n = m, m+1, \cdots$$

Conversely, these conditions fully determine the matrices $\mathbf{D}_{n,m}$, and the vectors (3-111), formed with the matrices thus defined, will be found to satisfy (3-110).

The matrices $\mathbf{D}_{n,m}$ obviously take the place of the scalar coefficients $d_{n,m}$ considered in §§1.4-3 and 2.3-2. It is also possible to show that for $h \to 0$ the elements of $\mathbf{D}_{n,m}$ approach the solution of a certain system of differential equations. Let $\mathbf{d}_{n,m}$ denote the kth column of the matrix $\mathbf{D}_{n,m}$ $(k = 1, 2, \cdots, s)$, and let $\mathbf{e}_k$ be the vector whose kth component is 1 and whose other components are 0. It then follows from (3-113) that

$$\begin{aligned} \mathbf{d}_{n,n} &= \mathbf{e}_k \\ \mathbf{d}_{n+1,m} &= \mathbf{d}_{n,m} + h\mathbf{G}(x_n)\mathbf{d}_{n,m} \end{aligned} \tag{3-114}$$

Exactly the same system of equations would result if the initial value problems

$$\begin{aligned} \mathbf{d}_m(x_m) &= \mathbf{e}_k \\ \mathbf{d}_m'(x) &= \mathbf{G}(x)\mathbf{d}_n(x), \qquad x \geq x_m \end{aligned} \tag{3-115}$$

were approximately solved by Euler's method. Because Euler's method converges and is of order 1, it follows that

$$\mathbf{d}_{n,m} = \mathbf{d}_m(x_n) + O(h)$$

Collecting the s vectors $\mathbf{d}_m(x)$ corresponding to the s different values of k in a matrix $\mathbf{D}_m(x)$, we find that

$$\mathbf{D}_{n,m} = \mathbf{D}_m(x_n) + \mathbf{\Theta}_{m,n}Kh \tag{3-116}$$

where $\mathbf{\Theta}_{n,m}$ is a matrix whose elements are bounded by 1, K is a constant, and the matrices $\mathbf{D}_m(x)$ are the solutions of the initial value problems

$$\begin{aligned} \mathbf{D}_m(x_m) &= \mathbf{I} \\ \mathbf{D}_m'(x) &= \mathbf{G}(x)\mathbf{D}_m(x) \end{aligned} \tag{3-117}$$

By applying this result to a simple special matrix $\mathbf{G}(x)$ such as

$$\mathbf{G}(x) = \begin{pmatrix} 0 & 1 \\ -1 & 0 \end{pmatrix}$$

we conclude that for $s > 1$ the elements of the matrices $\mathbf{D}_{n,m}$ are no longer positive even for arbitrarily small values of h.

3.4-3. Dependence of the accumulated round-off error on the local errors: the general case. In order to derive a formula similar to (3-111) for the approximate solution of an arbitrary differential equation by an arbitrary method, we now shall assume that the bound ε for the local round-off error depends on h in such a manner that

$$Nh^{p+1} \leq \varepsilon \leq Kh^{q+1}, \qquad h \leq h_0 \tag{3-118}$$

Here p denotes the order of the method defined by $\mathbf{\Phi}$, N is the constant occurring in the bound (3-57) for the local discretization error, K is some constant which does not depend on h, and q is supposed to satisfy $q \geq 1$. Condition (3-118) may be interpreted by saying that the local round-off error is permitted to exceed the local discretization error, but still must tend to 0 at least as rapidly as h^2 as $h \to 0$. As a consequence of (3-118) we have from Theorems 3.3 and 3.5 the inequalities

$$\begin{aligned} \|\tilde{\mathbf{y}}_n - \mathbf{y}_n\| &\leq h^q KE \\ \|\tilde{\mathbf{y}}_n - \mathbf{y}(x_n)\| &\leq h^p NE, \qquad x_n \in [a, b] \end{aligned} \tag{3-119}$$

where $E = E_L(b - a)$.

We now try to write (3-107) in a form which resembles (3-110). By repeated use of Taylor's theorem (both for functions of one and of several variables) we have, using the first relation (3-119) and denoting by K_1 the bound for certain second derivatives of $\mathbf{\Phi}$ as given in (3-90),

$$\mathbf{\Phi}(x_n, \tilde{\mathbf{y}}_n; h) - \mathbf{\Phi}(x_n, \mathbf{y}(x_n); h) = \mathbf{\Phi}_j(x_n, \mathbf{y}(x_n); h)(\tilde{y}^j - y^j(x_n))$$
$$+ \tfrac{1}{2}\mathbf{\Phi}_{jk}(x_n, \mathbf{y}^+; h)(\tilde{y}^j - y^j(x_n))(\tilde{y}^k - y^k(x_n)) = \mathbf{G}(x_n)(\tilde{\mathbf{y}}_n - \mathbf{y}(x_n)) + \varepsilon\boldsymbol{\theta}_n K_2 h^q$$

Here $\mathbf{y}^+$ denotes a certain vector of intermediate values which may be different in the different components of $\mathbf{\Phi}_{jk}$. The matrix $\mathbf{G}(x)$ is defined as in §3.3-6, and

$$K_2 = K_1 E(1 + 2h_0^{q-1} KE)$$

Similarly we find

$$\mathbf{\Phi}(x_n, \mathbf{y}_n; h) - \mathbf{\Phi}(x_n, \mathbf{y}(x_n); h) = \mathbf{G}(x_n)(\mathbf{y}_n - \mathbf{y}(x_n)) + \varepsilon\boldsymbol{\theta}_n' K_3$$

where

$$K_3 = K_1 E(1 + \tfrac{1}{2}h_0^{q-1} KE)$$

Thus we have

(3-120) $$\mathbf{\Phi}(x_n, \tilde{\mathbf{y}}_n; h) - \mathbf{\Phi}(x_n, \mathbf{y}_n; h) = \mathbf{G}(x_n)\mathbf{r}_n + \boldsymbol{\theta}_n'' K_4$$

where $K_4 = K_2 + K_3$, $\|\boldsymbol{\theta}_n''\| \le 1$, and hence, in place of (3-107),

(3-121) $$\mathbf{r}_{n+1} = \mathbf{r}_n + h\mathbf{G}(x_n)\mathbf{r}_n + \boldsymbol{\epsilon}_{n+1} + \varepsilon\boldsymbol{\theta}_{n+1} hK_4$$

This relation is identical with (3-110), except that the place of $\boldsymbol{\epsilon}_{n+1}$ is now taken by $\boldsymbol{\epsilon}_{n+1} + \varepsilon\boldsymbol{\theta}_{n+1} h^{q+1} K_4$. It follows that

(3-122) $$\mathbf{r}_n = \mathbf{r}_n^{(1)} + \mathbf{r}_n^{(2)}$$

where

(3-122*a*) $$\mathbf{r}_n^{(1)} = \sum_{m=1}^{n} \mathbf{D}_{n,m}\boldsymbol{\epsilon}_m$$

(3-122*b*) $$\mathbf{r}_n^{(2)} = hK_4 \sum_{m=1}^{n} \mathbf{D}_{n,m}\boldsymbol{\theta}_m$$

As in §2.3-2, we call $\mathbf{r}_n^{(1)}$ the *primary* and $\mathbf{r}_n^{(2)}$ the *secondary* round-off error. If the differential equation were linear and would be integrated by the Euler method, the primary error alone would be present. The secondary error owes its existence to the possible nonlinearity of the equation and to the complexity of the integration method. From (3-116) and (3-117) it follows that the elements of the matrices $\mathbf{D}_{n,m}$ are bounded independently of h. If $\mathbf{D}_{n,m} = (d_{n,m}^{ij})$, it follows that

(3-123) $$\|\mathbf{r}_n^{(2)}\| \le K_5\varepsilon$$

where

$$K_5 = (b - a) \max_{x_n, x_m \in [a,b]} \left\{ \max_{1 \le j \le s} \sum_{i=1}^{s} |d_{n,m}^{ij}| \right\}$$

i.e., the secondary round-off error is of the order of magnitude of one local round-off error as $h \to 0$.

3.4-4. An a posteriori bound for the accumulated round-off error. The following notations will be convenient. For any vector $\mathbf{v} = (v^i)$ we denote by $\hat{\mathbf{v}}$ the vector with the components $|v^i|$. By an inequality such as $\mathbf{v} \le \mathbf{w}$ we mean that the inequality $v^i \le w^i$ holds for all components. Similar notations will be employed for matrices.

We shall in this section derive a bound for $\hat{\mathbf{r}}_n$ under the assumption that

(3-124)
$$\hat{\boldsymbol{\epsilon}}_n \le \mathbf{p}(x_n)\varepsilon$$

where $\mathbf{p}(x)$ is a vector function, assumed to be known, with nonnegative elements which are piecewise continuous functions of x. The constant ε is assumed to satisfy condition (3-118). Since for the secondary error the satisfactory bound (3-123) has already been found, we shall concentrate on the primary error. By (3-122*a*), taking into account the fact that the elements of $\mathbf{D}_{n,m}$ are no longer positive as in the case of a single equation, we have

$$\hat{\mathbf{r}}_n^{(1)} \le \frac{\varepsilon}{h} \mathbf{m}_n$$

where, writing $\mathbf{p}_n = \mathbf{p}(x_n)$,

(3-125)
$$\mathbf{m}_n = h \sum_{m=1}^{n} \hat{\mathbf{D}}_{n,m} \mathbf{p}_m$$

It follows from (3-113) that

$$\hat{\mathbf{D}}_{n,n} = \mathbf{I}, \qquad \hat{\mathbf{D}}_{n+1,m} \le \hat{\mathbf{D}}_{n,m} + h\hat{\mathbf{G}}(x_n)\hat{\mathbf{D}}_{n,m}$$

By a procedure similar to the one used in deriving (2-77), we thus find

$$\mathbf{m}_{n+1} - \mathbf{m}_n \le h(\mathbf{p}_{n+1} + \hat{\mathbf{G}}(x_n)\mathbf{m}_n)$$

By an application of Theorem 3.3, with Euler's method in mind, we conclude that

(3-126)
$$\mathbf{m}_n \le \mathbf{m}(x_n) + O(h)$$

where $\mathbf{m}(x)$ is the continuous solution of the initial value problem

(3-127)
$$\mathbf{m}(a) = \mathbf{0}$$
$$\mathbf{m}'(x) = \hat{\mathbf{G}}(x)\mathbf{m}(x) + \mathbf{p}(x)$$

We thus find that

$$\hat{\mathbf{r}}_n^{(1)} \le \frac{\varepsilon}{h}\{\mathbf{m}(x_n) + O(h)\} \tag{3-128}$$

and by virtue of (3-123) a similar inequality holds for $\hat{\mathbf{r}}_n$. We have thus proved:

THEOREM 3.6. *If the increment function has continuous and bounded second derivatives with respect to all its arguments, and if the local round-off errors are bounded by* (3-124), *then the accumulated round-off error satisfies*

$$\hat{\mathbf{r}}_n \le \frac{\varepsilon}{h}\{\mathbf{m}(x_n) + O(h)\} \tag{3-129}$$

where $\mathbf{m}(x)$ *is defined by* (3-127).

For the simple example $a = 0$,

$$y^{1\prime} = y^2, \qquad y^{2\prime} = -y^1 \tag{3-130}$$

assuming that $p^i(x) = 1$, $i = 1, 2$, we find for $\mathbf{m}(x)$ the system

$$m^{1\prime} = m^2 + 1, \qquad m^1(0) = 0$$
$$m^{2\prime} = m^1 + 1, \qquad m^2(0) = 0$$

which has the solution

$$m^1(x) = m^2(x) = e^x - 1$$

Thus the estimate (3-129) in this case permits the round-off error to grow exponentially, in spite of the fact that the solutions of (3-130) remain bounded. Such a situation cannot arise in the case of a single linear equation. We shall also see that this result is not confirmed by the statistical theory to be presented below.

3.4-5. Statistical theory. In order to obtain realistic statements about the behavior of the accumulated round-off errors, we now shall once again assume that the components of the local round-off error vector $\boldsymbol{\epsilon}_n$ are random variables. The following assumptions on the random nature of $\boldsymbol{\epsilon}_n$ will be made:

(i) Writing

$$\boldsymbol{\mu}_n = E(\boldsymbol{\epsilon}_n) \tag{3-131}$$

we shall assume that

$$\hat{\boldsymbol{\mu}}_n \le \mu\mathbf{p}(x_n) \tag{3-132}$$

where $\mathbf{p}(x)$ is a known vector with nonnegative components which are piecewise smooth functions of x, and where $\mu \ge 0$ is a constant independent of n (but possibly dependent on h).

(ii) If $m \neq n$, then, denoting by a superscript T the transposed vector or matrix,

$$E[(\boldsymbol{\epsilon}_n - \boldsymbol{\mu}_n)(\boldsymbol{\epsilon}_m^T - \boldsymbol{\mu}_m^T)] = \mathbf{0} \tag{3-133}$$

(iii) If $m = n$, then

$$E[(\boldsymbol{\epsilon}_n - \boldsymbol{\mu}_n)(\boldsymbol{\epsilon}_n^T - \boldsymbol{\mu}_n^T)] = \sigma^2 \mathbf{C}(x_n) \tag{3-134}$$

where $\mathbf{C}(x)$ is a known positive semidefinite symmetric matrix whose elements are piecewise smooth functions of x, and where σ^2 is independent of n.

Condition (ii) holds if the local round-off errors at different points x_n are considered as independent random variables. The matrix on the left of (3-134) is called the covariance matrix of the random variables making up the vector $\boldsymbol{\epsilon}_n$. Its elements are the expected values of the products $(\varepsilon_n^i - \mu_n^i)(\varepsilon_n^j - \mu_n^j)$. If the variables ε_n^i and ε_n^j are independent for $i \neq j$, then the covariance matrix is a diagonal matrix. It is shown in §3.4-8 that the variables ε_n^i cannot always be assumed to be independent. We do not therefore assume $\mathbf{C}(x)$ to be diagonal.

In this section we are concerned only with the primary error $\mathbf{r}_n^{(1)}$, since statistical assumptions on the vectors $\boldsymbol{\theta}_n$ appearing in the definition of $\mathbf{r}_n^{(2)}$ are difficult to justify. For the expected value of $\mathbf{r}_n^{(1)}$ we find as a consequence of (1-94),

$$E(\mathbf{r}_n^{(1)}) = \sum_{m=1}^{n} \mathbf{D}_{n,m} \boldsymbol{\mu}_m \tag{3-135}$$

By (3-132), the absolute values of the components of this vector are bounded by those of $(\mu/h)\mathbf{m}_n$, where $\mathbf{m}_n$ is defined by (3-125). Using a result of the preceding section, we thus find

$$\hat{E}(\mathbf{r}_n^{(1)}) \leq \frac{\mu}{h} (\mathbf{m}(x_n) + O(h)) \tag{3-136}$$

where the vector $\mathbf{m}(x)$ is defined by (3-127).

We now turn to the slightly more complicated task of determining the covariance matrix of $\mathbf{r}_n^{(1)}$. If

$$E[(\mathbf{r}_n^{(1)} - E(\mathbf{r}_n^{(1)}))(\mathbf{r}_n^{(1)} - E(\mathbf{r}_n^{(1)}))^T] = \text{covar}\,(\mathbf{r}_n^{(1)})$$

then, by (3-122*a*) and (3-135),

$$\begin{aligned} \text{covar}\,(\mathbf{r}_n^{(1)}) &= E\left[\sum_{m=1}^{n} \mathbf{D}_{nm}(\boldsymbol{\epsilon}_m - \boldsymbol{\mu}_m) \sum_{l=1}^{n} (\boldsymbol{\epsilon}_l^T - \boldsymbol{\mu}_l^T)\mathbf{D}_{n\,l}^T\right] \\ &= \sum_{m=1}^{n} \sum_{l=1}^{n} \mathbf{D}_{nm} E[(\boldsymbol{\epsilon}_m - \boldsymbol{\mu}_m)(\boldsymbol{\epsilon}_l^T - \boldsymbol{\mu}_l^T)]\mathbf{D}_{n\,l}^T \end{aligned}$$

By (3-133) and (3-134) this reduces to

$$\text{(3-137)} \qquad \operatorname{covar}(\mathbf{r}_n^{(1)}) = \frac{\sigma^2}{h}\mathbf{V}_n$$

where, upon writing $\mathbf{C}_m = \mathbf{C}(x_m)$,

$$\text{(3-138)} \qquad \mathbf{V}_n = h\sum_{m=1}^{n}\mathbf{D}_{nm}\mathbf{C}_m\mathbf{D}_{nm}^T$$

As on earlier occasions, we shall compare $\mathbf{V}_n$ to the solution of a differential equation by establishing a difference equation. Writing $\mathbf{G}_n = \mathbf{G}(x_n)$, we have, using the recurrence relations (3-113) repeatedly,

$$\begin{aligned}\mathbf{V}_{n+1} - \mathbf{V}_n &= h\left\{\sum_{m=1}^{n+1}\mathbf{D}_{n+1,m}\mathbf{C}_m\mathbf{D}_{n+1,m}^T - \sum_{m=1}^{n}\mathbf{D}_{n,m}\mathbf{C}_m\mathbf{D}_{n,m}^T\right\} \\ &= h\left\{\mathbf{C}_{n+1} + \sum_{m=1}^{n}[\mathbf{D}_{n+1,m}\mathbf{C}_m\mathbf{D}_{n+1,m} - \mathbf{D}_{nm}\mathbf{C}_m\mathbf{D}_{nm}^T]\right\} \\ &= h\left\{\mathbf{C}_{n+1} + \sum_{m=1}^{n}[(\mathbf{I} + h\mathbf{G}_n)\mathbf{D}_{nm}\mathbf{C}_m\mathbf{D}_{nm}^T(\mathbf{I} + h\mathbf{G}_n^T) \right. \\ &\qquad \left. - \mathbf{D}_{nm}\mathbf{C}_m\mathbf{D}_{nm}^T]\right\}\end{aligned}$$

At points of continuity of $\mathbf{C}(x)$, using the fact that

$$h^2\sum_{m=1}^{n}\mathbf{G}_n\mathbf{D}_{nm}\mathbf{C}_m\mathbf{D}_{nm}^T\mathbf{G}_n^T = O(h)$$

[which follows from (3-116)], we thus find

$$\text{(3-139)} \qquad \mathbf{V}_{n+1} - \mathbf{V}_n = h\{\mathbf{C}_n + \mathbf{G}_n\mathbf{V}_n + \mathbf{V}_n^T\mathbf{G}_n^T\} + O(h^2)$$

The same equation [without the $O(h^2)$ term] would result if the following initial value problem for a matrix function $\mathbf{V}(x)$ were solved approximately by Euler's method:

$$\text{(3-140)} \qquad \begin{aligned}\mathbf{V}(a) &= \mathbf{0} \\ \mathbf{V}'(x) &= \mathbf{C}(x) + \mathbf{G}(x)\mathbf{V}(x) + \mathbf{V}^T(x)\mathbf{G}^T(x)\end{aligned}$$

By Theorem 3.3 the presence of the $O(h^2)$ term in (3-139) does not invalidate the fact that

$$\text{(3-141)} \qquad \mathbf{V}_n = \mathbf{V}(x_n) + O(h)$$

where $\mathbf{V}(x)$ is defined by (3-140). It thus follows that

$$\text{(3-142)} \qquad \operatorname{covar}(\mathbf{r}_n^{(1)}) = \frac{\sigma^2}{h}(\mathbf{V}(x_n) + O(h))$$

We summarize the results of this section in the following statement:

THEOREM 3.7. *Suppose the local round-off errors* $\boldsymbol{\epsilon}_m$ *are random variables satisfying conditions* (*i*), (*ii*), (*iii*) *stated at the beginning of* §3.4-5. *Then the expected value and covariance matrix of the primary component* $\mathbf{r}_n^{(1)}$ *of the accumulated round-off error satisfy* (3-136) *and* (3-142), *where* $\mathbf{m}(x)$ *and* $\mathbf{V}(x)$ *are defined by* (3-127) *and* (3-140).

A statement about the distribution of the accumulated error is made in §3.4-7.

3.4-6. The matrix $\mathbf{V}(x)$**.** The second equation (3-140) represents a system of s^2 simultaneous differential equations for the s^2 elements of the matrix $\mathbf{V}(x)$. The calculation of $\mathbf{V}(x)$ directly from (3-140) may therefore be a formidable task. The results presented in this section may facilitate the problem of calculating $\mathbf{V}(x)$.

LEMMA 3.3. *The matrix* $\mathbf{V}(x)$ *is symmetric.*

Proof. By transposing the system (3-140), we find, since $\mathbf{C}^T(x) = \mathbf{C}(x)$,

$$\mathbf{V}^{T\prime}(x) = \mathbf{C}(x) + \mathbf{V}^T(x)\mathbf{G}^T(x) + \mathbf{G}(x)\mathbf{V}(x)$$

Thus $\mathbf{V}^{T\prime}(x) = \mathbf{V}'(x)$, and since $\mathbf{V}^T(0) = \mathbf{V}(0) = \mathbf{0}$, the assertion follows.

By the above result, the integration of s^2 simultaneous equations is reduced to the integration of $\frac{1}{2}s(s+1)$ equations. The results to be given below reduce the problem to one involving the integration of s systems, each involving s equations.

LEMMA 3.4. *Denote by* $\mathbf{Y}(x)$ *the solution of the initial value problem*

$$\mathbf{Y}(a) = \mathbf{I}, \qquad \mathbf{Y}'(x) = \mathbf{G}(x)\mathbf{Y}(x), \qquad a \le x \le b \tag{3-143}$$

Then $\mathbf{Y}^{-1}(x)$ *exists for* $a \le x \le b$.

Note that the system (3-143) splits into s systems

$$\mathbf{y}'(x) = \mathbf{G}(x)\mathbf{y}(x) \tag{3-144}$$

for the s columns of $\mathbf{Y}(x)$, each involving only s unknown functions. The matrix $\mathbf{Y}(x)$ is called the *complete solution* of the system (3-143).

Proof of Lemma 3.4. Define the matrix $\mathbf{Z}(x)$ as the solution of the initial value problem

$$\mathbf{Z}(a) = \mathbf{I}, \qquad \mathbf{Z}'(x) = -\mathbf{Z}(x)\mathbf{G}(x), \qquad a \le x \le b \tag{3-145}$$

Then

$$\begin{aligned}(\mathbf{Z}(x)\mathbf{Y}(x))' &= \mathbf{Z}'(x)\mathbf{Y}(x) + \mathbf{Z}(x)\mathbf{Y}'(x)\\ &= -\mathbf{Z}(x)\mathbf{G}(x)\mathbf{Y}(x) + \mathbf{Z}(x)\mathbf{G}(x)\mathbf{Y}(x)\\ &= \mathbf{0}\end{aligned}$$

or variable stepsize, provided that the definitions (3-127) and (3-140) of he vector $\mathbf{m}(x)$ and the matrix $\mathbf{V}(x)$ are replaced by

(3-155) $$\mathbf{m}(a) = \mathbf{0}$$
$$\mathbf{m}'(x) = \hat{\mathbf{G}}(x)\mathbf{m}(x) + [\vartheta(x)]^{-1}\mathbf{p}(x)$$

and

(3-156) $$\mathbf{V}(a) = \mathbf{0}$$
$$\mathbf{V}'(x) = [\vartheta(x)]^{-1}\mathbf{C}(x) + \mathbf{G}(x)\mathbf{V}(x) + \mathbf{V}(x)\mathbf{G}^T(x)$$

respectively.

3.4-8. The local round-off error. The local round-off error was defined by (3-104). Its nature depends on how the arithmetical operations are performed.

(i) Single precision, fixed point. It has already been stated that

$$\boldsymbol{\epsilon}_{n+1} = [h\tilde{\boldsymbol{\Phi}}(x_n, \tilde{\mathbf{y}}_n; h)]^* - h\boldsymbol{\Phi}(x_n, \tilde{\mathbf{y}}_n; h)$$

where $\tilde{\boldsymbol{\Phi}}$ denotes a numerical approximation to $\boldsymbol{\Phi}$, and where, for any vector $\mathbf{z}$, $\mathbf{z}^*$ denotes the vector whose components are rounded values of the corresponding components of $\mathbf{z}$. As on previous occasions,

$$\boldsymbol{\epsilon}_{n+1} = \boldsymbol{\pi}_{n+1} + \boldsymbol{\rho}_{n+1}$$

where

$$\boldsymbol{\pi}_{n+1} = [h\tilde{\boldsymbol{\Phi}}(x_n, \tilde{\mathbf{y}}_n; h)]^* - h\tilde{\boldsymbol{\Phi}}(x_n, \tilde{\mathbf{y}}_n; h)$$

is the *induced* error (induced by rounding the product $h\tilde{\boldsymbol{\Phi}}$), and

$$\boldsymbol{\rho}_{n+1} = h[\tilde{\boldsymbol{\Phi}}(x_n, \tilde{\mathbf{y}}_n; h) - \boldsymbol{\Phi}(x_n, \tilde{\mathbf{y}}_n; h)]$$

is the *inherent* error (inherent in the evaluation of $\boldsymbol{\Phi}$). If the basic unit is called u, then each component of $\boldsymbol{\pi}_n$ is bounded by $\frac{1}{2}u$, whereas the components of $\boldsymbol{\rho}_{n+1}$ are of the order of hu. Thus for h sufficiently small most of the local round-off error is due to the induced error. If the components of $\boldsymbol{\pi}_{n+1}$ are considered random variables with the rectangular distributions $F_u(x)$ [see (1-109)] and if $\tilde{\boldsymbol{\Phi}} - \boldsymbol{\Phi}$ is considered a random vector such that

$$E(\tilde{\boldsymbol{\Phi}} - \boldsymbol{\Phi}) = \mathbf{m}u, \qquad \operatorname{covar}(\tilde{\boldsymbol{\Phi}} - \boldsymbol{\Phi}) = \mathbf{D}u^2$$

when $\mathbf{D}$ is a diagonal matrix with nonnegative elements, and if the variables $\boldsymbol{\pi}_{n+1}$ and $\tilde{\boldsymbol{\Phi}} - \boldsymbol{\Phi}$ are independent, then the variable $\boldsymbol{\epsilon}_{n+1}$ satisfies

(3-157) $$E(\boldsymbol{\epsilon}_{n+1}) = \mathbf{m}uh, \qquad \operatorname{covar}(\boldsymbol{\epsilon}_{n+1}) = (\tfrac{1}{12}\mathbf{I} + h^2\mathbf{D})u^2$$

(ii) Double precision, fixed point. From a theoretical point of view, working in double precision with a fixed decimal or binary point is equivalent to working with single precision using the basic unit u^2.

and hence,

$$\mathbf{Z}(x)\mathbf{Y}(x) = \mathbf{Z}(a)\mathbf{Y}(a) = \mathbf{I}, \qquad a \le x \le b$$

It follows that $\mathbf{Z}(x)$ is a left inverse of $\mathbf{Y}(x)$, and thus *the* inverse of $\mathbf{Y}(x)$.

LEMMA 3.5.* *Let* $\mathbf{Y}(x)$ *be defined by* (3-143), *and suppose that the matrix* $\mathbf{N}(x)$ *satisfies*

(3-146) $$\mathbf{N}(a) = \mathbf{0}, \qquad \mathbf{N}'(x) = \mathbf{N}(x)\mathbf{G}^T(x) + \mathbf{Y}^{-1}(x)\mathbf{C}(x), \qquad a \le x \le b$$

Then

(3-147) $$\mathbf{V}(x) = \mathbf{Y}(x)\mathbf{N}(x), \qquad a \le x \le b$$

Note that the integration of (3-146) can again be accomplished by integrating s systems with s unknown functions each for the s columns of $\mathbf{N}(x)$.

Proof of Lemma 3.5. Let $\mathbf{W}(x) = \mathbf{Y}(x)\mathbf{N}(x)$. It is obvious that $\mathbf{W}(a) = \mathbf{V}(a)$. For the derivative we find

$$\begin{aligned}\mathbf{W}'(x) &= \mathbf{Y}'(x)\mathbf{N}(x) + \mathbf{Y}(x)\mathbf{N}'(x)\\ &= \mathbf{G}(x)\mathbf{Y}(x)\mathbf{N}(x) + \mathbf{Y}(x)[\mathbf{N}(x)\mathbf{G}^T(x) + \mathbf{Y}^{-1}(x)\mathbf{C}(x)]\\ &= \mathbf{G}(x)\mathbf{W}(x) + \mathbf{W}(x)\mathbf{G}^T(x) + \mathbf{C}(x)\end{aligned}$$

By Lemma 3.3, the same differential equation is satisfied by $\mathbf{V}(x)$. It follows that $\mathbf{W}(x) = \mathbf{V}(x)$.

We next solve the system (3-146) in terms of the complete solution.

LEMMA 3.6. *The solution of* (3.146) *is given by*

(3-148) $$\mathbf{N}(x) = \left[\int_a^x \mathbf{Y}^{-1}(t)\mathbf{C}(t)\mathbf{Y}^{-1T}(t)\,dt\right]\mathbf{Y}^T(x)$$

Proof. It is evident that $\mathbf{N}(a) = \mathbf{0}$. Differentiating by using the product rule for matrices, we find

$$\begin{aligned}\mathbf{N}'(x) &= \mathbf{Y}^{-1}(x)\mathbf{C}(x)\mathbf{Y}^{-1T}(x)\mathbf{Y}^T(x) + \left[\int_a^x \mathbf{Y}^{-1}(t)\mathbf{C}(t)\mathbf{Y}^{-1T}(t)\,dt\right]\mathbf{Y}^{T\prime}(x)\\ &= \mathbf{Y}^{-1}(x)\mathbf{C}(x) + \mathbf{N}(x)\mathbf{G}^T(x)\end{aligned}$$

as required.

Putting together the statements of the Lemmas 3.5 and 3.6, we obtain the following result:

THEOREM 3.8. *If* $\mathbf{Y}(x)$ *denotes the complete solution of the system* (3-143), *then the matrix* $\mathbf{V}(x)$ *defined by* (3-140) *can be represented in the form*

(3-149) $$\mathbf{V}(x) = \mathbf{Y}(x)\left[\int_a^x \mathbf{Y}^{-1}(t)\mathbf{C}(t)\mathbf{Y}^{-1T}(t)\,dt\right]\mathbf{Y}^T(x)$$

* The author is indebted to Dr. G. Culler for this result.

Example. We shall determine $\mathbf{V}(x)$ in the case that

$$\mathbf{G}(x) = \begin{pmatrix} 0 & \omega \\ -\omega & 0 \end{pmatrix} \tag{3-150}$$

where ω is a constant, assuming that $a = 0$ and $\mathbf{C}(x) = \mathbf{I}$. It is easily verified that

$$\mathbf{Y}(x) = \begin{pmatrix} \cos \omega x & \sin \omega x \\ -\sin \omega x & \cos \omega x \end{pmatrix}$$

Since this matrix $\mathbf{Y}(x)$ is orthogonal,

$$\mathbf{Y}^{-1}(x) = \mathbf{Y}^T(x) \tag{3-151}$$

and

$$\int_0^x \mathbf{Y}^{-1}(t)\mathbf{C}(t)\mathbf{Y}^{-1^T}(t)\, dt = x\mathbf{I}$$

Using (3-151) once again, we find

$$\mathbf{V}(x) = x\mathbf{I} \tag{3-152}$$

3.4-7. The distribution of $\mathbf{r}_n^{(1)}$. We shall apply some of the results of the preceding section to the study of the distribution of the components of $\mathbf{r}_n^{(1)}$ as $n \to \infty$. It is evident from (3-122) that every component of $\mathbf{r}_n^{(1)}$ depends, in general, on all components of all vectors $\boldsymbol{\epsilon}_m$ with $m \le n$. In order to apply Theorem 1.8, it is therefore necessary to assume that the components of each vector $\boldsymbol{\epsilon}_m$ are independent among themselves or that the matrices $\mathbf{C}(x_n)$ are diagonal. If we put

$$\mathbf{V}_n = (v_n^{ij}), \qquad \mathbf{D}_{nm} = (d_{nm}^{ij})$$

condition (1-96) is found to be equivalent to the condition that for $x_n = x$, $a < x \le b$,

$$\lim h^{1/2} \frac{d_{nm}^{ij}}{(v_n^{ii})^{1/2}} = 0, \quad i, j = 1, \cdots, s \tag{3-153}$$

uniformly for all $m < n$.

Relation (3-116) shows that the elements of the matrices $\mathbf{D}_{n,m}$ are uniformly bounded for $a \le x \le b$. The proof of (3-153) thus hinges upon the fact that for all sufficiently large values of n,

$$v_n^{ii} \ge c > 0, \qquad i = 1, \cdots, s \tag{3-154}$$

where c is independent of n. We shall establish (3-154) by proving

LEMMA 3.7. *If the symmetric matrix $\mathbf{C}(x)$ is piecewise continuous and positive definite for $a \le x \le b$, then $\mathbf{V}(x)$ is positive definite for $a < x \le b$.*

The proof* is based on repeated applications of the fol terization of positive definite matrices:

LEMMA 3.8. *A symmetric matrix $\mathbf{A}$ is positive definite if an be represented in the form $\mathbf{A} = \mathbf{U}\mathbf{U}^T$, where $\mathbf{U}$ is nonsingula*

This is Theorem 20 of Chapter 9 of Birkhoff and Ma to which we refer for a proof.

It follows from Lemma 3.8 that for $a \le t \le b$ we can fin matrices $\mathbf{B} = \mathbf{B}(t)$ such that $\mathbf{C}(t) = \mathbf{B}\mathbf{B}^T$. The matrix $\mathbf{Y}^{-1}$ the product of two nonsingular matrices, is nonsingular. Hen

$$\mathbf{L}(t) = \mathbf{Y}^{-1}(t)\mathbf{C}(t)\mathbf{Y}^{-1^T}(t) = \mathbf{Y}^{-1}(t)\mathbf{B}(t)[\mathbf{Y}^{-1}(t)\mathbf{B}(t)]^T$$

is positive definite by Lemma 3.8. Thus, if $\mathbf{k}$ is any nonzero

$$\mathbf{k}^T\mathbf{L}(t)\mathbf{k} > 0, \qquad a \le t \le b$$

and thus, since $\mathbf{L}(t)$ is piecewise continuous,

$$\int_a^x \mathbf{k}^T\mathbf{L}(t)\mathbf{k}\, dt = \mathbf{k}^T\left[\int_a^x \mathbf{L}(t)\, dt\right]\mathbf{k} > 0, \qquad a \le x \le$$

It follows that the matrix

$$\mathbf{M}(x) = \int_a^x \mathbf{L}(t)\, dt$$

is positive definite for $a < x \le b$ and can thus be factored i $\mathbf{M}(x) = \mathbf{U}(x)\mathbf{U}^T(x)$, where $\mathbf{U}(x)$ is nonsingular. By a final app Lemma 3.8,

$$\mathbf{V}(x) = \mathbf{Y}(x)\mathbf{M}(x)\mathbf{Y}^T(x) = \mathbf{Y}(x)\mathbf{U}(x)[\mathbf{Y}(x)\mathbf{U}(x)]^T$$

is positive definite, and Lemma 3.7 is demonstrated.

Relation (3-154) now follows from the fact that the diagonal el a positive definite matrix are positive. Thus (3-153) is establishe may state the following result:

THEOREM 3.9. *Suppose the local round-off errors $\boldsymbol{\epsilon}_m$ satisfy c (i), (ii), and (iii) stated at the beginning of §3.4-5, where $\mathbf{C}(x)$ is wise continuous positive definite diagonal matrix. Then the distri the components of the primary round-off error $\mathbf{r}_n^{(1)}$ approaches th distribution as $n \to \infty$ and $x_n = x$, $a < x \le b$.*

Variable stepsize. The above developments are easily adapte case where the stepsize is varied according to (2-51). The only d arises from our insisting that the abscissas x_n are exact machine n This limits the step modifier $\vartheta(x)$ to the class of piecewise constant functions. By an analysis entirely analogous to that given in §2.2-8 be shown that the statements of Theorems 3.6, 3.7, and 3.9 remai

* Suggested by Dr. G. Culler.

(iii) Partial double precision, fixed point. As before, it is assumed that h and x_n are exact single precision numbers. Denoting by $\mathbf{z}^*$ a correctly rounded single precision approximation to a vector $\mathbf{z}$, the algorithm of partial double precision is described by the formulas

$$\text{(3-158)} \qquad \begin{aligned} \tilde{\mathbf{y}}_0 &= \mathbf{y}_0 \\ \tilde{\mathbf{y}}_{n+1} &= \tilde{\mathbf{y}}_n + h\tilde{\mathbf{\Phi}}(x_n, \tilde{\mathbf{y}}_n^*; h), \qquad n = 0, 1, 2, \cdots \end{aligned}$$

It is understood that $\tilde{\mathbf{\Phi}}$ is a digital single precision approximation to (but not necessarily a correctly rounded value of) $\mathbf{\Phi}$. The product $h\tilde{\mathbf{\Phi}}$ is not rounded. Since both h and $\tilde{\mathbf{\Phi}}$ are represented by exact single precision numbers, the components of $h\tilde{\mathbf{\Phi}}$, and thus those of $\tilde{\mathbf{y}}_n$, are exact double precision numbers. The local round-off error in this case is given by

$$\text{(3-159)} \qquad \boldsymbol{\epsilon}_{n+1} = h\{\tilde{\mathbf{\Phi}}(x_n, \tilde{\mathbf{y}}_n^*; h) - \mathbf{\Phi}(x_n, \tilde{\mathbf{y}}_n; h)\}$$

We have

$$\begin{aligned} \tilde{\mathbf{\Phi}}(x_n, \tilde{\mathbf{y}}_n^*; h) - \mathbf{\Phi}(x_n, \tilde{\mathbf{y}}_n; h) = \tilde{\mathbf{\Phi}}(x_n, \tilde{\mathbf{y}}_n^*; h) &- \mathbf{\Phi}(x_n, \tilde{\mathbf{y}}_n^*; h) \\ + \mathbf{\Phi}(x_n, \tilde{\mathbf{y}}_n^*; h) &- \mathbf{\Phi}(x_n, \tilde{\mathbf{y}}_n; h) \end{aligned}$$

Denoting by u the basic unit, we shall assume that

$$\|\tilde{\mathbf{\Phi}} - \mathbf{\Phi}\| \le ku$$

where k is a fixed constant. Using the Lipschitz condition (3-56) together with the fact that

$$\|\tilde{\mathbf{y}}_n^* - \tilde{\mathbf{y}}_n\| \le \tfrac{1}{2}su$$

we obtain

$$\|\boldsymbol{\epsilon}_{n+1}\| \le (k + \tfrac{1}{2}sL)hu$$

It follows that (3-118) is valid (and thus the splitting of $\mathbf{r}_n$ into a primary and a secondary component is possible) if $u = O(h)$. [For single precision operations $u = O(h^2)$ is necessary.] If this condition is met, the behavior of the accumulated error is essentially that of the primary error defined by (3-122*a*). This opens the way to the application of the statistical theory.

We shall assume that the components of the vector

$$\text{(3-160)} \qquad \boldsymbol{\delta}_n = \tilde{\mathbf{y}}_n^* - \tilde{\mathbf{y}}_n$$

are independent random variables with the distribution function $F_u(x)$, and that the components of the vector

$$\boldsymbol{\eta}_n = \tilde{\mathbf{\Phi}}(x_n, \tilde{\mathbf{y}}_n^*; h) - \boldsymbol{\phi}(x_n, \tilde{\mathbf{y}}_n^*; h)$$

are random variables satisfying

$$\hat{E}(\boldsymbol{\eta}_n) \le \mathbf{k}u, \qquad \text{covar}\,(\boldsymbol{\eta}_n) = \mathbf{W}u^2$$

The vector $\mathbf{k}$ and the matrix $\mathbf{W}$ may depend on x. In the ideal case where the only error in $\tilde{\mathbf{\Phi}}$ is due to rounding, $\mathbf{W} = \frac{1}{12}\mathbf{I}$. In order to find a bound for $\hat{E}(\boldsymbol{\epsilon}_{n+1})$ and an expression for covar $(\boldsymbol{\epsilon}_{n+1})$, we note that by the reasoning which led from (3-107) to (3-121) we have

$$\mathbf{\Phi}(x_n, \tilde{\mathbf{y}}_n^*; h) - \mathbf{\Phi}(x_n, \tilde{\mathbf{y}}_n; h) = \mathbf{G}(x_n)\boldsymbol{\delta}_n + O(h^2)$$

where $\mathbf{G}(x)$ denotes the functional matrix defined by (3-120). We thus have

$$\boldsymbol{\epsilon}_{n+1} = h\{\boldsymbol{\eta}_n + \mathbf{G}(x_n)\boldsymbol{\delta}_n + O(h^2)\}$$

and in view of $E(\boldsymbol{\delta}_n) = 0$ we find

$$\hat{E}(\boldsymbol{\epsilon}_{n+1}) \leq hu\mathbf{k} + O(h^3) \tag{3-161}$$

Furthermore, since

$$\begin{aligned}\text{covar}\,(\mathbf{G}(x_n)\boldsymbol{\delta}_n) &= E(\mathbf{G}(x_n)\boldsymbol{\delta}_n\boldsymbol{\delta}_n^T\mathbf{G}^T(x_n))\\ &= \mathbf{G}(x_n)E(\boldsymbol{\delta}_n\boldsymbol{\delta}_n^T)\mathbf{G}^T(x_n)\\ &= \tfrac{1}{12}\mathbf{G}(x_n)\mathbf{G}^T(x_n)u^2\end{aligned}$$

and because the variables $\boldsymbol{\eta}_n$ and $\boldsymbol{\delta}_n$ are independent, we have

$$\text{covar}\,(\boldsymbol{\epsilon}_n) = h^2u^2\{\mathbf{W} + \tfrac{1}{12}\mathbf{G}(x_n)\mathbf{G}^T(x_n) + O(h)\} \tag{3-162}$$

The equations (3-161) and (3-162) define the quantities $\mathbf{p}(x)$ and $\mathbf{C}(x)$, which are necessary to perform the statistical analysis discussed in §3.4-5.

(iv) Floating point. Let x be any nonzero real number, and let b be a positive integer ≥ 2. Denoting, for any real z, by $[z]$ the greatest integer not exceeding z, and by $\log_b$ the logarithm to the base b, we set

$$m = [\log_b |x|] + 1, \qquad a = b^{-m}x \tag{3-163}$$

We then can represent x in the form

$$x = ab^m \tag{3-164}$$

called the *floating representation of x to the base b*. The integer m is called the *exponent* and the real number a the *mantissa* of the *floating number x*. The absolute value of the mantissa by definition belongs to the interval $[b^{-1}, 1)$.

If a floating number is represented in a computing machine, the integer m can (within wide limits) be represented exactly. The real number a, on the other hand, can in general be represented only approximately; it is usually approximated by a fixed precision number, whose basic unit we shall again denote by u. The fixed precision approximation of a will be denoted by a^*. If a^* is selected such that $|a - a^*| \leq \frac{1}{2}u$, we shall say that the floating number is *rounded symmetrically*. (Unfortunately, some computing systems do not round symmetrically.) A floating number is called an *exact machine number* if $a = a^*$.

Operations with floating numbers are almost universally used today on big computing machines, because they make it unnecessary to predict the size of the numbers occurring in a computation and to "scale" a problem to the range of fixed point arithmetic. However, in floating operations not only products but also sums are rounded.

We shall call two floating numbers *nearly equal* if their exponents are identical. The sum of two floating numbers x and y will be called *regular* if the exponent of the floating representation of $x + y$ equals the larger of the exponents of x and y. A simple probability argument shows that the sum of two not nearly equal numbers is regular in the majority of cases. Let us now discuss the quantity $r = (x + y)^* - (x + y)$, assuming that x and y are two not nearly equal machine numbers, and that the sum $x + y$ is regular. If $|x| > |y|$, $x = ab^m$, and if rounding is symmetric, we clearly have

$$|r| \leq \tfrac{1}{2}ub^m = \tfrac{1}{2}ub^{[\log_b|x|]+1} \tag{3-165}$$

The above bound, although sharp, is difficult to work with analytically because of the appearance of the greatest integer function. Since $m = 1 + [\log_b |x|] = \log_b |x| - \log_b |a|$, we have

$$b^{[\log_b|x|]+1} = |x|\ b^{-\log_b|a|}$$

and thus may write

$$|r| \leq u\,|x|\,\theta \tag{3-166}$$

where $\theta = \frac{1}{2}b^{-\log_b|a|}$ and thus

$$\tfrac{1}{2} \leq \theta < \tfrac{1}{2}b \tag{3-167}$$

In practice, θ can be replaced by $\frac{1}{2}b$ or, in a long computation, by a suitable mean value. In the practically important case $b = 2$, θ is contained between the limits $\frac{1}{2}$ and 1.

Let us now apply the above to the particular problem on hand. In floating operations, successive vectors $\tilde{\mathbf{y}}_n$ are generated by the formulas

$$\begin{aligned} \tilde{\mathbf{y}}_0 &= \boldsymbol{\eta} \\ \tilde{\mathbf{y}}_{n+1} &= \{\tilde{\mathbf{y}}_n + [h\tilde{\mathbf{\Phi}}(x_n, \tilde{\mathbf{y}}_n;\ h)]^*\}^* \end{aligned} \tag{3-168}$$

The local round-off error is thus given by

$$\boldsymbol{\epsilon}_{n+1} = \{\tilde{\mathbf{y}}_n + [h\tilde{\mathbf{\Phi}}(x_n, \tilde{\mathbf{y}}_n;\ h)]^*\}^* - \{\tilde{\mathbf{y}}_n + h\mathbf{\Phi}(x_n, \tilde{\mathbf{y}}_n;\ h)\}$$

and may be decomposed into three components as follows:

$$\boldsymbol{\epsilon}_{n+1} = \boldsymbol{\alpha}_{n+1} + \boldsymbol{\pi}_{n+1} + \boldsymbol{\rho}_{n+1}$$

Here $\boldsymbol{\pi}_{n+1}$ and $\boldsymbol{\rho}_{n+1}$ are defined as under (i), while the quantity $\boldsymbol{\alpha}_{n+1}$, called the *adduced error*, is defined by

$$\text{(3-169)} \quad \boldsymbol{\alpha}_{n+1} = \{\tilde{\mathbf{y}}_{n+1} + [h\tilde{\boldsymbol{\Phi}}(x_n, \tilde{\mathbf{y}}_n; h)]^*\}^* - \{\tilde{\mathbf{y}}_n + [h\tilde{\boldsymbol{\Phi}}(x_n, \tilde{\mathbf{y}}_n; h)]^*\}$$

Let us discuss a typical component α_{n+1}^i of $\boldsymbol{\alpha}_{n+1}$. We make the assumptions that the floating numbers y_n^i and $h\Phi^i$ are not nearly equal, that their sum is regular, and that $|y_n^i| > |h\Phi^i|$. (These assumptions are particularly well justified here, because h is small.) The adduced error is then of the order of uy_n^i, whereas both the induced and the inherent error are of the order of $uh\Phi^i$. The significant contribution to the local round-off error thus arises from the adduced error. Neglecting inherent and induced error entirely, we may thus assume that

$$\hat{\boldsymbol{\epsilon}}_{n+1} \leq \boldsymbol{\Theta} u \hat{\mathbf{y}}_n$$

where $\boldsymbol{\Theta}$ is a diagonal matrix whose diagonal elements satisfy (3-167).

For statistical work the most natural assumption consists in regarding the components ε_{n+1}^i of $\boldsymbol{\epsilon}_{n+1}$ as independent random variables with the distribution function $F_w(x)$, where

$$w = 2u\theta \, |y_n^i|$$

It then follows that $E(\varepsilon_{n+1}^i) = 0$ and

$$\text{var}\,(\varepsilon_{n+1}^i) = \tfrac{1}{3}u^2\theta^2(y_n^i)^2$$

3.4-9. Numerical illustration. The numerical experiment described in §2.3-7 may be generalized to systems in the following manner: One finds a family of numerical solutions $\tilde{\mathbf{y}}_{n,q}$ ($q = 1, \cdots, Q$) of the linear system

$$\mathbf{y}' = \mathbf{G}(x)\mathbf{y}$$

corresponding to the Q initial conditions

$$\mathbf{y}_{0,q} = \mathbf{y}_{0,0}(1 + q\Delta) \qquad (q = 1, \cdots, Q)$$

Ascertaining the values of the theoretical approximation $\mathbf{y}_{n,q}$ by calculating with higher precision,* one thus obtains, for any desired value of n, Q samples of the accumulated round-off error,

$$\mathbf{r}_{n,q} = \tilde{\mathbf{y}}_{n,q} - \mathbf{y}_{n,q}$$

* Since $\mathbf{y}_{n,q} = (1 + q\Delta)\mathbf{y}_{n,0}$, the theoretical approximation has to be determined for one value of q only.

From there one may determine experimental values of the mean and of the covariance matrix of the values $\mathbf{r}_{n,q}$ by means of the relations

$$\mathbf{a}_n = E(\mathbf{r}_n)_e = \frac{1}{Q}\sum_{q=1}^{Q}\mathbf{r}_{n,q}$$

$$\text{covar }(\mathbf{r}_n)_e = \frac{1}{Q}\sum_{q=1}^{Q}(\mathbf{r}_{n,q} - \mathbf{a}_n)(\mathbf{r}_{n,q}^T - \mathbf{a}_n^T)$$

$$= \frac{1}{Q}\sum_{q=1}^{Q}\mathbf{r}_{n,q}\mathbf{r}_{n,q}^T - \mathbf{a}_n\mathbf{a}_n^T$$

We shall report below the results of this experiment as carried out for two special systems, each involving two unknown functions. In order to simplify the notation, we shall write

$$y^1 = y, \quad y^2 = z$$

and also

$$E(\mathbf{r}_n) = \begin{pmatrix} p \\ q \end{pmatrix}$$

$$\text{covar }(\mathbf{r}_n) = \begin{pmatrix} a & b \\ b & c \end{pmatrix}, \qquad \mathbf{V}(x) = \begin{pmatrix} u & v \\ v & w \end{pmatrix}$$

We content ourselves with a concise description of the conditions of the experiment and of the numerical results.

Experiment I

Differential equation: $y' = -\pi z$, $z' = \pi y$.

Initial condition: $x_0 = 0$, $y_{0,0} = 10^{-1} \times 2^{-12}$, $z_{0,0} = 0$.

Method: Modified Euler [(3-47) with $\alpha = 1$], single precision, fixed binary point.

$$h = 2^{-6}$$
$$\Delta = \tfrac{1}{3} \cdot 2^{-8}$$
$$Q = 100$$
$$u = 2^{-36}$$

Assumption on local round-off error: Uniformly distributed in $(-\frac{1}{2}u, \frac{1}{2}u)$, hence $\sigma^2 = \frac{1}{12}u^2$, $\mathbf{C} = \mathbf{I}$.

Theoretical expected value of $\mathbf{r}_n$: $p = q = 0$.

Theoretical value of covariance matrix: For the case under consideration, the matrix $\mathbf{V}(x)$ was calculated in §3.4-6 with the result

$$u = w = x, \qquad v = 0$$

We thus expect

$$a = c = \tfrac{1}{12} \cdot 2^6 \cdot u^2 x, \quad b = 0 \tag{3-170}$$

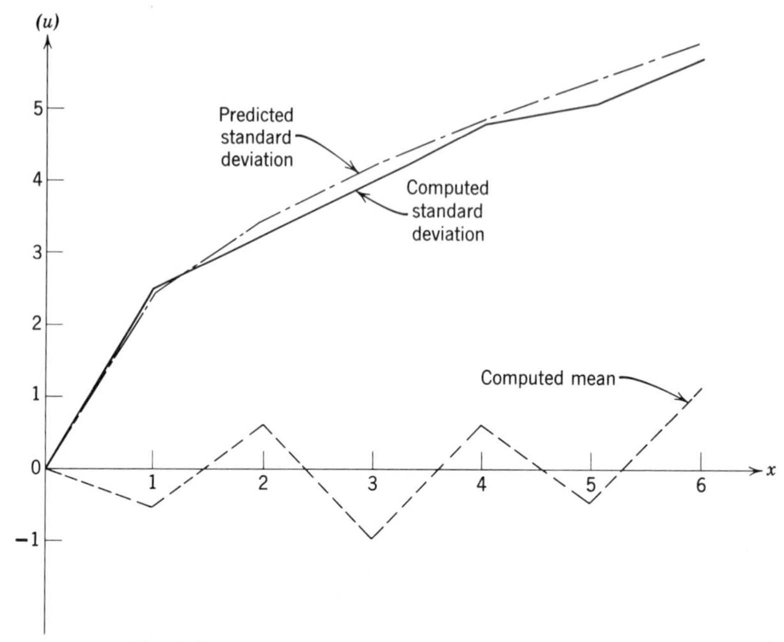

Fig. 3.1. Rounding error in y_n values of Experiment 1.

Experimental and predicted values of p, q, a, b, and c are shown in Table 3.2. The values of p and of the standard deviation $a^{1/2}$ are also shown graphically on Fig. 3.1. The good agreement between predicted and experimental values is evident.

Table 3.2

	x	1	2	3	4	5	6
Experimental values	$u^{-1}p$	−0.5	0.6	−0.9	0.6	−0.4	1.1
	$u^{-1}q$	−0.7	0.7	−1.8	2.2	−2.9	3.8
	$u^{-2}a$	5.7	9.7	14.4	20.8	23.2	29.8
	$u^{-2}b$	0.1	−1.3	−2.3	−2.1	−1.5	−4.2
	$u^{-2}c$	5.8	12.7	17.0	21.6	26.9	33.9
Predicted values	$p = q = b$	0	0	0	0	0	0
	$u^{-2}a = u^{-2}c$	5.3	10.7	16.0	21.3	26.7	32.0

Experiment 2

Differential equation: $y' = \frac{1}{2}z$, $z' = \frac{1}{2}y$.

Initial condition: $x_0 = 0$, $y_{0,0} = 2^{-17}$, $z_{0,0} = 0$.

Method: Runge-Kutta, single precision, fixed binary point. h, Δ, Q, and u as in Experiment 1.

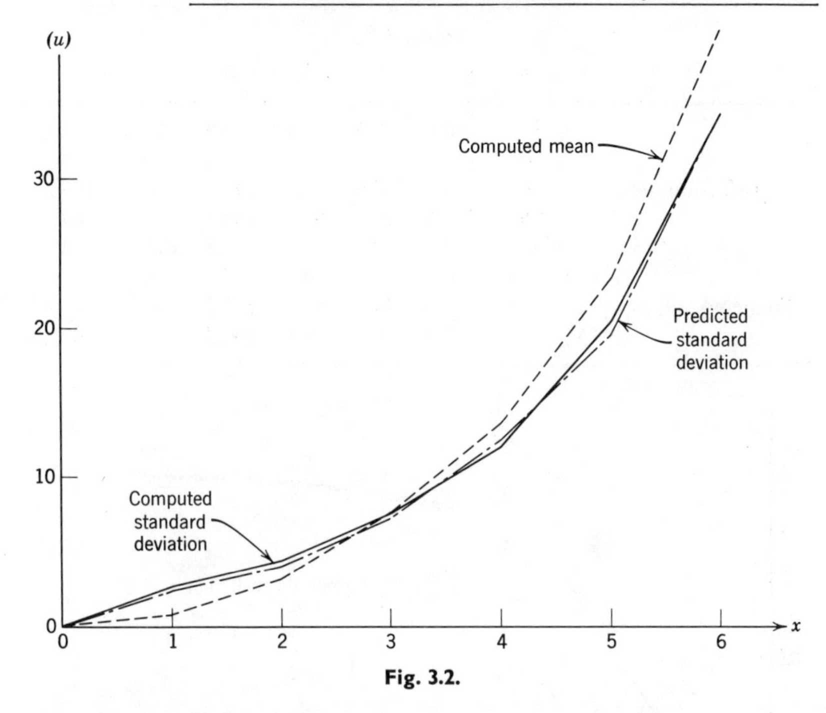

Fig. 3.2.

Assumption on local round-off error: as in Experiment 1.

Theoretical expected value of $\mathbf{r}_n$: $p = q = 0$.

Theoretical value of covariance matrix: Equation (3-140) reduces to the following system for u, v, and w:

$$u' = 1 + v, \qquad v' = u, \qquad w' = 1 + v$$

The solution [under the initial condition $\mathbf{V}(0) = \mathbf{0}$] is (see Problem 13)

$$u = w = \sinh x, \quad v = \cosh x - 1 \tag{3-171}$$

Experimental and predicted values of p, q, a, b, and c are shown in Table 3.3; values of p and $a^{1/2}$ also in Fig. 3.2.

It is obvious that the prediction of zero mean is not justified here. It can be verified that the values of the mean error closely agree with the values

$$p = q = \frac{\mu}{h} 2(e^{x/2} - 1)$$

predicted by (3-136), if we assume that $\mu = uh$. One would like to justify this value by assuming the distribution $F_{u,v}(x)$ in place of the symmetric distribution $F_u(x)$ for the local round-off error, but the value of μ predicted by §1.6-1 is only $\mu = \frac{1}{2}uh$. The reasons for this discrepancy are not clear.

Table 3.3

	x	1	2	3	4	5	6
Experimental values	$u^{-1}p$	0.9	3.2	7.3	13.3	23.0	39.8
	$u^{-1}q$	1.6	3.4	6.8	13.5	22.7	39.0
	$u^{-2}a$	6.8	20.0	57.9	143.1	400.6	1075.1
	$u^{-2}b$	2.3	15.2	51.0	140.8	399.6	1075.2
	$u^{-2}c$	7.0	19.2	53.4	149.7	410.8	1087.3
Predicted values	$p = q$	0	0	0	0	0	0
	$u^{-2}a = u^{-2}c$	6.3	19.3	53.4	145.6	395.8	1076.3
	$u^{-2}b$	2.9	14.7	48.4	140.3	390.4	1071.0

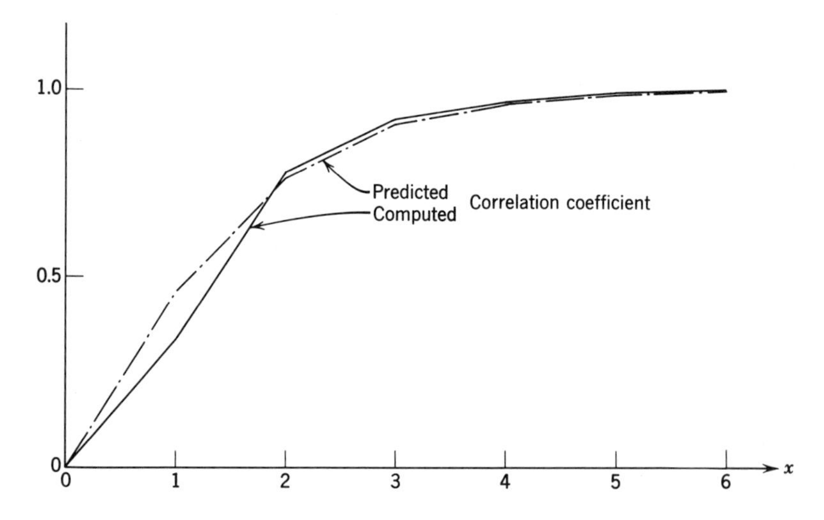

Fig. 3.3. Correlation between the components of $\mathbf{r}_n$ in Experiment 2.

From (3-171) we expect the two components of $\mathbf{r}_n$ to be very strongly correlated. As shown by Fig. 3.3, where predicted and experimental values of the correlation coefficient $b/(ac)^{1/2}$ are given, this expectation is confirmed accurately by the experiment.

3.5. Problems for Solution

Section 3.2

1. Verify that for $s = 1$ the expressions for $\mathbf{f}'(\mathbf{y})$ and $\mathbf{f}''(\mathbf{y})$ given by (3-42) agree with the corresponding expressions given in Chapter 2 (see in particular §2.2-4).

2. Assuming $s = 3$ and setting $y^1 = x$, $y^2 = y$, $y^3 = z$; $f^1 = 1$, $f^2 = f$, $f^3 = g$, write out in full the quantities A^i, B^i, C^i, and D^i defined in (3-42).

3. Verify that for $s = 2$, setting $y^1 = x$, $y^2 = y$, and $f^1 = 1$, $f^2 = f$:

$$E^2 = f_{xxx} + 3f_{xxy}f + 3f_{xyy}f^2 + f_{yyy}f^3$$
$$F^2 = (f_x + f_y f)(f_{yx} + f_{yy}f)$$
$$G^2 = f_y(f_{xx} + 2f_{xy}f + f_{yy}f^2)$$
$$H^2 = f_y^2(f_x + f_y f)$$

4. Determine the most general Runge-Kutta method of order 3 which is expressible in the form

$$\mathbf{\Phi}(\mathbf{y};\ h) = a_1\mathbf{k}_1 + a_2\mathbf{k}_2 + a_3\mathbf{k}_3$$

where $\mathbf{k}_1$, $\mathbf{k}_2$, and $\mathbf{k}_3$ are defined as in (3-48). Obtain the method described in §3.3-7 as a special case.

Section 3.3

5. Using tensor notation, determine the order and principal error function for the method where $\mathbf{\Phi}$ is defined implicitly by the relation

$$\mathbf{y}_{n+1} = \mathbf{y}_n + \tfrac{1}{2}h[\mathbf{f}(\mathbf{y}_n) + \mathbf{f}(\mathbf{y}_{n+1})]$$

6*. Repeat Problem 5 for the method defined by

$$\mathbf{y}_{n+1} = \mathbf{y}_n + \tfrac{1}{2}h[\mathbf{f}(\mathbf{y}_n) + \mathbf{f}(\mathbf{y}_{n+1})] + \tfrac{1}{12}h^2[\mathbf{f}'(\mathbf{y}_n) - \mathbf{f}'(\mathbf{y}_{n+1})]$$

7*. *A special method based on Gaussian quadrature.* The integration method based on the Gaussian fourth order formula may be written

$$\mathbf{y}_{n+1} - \mathbf{y}_n = \tfrac{1}{2}h[\mathbf{f}(\mathbf{y}_{n+p}) + \mathbf{f}(\mathbf{y}_{n+q})] \tag{3-172}$$

Here $\mathbf{y}_{n+p}$ and $\mathbf{y}_{n+q}$ are predicted values of the solution at the points $x_n + ph$ and $x_n + qh$, respectively, where

$$p = \frac{3 - \sqrt{3}}{6}, \qquad q = \frac{3 + \sqrt{3}}{6}$$

Hammer and Hollingsworth [1955] suggest that $\mathbf{y}_{n+p}$ and $\mathbf{y}_{n+q}$ be determined as solutions of the system

$$\begin{aligned} \mathbf{y}_{n+p} &= \mathbf{y}_n + \frac{ph}{2(q-p)}[(2q - p)\mathbf{f}(\mathbf{y}_{n+p}) - p\mathbf{f}(\mathbf{y}_{n+q})] \\ \mathbf{y}_{n+q} &= \mathbf{y}_n + \frac{qh}{2(q-p)}[q\mathbf{f}(\mathbf{y}_{n+p}) - (2p - q)\mathbf{f}(\mathbf{y}_{n+q})] \end{aligned} \tag{3-173}$$

(a) Setting

$$\mathbf{y}_{n+p} = \mathbf{y}_n + ph\bar{\mathbf{\Phi}}_{(1)}(\mathbf{y}_n;\ ph)$$
$$\mathbf{y}_{n+q} = \mathbf{y}_n + qh\bar{\mathbf{\Phi}}_{(2)}(\mathbf{y}_n;\ qh)$$

show that

$$\begin{aligned} \bar{\mathbf{\Phi}}_{(1)}(\mathbf{y};\ h) &= \mathbf{A} + \tfrac{1}{2}h\mathbf{B} + \tfrac{1}{6}h^2\gamma_p(\mathbf{C} + \mathbf{D}) \\ &\quad + \tfrac{1}{24}h^3[\varepsilon_p(\mathbf{E} + 3\mathbf{F}) + \kappa_p(\mathbf{G} + \mathbf{H})] + O(h^4) \\ \bar{\mathbf{\Phi}}_{(2)}(\mathbf{y};\ h) &= \mathbf{A} + \tfrac{1}{2}h\mathbf{B} + \tfrac{1}{6}h^2\gamma_q(\mathbf{C} + \mathbf{D}) \\ &\quad + \tfrac{1}{24}h^3[\varepsilon_q(\mathbf{E} + 3\mathbf{F}) + \kappa_q(\mathbf{G} + \mathbf{H})] + O(h^4) \end{aligned} \tag{3-174}$$

where

$$\gamma_p = \frac{3}{2}\frac{p-q}{p}, \qquad \gamma_q = \frac{3}{2}\frac{q-p}{2}$$

$$\varepsilon_p = 2\frac{p^2 - pq - q^2}{p^2}, \qquad \varepsilon_q = 2\frac{q^2 - pq - p^2}{q^2}$$

$$\kappa_p = 2\frac{p^2 - 2pq - q^2}{p^2}, \qquad \kappa_q = 2\frac{q^2 - 2pq - p^2}{q^2}$$

Conclude that the predictor formulas are of order 2.

(b) Calculate the increment function of the method defined by (3-172) and (3-173) and show that this method is of order 4 [and not 3, as might be expected from the result obtained in (a)].

(c) Show that the principal error function is given by

$$\begin{aligned} \boldsymbol{\varphi}(\mathbf{y}) &= -\tfrac{1}{4320}\mathbf{f}^{\text{iv}} - \tfrac{1}{108}(\mathbf{K} + \mathbf{L}) + \tfrac{1}{864}(\mathbf{M} + 3\mathbf{N}) - \tfrac{1}{288}(\mathbf{P} + \mathbf{Q}) \\ &= -\tfrac{1}{4320}\mathbf{f}^{\text{iv}} + \tfrac{1}{864}\mathbf{f}_j f'''^j + \tfrac{1}{432}(\mathbf{f}_{jk}f^j - \mathbf{f}_j f^j_k)f''^k \end{aligned} \tag{3-175}$$

(d) Apply the results to the integration of the system*

$$\begin{aligned} y' &= z, & y(0) &= 0 \\ z' &= 4y + \tfrac{1}{10}x^2, & z(0) &= \tfrac{1}{10} \end{aligned} \tag{3-176}$$

the exact solution of which is given by

$$\begin{aligned} y &= \tfrac{1}{160}(5e^{2x} - 3e^{-2x} - 4x^2 - 2) \\ z &= \tfrac{1}{160}(10e^{2x} + 6e^{-2x} - 8x) \end{aligned}$$

In particular, show that the components $e = e^1$ and $f = e^2$ of the magnified error function are given by

$$\begin{aligned} e(x) &= -x(\tfrac{1}{720}e^{2x} + \tfrac{1}{1200}e^{-2x}) \\ f(x) &= -x(\tfrac{1}{360}e^{2x} - \tfrac{1}{600}e^{-2x}) \end{aligned}$$

[Numerical integrations over the interval (0, 1), using the mesh sizes $h = 2^{-p}$, $p = 2, 3, \cdots, 7$, show that the errors at $x_n = 1$ follow very accurately the h^4 law

$$e_n \approx -0.010h^4, \qquad f_n \approx -0.020h^4$$

in good agreement with the theory.]

8. The numerical values given in Table 3.4 were obtained in the numerical integration of the system

$$\begin{aligned} y' &= -\pi z, & y(0) &= 0.1 \\ z' &= \pi y, & z(0) &= 0 \end{aligned}$$

by the method described in §3.3-7, working with the steps $h = 2^{-p}, p = 2, 3, \cdots$. Test the actual errors against the errors predicted by the magnified error vector.

* The author is indebted to Prof. C. B. Tompkins and Dr. D. Pope for suggesting this example and performing the numerical calculations involving it.

Table 3.4

	x_n	1	2	3	4	5
y_n	$p = 2$	−0.094958995	0.090045421	−0.085265627	0.080624836	−0.076127235
	3	− .099249852	.098504742	− .097764639	.097029517	− .096299345
	4	− .099902224	.099804542	− .099706952	.099609456	− .099512053
	5	− .099987654	.099975310	− .099962967	.099950626	− .099938286
	6	− .099998453	.099996906	− .099995359	.099993812	− .099992265
	7	− .099999806	.099999613	− .099999419	.099999226	− .099999032
	8	− .099999975	.099999951	− .099999927	.099999903	− .099999879
	9	− .099999996	.099999994	− .099999991	.099999988	− .099999985
z_n	$p = 2$	−0.003559302	0.006759755	−0.009623984	0.012173700	−0.014429704
	3	− .000242890	.000482136	− .000717777	.000949854	− .001178404
	4	− .000015479	.000030929	− .000046349	.000061738	− .000077097
	5	− .000000972	.000001944	− .000002915	.000003887	− .000004858
	6	− .000000061	.000000122	− .000000184	.000000245	− .000000306
	7	− .000000004	.000000008	− .000000013	.000000017	− .000000021
	8	− .000000000	.000000001	− .000000002	.000000003	− .000000004
	9	− .000000000	.000000001	− .000000001	.000000002	− .000000002

9. *Step modifier not bounded away from zero.* Show that for $a = 0$, $b = 1$ the step modifier $\vartheta = 1 - x$ requires an infinite number of steps to go from $x = 0$ to $x = 1$, even if h is as large as $\frac{1}{2}$.

Section 3.4

10. Show that the matrix function $\mathbf{D}_m(x)$ defined by (3-117) can be represented in the form $\mathbf{Y}(x)\mathbf{Y}^{-1}(x_m)$, where $\mathbf{Y}(x)$ satisfies

$$\mathbf{Y}'(x) = \mathbf{G}(x)\mathbf{Y}(x), \qquad \mathbf{Y}(a) = \mathbf{I}$$

Hence conclude that

$$\mathbf{D}_{nm} = \mathbf{Y}(x_n)\mathbf{Y}^{-1}(x_m) + O(h)$$

11. Give an alternative, direct proof of the representation (3-149) for $\mathbf{V}(x)$ by considering the sum (3-138) a Riemann sum approximating a definite integral and using the result of Problem 10.

12. *Covariance of round-off in floating operations.* If $\mathbf{C}(x) = \mathbf{y}(x)\mathbf{y}^T(x)$, where $\mathbf{y}(x)$ denotes a solution of $\mathbf{y}'(x) = \mathbf{G}(x)\mathbf{y}(x)$, show that

$$\mathbf{V}(x) = (x - a)\mathbf{y}(x)\mathbf{y}^T(x)$$

As a special case, obtain (3-152).

13. Let $s = 2$, $\mathbf{C} = \mathbf{I}$, and

$$\mathbf{G}(x) = \begin{pmatrix} 0 & g(x) \\ g(x) & 0 \end{pmatrix}, \qquad \mathbf{V} = \begin{pmatrix} u & v \\ v & w \end{pmatrix}$$

Show that

$$u = w = \int_a^x \cosh\left(\int_\xi^x g(t)\,dt\right) d\xi$$

$$v = \int_a^x \sinh\left(\int_\xi^x g(t)\,dt\right) d\xi$$

As a special case, obtain (3-171).

14. Let $s = 2$, $\mathbf{C} = \mathbf{I}$, and

$$\mathbf{G}(x) = \begin{pmatrix} 0 & g(x) \\ -g(x) & 0 \end{pmatrix}$$

Show that

$$\mathbf{V}(x) = (x - a)\mathbf{I}$$

As a special case, obtain once again (3-152).

15. Assuming a symmetric distribution of the local round-off error and fixed decimal point operations, study the propagation of round-off error in the solution of the system

$$y' = -z, \qquad y(0) = a^{-1}$$

$$z' = -\frac{2}{(x+a)^2}y, \qquad z(0) = a^{-2}$$

by a one-step method ($a > 0$).

16. Repeat Problem 15 for the system

$$y' = -z, \qquad y(0) = a^{-1}$$

$$z' = -2y^3, \qquad z(0) = a^{-2}$$

17. Repeat Problem 15 for the system

$$y' = 1/z, \qquad y(0) = 1$$

$$z' = -1/y, \qquad z(0) = 1$$

18. Assuming $s = 2$, put

$$\mathbf{V} = \begin{pmatrix} u & v \\ v & w \end{pmatrix}, \qquad \mathbf{G} = \begin{pmatrix} a & b \\ c & d \end{pmatrix}$$

where a, b, c, and d are constants. Introducing the vector

$$\mathbf{v} = \begin{pmatrix} u \\ v \\ w \end{pmatrix}$$

the system (3-140) can be written in the form

$$\mathbf{v}' = \mathbf{a} + \mathbf{B}\mathbf{v}$$

where

$$\mathbf{a} = \begin{pmatrix} 1 \\ 0 \\ 1 \end{pmatrix}, \qquad \mathbf{B} = \begin{pmatrix} 2a & 2b & 0 \\ c & a+d & b \\ 0 & 2c & 2d \end{pmatrix}$$

Show that the corresponding homogeneous equation

$$\mathbf{v}' = \mathbf{B}\mathbf{v}$$

has solutions of the form $\mathbf{v} = \mathbf{r}e^{(\lambda_i+\lambda_j)x}$, $i, j = 1, 2$, where $\mathbf{r}$ is a suitable vector and where λ_1 and λ_2 are the eigenvalues of $\mathbf{G}$.

19. Let $\mathbf{G}$ be a constant matrix of order s with eigenvalues $\lambda_1, \cdots, \lambda_s$, and corresponding eigenvectors $\mathbf{a}_1, \cdots, \mathbf{a}_s$. Show that the differential equation for the square matrix $\mathbf{Y}(x)$ of order s,

$$\mathbf{Y}' = \mathbf{G}\mathbf{Y} + \mathbf{Y}\mathbf{G}^T$$

has the solutions $\mathbf{Y} = \mathbf{a}_i\mathbf{a}_j^T e^{(\lambda_i+\lambda_j)x}$, $i, j = 1, \cdots, s$. (It is not claimed that these solutions are independent.)

Notes

§3.2-3. The analysis of this section is patterned after Gill [1951]. For a different treatment see Albrecht [1955]. Runge-Kutta formulas of higher order are given by Huta [1956, 1957]. For further theoretical contributions on the Runge-Kutta method refer to Kuntzmann [1953, 1959a] and Ionescu [1954, 1956]. Discussions of computational procedure are given by Murray [1950], Gill [1951], Blum [1957], Martin [1958], Romanelli [1960], Anderson [1960]. A modification of the Runge-Kutta method which uses different stepsizes in different components of $\mathbf{y}_n$ is proposed by Rice [1960]. The application of Chaplygin's method (see notes on §2.1-2) is discussed by Babkin [1954] and Artemov [1955].

§3.3-2. General a priori bounds for the discretization error are given by Lozinskii [1953] and Capra [1956].

§3.3-4. A different value for N for the classical Runge-Kutta method is given by Bieberbach [1951].

§3.4-5. For the case where a system of two equations is integrated by the Heun method the statistical theory given here is outlined by Rademacher [1948].

chapter

4 One-step Methods for Systems of Equations of Higher Order

4.1. Introduction

4.1-1. The general system of s equations of order q. In this section q denotes an integer ≥ 1, and $\mathbf{f}, \mathbf{y}^1, \mathbf{y}^2, \cdots, \mathbf{y}^q$ denote vectors with s components ($s \geq 1$). We assume that the vector-valued function $\mathbf{f}(x, \mathbf{y}^1, \mathbf{y}^2, \cdots, \mathbf{y}^q)$ is defined for $x \in [a, b]$ and arbitrary vectors $\mathbf{y}^1, \cdots, \mathbf{y}^q$. By a solution of the system of differential equations of order q

$$\mathbf{y}^{(q)} = \mathbf{f}(x, \mathbf{y}, \mathbf{y}', \cdots, \mathbf{y}^{(q-1)}) \tag{4-1}$$

is meant any vector-valued function $\mathbf{y}(x)$ which is q times differentiable and satisfies the identity

$$\mathbf{y}^{(q)}(x) = \mathbf{f}(x, \mathbf{y}(x), \mathbf{y}'(x), \cdots, \mathbf{y}^{(q-1)}(x)), \qquad x \in [a, b] \tag{4-2}$$

The following *initial value problem* is frequently encountered in practice: to find a solution $\mathbf{y}(x)$ of the system (4-1) which satisfies the conditions

$$\mathbf{y}(a) = \boldsymbol{\eta}, \qquad \mathbf{y}'(a) = \boldsymbol{\eta}', \qquad \cdots \qquad \mathbf{y}^{(q-1)}(a) = \boldsymbol{\eta}^{(q-1)} \tag{4-3}$$

where $\boldsymbol{\eta}, \boldsymbol{\eta}', \cdots, \boldsymbol{\eta}^{(q-1)}$ are preassigned vectors.

The existence of a unique solution of this initial value problem can, under certain conditions, be inferred from Theorem 3.1. We shall assume that

(A) $\mathbf{f}(x, \mathbf{y}^1, \mathbf{y}^2, \cdots, \mathbf{y}^q)$ is defined and continuous for $x \in [a, b]$ and for arbitrary finite values of the components of the vectors $\mathbf{y}^1, \mathbf{y}^2, \cdots, \mathbf{y}^q$;

(B) there exists a constant Λ such that for arbitrary $\mathbf{y}^1, \cdots, \mathbf{y}^q$ and $\mathbf{y}^{1*}, \cdots, \mathbf{y}^{q*}$ and all $x \in [a, b]$

$$\|\mathbf{f}(x, \mathbf{y}^{1*}, \cdots, \mathbf{y}^{q*}) - \mathbf{f}(x, \mathbf{y}^1, \cdots, \mathbf{y}^q)\| \leq \Lambda(\|\mathbf{y}^{1*} - \mathbf{y}^1\| + \cdots + \|\mathbf{y}^{q*} - \mathbf{y}^q\|) \tag{4-4}$$

i.e., $\mathbf{f}$ satisfies a uniform Lipschitz condition with respect to the arguments $\mathbf{y}^1, \cdots, \mathbf{y}^q$.

We then can prove:

THEOREM 4.1. *Let the function* $\mathbf{f}(x, \mathbf{y}^1, \cdots, \mathbf{y}^q)$ *satisfy conditions (A) and (B) above, and let the vectors* $\boldsymbol{\eta}, \boldsymbol{\eta}', \cdots, \boldsymbol{\eta}^{(q-1)}$ *be arbitrary. Then there exists exactly one function* $\mathbf{y}(x)$ *which is continuous and has continuous derivatives up to the order* q *in* $[a, b]$, *and which satisfies the relations* (4-2) *and* (4-3).

Proof. We define the composite sq-dimensional vectors

$$\mathsf{y} = \begin{pmatrix} \mathbf{y}^1 \\ \mathbf{y}^2 \\ \cdot \\ \cdot \\ \cdot \\ \mathbf{y}^q \end{pmatrix}, \qquad \underline{\eta} = \begin{pmatrix} \boldsymbol{\eta} \\ \boldsymbol{\eta}' \\ \cdot \\ \cdot \\ \cdot \\ \boldsymbol{\eta}^{(q-1)} \end{pmatrix}$$

and, letting $\mathbf{f}(x, \mathbf{y}^1, \mathbf{y}^2, \cdots, \mathbf{y}^q) = \mathbf{f}(x, \mathsf{y})$,

$$\mathsf{f}(x, \mathsf{y}) = \begin{pmatrix} \mathbf{y}^2 \\ \mathbf{y}^3 \\ \cdot \\ \cdot \\ \cdot \\ \mathbf{y}^q \\ \mathbf{f}(x, \mathsf{y}) \end{pmatrix}$$

We now consider the following initial value problem involving a system of sq differential equations of the first order:

$$\mathsf{y}' = \mathsf{f}(x, \mathsf{y}) \tag{4-5a}$$

$$\mathsf{y}(a) = \underline{\eta} \tag{4-5b}$$

To this problem we apply Theorem 3.1. Clearly condition (A) of that theorem is satisfied, and from

$$\|\mathsf{f}(x, \mathsf{y}^*) - \mathsf{f}(x, \mathsf{y})\| \leq \|\mathbf{y}^{2*} - \mathbf{y}^2\| + \cdots + \|\mathbf{y}^{q*} - \mathbf{y}^q\| + \Lambda(\|\mathbf{y}^{1*} - \mathbf{y}^1\| + \cdots + \|\mathbf{y}^{q*} - \mathbf{y}^q\|) \leq (1 + \Lambda)\|\mathsf{y}^* - \mathsf{y}\|$$

it follows that condition (B) of Theorem 3.1 is satisfied with $L = 1 + \Lambda$. Hence the initial value problem (4-5) has a unique solution $\mathsf{y}(x)$. If $\mathbf{y}^1(x)$ denotes the first subvector of that solution, and if we set $\mathbf{y}(x) = \mathbf{y}^1(x)$, then it follows from the first $(q - 1)s$ components of the differential equation (4-5a) that $\mathbf{y}^{i+1}(x) = \mathbf{y}^{i\prime}(x)$, $i = 1, 2, \cdots, q-1$, and hence $\mathbf{y}^{i+1}(x) = \mathbf{y}^{(i)}(x)$, $i = 0, 1, \cdots, q-1$.

Relation (4-5*b*) now implies (4-3). By considering the last s components of (4-5*a*) we conclude that $\mathbf{y}(x)$ satisfies (4-2). The existence of a solution of the initial value problem defined by (4-1) and (4-3) has thus been demonstrated. In order to prove the uniqueness of the solution, let $\mathbf{z}(x)$ denote an arbitrary solution of the initial value problem. The vector $\mathbf{z}(x)$ defined by

$$\mathbf{z}(x) = \begin{pmatrix} \mathbf{z}(x) \\ \mathbf{z}'(x) \\ \cdot \\ \cdot \\ \mathbf{z}^{(q-1)}(x) \end{pmatrix}$$

is readily shown to be a solution of the initial value problem (4-5). In view of the uniqueness of the solution of this latter problem, $\mathbf{z}(x)$ is identical with the solution $\mathbf{y}(x)$ defined above. In particular, the first s components are identical, whence it follows that $\mathbf{z}(x) = \mathbf{y}(x)$, $x \in [a, b]$. This completes the proof of Theorem 4.1.

We shall require a notation for the exact relative increments of the derivatives $\mathbf{z}(t), \mathbf{z}'(t), \cdots, \mathbf{z}^{(q-1)}(t)$ of the solution $\mathbf{z}(t)$ of the equation

$$\mathbf{z}^{(q)} = \mathbf{f}(t, \mathbf{z}, \mathbf{z}', \cdots, \mathbf{z}^{(q-1)}) \tag{4-6a}$$

satisfying

$$\mathbf{z}(x) = \mathbf{y}^1, \qquad \mathbf{z}'(x) = \mathbf{y}^2, \cdots, \mathbf{z}^{(q-1)}(x) = \mathbf{y}^q \tag{4-6b}$$

where x is a fixed point in $[a, b]$ and

$$\mathbf{y} = \begin{pmatrix} \mathbf{y}^1 \\ \cdot \\ \cdot \\ \cdot \\ \mathbf{y}^q \end{pmatrix}$$

is a given sq vector. If

$$\mathbf{z}(t) = \begin{pmatrix} \mathbf{z}(t) \\ \mathbf{z}'(t) \\ \cdot \\ \cdot \\ \mathbf{z}^{(q-1)}(t) \end{pmatrix} \tag{4-7}$$

we shall put

$$\underline{\mathbf{\Delta}}(x, \mathbf{y}; h) = \begin{pmatrix} \mathbf{\Delta}^1(x, \mathbf{y}; h) \\ \cdot \\ \cdot \\ \cdot \\ \mathbf{\Delta}^q(x, \mathbf{y}; h) \end{pmatrix} = \frac{1}{h}[\mathbf{z}(x+h) - \mathbf{z}(x)] \tag{4-8}$$

4.1-2. Systems of special equations of the second order. The following special case of the general initial value problem considered in §4.1-1 occurs frequently, especially in problems of celestial mechanics:

$$\mathbf{y}(a) = \boldsymbol{\eta}, \qquad \mathbf{y}'(a) = \boldsymbol{\eta}', \qquad \mathbf{y}'' = \mathbf{f}(x, \mathbf{y}) \tag{4-9}$$

The characteristic feature here is that the function $\mathbf{f}$ does not depend on the first derivative $\mathbf{y}'$. A differential equation of the form (4-1) in which $\mathbf{f}$ does not depend on any derivatives is referred to as a *special* differential equation.

In view of various simplifications which occur, special differential equations of the second order merit attention also from the numerical point of view. In order to save indices, we shall introduce some notations which will be used only in connection with such special equations. We shall consistently write $\mathbf{y}' = \mathbf{z}$. Introducing the $2s$ component vectors

$$\mathsf{y} = \begin{pmatrix} \mathbf{y} \\ \mathbf{z} \end{pmatrix}, \qquad \mathsf{f}(x, \mathsf{y}) = \begin{pmatrix} \mathbf{z} \\ \mathbf{f}(x, \mathbf{y}) \end{pmatrix}$$

the system of $2s$ first order equations equivalent to the differential equations (4-9) is again given by (4-5*a*). Denoting by $\mathbf{u}(t)$ the solution of $\mathbf{u}'' = \mathbf{f}(t, \mathbf{u})$ satisfying $\mathbf{u}(x) = \mathbf{y}$, $\mathbf{u}'(x) = \mathbf{z}$, where $\mathbf{y}$ and $\mathbf{z}$ are given vectors, we put

$$\begin{aligned} \boldsymbol{\Delta}(x, \mathsf{y};\ h) &= \frac{\mathbf{u}(x + h) - \mathbf{u}(x)}{h} \\ \boldsymbol{\Theta}(x, \mathsf{y};\ h) &= \frac{\mathbf{u}'(x + h) - \mathbf{u}'(x)}{h} \end{aligned} \tag{4-10}$$

Thus the subvectors $\boldsymbol{\Delta}^1$ and $\boldsymbol{\Delta}^2$ of the exact relative increment are now denoted by $\boldsymbol{\Delta}$ and $\boldsymbol{\Theta}$.

4.2. Numerical Methods for Systems of Equations of Higher Order

Notwithstanding the fact that systems of equations of higher order can always be treated numerically by reducing them to a larger system of first order equations, a number of integration methods have been devised which attack such systems directly, without reduction. Based on the fact that some of these methods have a small local discretization error, it has been claimed that these methods also have a small accumulated error. The theoretical analysis in §4.3 will show that these claims are unjustified. There is no reason not to reduce a high order equation to a system of first order equations from the point of view of discretization error. In view of this negative result we shall restrict ourselves to a description without proofs of some of the methods in question. Some attention will be given to systems of special equations of order 2, because, although the above statements about the error still hold, it is possible here to save some computational work by taking advantage of the special form of the differential equation.

4.2-1. The increment function for a general system. A one-step method for the numerical integration of the system (4-1) is defined by a vector-valued function

$$\underline{\mathbf{\Phi}}(x, \mathbf{y};\ h) = \begin{pmatrix} \mathbf{\Phi}^1(x, \mathbf{y};\ h) \\ \vdots \\ \mathbf{\Phi}^q(x, \mathbf{y};\ h) \end{pmatrix} \tag{4-11}$$

with sq components, which enables one to calculate vectors $\mathbf{y}_n$, intended as approximations to $\mathbf{y}(x_n)$, by means of the formulas

$$\begin{aligned} \mathbf{y}_0 &= \underline{\boldsymbol{\eta}} \\ \mathbf{y}_{n+1} &= \mathbf{y}_n + h\underline{\mathbf{\Phi}}(x_n, \mathbf{y}_n;\ h), \qquad n = 0, 1, 2, \cdots \end{aligned} \tag{4-12}$$

For each of the functions $\mathbf{y}^{(i)}(x)$ $(i = 0, 1, \cdots, q - 1)$ the increment has to be calculated separately by means of the corresponding component $\mathbf{\Phi}^{i+1}$ of the increment function. There is no way of obtaining approximate values of the solution $\mathbf{y}(x)$ by a one-step method without also obtaining values of its first $q - 1$ derivatives. For multistep methods the situation is different; see Chapter 6.

For the numerical integration of systems of special second order equations we shall write the increment function in the form

$$\underline{\mathbf{\Phi}}(x, \mathbf{y};\ h) = \begin{pmatrix} \mathbf{\Phi}(x, \mathbf{y};\ h) \\ \mathbf{\Psi}(x, \mathbf{y};\ h) \end{pmatrix} \tag{4-13}$$

4.2-2. Taylor expansion. As in earlier chapters, the general aim in setting up the increment function $\underline{\mathbf{\Phi}}$ is to produce the closest possible agreement with the exact relative increment $\underline{\mathbf{\Delta}}$ in all components. This agreement can be realized by explicit or implicit use of Taylor's expansion. We shall continue to denote by $\mathbf{f}', \mathbf{f}'', \cdots$ the *total* derivatives of the function $\mathbf{f}$. These derivatives now depend not only on $\mathbf{y} = \mathbf{y}^1$, but in general also on $\mathbf{y}^2, \cdots, \mathbf{y}^q$. For instance, if the differential equation $\mathbf{y}'' = \mathbf{f}(x, \mathbf{y}, \mathbf{y}')$ is given, then, denoting by $y^{\nu i}$ the components of $\mathbf{y}^\nu$ $(\nu = 1, 2)$ and adopting the summation convention,

$$\mathbf{f}' = \mathbf{f}_x + \mathbf{f}_{y^{1i}} y^{2i} + \mathbf{f}_{y^{2i}} f^i$$

In the case of a system of special second order equations $\mathbf{y}'' = \mathbf{f}(x, \mathbf{y})$ the derivatives of $\mathbf{f}$ with respect to the components of $\mathbf{y}$ may unambiguously be denoted by $\mathbf{f}_i$, and, writing $\mathbf{y}^2 = \mathbf{z}$, we have

$$\mathbf{f}' = \mathbf{f}_x + \mathbf{f}_i z^i$$

We shall denote by $\mathbf{T}^\nu(x, \mathbf{y};\ h)$ that part of the Taylor expansion of the subvector $\mathbf{\Delta}^\nu$ of $\underline{\mathbf{\Delta}}$ which can be calculated without using the function $\mathbf{f}$. Thus,

$$\mathbf{T}^\nu(x, \mathbf{y};\ h) = \mathbf{y}^{\nu+1} + \frac{h}{2!}\mathbf{y}^{\nu+2} + \cdots + \frac{h^{q-\nu-1}}{(q-\nu)!}\mathbf{y}^q, \qquad \nu = 1, 2, \cdots, q-1$$

$$\mathbf{T}^q(x, \mathbf{y};\ h) = 0$$

We then can write

$$\text{(4-14)} \quad \mathbf{\Delta}^\nu(x, \mathbf{y};\ h) = \mathbf{T}^\nu(x, \mathbf{y};\ h) + \frac{h^{q-\nu}}{(q-\nu+1)!}\mathbf{f}(x, \mathbf{y}) + \frac{h^{q-\nu+1}}{(q-\nu+2)!}\mathbf{f}'(x, \mathbf{y}) + \cdots, \qquad \nu = 1, 2, \cdots, q$$

For a special second order system we have

$$\text{(4-15)} \quad \begin{aligned} \mathbf{\Delta}(x, \mathbf{y}, \mathbf{z};\ h) &= \mathbf{z} + \frac{h}{2}\mathbf{f}(x, \mathbf{y}) + \frac{h^2}{3!}\mathbf{f}'(x, \mathbf{y}, \mathbf{z}) + \cdots \\ \mathbf{\Theta}(x, \mathbf{y}, \mathbf{z};\ h) &= \mathbf{f}(x, \mathbf{y}) + \frac{h}{2}\mathbf{f}'(x, \mathbf{y}, \mathbf{z}) + \cdots \end{aligned}$$

If some of the derivatives $\mathbf{f}', \mathbf{f}'', \cdots$ are easily calculated, accurate increment functions can be obtained by truncating the expansions (4-14) and (4-15) after a suitable number of terms. In general, this method is not recommended.

4.2-3. Runge-Kutta formulas. Application of the ordinary Runge-Kutta method (3-48) to the system $\mathbf{y}' = \mathbf{f}(x, \mathbf{y})$ seems wasteful, because only the last s of the sq components of $\mathbf{f}$ involve the given function $\mathbf{f}$; all other components are trivial. One is therefore tempted to calculate only the last subvector $\mathbf{k}_\rho$ of the Runge-Kutta coefficients $\mathbf{k}_\rho$ $(\rho = 1, 2, \cdots, \kappa)$ and to represent all subvectors $\mathbf{\Phi}^\nu$ $(\nu = 1, 2, \cdots, q)$ of the increment function $\underline{\mathbf{\Phi}}$ in the form

$$\text{(4-16)} \quad \mathbf{\Phi}^\nu(x, \mathbf{y};\ h) = \mathbf{T}^\nu(x, \mathbf{y};\ h) + \frac{h^{q-\nu}}{(q-\nu+1)!}\sum_{\rho=1}^{\kappa}\gamma_{\nu\rho}\mathbf{k}_\rho, \qquad \nu = 1, 2, \cdots, q$$

Here the $\mathbf{k}_\rho$ denote certain values of the function $\mathbf{f}$ which do not depend on ν. Agreement (up to a certain degree) with the Taylor expansion (4-14) will be enforced by choosing suitable numerical coefficients $\gamma_{\nu\rho}$.

The idea thus outlined was pursued by Zurmühl [1948], who, in the case of a single equation of the form (4-1) ($s = 1$) arrived at very complicated formulas for the $\mathbf{k}_\rho$. In view of their more symmetrical appearance we prefer the slightly different formulas given below, which hold for an arbitrary number of equations of the form (4-1). For formal simplification we assume that $\mathbf{f}$ does not depend explicitly on x. This can always be achieved by adjoining to the vector $\mathbf{y}$ a component y^0 which is subjected to the differential equation

$$y^{0(q)} = 0$$

and to the initial conditions

$$y^0(a) = a, \qquad y^{0\prime}(a) = 1, \qquad y^{0(\nu)}(a) = 0 \qquad (\nu = 2, \ldots, q - 1)$$

For $q \geq 2$ it then follows that $y^0(x) = x$. The system (4-1) is thus equivalent to

$$\mathbf{y}^{(q)} = \mathbf{f}(\mathbf{y}, \mathbf{y}', \cdots, \mathbf{y}^{(q-1)})$$

where $\mathbf{y}$ now denotes the augmented vector, and where the component $f^0 = 0$ has been added to $\mathbf{f}$.

Choosing, as Zurmühl does, $\kappa = 4$ in (4-16), we now define the vectors $\mathbf{k}_\rho$ by

$$\mathbf{k}_\rho = \mathbf{k}_\rho(\mathbf{y};\, h) = \mathbf{f}(\mathbf{Y}^1, \mathbf{Y}^2, \cdots, \mathbf{Y}^q) \tag{4-17}$$

where the values of the arguments $\mathbf{Y}^1, \mathbf{Y}^2, \cdots$ are given in Table 4.1.

Table 4.1

Arguments of $\mathbf{f}$ in (4-17)

ρ	1	2	3	4
$\mathbf{Y}^1$	$\mathbf{y}^1$	$\mathbf{y}^1 + \frac{1}{2}h\mathbf{y}^2$	$\mathbf{y}^1 + \frac{1}{2}h\mathbf{y}^2 + \frac{1}{4}h^2\mathbf{y}^3$	$\mathbf{y}^1 + h\mathbf{y}^2 + \frac{1}{2}h^2\mathbf{y}^3 + \frac{1}{4}h^3\mathbf{y}^4$
$\mathbf{Y}^2$	$\mathbf{y}^2$	$\mathbf{y}^2 + \frac{1}{2}h\mathbf{y}^3$	$\mathbf{y}^2 + \frac{1}{2}h\mathbf{y}^3 + \frac{1}{4}h^2\mathbf{y}^4$	$\mathbf{y}^2 + h\mathbf{y}^3 + \frac{1}{2}h^2\mathbf{y}^4 + \frac{1}{4}h^3\mathbf{y}^5$
.	.	.	.	.
.	.	.	.	.
.	.	.	.	.
$\mathbf{Y}^{q-3}$	$\mathbf{y}^{q-3}$	$\mathbf{y}^{q-3} + \frac{1}{2}h\mathbf{y}^{q-2}$	$\mathbf{y}^{q-3} + \frac{1}{2}h\mathbf{y}^{q-2} + \frac{1}{4}h^2\mathbf{y}^{q-1}$	$\mathbf{y}^{q-3} + h\mathbf{y}^{q-2} + \frac{1}{2}h^2\mathbf{y}^{q-1} + \frac{1}{4}h^3\mathbf{y}^q$
$\mathbf{Y}^{q-2}$	$\mathbf{y}^{q-2}$	$\mathbf{y}^{q-2} + \frac{1}{2}h\mathbf{y}^{q-1}$	$\mathbf{y}^{q-2} + \frac{1}{2}h\mathbf{y}^{q-1} + \frac{1}{4}h^2\mathbf{y}^q$	$\mathbf{y}^{q-2} + h\mathbf{y}^{q-1} + \frac{1}{2}h^2\mathbf{y}^q + \frac{1}{4}h^3\mathbf{k}_1$
$\mathbf{Y}^{q-1}$	$\mathbf{y}^{q-1}$	$\mathbf{y}^{q-1} + \frac{1}{2}h\mathbf{y}^q$	$\mathbf{y}^{q-1} + \frac{1}{2}h\mathbf{y}^q + \frac{1}{4}h^2\mathbf{k}_1$	$\mathbf{y}^{q-1} + h\mathbf{y}^q + \frac{1}{2}h^2\mathbf{k}_2$
$\mathbf{Y}^q$	$\mathbf{y}^q$	$\mathbf{y}^q + \frac{1}{2}h\mathbf{k}_1$	$\mathbf{y}^q + \frac{1}{2}h\mathbf{k}_2$	$\mathbf{y}^q + h\mathbf{k}_3$

The numerical coefficients $\gamma_{\nu\rho}$ in (4-16) are given by

$$\begin{aligned} \gamma_{\nu 1} &= \frac{(q - \nu + 1)^2}{(q - \nu + 2)(q - \nu + 3)} \\ \gamma_{\nu 2} = \gamma_{\nu 3} &= \frac{2(q - \nu + 1)}{(q - \nu + 2)(q - \nu + 3)} \\ \gamma_{\nu 4} &= \frac{1 - q + \nu}{(q - \nu + 2)(q - \nu + 3)}, \qquad \nu = 1, 2, \cdots, q \end{aligned} \tag{4-18}$$

It will be noted that the coefficients $\gamma_{\nu\rho}$ depend only on ρ and $q - \nu$. Some numerical values are given in Table 4.2.

Table 4.2

Values of the Coefficients $\gamma_{\nu\rho}$ in (4-16)

$q - \nu$	$\gamma_{\nu 1}$	$\gamma_{\nu 2} = \gamma_{\nu 3}$	$\gamma_{\nu 4}$
0	$\frac{1}{6}$	$\frac{1}{3}$	$\frac{1}{6}$
1	$\frac{1}{3}$	$\frac{1}{3}$	0
2	$\frac{9}{20}$	$\frac{6}{20}$	$-\frac{1}{20}$
3	$\frac{8}{15}$	$\frac{4}{15}$	$-\frac{1}{15}$

For the system of nonspecial second order equations

$$\mathbf{y}'' = \mathbf{f}(x, \mathbf{y}, \mathbf{z})$$

where $\mathbf{z} = \mathbf{y}'$, (4-16) reduces to

$$\begin{aligned} \mathbf{\Phi}^1(x, \mathbf{y}, \mathbf{z};\ h) &= z + \tfrac{1}{6}h(\mathbf{k}_1 + \mathbf{k}_2 + \mathbf{k}_3) \\ \mathbf{\Phi}^2(x, \mathbf{y}, \mathbf{z};\ h) &= \tfrac{1}{6}(\mathbf{k}_1 + 2\mathbf{k}_2 + 2\mathbf{k}_3 + \mathbf{k}_4) \end{aligned} \tag{4-19a}$$

where

$$\begin{aligned} \mathbf{k}_1 &= \mathbf{f}(x, \mathbf{y}, \mathbf{z}) \\ \mathbf{k}_2 &= \mathbf{f}(x + \tfrac{1}{2}h, \mathbf{y} + \tfrac{1}{2}h\mathbf{z};\ \mathbf{z} + \tfrac{1}{2}h\mathbf{k}_1) \\ \mathbf{k}_3 &= \mathbf{f}(x + \tfrac{1}{2}h, \mathbf{y} + \tfrac{1}{2}h\mathbf{z} + \tfrac{1}{4}h^2\mathbf{k}_1, \mathbf{z} + \tfrac{1}{2}h\mathbf{k}_2) \\ \mathbf{k}_4 &= \mathbf{f}(x + h, \mathbf{y} + h\mathbf{z} + \tfrac{1}{2}h^2\mathbf{k}_2, \mathbf{z} + h\mathbf{k}_3) \end{aligned} \tag{4-19b}$$

It can be shown that for the increment functions $\mathbf{\Phi}^\nu(\mathbf{y};\ h)$ defined by (4-16)

$$\begin{aligned} \mathbf{\Phi}^q(\mathbf{y};\ h) - \mathbf{\Delta}^q(\mathbf{y};\ h) &= O(h^4) \\ \mathbf{\Phi}^\nu(\mathbf{y};\ h) - \mathbf{\Delta}^\nu(\mathbf{y};\ h) &= O(h^{q-\nu+3}), \qquad \nu = 1, 2, \cdots, q - 1 \end{aligned} \tag{4-20}$$

Thus, while the values of $\mathbf{y}$ are calculated with considerable accuracy, the accuracy of the increment functions for $\mathbf{y}^{(q-2)}$ and $\mathbf{y}^{(q-1)}$ is not greater than for the ordinary Runge-Kutta method.

4.2-4. Nyström's formulas for special equations. For special second order equations the formulas for the total derivatives of **f** simplify somewhat, and it becomes possible to achieve in the Runge-Kutta technique a higher degree of agreement with a given number of substitutions than can be expected from the general case. Thus, with two substitutions, we can achieve agreement in $\underline{\mathbf{\Phi}}$ and $\underline{\mathbf{\Delta}}$ with an error of $O(h^3)$, of $O(h^4)$ with three substitutions, and of $O(h^5)$ with four substitutions.

We again assume that $\mathbf{f}$ does not depend explicitly on x, $\mathbf{f} = \mathbf{f}(\mathbf{y})$. The increment functions will then also be independent of x, although they depend on $\mathbf{z} = \mathbf{y}'$ in addition to $\mathbf{y}$. The special methods considered below are special cases of the formulas

$$\begin{aligned}\mathbf{\Phi}(\mathbf{y}, \mathbf{z}; h) &= \mathbf{z} + h \sum_{\rho=1}^{\kappa} a_\rho \mathbf{k}_\rho \\ \mathbf{\Psi}(\mathbf{y}, \mathbf{z}; h) &= \sum_{\rho=1}^{\kappa} b_\rho \mathbf{k}_\rho\end{aligned} \tag{4-21a}$$

where

$$\begin{aligned}\mathbf{k}_1 &= \mathbf{f}(\mathbf{y}) \\ \mathbf{k}_2 &= \mathbf{f}(\mathbf{y} + hp_1\mathbf{z} + h^2 p_1 q_1 \mathbf{k}_1) \\ \mathbf{k}_3 &= \mathbf{f}(\mathbf{y} + hp_2\mathbf{z} + h^2 p_2[(q_2 - q_3)\mathbf{k}_1 + q_3\mathbf{k}_3]) \\ \mathbf{k}_4 &= \mathbf{f}(\mathbf{y} + hp_3\mathbf{z} + h^2 p_3[(q_4 - q_5 - q_6)\mathbf{k}_1 + q_5\mathbf{k}_2 + q_6\mathbf{k}_3])\end{aligned} \tag{4-21b}$$

The number of substitutions κ and the values of the parameters a_ρ, b_ρ ($\rho = 1, \cdots, \kappa$), and $p_1, p_2, \cdots, q_6$ depend on the desired accuracy.

If p is the largest integer such that, assuming sufficient differentiability of the functions involved,

$$\underline{\Phi}(\mathbf{y}; h) - \underline{\Delta}(\mathbf{y}; h) = O(h^p)$$

for all vectors $\mathbf{y}$, we shall call p the *order* of the method defined by $\underline{\Phi}$. It can be shown (see Problem 5) that $p = 3$ if $\kappa = 2$, $a_1 = a_2 = \frac{1}{4}$, $b_1 = \frac{1}{4}$, $b_2 = \frac{3}{4}$, $p_1 = \frac{2}{3}$, $q_1 = \frac{1}{2}$. Formulas (4-21) then read

$$\begin{aligned}\mathbf{\Phi}(\mathbf{y}, \mathbf{z}; h) &= \mathbf{z} + \tfrac{1}{4}h(\mathbf{k}_1 + \mathbf{k}_2) \\ \mathbf{\Psi}(\mathbf{y}, \mathbf{z}; h) &= \tfrac{1}{4}(\mathbf{k}_1 + 3\mathbf{k}_2)\end{aligned} \tag{4-22a}$$

where

$$\begin{aligned}\mathbf{k}_1 &= \mathbf{f}(\mathbf{y}) \\ \mathbf{k}_2 &= \mathbf{f}(\mathbf{y} + \tfrac{2}{3}h\mathbf{z} + \tfrac{1}{3}h^2\mathbf{k}_1)\end{aligned} \tag{4-22b}$$

We obtain $p = 4$ (see Problem 7) if $\kappa = 3$, and the parameters are chosen such that

$$\begin{aligned}\mathbf{\Phi}(\mathbf{y}, \mathbf{z}; h) &= \mathbf{z} + \tfrac{1}{6}h(\mathbf{k}_1 + 2\mathbf{k}_2) \\ \mathbf{\Psi}(\mathbf{y}, \mathbf{z}; h) &= \tfrac{1}{6}(\mathbf{k}_1 + 4\mathbf{k}_2 + \mathbf{k}_3)\end{aligned} \tag{4-23a}$$

where

$$\begin{aligned}\mathbf{k}_1 &= \mathbf{f}(\mathbf{y}) \\ \mathbf{k}_2 &= \mathbf{f}(\mathbf{y} + \tfrac{1}{2}h\mathbf{z} + \tfrac{1}{8}h^2\mathbf{k}_1) \\ \mathbf{k}_3 &= \mathbf{f}(\mathbf{y} + h\mathbf{z} + \tfrac{1}{2}h^2\mathbf{k}_2)\end{aligned} \tag{4-23b}$$

It is finally possible to achieve a method of order $p = 5$ (see Problem 8) by putting

$$\text{(4-24}a\text{)} \qquad \begin{aligned} \mathbf{\Phi}(\mathbf{y}, \mathbf{z};\ h) &= \mathbf{z} + \tfrac{1}{192}h[23\mathbf{k}_1 + 75\mathbf{k}_2 - 27\mathbf{k}_3 + 25\mathbf{k}_4] \\ \mathbf{\Psi}(\mathbf{y}, \mathbf{z};\ h) &= \tfrac{1}{192}[23\mathbf{k}_1 + 125\mathbf{k}_2 - 81\mathbf{k}_3 + 125\mathbf{k}_4] \end{aligned}$$

where

$$\text{(4-24}b\text{)} \qquad \begin{aligned} \mathbf{k}_1 &= \mathbf{f}(\mathbf{y}) \\ \mathbf{k}_2 &= \mathbf{f}(\mathbf{y} + \tfrac{2}{5}h\mathbf{z} + \tfrac{2}{25}h^2\mathbf{k}_1) \\ \mathbf{k}_3 &= \mathbf{f}(\mathbf{y} + \tfrac{2}{3}h\mathbf{z} + \tfrac{2}{9}h^2\mathbf{k}_1) \\ \mathbf{k}_4 &= \mathbf{f}(\mathbf{y} + \tfrac{4}{5}h\mathbf{z} + \tfrac{4}{25}h^2(\mathbf{k}_1 + \mathbf{k}_2)) \end{aligned}$$

Formulas (4-22), (4-23), and (4-24) have been given by Nyström [1925] in his fundamental memoir on the numerical solution of differential equations.

4.3. Discretization Error

4.3-1. Convergence, consistency, a priori bounds. The method defined by the increment function $\underline{\mathbf{\Phi}}$ is called *convergent*, if for every choice of the starting vector $\underline{\boldsymbol{\eta}}$ the components $\mathbf{y}_n^1, \mathbf{y}_n^2, \cdots, \mathbf{y}_n^q$ of the vectors y_n defined by (4-12) satisfy for $x \in [a, b]$

$$\text{(4-25)} \qquad \lim_{\substack{h\to 0\\ x_n = x}} \mathbf{y}_n^\nu = \mathbf{y}^{(\nu-1)}(x), \qquad \nu = 1, 2, \cdots, q$$

where $\mathbf{y}(x)$ denotes the exact solution of the initial value problem. The essence of the convergence behavior of the vectors y_n may be stated briefly by saying that these vectors act as if they were to approximate the solution of a system of sq first order equations. Indeed, if, as we shall always assume, the function $\underline{\mathbf{\Phi}}(x, \mathsf{y};\ h)$ is continuous in the appropriate region and satisfies a Lipschitz condition with respect to y, then, according to the proof of Theorem 3.1, the vectors y_n defined in (4-12) satisfy

$$\text{(4-26)} \qquad \lim_{\substack{h\to 0\\ x_n = x}} \mathsf{y}_n = \mathsf{z}(x)$$

where $\mathsf{z}(x)$ denotes the solution of the initial value problem

$$\text{(4-27)} \qquad \mathsf{z}' = \underline{\mathbf{\Phi}}(x, \mathsf{z};\ 0), \qquad \mathsf{z}(a) = \underline{\boldsymbol{\eta}}$$

This solution agrees with $\mathsf{y}(x)$ for all choices of the starting vector $\underline{\boldsymbol{\eta}}$ if and only if

$$\underline{\mathbf{\Phi}}(x, \mathsf{z};\ 0) \equiv \mathsf{f}(x, \mathsf{z})$$

identically in x and z, i.e., if

$$\text{(4-28)} \qquad \begin{aligned} \mathbf{\Phi}^\nu(x, \mathsf{y};\ 0) &= \mathbf{y}^{\nu+1}, \qquad \nu = 1, 2, \cdots, q - 1 \\ \underline{\mathbf{\Phi}}(x, \mathsf{y};\ 0) &= \mathbf{f}(x, \mathsf{y}) \end{aligned}$$

These relations thus represent the *conditions of consistency* of the method defined by $\underline{\mathbf{\Phi}}$. They are obviously satisfied for the special methods considered in §4.2. For systems of special second order equations the consistency conditions reduce to

$$\mathbf{\Phi}(x, \mathbf{y};\ 0) = \mathbf{z}, \qquad \mathbf{\Psi}(x, \mathbf{y};\ 0) = \mathbf{f}(x, \mathbf{y}) \tag{4-29}$$

By applying Theorem 3.3 to the vectors $\mathbf{y}_n$ we obtain the following a priori bound for the errors $\mathbf{e}_n = \mathbf{y}_n - \mathbf{y}(x_n)$:

THEOREM 4.2. *Let there exist a constant L such that for arbitrary vectors* $\mathbf{y}$ *and* $\mathbf{y}^*$, *for all* $x \in [a, b]$ *and for* $h \le h_0$, *where* $h_0 > 0$,

$$\|\underline{\mathbf{\Phi}}(x, \mathbf{y}^*;\ h) - \underline{\mathbf{\Phi}}(x, \mathbf{y};\ h)\| \le L\|\mathbf{y}^* - \mathbf{y}\| \tag{4-30}$$

and let N and p be constants such that

$$\|\underline{\mathbf{\Phi}}(x, \mathbf{y}(x);\ h) - \underline{\mathbf{\Delta}}(x, \mathbf{y}(x);\ h)\| \le Nh^p \tag{4-31}$$

for $x \in [a, b]$ *and* $h \le h_0$ *where* $\mathbf{y}(x)$ *denotes the exact solution of the initial value problem* (4-5). *Then the vectors* $\mathbf{y}_n$ *determined by* (4-12) *satisfy*

$$\|\mathbf{y}_n - \mathbf{y}(x_n)\| \le Nh^p E_L(x_n - a), \qquad x_n \in [a, b],\ h \le h_0 \tag{4-32}$$

It will be noted that the exponent p in (4-31) is governed by the *worst* degree of approximation of the components of $\underline{\mathbf{\Phi}} - \underline{\mathbf{\Delta}}$, i.e., by the smallest of the numbers p_ν such that

$$\|\mathbf{\Phi}^\nu(x, \mathbf{y}(x);\ h) - \mathbf{\Delta}^\nu(x, \mathbf{y}(x);\ h)\| \le N_\nu h^{p_\nu}, \qquad h \le h_0 \tag{4-33}$$

Since (4-31) is only a bound and not an asymptotic formula, this fact in itself does not guarantee that the error in every component of $\mathbf{y}_n$ is of the order of h^p and not of any higher order. However, we see in §4.3-3 that, except in very special instances, such is indeed the case.

In order to apply the result of Theorem 4.2 to special methods, one would have to know numerical values for L and N. Bounds for these numbers in terms of bounds for derivatives of $\mathbf{f}$ can be found by the methods of §§3.3-3 and 3.3-4. See also Gautschi [1955] for bounds for N for the Zurmühl formulas.

4.3-2. Principal error function. Assuming that $\mathbf{f}$ and $\underline{\mathbf{\Phi}}$ are sufficiently differentiable, we can expand both $\underline{\mathbf{\Delta}}(x, \mathbf{y};\ h)$ and $\underline{\mathbf{\Phi}}(x, \mathbf{y};\ h)$ in powers of h at an arbitrary point $(x, \mathbf{y})$ such that $x \in [a, b]$. If p is the largest integer such that

$$\underline{\mathbf{\Phi}}(x, \mathbf{y};\ h) - \underline{\mathbf{\Delta}}(x, \mathbf{y};\ h) = O(h^p) \tag{4-34}$$

for all $x \in [a, b]$ and all $\mathbf{y}$, then p is called the order of the method defined by $\underline{\mathbf{\Phi}}$. The function

$$\underline{\boldsymbol{\varphi}}(x, \mathbf{y}) = \begin{pmatrix} \boldsymbol{\varphi}^1(x, \mathbf{y}) \\ \cdot \\ \cdot \\ \cdot \\ \boldsymbol{\varphi}^q(x, \mathbf{y}) \end{pmatrix} \tag{4-35}$$

defined by

$$\underline{\mathbf{\Phi}}(x, \mathbf{y};\, h) - \underline{\mathbf{\Delta}}(x, \mathbf{y};\, h) = \underline{\boldsymbol{\varphi}}(x, \mathbf{y})h^p + O(h^{p+1}) \tag{4-36}$$

is called the *principal error function* of the method. As in previous chapters, the principal error function serves to estimate asymptotically both the local and the accumulated discretization error.

For any given method the principal error function is easily calculated as the discrepancy between the first disagreeing terms in the Taylor expansions of $\underline{\mathbf{\Phi}}$ and $\underline{\mathbf{\Delta}}$. We shall not carry out this calculation for the methods described in §4.2-2. We only note that, although $\underline{\boldsymbol{\varphi}}$ can never vanish identically, it may very well happen that some of the subvectors $\boldsymbol{\varphi}^i$ in (4-35) are zero; for the method (4-16), for instance, we have $\boldsymbol{\varphi}^1 = \boldsymbol{\varphi}^2 = \cdots = \boldsymbol{\varphi}^{q-2} = 0$.

For methods appropriate for systems of special second order equations we write

$$\underline{\boldsymbol{\varphi}}(x, \mathbf{y}) = \begin{pmatrix} \boldsymbol{\varphi}(x, \mathbf{y}, \mathbf{z}) \\ \boldsymbol{\psi}(x, \mathbf{y}, \mathbf{z}) \end{pmatrix}$$

We shall give without proof the principal error functions for some of Nyströms formulas presented in §4.2-4. Assuming, as we may, that $\mathbf{f}$ does not depend on x, and using the summation convention, we have for the method (4-22)

$$\begin{aligned} \boldsymbol{\varphi}(\mathbf{y}, \mathbf{z}) &= \tfrac{1}{72}\mathbf{f}'' \\ \boldsymbol{\psi}(\mathbf{y}, \mathbf{z}) &= -\tfrac{1}{648}\mathbf{f}''' - \tfrac{13}{324}\mathbf{f}_i f^{i\prime} \end{aligned} \tag{4-37}$$

and for the method (4-23):

$$\begin{aligned} \boldsymbol{\varphi}(\mathbf{y}, \mathbf{z}) &= -\tfrac{1}{720}\mathbf{f}''' - \tfrac{1}{144}\mathbf{f}_i f^{i\prime} \\ \boldsymbol{\psi}(\mathbf{y}, \mathbf{z}) &= +\tfrac{1}{2880}\mathbf{f}^{\text{iv}} + \tfrac{1}{576}\mathbf{f}_i f^{i\prime\prime} + \tfrac{1}{144}\mathbf{f}_{ij} f^{i\prime} z^j - \tfrac{1}{128}\mathbf{f}_{ijk} f^i z^j z^k \end{aligned} \tag{4-38}$$

For the fifth order method (4-24) the principal error function has not been computed.

It is of interest to see how the above formulas work out in a simple special case such as $\mathbf{y}'' = -\mathbf{y}$. Here we easily find $\mathbf{f}' = -\mathbf{f}''' = -\mathbf{z}$ and $\mathbf{f}'' = -\mathbf{f}^{\text{iv}} = -\mathbf{y}$; furthermore, $\mathbf{f}_i = -\mathbf{e}_i$, $\mathbf{f}_{ij} = \mathbf{f}_{ijk} = \cdots = 0$. We thus obtain

$\boldsymbol{\varphi} = \tfrac{1}{72}\mathbf{y}$ and $\boldsymbol{\psi} = \tfrac{1}{24}\mathbf{z}$ for (4-22)

and

$\boldsymbol{\varphi} = -\tfrac{1}{120}\mathbf{z}$ and $\boldsymbol{\psi} = -\tfrac{1}{480}\mathbf{y}$ for (4-23)

4.3-3. Asymptotic formulas for the accumulated discretization error. In order to apply Theorem 3.4 to the system $\mathbf{y}' = \mathbf{f}(x, \mathbf{y})$ we need to know the functional matrix $\mathbf{G}(x)$ of this system. From the definition of $\mathbf{f}(x, \mathbf{y})$ it follows that $\mathbf{G}(x)$ is given by the $sq \times sq$ matrix

$$\mathbf{G}(x) = \begin{pmatrix} \mathbf{0} & \mathbf{I} & & & \\ & \mathbf{0} & \mathbf{I} & \cdot & \\ & & & \cdot & \\ & & & & \cdot \\ & & \mathbf{0} & & \mathbf{I} \\ \mathbf{G}_1(x) & \mathbf{G}_2(x) & \mathbf{G}_3(x) & \cdots & \mathbf{G}_q(x) \end{pmatrix} \tag{4-39}$$

Here $\mathbf{I}$ denotes the $s \times s$ unit matrix, and

$$\mathbf{G}_\nu(x) = (g_\nu^{ij}(x)), \qquad \nu = 1, 2, \cdots, q$$

where

$$g_\nu^{ij}(x) = \frac{\partial f^i}{\partial y^{\nu j}}(x, \mathbf{y}(x)), \qquad i, j = 1, \cdots, s$$

with $y^{\nu j}$ denoting the jth component of $\mathbf{y}^\nu$. For systems of special second order equations we have, since $\mathbf{f}$ does not depend on $\mathbf{y}^2$,

$$\mathbf{G}(x) = \begin{pmatrix} \mathbf{0} & \mathbf{I} \\ \mathbf{G}(x) & \mathbf{0} \end{pmatrix} \tag{4-39a}$$

where $\mathbf{G}(x) = \mathbf{G}_1(x)$.

It then follows from Theorem 3.4 that

$$\mathbf{y}_n - \mathbf{y}(x_n) = h^p\mathbf{e}(x_n) + O(h^{p+1}) \tag{4-40}$$

where $\mathbf{e}(x)$ denotes the solution of the initial value problem

$$\begin{aligned} \mathbf{e}'(x) &= \mathbf{G}(x)\mathbf{e}(x) + \underline{\boldsymbol{\varphi}}(x, \mathbf{y}(x)) \\ \mathbf{e}(a) &= \mathbf{0} \end{aligned} \tag{4-41}$$

It is clear that in general no component of $\mathbf{e}(x)$ vanishes identically, even if some components of $\underline{\boldsymbol{\varphi}}$ vanish. For instance, if $\boldsymbol{\varphi}^1 = \boldsymbol{\varphi}^2 = \cdots = \boldsymbol{\varphi}^{q-1} = \mathbf{0}$, but $\boldsymbol{\varphi}^q \neq \mathbf{0}$, then, writing

$$\mathbf{e}(x) = \begin{pmatrix} \mathbf{e}^1(x) \\ \mathbf{e}^2(x) \\ \cdot \\ \cdot \\ \cdot \\ \mathbf{e}^q(x) \end{pmatrix}$$

we can eliminate $\mathbf{e}^2(x), \cdots, \mathbf{e}^q(x)$ from (4-41) and define $\mathbf{e}^1(x)$ as the solution of the initial value problem

(4-42)
$$\mathbf{e}^{1(q)}(x) = \mathbf{G}_1(x)\mathbf{e}^1(x) + \mathbf{G}_2(x)\mathbf{e}^{1'}(x) + \cdots + \mathbf{G}_q(x)\mathbf{e}^{1(q-1)}(x) + \boldsymbol{\varphi}^q(x, \mathbf{y}(x))$$
$$\mathbf{e}^1(a) = \mathbf{e}^{1'}(a) = \cdots = \mathbf{e}^{1(q-1)}(a) = \mathbf{0}$$

It thus follows in this case that asymptotically the error in $\mathbf{y}^1(x)$ is $O(h^p)$, even though $\mathbf{\Phi}^1 - \mathbf{\Delta}^1$ is of some higher order. This important negative result has in the special case of the Zurmühl formulas and for a particular differential equation been stated by Rutishauser [1955].

For systems of special differential equations of the second order a more complete analysis is possible. If we put

$$\mathsf{e}(x) = \begin{pmatrix} \mathbf{e}(x) \\ \mathbf{d}(x) \end{pmatrix}$$

then the system (4-41) can be written in the form

$$\begin{aligned} \mathbf{e}'(x) &= \mathbf{d}(x) + \boldsymbol{\varphi}(x, \mathbf{y}(x)) \\ \mathbf{d}'(x) &= \mathbf{G}(x)\mathbf{e}(x) + \boldsymbol{\psi}(x, \mathbf{y}(x)) \\ \mathbf{e}(a) &= \mathbf{d}(a) = \mathbf{0} \end{aligned} \tag{4-43}$$

Elimination of $\mathbf{d}(x)$ yields

$$\begin{aligned} \mathbf{e}''(x) &= \mathbf{G}(x)\mathbf{e}(x) + \boldsymbol{\psi}(x, \mathbf{y}(x)) + \boldsymbol{\varphi}'(x, \mathbf{y}(x)) \\ \mathbf{e}(a) &= \mathbf{0}, \qquad \mathbf{e}'(a) = \boldsymbol{\varphi}(a, \mathbf{y}(a)) \end{aligned} \tag{4-44}$$

It follows that $\mathbf{e}(x) = \mathbf{0}$ only in the exceptional case that $\boldsymbol{\psi}(x, \mathbf{y}(x)) + \boldsymbol{\varphi}'(x, \mathbf{y}(x)) = \mathbf{0}$ and $\boldsymbol{\varphi}(a, \mathbf{y}(a)) = \mathbf{0}$.

The above results may be formally stated as follows:

THEOREM 4.3. *If the method defined by $\underline{\mathbf{\Phi}}(x, \mathsf{y}; h)$ is of order p, and if the functions f and $\underline{\mathbf{\Phi}}$ have continuous derivatives of order $p + 2$ for $x \in [a, b]$ and arbitrary y, then the error $\mathsf{e}_n = \mathsf{y}_n - \mathsf{y}(x_n)$ in the solution of the initial value problem* (4-5) *satisfies*

$$\mathsf{e}_n = h^p\mathsf{e}(x_n) + O(h^{p+1}) \tag{4-45}$$

for $x_n \in [a, b]$, $h \to 0$, where $\mathsf{e}(x)$ is the solution of the initial value problem (4-41).

For the sake of illustration we shall apply the above result to the trivial initial value problem $y'' = -y$, $y(0) = 1$, $y'(0) = 0$ where $s = 1$. For the special Nyström formula (4-23) we have $p = 4$ and, using (4-44),

$$e''(x) = -e(x) + \tfrac{1}{160}y(x) = -e(x) + \tfrac{1}{160}\cos x,$$
$$e(0) = e'(0) = 0$$

hence

$$e(x) = \tfrac{1}{320}(x \cos x - \sin x)$$

For the ordinary Runge-Kutta method (3-48), as applied to the system $y' = z$, $z' = -y$, $y(0) = 1$, $z(0) = 0$ we would have (in the notation of the present chapter) $\varphi = -\frac{1}{120}z$, $\psi = \frac{1}{120}y$, hence

$$\mathbf{e}''(x) = -e(x) + \tfrac{1}{60} \cos x,$$

$$e(0) = e'(0) = 0$$

This leads to the slightly less favorable result

$$e(x) = \tfrac{1}{120}(x \cos x - \sin x)$$

4.4. Propagation of Round-off Error

4.4-1. Nonstatistical theory. Let the numerical values $\tilde{\mathbf{y}}_n$ satisfy

$$\tilde{\mathbf{y}}_{n+1} = \tilde{\mathbf{y}}_n + h\underline{\mathbf{\Phi}}(x_n, \tilde{\mathbf{y}}_n;\ h) + \underline{\boldsymbol{\epsilon}}_{n+1} \tag{4-46}$$

where the vectors $\underline{\boldsymbol{\epsilon}}_{n+1}$ represent the local round-off errors. A crude bound for the accumulated round-off error $\mathbf{r}_n = \tilde{\mathbf{y}}_n - \mathbf{y}_n$ can be obtained from Theorem 3.5: If the local round-off errors satisfy

$$\|\underline{\boldsymbol{\epsilon}}_n\| \leq \varepsilon, \qquad n = 1, 2, \cdots \tag{4-47}$$

and if L denotes the Lipschitz constant defined by (4-30), then we clearly have

$$\|\mathbf{r}_n\| \leq \frac{\varepsilon}{h} E_L(x_n - a), \qquad x_n \in [a, b] \tag{4-48}$$

More refined bounds can be obtained by assuming that

$$\hat{\underline{\boldsymbol{\epsilon}}}_n \leq \varepsilon \mathbf{p}(x_n) \tag{4-49}$$

where $\mathbf{p}(x)$ is a fixed, piecewise continuous and continuously differentiable vector function with nonnegative components, and where

$$Nh^{p+1} \leq \varepsilon \leq Kh^{q+1} \tag{4-50}$$

Here N, K, and q are constants which do not depend on h, $1 \leq q \leq p$. The hypothesis (4-50) is natural when the local round-off error is allowed to dominate the local discretization error. It then follows from Theorem 3.6 and the discussion preceding it that

$$\mathbf{r}_n = \mathbf{r}_n^{(1)} + \mathbf{r}_n^{(2)}$$

where $\mathbf{r}_n^{(2)} = O(\varepsilon)$, and

$$\hat{\mathbf{r}}_n^{(1)} \leq \frac{\varepsilon}{h}\{\mathbf{m}(x_n) + O(h)\} \tag{4-51}$$

The function $\mathbf{m}(x)$ is defined as the solution of the initial value problem

$$\mathbf{m}' = \hat{\mathbf{G}}(x)\mathbf{m} + \mathbf{p}(x), \qquad \mathbf{m}(a) = \mathbf{0} \tag{4-52}$$

In the case of a system of special second order equations we write

$$\mathbf{r}_n^{(1)} = \begin{pmatrix} \mathbf{r}_n^{(1)} \\ \mathbf{s}_n^{(1)} \end{pmatrix}, \qquad \mathbf{m} = \begin{pmatrix} \mathbf{m} \\ \mathbf{n} \end{pmatrix}, \qquad \mathbf{p}(x) = \begin{pmatrix} \mathbf{p}(x) \\ \mathbf{q}(x) \end{pmatrix}$$

In view of the special form (4-39a) of the matrix $\mathbf{G}$ the system (4-52) then takes the form

$$\begin{aligned} \mathbf{m}' &= \mathbf{n} + \mathbf{p}(x) \\ \mathbf{n}' &= \hat{\mathbf{G}}(x)\mathbf{m} + \mathbf{q}(x) \\ \mathbf{m}(a) &= \mathbf{n}(a) = \mathbf{0} \end{aligned} \tag{4-53}$$

If only the growth of $\mathbf{r}_n^{(1)}$ (i.e., the round-off error in $\mathbf{y}_n$) is of interest, $\mathbf{n}(x)$ can be eliminated from (4-53), and we find that

$$\hat{\mathbf{r}}_n^{(1)} \le \frac{\varepsilon}{h}\{\mathbf{m}(x_n) + O(h)\} \tag{4-54}$$

where

$$\begin{aligned} \mathbf{m}'' &= \hat{\mathbf{G}}(x)\mathbf{m} + \mathbf{p}'(x) + \mathbf{q}(x) \\ \mathbf{m}(a) &= \mathbf{0}, \qquad \mathbf{m}'(a) = \mathbf{p}(a) \end{aligned} \tag{4-55}$$

4.4-2. Statistical theory. The results of §3.4-5 are readily adapted to the present situation. We consider the round-off errors $\boldsymbol{\epsilon}_n$ as random variables with the expected values

$$\boldsymbol{\mu}_n = E(\boldsymbol{\epsilon}_n)$$

and assume that

$$\hat{\boldsymbol{\mu}}_n \le \mu\mathbf{p}(x_n), \tag{4-56}$$

$$E[(\boldsymbol{\epsilon}_n - \boldsymbol{\mu}_n)(\boldsymbol{\epsilon}_n^T - \boldsymbol{\mu}_n^T)] = \begin{cases} \mathbf{0}, & n \ne m \\ \sigma^2\mathbf{C}(x_n) \end{cases}$$

where $\mathbf{p}(x)$ is a known, piecewise continuous and continuously differentiable vector function, and where $\mathbf{C}(x)$ is a known, positive semidefinite matrix whose elements are functions of x with similar properties. By applying Theorem 3.7 to the initial value problem (4-5) we then find that

$$\widehat{E(\mathbf{r}_n^{(1)})} \le \frac{\mu}{h}\{\mathbf{m}(x_n) + O(h)\} \tag{4-57}$$

where $\mathbf{m}(x)$ is defined by (4-52). In particular, for systems of special second order equations,

$$\widehat{E(\mathbf{r}_n^{(1)})} \le \frac{\mu}{h}\{\mathbf{m}(x_n) + O(h)\} \tag{4-58}$$

with $\mathbf{m}(x)$ being defined by (4-55).

For the covariance matrix the results are more complicated. Theorem 3.7 states that

$$\text{covar}\,(\mathbf{r}_n^{(1)}) = \frac{\sigma^2}{h}\{\mathbf{V}(x_n) + O(h)\} \tag{4-59}$$

where $\mathbf{V}(x)$ is defined as the solution of the system of s^2q^2 differential equations

$$\mathbf{V}' = \mathbf{C} + \mathbf{G}\mathbf{V} + \mathbf{V}\mathbf{G}^T, \qquad \mathbf{V}(a) = \mathbf{0} \tag{4-60}$$

For systems of special second order equations we put

$$\text{covar}\,(\mathbf{r}_n^{(1)}) = \begin{pmatrix} \mathbf{R} & \mathbf{S} \\ \mathbf{S}^T & \mathbf{T} \end{pmatrix}, \qquad \mathbf{C} = \begin{pmatrix} \mathbf{C} & \mathbf{D} \\ \mathbf{D}^T & \mathbf{E} \end{pmatrix}, \qquad \mathbf{V} = \begin{pmatrix} \mathbf{V} & \mathbf{W} \\ \mathbf{W}^T & \mathbf{U} \end{pmatrix}$$

The system (4-60) can then be written in the form

$$\begin{aligned} \mathbf{V}' &= \mathbf{C} + \mathbf{W}^T + \mathbf{W}, \\ \mathbf{W}' &= \mathbf{D} + \mathbf{U} + \mathbf{V}\mathbf{G}^T, \qquad \mathbf{V}(a) = \mathbf{W}(a) = \mathbf{U}(a) = \mathbf{0} \\ \mathbf{U}' &= \mathbf{E} + \mathbf{G}\mathbf{W} + \mathbf{W}^T\mathbf{G}^T, \end{aligned} \tag{4-61}$$

If only the covariance of $\mathbf{r}_n^{(1)}$ is of interest, then only the matrix $\mathbf{V}$ needs to be determined. It does not seem possible, however, to eliminate $\mathbf{W}$ and $\mathbf{U}$ in any simple manner from the system except in the case $s = 1$, where the matrices $\mathbf{C}, \mathbf{D}, \cdots$ reduce to scalars. If

$$\mathbf{C} = \begin{pmatrix} c & d \\ d & e \end{pmatrix}, \qquad \mathbf{V} = \begin{pmatrix} v & w \\ w & u \end{pmatrix}$$

the system (4-61) reduces to

$$\begin{aligned} v' &= c + 2w \\ w' &= d + u + gv \\ u' &= e + 2gw \end{aligned} \tag{4-62}$$

where $g = g(x) = f_y(x, y(x))$. Assuming that $g(x)$ and the elements of $\mathbf{C}$ are sufficiently differentiable, we find

$$\begin{aligned} v'' &= c' + 2w' \\ &= c' + 2(d + u + gv) \\ v''' &= c'' + 2(d' + u' + (gv)') \\ &= c'' + 2(d' + e + g(v' - c) + (gv)') \end{aligned}$$

Thus $v(x)$ satisfies the single differential equation

$$v''' - 4gv' - 2g'v = c'' + 2d' + 2e - 2gc \tag{4-63a}$$

and the initial conditions

$$v(a) = 0, \qquad v'(a) = c(a), \qquad v''(a) = c'(0) + 2d(0) \tag{4-63b}$$

For the initial value problem $y'' = -y$, $y(0) = 1$, $y'(0) = 0$, which we considered already in §4.3-3, we have $g(x) = -1$. Assuming $c = e = 1$, $d = 0$ (constant diagonal local covariance matrix, appropriate to fixed point operations), the initial problem (4-63) reduces to

$$v''' + 4v' = 4, \qquad v(0) = v''(0) = 0, \qquad v'(0) = 1$$

It is easily verified that the solution is $v(x) = x$, in accordance with (3-152).

4.5. Problems for Solution

Section 4.2

1. Using the Nyström formula (4-22) with the steps $h = 0.4$ and $h = 0.2$, find the solution $y(x)$ of the initial value problem

$$y'' = -\sin y, \qquad y(0) = 1, \qquad y'(0) = 0$$

and determine by inverse interpolation (see Hildebrand [1956], p. 50) the smallest positive root ξ of $y(x)$.

[For $y(0) = \alpha$, an analytical solution is given by

$$\xi = \frac{\pi}{2}\left\{1 + \left(\frac{1}{2}\right)^2\left(\sin\frac{\alpha}{2}\right)^2 + \left(\frac{1.3}{2.4}\right)^2\left(\sin\frac{\alpha}{2}\right)^4 + \cdots\right\}\Bigg]$$

2. Find by numerical integration, using the Nyström formula (4-23), the asymptotic amplitude and phase shift of the sine-like curve defined by

$$y'' + \frac{4x^2}{1 + x^2}y = 0, \qquad y(0) = 0, \qquad y'(0) = 1$$

3. Determine the parameter p_1 in such a manner that the formulas (requiring *one* substitution!)

$$\begin{aligned} \mathbf{\Phi}(\mathbf{y}, \mathbf{z};\, h) &= \mathbf{z} + \tfrac{1}{2}h\mathbf{k} \\ \mathbf{\Psi}(\mathbf{y}, \mathbf{z};\, h) &= \mathbf{k} \end{aligned} \tag{4-64}$$

where $\mathbf{k} = \mathbf{f}(\mathbf{y} + hp_1\mathbf{z})$, represent a second order method for systems of special second order equations.

4. Let $\mathbf{f} = \mathbf{f}(\mathbf{y})$, and put, setting $\mathbf{f}_i = \partial\mathbf{f}/\partial y^i$ and adopting the summation convention,

$$\begin{aligned}
&\mathbf{A} = \mathbf{f} && \mathbf{H} = \mathbf{f}_{ijkm}z^iz^jz^kz^m \\
&\mathbf{B} = \mathbf{f}_iz^i && \mathbf{I} = \mathbf{f}_{ijk}f^iz^jz^k \\
&\mathbf{C} = \mathbf{f}_{ij}z^iz^j && \mathbf{J} = \mathbf{f}_{ij}f^i_kz^jz^k \\
&\mathbf{D} = \mathbf{f}_if^i && \mathbf{K} = \mathbf{f}_{ij}f^if^j \\
&\mathbf{E} = \mathbf{f}_{ijk}z^iz^jz^k && \mathbf{L} = \mathbf{f}_if^i_{jk}z^jz^k \\
&\mathbf{F} = \mathbf{f}_{ij}f^iz^j && \mathbf{M} = \mathbf{f}_if^i_jf^j \\
&\mathbf{G} = \mathbf{f}_if^i_jz^j &&
\end{aligned}$$

If $\mathbf{y}' = \mathbf{z}$, $\mathbf{y}'' = \mathbf{f}(\mathbf{y})$, show that the total derivatives of $\mathbf{f}$ with respect to x are given by

$$\begin{aligned} \mathbf{f}' &= \mathbf{B} \\ \mathbf{f}'' &= \mathbf{C} + \mathbf{D} \\ \mathbf{f}''' &= \mathbf{E} + 3\mathbf{F} + \mathbf{G} \\ \mathbf{f}^{\text{iv}} &= \mathbf{H} + 6\mathbf{I} + 4\mathbf{J} + 3\mathbf{K} + \mathbf{L} + \mathbf{M} \end{aligned}$$

5. If the vectors $\mathbf{k}_\rho$ are defined by (4-21), show that, in the notation of Problem 4,

$$\begin{aligned} \mathbf{k}_2 = \mathbf{A} &+ hp_1\mathbf{B} + h^2[\tfrac{1}{2}p_1^2\mathbf{C} + p_1q_1\mathbf{D}] + h^3[\tfrac{1}{6}p_1^3\mathbf{E} + p_1^2q_1\mathbf{F}] \\ &+ h^4[\tfrac{1}{24}p_1^4\mathbf{H} + \tfrac{1}{2}p_1^2q_1^2\mathbf{K} + \tfrac{1}{2}p_1^3q_1\mathbf{I}] + O(h^5) \end{aligned}$$

Verify that the method defined by (4-22) is the only method of the form (4-21) which achieves $p = 3$ with $\kappa = 2$.

6. Verify that

$$\begin{aligned} \mathbf{k}_3 = \mathbf{A} &+ hp_2\mathbf{B} + h^2[\tfrac{1}{2}p_2^2\mathbf{C} + p_2q_2\mathbf{D}] + h^3[\tfrac{1}{6}p_2^3\mathbf{E} + p_2^2q_2\mathbf{F} + p_1p_2q_3\mathbf{G}] \\ &+ h^4[\tfrac{1}{24}p_2^4\mathbf{H} + \tfrac{1}{2}p_2^3q_2\mathbf{I} + \tfrac{1}{2}p_2^2q_2^2\mathbf{K} + p_1p_2^2q_3\mathbf{J} \\ &+ \tfrac{1}{2}p_1^2p_2q_3\mathbf{L} + p_1p_2q_1q_3\mathbf{M}] + O(h^5) \end{aligned}$$

7. Using the results of the preceding problems, show that the requirement that the method (4-21) with $\kappa = 3$ has order $p = 4$ leads to the following one-parameter family of values of the coefficients a_i, b_i, p_i, and q_i:

$$p_1 = \frac{4t - 1}{6t}, \qquad p_2 = \frac{2 + t}{3}$$

$$q_1 = \frac{4t - 1}{12t}, \qquad q_2 = \frac{2 + t}{6}, \qquad q_3 = \frac{t(t + 2t^2)}{2(4t - 1)}$$

$$a_1 = \frac{2t - 1 + 4t^3(1 + t)}{(4t - 1)(4 + 2t)(1 + 2t^2)}, \qquad a_2 = \frac{t^2(1 + 2t)}{(4t - 1)(1 + 2t^2)},$$

$$a_3 = \frac{1 - t}{(4 + 2t)(1 + 2t^2)}$$

$$b_1 = \frac{2t - 1 + 4t^3(1 + t)}{(4t - 1)(4 + 2t)(1 + 2t^2)}, \qquad b_2 = \frac{6t^3}{(4t - 1)(1 + 2t^2)},$$

$$b_3 = \frac{3}{(4 + 2t)(1 + 2t^2)}$$

[Formulas (4-23) are obtained for $t = 1$.]

8*. Show that

$$\begin{aligned}\mathbf{k}_4 = \mathbf{A} &+ hp_3\mathbf{B} + h^2[\tfrac{1}{2}p_3^2\mathbf{C} + p_2q_4\mathbf{D}] + h^3[\tfrac{1}{6}p_3^3\mathbf{E} + p_3^2q_4\mathbf{F} + p_3(q_5p_1 + q_6p_2)\mathbf{G}] \\ &+ h^4[\tfrac{1}{24}p_3^4\mathbf{H} + \tfrac{1}{2}p_3^3q_4\mathbf{I} + p_3^2(q_5p_1 + q_6p_2)\mathbf{J} + \tfrac{1}{2}p_3(q_5p_1^2 + q_6p_2^2)\mathbf{L} \\ &+ p_3(p_1q_1q_5 + p_2q_2q_6)\mathbf{M} + \tfrac{1}{2}p_3^2q_4^2\mathbf{K}] + O(h^5)\end{aligned}$$

and verify that the Nyström formulas (4-24) are of order 5.

9. Determine all third order methods for special second order equations which are of the form

$$\begin{aligned}\mathbf{\Phi}(\mathbf{y}, \mathbf{z};\ h) &= \mathbf{z} + h(a_1\mathbf{k}_1 + a_2\mathbf{k}_2) \\ \mathbf{\Psi}(\mathbf{y}, \mathbf{z};\ h) &= b_1\mathbf{k}_1 + b_2\mathbf{k}_2\end{aligned}$$

where

$$\begin{aligned}\mathbf{k}_1 &= \mathbf{f}(\mathbf{y} + hp_0\mathbf{z}) \\ \mathbf{k}_2 &= \mathbf{f}(\mathbf{y} + hp_1\mathbf{z} + h^2p_1q_1\mathbf{k}_1)\end{aligned}$$

[Formulas (4-22) are the special case $p_0 = 0$.]

Section 4.3

10. Determine bounds for the constants L and N in Theorem 4.2 for the third order method (4-22).

11*. Repeat Problem 10 for the fourth order method (4-23).

12. Express the components $\boldsymbol{\varphi}$ and $\boldsymbol{\psi}$ of the principal error function of the method (4-22) in terms of the vectors **A**, **B**, **C**, $\cdots$ defined in Problem 4, and check the relations (4-37).

13. Show that for the method (4-23) the components of the principal error function are given by

$$\begin{aligned}\boldsymbol{\varphi}(\mathbf{y}, \mathbf{z}) &= -\tfrac{1}{720}\mathbf{E} - \tfrac{1}{240}\mathbf{F} - \tfrac{1}{120}\mathbf{G} \\ \boldsymbol{\psi}(\mathbf{y}, \mathbf{z}) &= \tfrac{1}{2880}\mathbf{H} + \tfrac{11}{1920}\mathbf{I} + \tfrac{1}{120}\mathbf{J} + \tfrac{1}{720}\mathbf{K} + \tfrac{1}{480}\mathbf{L} + \tfrac{1}{480}\mathbf{M}\end{aligned}$$

and verify formulas (4-38).

14. Calculate the principal error function $\underline{\boldsymbol{\varphi}}$ for the family of methods constructed in Problem 9. Is it possible to select the parameter in such a manner that $\boldsymbol{\psi} + \boldsymbol{\varphi}' = 0$? What would be the significance of this result?

15. Determine the magnified error function $\mathbf{e}(x)$, if the initial value problem (3-99) is solved by the third order Nyström formula (4-23). How does the result compare with the third-order Radau formula?

Section 4.4

16. Study the propagation of round-off error in the solution of the initial value problem $y'' = y$, $y(0) = 1$, $y'(0) = -1$, assuming symmetric distribution of local round-off and (a) fixed decimal point operations, (b) floating decimal point operations. [In the latter case, set $c(x) = (y(x))^2$, $e(x) = (y'(x))^2$, $d(x) = 0$ in (4-62).]

17*. Formulate results analogous to those implied by §§4.4-1 and 4.4-2 for the numerical integration of a differential equation $y''' = f(x, y)$ and, more generally, $y^{(p)} = f(x, y)$, by a special one-step method.

Notes

§4.1. The advantages and disadvantages of the reduction of an nth order equation to a system of first order equations are argued by Collatz and Zurmühl [1942] (see also Collatz [1960], p. 117), Rutishauser [1955], Ceschino and Kuntzmann [1960].

§4.2. For further methods for the direct integration of equations of higher order, see Nyström [1925], Babkin [1948], Rapoport [1952], de Vogelaere [1955], Vorobev [1956], and the methods discussed in Chapter 6.

part II MULTISTEP METHODS FOR INITIAL VALUE PROBLEMS

IN THE methods considered so far the increment $y_{n+1} - y_n$ of the approximate solution at the point x_n was calculated by means of a function which depended (in the case of a single equation) only on x_n, y_n, and the step h. The methods thus obtained are conceptually simple, versatile, and easy to code. On the other hand, they are inefficient in that they do not make full use of the available information. It seems plausible that more accuracy can be obtained if the value of y_{n+1} is made to depend not only on y_n but also, say, on y_{n-1} and y_{n-2}. Some of the methods based on this idea have become very popular. For comparable accuracy, they require in general less work than one-step methods. This advantage is obtained at the expense of some increase in complexity (a special starting procedure is necessary) and, in some cases, of the possibility of a dangerous build-up of both discretization and round-off error (numerical instability).

chapter

5 Multistep Methods for Equations of the First Order

In this chapter we shall study the application of multistep methods to the initial value problem

$$y' = f(x, y), \qquad y(a) = \eta \tag{5-1}$$

where $f(x, y)$ satisfies the conditions of the existence theorem of Chapter 1. Only problems involving a single differential equation will be considered. All essential features are present in this special case, although the methods to be discussed lend themselves with equal ease to the solution of systems of differential equations.

5.1. Special Multistep Methods

5.1-1. The interpolating polynomial. Since most of the commonly used multistep methods are based on interpolation, we shall review in this section some properties of the Lagrangian interpolation polynomial. Let the function $z(x)$ be defined on an interval J containing the $q + 1$ *distinct* points $x_0, x_1, \cdots, x_q$. It is well known that among all polynomials in x of degree not exceeding q there exists exactly one polynomial $P(x)$ which satisfies the relations

$$P(x_\nu) = z(x_\nu), \qquad \nu = 0, 1, \cdots, q$$

or which, in other words, *interpolates* the function $z(x)$ at the points x_ν. The uniqueness of this so-called interpolating polynomial follows from the fact that the difference of any two such polynomials is a polynomial of degree $\leq q$ which has $q + 1$ zeros (namely at the points x_ν) and therefore vanishes identically. The existence can be proved by exhibiting the polynomial explicitly, for instance in the form

$$P(x) = L(x) \sum_{\nu=0}^{q} \frac{z(x_\nu)}{L'(x_\nu)(x - x_\nu)} \tag{5-2}$$

where

$$L(x) = (x - x_0)(x - x_1) \cdots (x - x_q) \tag{5-3}$$

It is natural to expect that $P(x)$ reproduces $z(x)$ not only exactly at the points $x = x_\nu$ but also approximately for other values of x. The following lemma estimates the error committed in this approximation.

LEMMA 5.1. *Let $z(x)$ have a continuous derivative of order $q + 1$ in J. Then for every point x in J there exists a point ξ in the smallest interval I containing both x and the points x_ν ($\nu = 0, \cdots, q$) such that*

$$z(x) - P(x) = \frac{1}{(q + 1)!} L(x) z^{(q+1)}(\xi) \tag{5-4}$$

Proof. If x is one of the points x_ν, there is nothing to prove, since then both sides of (5-4) vanish for arbitrary ξ. If $x \neq x_\nu$, $\nu = 0, \cdots, q$, consider the auxiliary function of the variable t,

$$Z(t) = z(t) - P(t) - \lambda L(t)$$

where

$$\lambda = \frac{z(x) - P(x)}{L(x)} \tag{5-5}$$

so that $Z(x) = 0$. Obviously the function $Z(t)$ also vanishes at all points x_ν, so that it vanishes at $q + 2$ distinct points of the interval I. By repeated applications of Rolle's theorem, its μth derivative thus vanishes at least at $q + 2 - \mu$ distinct points of I ($\mu = 1, 2, \cdots, q + 1$). Now let ξ be a point where the $(q + 1)$st derivative vanishes. Since $P^{(q+1)}(\xi) \equiv 0$ and $L^{(q+1)}(\xi) = (q + 1)!$, we have at this point

$$Z^{(q+1)}(\xi) = z^{(q+1)}(\xi) - \lambda(q + 1)! = 0$$

Using (5-5) and transposing, we get the desired result.

Remark. Although the quantity ξ as a function of x is in general not known, we can assert that $z^{(q+1)}(\xi)$ is a continuous function of x for $x \in J$. This is readily seen by solving (5-4) for $z^{(q+1)}(\xi)$. It is then clear that $z^{(q+1)}(\xi)$ is continuous for $x \neq x_\nu$; at the points $x = x_\nu$ the continuity follows from L'Hospital's rule, since $L'(x_\nu) = \prod_{\mu \neq \nu}(x_\mu - x_\nu) \neq 0$.

We shall also use the derivative of $P(x)$ to approximate $z'(x)$. The quality of this approximation *at the points* x_ν is estimated in a simple manner by the following result.*

LEMMA 5.2. *Let $z(x)$ satisfy the same hypothesis as in Lemma 5.1. Then for every x_μ $(\mu = 0, 1, \cdots, q)$ there exists a number $\xi \in J$ such that*

$$z'(x_\mu) - P'(x_\mu) = \frac{1}{(q+1)!} z^{(q+1)}(\xi) L'(x_\mu) \tag{5-6}$$

Proof. Without loss of generality we may assume that $x_0 < x_1 < \cdots < x_q$. For every choice of the constant λ the function

$$Z(x) = z(x) - P(x) - \lambda L(x)$$

has $q + 1$ zeros at the points $x = x_\nu$, $\nu = 0, \cdots, q$. By Rolle's theorem, $Z'(x)$ has q zeros, namely one in the interior of each interval $(x_\nu, x_{\nu+1})$ $(\nu = 0, \cdots, q - 1)$. Now choose λ such that $Z'(x)$ has an additional zero at $x = x_\mu$. This leads to the condition

$$\lambda = \frac{z'(x_\mu) - P'(x_\mu)}{L'(x_\mu)} \tag{5-7}$$

which can be satisfied in view of $L'(x_\mu) \neq 0$. The function $Z'(x)$ now has $q + 1$ distinct zeros in J. By repeated applications of Rolle's theorem it follows that $Z^{(q+1)}(x)$ has at least one zero in J, say at $x = \xi$. Thus,

$$Z^{(q+1)}(\xi) = z^{(q+1)}(\xi) - \lambda(q + 1)!$$

and the conclusion follows by (5-7).

In applications to differential equations the points x_ν are usually equally spaced. The interpolating polynomial is then most conveniently represented in terms of finite differences. There exist a bewildering variety of definitions of finite difference symbols and of formulas that can be built up from them.† In the interest of formal economy we shall use only one difference symbol. Since the interpolating polynomial itself is unique and only the various shapes into which it can be put are different, nothing is lost by this except occasionally a touch of formal elegance.

We now assume that $x_p = x_0 + ph$, where h is a constant and p is an integer, and write $z_p = z(x_p)$. The *first backward difference* of the function $z(x)$ at the point $x = x_p$ is defined by

$$\nabla z_p = z_p - z_{p-1} \tag{5-8}$$

* At an arbitrary point x the estimate is more complicated (see Hildebrand [1956] p. 66).

† See, e.g., Hildebrand [1956], Chapters 4 and 5.

Higher backward differences are defined by $\nabla^q z_p = \nabla(\nabla^{q-1} z_p)$, so that for instance

$$\nabla^2 z_p = (z_p - z_{p-1}) - (z_{p-1} - z_{p-2}) = z_p - 2z_{p-1} + z_{p-2}$$

For symmetry one also puts $\nabla^0 z_p = z_p$. One easily verifies by induction that

$$\nabla^q z_p = \sum_{m=0}^{q} (-1)^m \binom{q}{m} z_{p-m}, \qquad q = 0, 1, \cdots \tag{5-9}$$

where $\binom{q}{m}$ denotes the binomial coefficient

$$\begin{aligned} \binom{q}{0} &= 1 \\ \binom{q}{m} &= \frac{q(q-1)\cdots(q-m+1)}{1 \cdot 2 \cdots m}, \qquad m = 1, 2, \cdots \end{aligned} \tag{5-10}$$

which can be defined also when q is not a natural number.

Formula (5-9) expresses the differences in terms of function values or *ordinates*, as they are sometimes called in this connection. It will also be necessary to express the ordinates in terms of differences. We have

$$\begin{aligned} z_p &= \nabla^0 z_p \\ z_{p-1} &= z_p - (z_p - z_{p-1}) = \nabla^0 z_p - \nabla^1 z_p \\ z_{p-2} &= z_p - 2(z_p - z_{p-1}) + (z_p - 2z_{p-1} + z_{p-2}) \\ &= \nabla^0 z_p - 2\nabla^1 z_p + \nabla^2 z_p \end{aligned}$$

and suspect the validity of the formula

$$z_{p-q} = \sum_{m=0}^{q} (-1)^m \binom{q}{m} \nabla^m z_p, \qquad q = 0, 1, \cdots \tag{5-11}$$

which is indeed easily verified by induction.

In order to express the interpolating polynomial in terms of differences, it is convenient to introduce the variable

$$s = \frac{x - x_p}{h}$$

which measures x in units of h, starting at $x = x_p$. It is now an easy matter to show that the polynomial $P(x)$ of degree $\leq q$ interpolating the function $z(x)$ at the points $x_p, x_{p-1}, \cdots, x_{p-q}$ can be represented in the form

$$P(x) = \sum_{m=0}^{q} (-1)^m \binom{-s}{m} \nabla^m z_p \tag{5-12}$$

Since $\binom{-s}{m}$ is a polynomial of degree m in s and since s is a linear function of x, this is indeed a polynomial of degree $\leq q$. Also, for $r = 0, 1, \cdots, q$,

$$P(x_{p-r}) = \sum_{m=0}^{q} (-1)^m \binom{r}{m} \nabla^m z_p = \sum_{m=0}^{r} (-1)^m \binom{r}{m} \nabla^m z_p = z_{p-r}$$

by (5-11), since $\binom{r}{m} = 0$ for r an integer $<m$.

The representation (5-12) of the interpolating polynomial is called the *Newtonian* backward-difference formula and will be used frequently in the following text.

5.1-2. Methods based on numerical integration. An exact solution of the differential equation (5-1) by definition satisfies the identity

$$y(x + k) - y(x) = \int_x^{x+k} f(t, y(t))\, dt \tag{5-13}$$

for any two points x and $x + k$ in the interval $[a, b]$. The methods now to be discussed are based on replacing the function $f(x, y(x))$, which is unknown, by an interpolating polynomial having the values $f_n = f(x_n, y_n)$ on a set of points x_n where y_n has already been computed or is just about to be computed, evaluating the integral and accepting its value as the increment of the approximate values y_n between x and $x + k$. We shall assume that the interpolating points are $x_p, x_{p-1}, \cdots, x_{p-q}$; the polynomial replacing $f(x, y(x))$ is then given by

$$P(x) = \sum_{m=0}^{q} (-1)^m \binom{-s}{m} \nabla^m f_p, \qquad s = \frac{x - x_p}{h}$$

The positive integer q is arbitrary in principle, but in practice rarely exceeds 6, say. Several classes of methods can now be distinguished

Table 5.1

Method	x	$x + k$
Adams-Bashforth	x_p	x_{p+1}
Adams-Moulton	x_{p-1}	x_p
Nyström	x_{p-1}	x_{p+1}
Milne-Simpson	x_{p-2}	x_p

according to the position of x and $x + k$ relative to the interpolating points; see Table 5.1. We shall consider each of these methods in greater detail.

(i) The Adams-Bashforth method (Bashforth and Adams [1883]). Here we have

$$y_{p+1} - y_p = \int_{x_p}^{x_{p+1}} P(x)\,dx = h \sum_{m=0}^{q} \gamma_m \nabla^m f_p \tag{5-14}$$

where the constants

$$\gamma_m = (-1)^m \frac{1}{h} \int_{x_p}^{x_{p+1}} \binom{-s}{m} dx = (-1)^m \int_0^1 \binom{-s}{m} ds \tag{5-15}$$

are independent of f and will be calculated numerically below. If the values $y_p, y_{p-1}, \cdots, y_{p-q}$ are known, the corresponding values $f_n = f(x_n, y_n)$ can be calculated and the differences $\nabla^m f_p$ are easily formed. The expression on the right of (5-14) is thus known, and $y_{p+1} - y_p$ and hence y_{p+1} can be calculated. The index p is then increased by 1, and the same formula is used to calculate y_{p+2}, etc. The method breaks down if some of the values $y_p, y_{p-1}, \cdots, y_{p-q}$ are not known. Such is the case at the beginning of the computation, where the initial condition furnishes only one of the required $q + 1$ values, or at places where the step h is changed. In such cases the missing values have to be secured by an independent method. One-step methods such as the Runge-Kutta method or Taylor's expansion are frequently used; other starting methods, which are more within the spirit of difference methods, are also known (Collatz [1960], p. 81). *The difficulty of missing starting values is typical for all multistep methods.*

In order to find a recurrence relation, numerical values, and other useful properties of the coefficients γ_m we use the method of generating functions. Such we call the function $G(t)$ which has the γ_m as coefficients in its Maclaurin expansion. Using the general form of the binomial theorem, we find

$$G(t) = \sum_{m=0}^{\infty} \gamma_m t^m = \sum_{m=0}^{\infty} (-t)^m \int_0^1 \binom{-s}{m} ds$$

$$= \int_0^1 \sum_{m=0}^{\infty} (-t)^m \binom{-s}{m} ds = \int_0^1 (1-t)^{-s}\,ds$$

The integral is easily evaluated by writing $(1 - t)^{-s} = e^{-s \log(1-t)}$. We thus find*

$$G(t) = -\frac{t}{(1-t)\log(1-t)} \tag{5-16}$$

which we may write as

$$-\frac{\log(1-t)}{t} G(t) = \frac{1}{1-t}$$

* The formal manipulation can be justified on the grounds that $G(t)$ is a holomorphic function of the complex variable t for $|t| < 1$.

Using the well-known expansions

$$\frac{1}{1-t} = 1 + t + t^2 + \cdots$$

(5-17)

$$-\frac{\log(1-t)}{t} = 1 + \tfrac{1}{2}t + \tfrac{1}{3}t^2 + \cdots$$

we obtain the following identity between power series:

$$(1 + \tfrac{1}{2}t + \tfrac{1}{3}t^2 + \cdots)(\gamma_0 + \gamma_1 t + \gamma_2 t^2 + \cdots) = 1 + t + t^2 + \cdots$$

By comparing the coefficients of corresponding powers of t, we find the relation

$$\gamma_m + \tfrac{1}{2}\gamma_{m-1} + \tfrac{1}{3}\gamma_{m-2} + \cdots + \frac{1}{m+1}\gamma_0 = 1, \qquad m = 0, 1, 2, \cdots$$

which makes it possible to calculate the γ_m recursively. The values given in Table 5.2 have been calculated in this manner.

Table 5.2

m	0	1	2	3	4	5	6
γ_m	1	$\frac{1}{2}$	$\frac{5}{12}$	$\frac{3}{8}$	$\frac{251}{720}$	$\frac{95}{288}$	$\frac{19087}{60480}$

Explicit use of the differences $\nabla^m f_p$, as indicated in (5-14), is especially suitable for computation by hand, where the computer at every moment can exert his free will and judgment to make changes in the computational procedure. For instance, if the difference of highest order $\nabla^q f_p$ is large, this may be considered an indication that the accuracy is insufficient and that the number of terms in (5-14) should be increased. However, if the number q is fixed once and for all, as is frequently the case if the Adams method is programmed for an automatic computer, there is no particular reason why the differences should be carried along explicitly. Expressing the differences in terms of ordinates by (5-9) and collecting the coefficients of equal ordinates, the Adams-Bashforth formula appears in the form

(5-18)
$$y_{p+1} - y_p = h \sum_{\rho=0}^{q} \beta_{q\rho} f_{p-\rho}$$

where the coefficients $\beta_{q\rho}$ are given by

$$\beta_{q\rho} = (-1)^\rho \left\{ \binom{\rho}{\rho}\gamma_\rho + \binom{\rho+1}{\rho}\gamma_{\rho+1} + \cdots + \binom{q}{\rho}\gamma_q \right\},$$

$$\rho = 0, 1, \cdots, q; \; q = 0, 1, \cdots$$

It should be noted that the coefficients $\beta_{q\rho}$ depend on q as well as ρ, which makes it more difficult to change the number of differences employed. Some numerical values are given in Table 5.3. The numerically large values of the coefficients and the alternating signs are a disadvantage of the method; see §5.3-4.

Table 5.3

Coefficients $\beta_{q\rho}$

ρ	0	1	2	3	4	5
$\beta_{0\rho}$	1					
$2\beta_{1\rho}$	3	−1				
$12\beta_{2\rho}$	23	−16	5			
$24\beta_{3\rho}$	55	−59	37	−9		
$720\beta_{4\rho}$	1901	−2774	2616	−1274	251	
$1440\beta_{5\rho}$	4227	−7673	9482	−6798	2627	−425

(ii) The Adams-Moulton method (Moulton [1926]). Equation (5-13) is now used in the form

(5-19)
$$y_p - y_{p-1} = \int_{x_{p-1}}^{x_p} P(x)\,dx = h \sum_{m=0}^{q} \gamma_m^* \nabla^m f_p$$

where

(5-20)
$$\gamma_m^* = (-1)^m h^{-1} \int_{x_{p-1}}^{x_p} \binom{-s}{m} dx = (-1)^m \int_{-1}^{0} \binom{-s}{m} ds$$

By operations which are entirely analogous to those which led to (5-16), the generating function of the coefficients γ_m^* is determined as follows:

(5-21)
$$G^*(t) = \sum_{m=0}^{\infty} \gamma_m^* t^m = -\frac{t}{\log(1-t)}$$

We thus have

$$-\frac{\log(1-t)}{t} G^*(t) = 1$$

or, using the expansion (5-17),

$$(1 + \tfrac{1}{2}t + \tfrac{1}{3}t^2 + \cdots)(\gamma_0^* + \gamma_1^* t + \gamma_2^* t^2 + \cdots) = 1$$

It follows that

$$\gamma_m^* + \tfrac{1}{2}\gamma_{m-1}^* + \tfrac{1}{3}\gamma_{m-2}^* + \cdots + \frac{1}{m+1}\gamma_0^* = \begin{cases} 1, & m = 0 \\ 0, & m = 1, 2, 3, \cdots \end{cases}$$

The numerical values given in Table 5.4 are easily found from these recurrence relations.

Table 5.4

m	0	1	2	3	4	5	6
γ_m^*	1	$-\frac{1}{2}$	$-\frac{1}{12}$	$-\frac{1}{24}$	$-\frac{19}{720}$	$-\frac{3}{160}$	$-\frac{863}{60480}$

We also note the relation

$$\frac{1}{1-t}G^*(t) = G(t)$$

or

$$(1 + t + t^2 + \cdots)(\gamma_0^* + \gamma_1^* t + \cdots) = \gamma_0 + \gamma_1 t + \gamma_2 t^2 + \cdots$$

Comparing the coefficients of t^m, we obtain

$$\gamma_0^* + \gamma_1^* + \cdots + \gamma_m^* = \gamma_m, \qquad m = 0, 1, 2, \cdots \tag{5-22}$$

Formula (5-19) is used like the Adams-Bashforth formula except that now only the values $y_{p-1}, y_{p-2}, \cdots, y_{p-q}$ are known and (5-19) is used to determine y_p. Since y_p occurs as an argument in $f_p = f(x_p, y_p)$ in the right-hand term of (5-19), this equation now represents a nontrivial equation for y_p. In general, it will not be possible to solve this equation explicitly. Fortunately, the special form of the equation suggests an iteration procedure which furnishes the solution very rapidly if h is sufficiently small.

Assuming that from some source an approximation $y_p^{(0)}$ of a solution of (5-19) has been obtained, we calculate $f_p^{(0)} = f(x_p, y_p^{(0)})$ and form the differences $\nabla f_p^{(0)} = f_p^{(0)} - f_{p-1}$, $\nabla^2 f_p^{(0)} = \nabla f_p^{(0)} - \nabla f_{p-1}, \cdots$. A better approximation $y_p^{(1)}$ is then obtained from

$$y_p^{(1)} = y_{p-1} + h \sum_{m=0}^{q} \gamma_m^* \nabla^m f_p^{(0)} \tag{5-23a}$$

Calculating $f_p^{(1)} = f(x_p, y_p^{(1)})$ and re-evaluating the differences, a still better value $y_p^{(2)}$ is

$$y_p^{(2)} = y_{p-1} + h \sum_{m=0}^{q} \gamma_m^* \nabla^m f_p^{(1)} \tag{5-23b}$$

Generally a sequence $y_p^{(\nu)}$ $(\nu = 0, 1, 2, \cdots)$ of approximations is obtained recursively from the relation

$$y_p^{(\nu+1)} = y_{p-1} + h \sum_{m=0}^{q} \gamma_m^* \nabla^m f_p^{(\nu)} \tag{5-23c}$$

where $f_p^{(\nu)} = f(x_p, y_p^{(\nu)})$. It follows from Theorem 5.4, to be proved in §5.2-2, that the sequence of numbers $y_p^{(0)}, y_p^{(1)}, y_p^{(2)}, \cdots$ thus defined converges for sufficiently small values of h to a solution y_p of (5-19) and that this solution is unique.

In actual numerical computation it is not necessary† to evaluate the full formula (5-23*c*) at each iteration step. Subtracting from (5-23*c*) the corresponding relation with ν replaced by $\nu - 1$, we get

$$y_p^{(\nu+1)} - y_p^{(\nu)} = h \sum_{m=0}^{q} \gamma_m^* \, (\nabla^m f_p^{(\nu)} - \nabla^m f_p^{(\nu-1)})$$

The differences $\nabla^m f_p^{(\nu)}$ and $\nabla^m f_p^{(\nu-1)}$ differ only in the foremost ordinates $f_p^{(\nu)}$ and $f_p^{(\nu-1)}$. It thus follows that

$$\nabla^m f_p^{(\nu)} - \nabla^m f_p^{(\nu-1)} = f_p^{(\nu)} - f_p^{(\nu-1)}$$

We thus have

$$y_p^{(\nu+1)} - y_p^{(\nu)} = h \sum_{m=0}^{q} \gamma_m^* (f_p^{(\nu)} - f_p^{(\nu-1)})$$

or, using (5-22),

(5-23*d*) $$y_p^{(\nu+1)} - y_p^{(\nu)} = h\gamma_q (f_p^{(\nu)} - f_p^{(\nu-1)})$$

The solution y_p of (5-19) is now obtained by summing up the terms of the series

$$y_p^* = y_p^{(1)} + (y_p^{(2)} - y_p^{(1)}) + (y_p^{(3)} - y_p^{(2)}) + \cdots$$

which are obtained by repeated use of (5-23*d*). Practically, the computation is stopped as soon as $f_p^{(\nu)} - f_p^{(\nu-1)}$ is negligible. The value of $f_p^{(\nu)}$ calculated last is accepted as the final value of f_p. With it are formed the definitive values of the differences $\nabla^m f_p$. Since the value y_p^* obtained by summing the series may be affected with round-off error, it is advisable to evaluate formula (5-19) now once more to obtain the final value of y_p.

A formula which furnishes a first approximation $y_p^{(0)}$ for the iteration procedure described above is called a *predictor formula.* Formula (5-23*c*) in this connection is called a *corrector formula.* The above iteration process thus consists in *predicting* a tentative value $y_p^{(0)}$ of y_p and correcting it (possibly a number of times) by means of the corrector formula. In the interest of minimizing the number of corrections it is clearly desirable to predict $y_p^{(0)}$ as accurately as possible. If the predictor formula is sufficiently accurate, it may be possible to draw a conclusion about the local discretization error from the difference between the predicted value $y_p^{(0)}$ and the final value y_p (see §5.3-6). Also, a sufficiently accurate predictor may make it unnecessary to correct more than once (see §5.3-7). For these reasons, the Adams-Bashforth formula [(5-14) with p diminished by 1] is recommended as an accurate predictor formula in connection with the Adams-Moulton method.

† The following simplification is due to Stohler [1943].

As a numerical illustration, we shall perform the first few integration steps in the solution of the initial value problem

$$y' = x - y^2, \qquad y(0) = 0$$

by the Adams-Moulton method with $h = 0.1$, using $q = 2$ in (5-19). The formula then reads

$$y_{p-1} - y_p = h\{f_p - \tfrac{1}{2}\nabla f_p - \tfrac{1}{12}\nabla^2 f_p\}$$

The Adams-Bashforth formula

$$y_p - y_{p-1} = h\{f_{p-1} + \tfrac{1}{2}\nabla f_{p-1} + \tfrac{5}{12}\nabla^2 f_{p-1}\}$$

is used as a predictor formula. The three starting values required are found from the series expansion

$$y(x) = \tfrac{1}{2}x^2 - \tfrac{1}{20}x^5 + \tfrac{1}{160}x^8 - \cdots$$

of the exact solution. The abscissas of the starting values are chosen symmetrically with respect to the "true" starting value $x_0 = 0$, so that the series is evaluated with $x = -h$ and $x = h$.

We now are ready to fill in the entries corresponding to x_{-1}, x_0, and x_1 in Table 5.5; the remaining work follows the procedure outlined above. Arrows indicate the order in which values are computed; an arrow of the form ↙‾‾ indicates that all values in the row above and to the right of the arrow have been used to calculate the result to the left. For $x = x_3$ the entries are given in abbreviated form only.

Expressing differences in terms of ordinates, the Adams-Moulton formula (5-19) can be written in the form

$$y_p - y_{p-1} = h \sum_{\rho=0}^{q} \beta^*_{q\rho} f_{p-\rho} \tag{5-24}$$

where

$$\beta^*_{q\rho} = (-1)^\rho \left\{ \binom{\rho}{\rho}\gamma^*_\rho + \binom{\rho+1}{\rho}\gamma^*_{\rho+1} + \cdots + \binom{q}{\rho}\gamma^*_q \right\},$$

$$\rho = 0, 1, \cdots, q; \; q = 0, 1, \cdots$$

Some numerical values are given in Table 5.6.

Figure 5.1 is a flow diagram for the solution of an initial value problem by the Adams-Moulton method, using the Adams-Bashforth formula as a predictor and working with ordinates. It is assumed that the starting values $y_0, y_1, \cdots, y_{q-1}$ are generated by a one-step method defined by the increment function Φ.

Because of the necessity of determining the new value y_p by solving an equation, the Adams-Moulton method is called an *implicit* method. In contrast, the Adams-Bashforth procedure may be called *explicit*. The greater computational complexity of the Adams-Moulton algorithm is tolerated because it leads to more accurate results. This is intuitively clear

Table 5.5

Example for Adams-Moulton Method

x_n	y_n		f_n	∇f_n	$\nabla^2 f_n$
−0.1	**0.00500 05000**	→	**−0.10002 50050**		
0	**0**	→	**0**	→ **0.10002 50050**	
0.1	**0.00499 95000** ↓	→	**0.09997 50050**	→ **0.09997 50050**	→ **−0.00005 00000**
0.2	$y_2^{(0)}$ = 0.01999 36674	↙ →	$f_2^{(0)}$ = 0.19960 02533	→ 0.09962 52483	→ −0.00034 97567
	$y_2^{(1)}$ = 0.01998 11776	↙ →	$f_2^{(1)}$ = 0.19960 07525 ↓	tentative differences formed with	
	$y_2^{(2)} - y_2^{(1)}$ = 208 ↓	←	$f_2^{(1)} - f_2^{(0)}$ = 4992	$f(x_2, y_2^{(0)})$	
	$y_2^{(2)}$ = 0.01998 11984	→	$f_2^{(2)}$ = 0.19960 07517 ↓		
	−0 ↓	←	$f_2^{(2)} - f_2^{(1)}$ = −8		
	$y_2^* = y_2^{(3)}$ = 0.01998 11984	→	$f_2^{(3)}$ = **0.19960 07517**	→ **0.09962 57467**	→ **−0.00034 92583**
	y_2 = **0.01998 11983** ↓	↙			
0.3	$y_3^{(0)}$ = 0.04490 80084	↙ →	$f_3^{(0)}$ = 0.29798 32708	0.09838 25919	−0.00124 32276
	$y_3^{(1)}$ = 0.04487 07597	↙ →	$f_3^{(1)}$ = 0.29798 66149	tentative differences	
	$y_3^{(2)}$ = 0.04487 08990	↙ →	$f_3^{(2)}$ = 0.29798 66024		
	$y_3^* = y_3^{(3)}$ = 0.04487 08985	↙ →	$f_3^{(3)}$ = **0.29798 66024**	**0.09838 58507**	**−0.00123 98960**
	y_3 = **0.04487 08985**	↙			

Table 5.6

Coefficients β^*_{qp}

p	0	1	2	3	4	5
β^*_{0p}	1					
$2\beta^*_{1p}$	1	1				
$12\beta^*_{2p}$	5	8	−1			
$24\beta^*_{3p}$	9	19	−5	1		
$720\beta^*_{4p}$	251	646	−264	106	−19	
$1440\beta^*_{5p}$	475	1427	−798	482	−173	27

because it uses the interpolating polynomial where it is more accurate; for a formal discussion see §5.1-3. The special case $q = 1$ of the Adams-Moulton method is known as the *trapezoidal rule.*

(iii) Nyström's method (Nyström [1925]). Here we set

$$y_{p+1} - y_{p-1} = \int_{x_{p-1}}^{x_{p+1}} P(x)\,dx = h \sum_{m=0}^{q} \kappa_m \nabla^m f_p \tag{5-25}$$

where

$$\kappa_m = (-1)^m \frac{1}{h} \int_{x_{p-1}}^{x_{p+1}} \binom{-s}{m} dx = (-1)^m \int_{-1}^{1} \binom{-s}{m} ds \tag{5-26}$$

This method is explicit and works much like the Adams-Bashforth method, except that now the increment of y_n is calculated over two steps in place of one. The method is only *weakly stable,* a property* which it shares with all methods which calculate $y_n - y_m$ where $n - m > 1$.

The generating function of the coefficients κ_m is found to be

$$K(t) = \sum_{m=0}^{\infty} \kappa_m t^m = -\frac{t}{\log(1-t)} \cdot \frac{2-t}{1-t} \tag{5-27}$$

Using the expansions (5-17) and

$$\frac{2-t}{1-t} = 1 + \frac{1}{1-t} = 2 + t + t^2 + \cdots$$

we thus have the identity

$$(1 + \tfrac{1}{2}t + \tfrac{1}{3}t^2 + \cdots)(\kappa_0 + \kappa_1 t + \kappa_2 t^2 + \cdots) = 2 + t + t^2 + \cdots$$

from which we may derive the recurrence relations

$$\kappa_m + \tfrac{1}{2}\kappa_{m-1} + \tfrac{1}{3}\kappa_{m-2} + \cdots + \frac{1}{m+1}\kappa_0 = \begin{cases} 2, & m = 0 \\ 1, & m = 1, 2, \cdots \end{cases}$$

* The phenomenon of weak stability is studied in §§5.3 and 5.4.

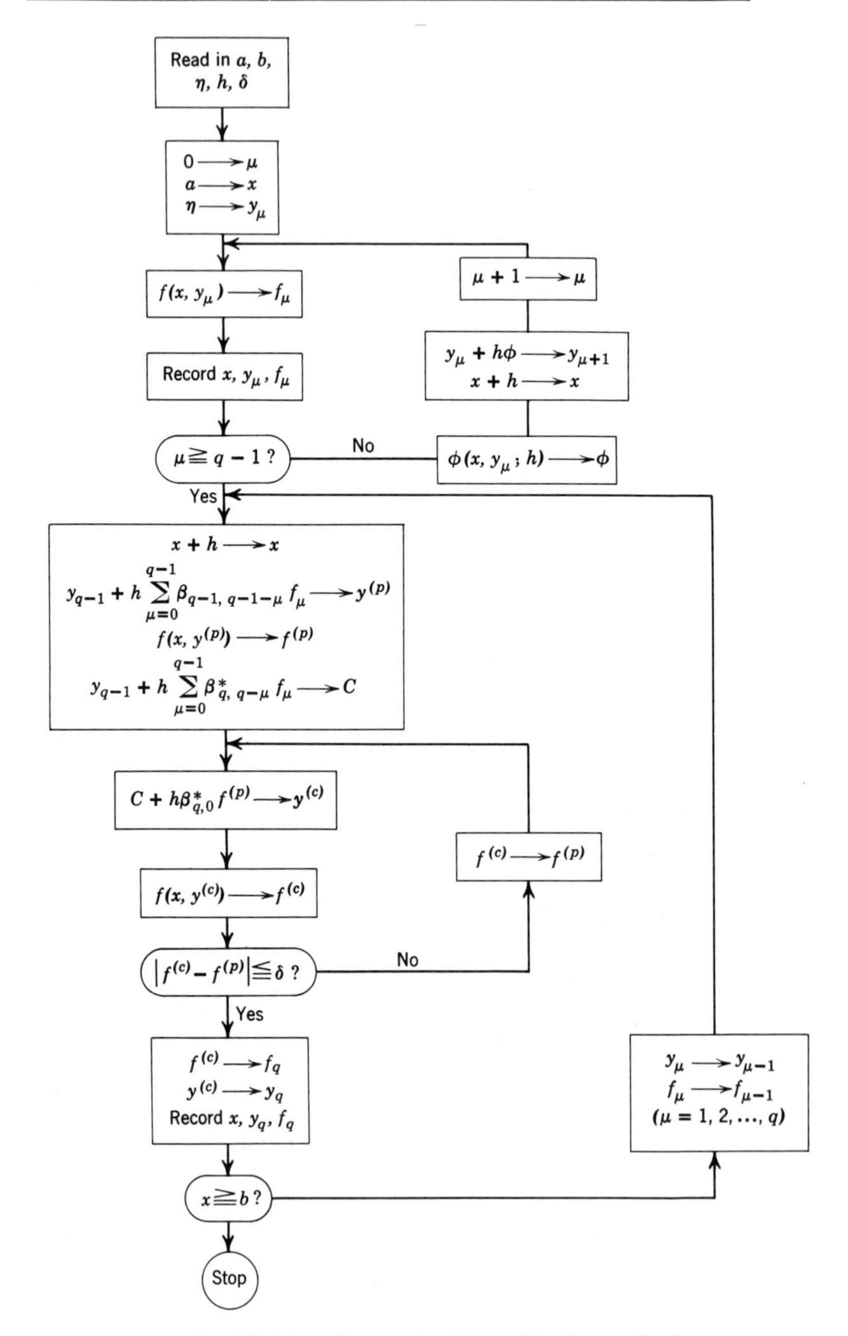

Fig. 5.1. Flow diagram for Adams-Moulton method.

The numerical values given in Table 5.7 are found.

Table 5.7

m	0	1	2	3	4	5	6
κ_m	2	0	$\frac{1}{3}$	$\frac{1}{3}$	$\frac{29}{90}$	$\frac{14}{45}$	$\frac{1139}{3780}$

The special case $q = 0$ of Nyström's method is called the *mid-point rule.* Since $\kappa_1 = 0$, it yields the same accuracy as could ordinarily be expected from taking $q = 1$, which accounts for some of the popularity of the method. The mid-point-rule has also been used as a predictor formula for the trapezoidal rule (Adams-Moulton with $q = 1$). For general q the Nyström method may be used as a predictor formula for the generalized Milne-Simpson method to be discussed now.

(iv) The generalized Milne-Simpson method. This results from taking $x = x_{p-2}$, $k = 2h$ in (5-13). Thus we have the formula

$$y_p - y_{p-2} = h\int_{x_{p-2}}^{x_p} P(x)\,dx = h\sum_{m=0}^{q} \kappa_m^*\nabla^m f_p \tag{5-28}$$

where

$$\kappa_m^* = (-1)^m \frac{1}{h}\int_{x_{p-2}}^{x_p}\binom{-s}{m}dx = (-1)^m\int_{-2}^{0}\binom{-s}{m}ds \tag{5-29}$$

Formula (5-28) resembles the Adams-Moulton formula in being *implicit.* However, the integration is now over two steps, which circumstance may cause *weak stability.* On the other hand, the method is, for comparable $q \geq 2$, more accurate than any of the methods previously considered. This is true in particular in the case $q = 2$, which is known as the *Milne method* (Milne [1926], [1953], p. 66), strictly speaking. The reason why will be understood by studying the coefficients κ_m^*. Their generating function is given by

$$K^*(t) = \sum_{m=0}^{\infty} \kappa_m^* t^m = -\frac{t}{\log(1-t)}(2-t) \tag{5-30}$$

It follows that

$$(1 + \tfrac{1}{2}t + \tfrac{1}{3}t^2 + \cdots)(\kappa_0^* + \kappa_1^* t + \kappa_2^* t^2 + \cdots) = 2 - t$$

Comparing coefficients, we find

$$\kappa_m^* + \tfrac{1}{2}\kappa_{m-1}^* + \tfrac{1}{3}\kappa_{m-2}^* + \cdots + \frac{1}{m+1}\kappa_0^* = \begin{cases} 2, & m = 0 \\ -1, & m = 1 \\ 0, & m = 2, 3, \cdots \end{cases}$$

Numerical values are given in Table 5.8. We note that $\kappa_3^* = 0$, which

Table 5.8

m	0	1	2	3	4	5	6
κ_m^*	2	-2	$\frac{1}{3}$	0	$-\frac{1}{90}$	$-\frac{1}{90}$	$-\frac{37}{3780}$

shows that the Milne-Simpson formula with $q = 2$ produces an effect which normally would be expected only with $q = 3$. We also note that

$$\frac{1}{1-t} K^*(t) = K(t)$$

from which there follows the relation

$$\kappa_0^* + \kappa_1^* + \kappa_2^* + \cdots + \kappa_m^* = \kappa_m, \qquad m = 0, 1, \cdots \tag{5-31}$$

The numerical values given in Table 5.8 suggest that (5-28) be better written and used in the form

$$y_p - y_{p-2} = h[2f_{p-1} + \tfrac{1}{3}\nabla^2 f_p - \tfrac{1}{90}(\nabla^4 f_p + \nabla^5 f_p) + \cdots] \tag{5-32}$$

The Milne formula resulting from $q = 2$ or 3 may be written out in terms of ordinates as follows

$$y_p - y_{p-2} = \tfrac{1}{3}h(f_p + 4f_{p-1} + f_{p-2}) \tag{5-33}$$

If f depends on x only, this reduces to the familiar Simpson formula in the theory of numerical integration.

Equations (5-28), (5-32), and (5-33) represent implicit equations for y_p and are usually solved by iteration, starting with a predicted first approximation $y_p^{(0)}$. Any explicit formula can be used in principle to calculate $y_p^{(0)}$. In the interest of economizing the number of corrections, one will choose, if possible, a predictor formula whose accuracy is comparable to that of the corrector formula. Milne suggests the formula

$$y_p - y_{p-4} = \tfrac{4}{3}h(8f_{p-1} - 4f_{p-2} + 8f_{p-3}) \tag{5-34}$$

[resulting from integrating the quadratic polynomial interpolating $f(x, y(x))$ at $x_{p-1}, x_{p-2}, x_{p-3}$ between x_{p-4} and x_p] as a predictor formula. Using the fact that by virtue of (5-31)

$$y_p - y_{p-2} = h\kappa_m f(x_p, y_p) + \text{a linear combination of } f_{p-1}, \cdots, f_{p-q}$$

the iteration procedure may, for any choice of the predictor formula, be arranged in the manner explained in the section on the Adams-Moulton formula.

5.1-3. The local error of the formulas based on integration. In the preceding section we encountered a number of difference expressions which approximate a quantity (namely the value of a definite integral) that depends on a given function $f(x)$. Such a quantity is called a *functional.* The situation that such a functional, which we denote by the symbol Lf, is approximated by a discrete operator $L_h f$ occurs frequently in numerical computation. It appears reasonable to measure the accuracy of such a finite difference approximation by substituting into L_h a function on which L can act and measuring the discrepancy

$$R = Lf - L_h f$$

in terms of h and of properties of the function f. The quantity R may be considered the equivalent of the local discretization error studied in connection with one-step methods. We see in §5.3-4 how the size of R influences the accumulated discretization error.

For the Adams-Bashforth and the Adams-Moulton method expressions for R are easily found by expressing $y'(x) = f(x, y(x))$ by its interpolating polynomial plus remainder term. In any interval where $y^{(q+2)}(x)$ is continuous, we have by Lemma 5.1, letting $s = (x - x_p)/h$,

$$y'(x) = \sum_{m=0}^{q} (-1)^m \binom{-s}{m} \nabla^m y'(x_p) + (-1)^{q+1} \binom{-s}{q+1} h^{q+1} y^{(q+2)}(\xi) \tag{5-35}$$

where ξ is a point between the largest and the smallest of the values x, x_p, and x_{p-q}. Integrating between x_p and x_{p+1}, we get by the definition of the quantities γ_m

$$y(x_{p+1}) - y(x_p) = h \sum_{m=0}^{q} \gamma_m \nabla^m y'(x_p) + R_q^{AB} \tag{5-36}$$

which is the result of substituting into the Adams-Bashforth formula the function $y'(x)$, augmented by the remainder term

$$R_q^{AB} = (-1)^{q+1} h^{q+1} \int_{x_p}^{x_{p+1}} \binom{-s}{q+1} y^{(q+2)}(\xi)\, dx$$

We now make use of the following two facts: (a) $\binom{-s}{q+1}$ is of constant sign in the interval $x_p \le x \le x_{p+1}$; (b) as emphasized in the remark following Lemma 5.1, $y^{(q+2)}(\xi)$ is a continuous function of x. We are thus in a position to apply the second mean value theorem of the integral calculus,† with the result that

$$R_q^{AB} = (-1)^{q+1} h^{q+1} y^{(q+2)}(\xi') \int_{x_p}^{x_{p+1}} \binom{-s}{q+1}\, dx$$

† See Taylor [1955], p. 118, Theorem V.

where ξ' is one of the values of ξ corresponding to values of x in (x_p, x_{p+1}), $x_{p-q} < \xi' < x_{p+1}$. By the definition of γ_{q+1}, this may be written

(5-37) $$R_q^{AB} = h^{q+2}y^{(q+2)}(\xi')\gamma_{q+1}$$

This is the desired expression for the remainder in the Adams-Bashforth formula. It resembles the first neglected term in that formula, except that the difference is replaced by a corresponding derivative expression.

In a completely analogous manner we find for the Adams-Moulton formula

$$y(x_p) - y(x_{p-1}) = h \sum_{m=0}^{q} \gamma_m^* \nabla^m y'(x_p) + R_q^{AM}$$

where

(5-38) $$R_q^{AM} = h^{q+2}y^{(q+2)}(\xi)\gamma_{q+1}^*$$

for some ξ satisfying $x_{p-q} < \xi < x_p$.

If one attempts to determine the quantities R_q^{NY} and R_q^{MS} defined by the formulas

(5-39) $$y(x_{p+1}) - y(x_{p-1}) = h \sum_{m=0}^{q} \kappa_m \nabla^m y'(x_p) + R_q^{NY}$$

and

(5-40) $$y(x_p) - y(x_{p-2}) = h \sum_{m=0}^{q} \kappa_m^* \nabla^m y'(x_p) + R_q^{MS}$$

in a similar manner, the trick used above does not work, since $\binom{-s}{q+1}$ now changes sign in the interval of integration. We postpone the estimation of R_q^{NY} and R_q^{MS} in the general case to §5.3-4, where in Lemma 5.7 a universally applicable method will be presented. In the present section we confine ourselves to a derivation, again by special methods, of error formulas for the mid-point rule and for the special Milne formula (5-33). Specifically we shall show that

(5-41) $$R_0^{NY} = \tfrac{1}{3}h^3y'''(\xi), \qquad x_{p-1} < \xi < x_{p+1}$$

and

(5-42) $$R_2^{MS} = -\tfrac{1}{90}h^5y^{\mathrm{v}}(\xi), \qquad x_{p-2} < \xi < x_p$$

In order to prove (5-41), we may assume without loss of generality that $p = 0$ and we then have to show that (5-41) holds with $x_{p\pm1} = \pm h$ and

(5-43) $$R_0^{NY} = y(h) - y(-h) - 2hy'(0)$$

Using Taylor's expansion with remainder, we have

$$y(h) = y(0) + hy'(0) + \tfrac{1}{2}h^2y''(0) + \tfrac{1}{6}h^3y'''(\xi_1)$$

where $0 < \xi_1 < h$, and similarly

$$y(-h) = y(0) - hy'(0) + \tfrac{1}{2}h^2y''(0) - \tfrac{1}{6}h^3y'''(\xi_2)$$

where $-h < \xi_2 < 0$. Inserting in (5-43), we get

$$R_0^{NY} = \tfrac{1}{6}h^3[y'''(\xi_1) + y'''(\xi_2)]$$

Since the function $y'''(x)$ is continuous, it assumes all values between $y'''(\xi_1)$ and $y'''(\xi_2)$ in the interval (ξ_2, ξ_1). Therefore, for some $\xi \in (\xi_2, \xi_1)$,

$$y'''(\xi_1) + y'''(\xi_2) = 2y'''(\xi)$$

which establishes the desired result.

In order to prove (5-42), we may assume, again without loss of generality, that $p = 1$, $x_p = h$. We then have to show that (5-42) holds for some $\xi \in (-h, h)$ and

$$R_2^{MS} = y(h) - y(-h) - h[2y'(0) + \tfrac{1}{3}\nabla^2 y'(h)]$$

It is easily verified that R_2^{MS} can be represented in terms of the definite integral

$$R_2^{MS} = \int_0^h \left[y'(t) + y'(-t) - 2y'(0) - \frac{t^2}{h^2}\nabla^2 y'(h) \right] dt$$

We now consider the function

$$F(x) = \int_0^x \left[y'(t) + y'(-t) - 2y'(0) - \frac{t^2}{h^2}\nabla^2 y'(h) \right] dt$$
$$- \lambda \int_0^x \left[\binom{1 - t/h}{4} + \binom{1 + t/h}{4} \right] dt$$

We have $F(0) = 0$, $F(-x) = -F(x)$, and select λ such that $F(h) = 0$. This is possible in view of

$$\int_0^h \left[\binom{1 - t/h}{4} + \binom{1 + t/h}{4} \right] dt = h \int_{-2}^0 \binom{-s}{4} ds = h\kappa_4^* = -\tfrac{1}{90}h \neq 0$$

With this choice of λ we have

$$F(\pm h) = R_2^{MS} + \tfrac{1}{90}\lambda h = 0 \tag{5-44}$$

Since $F(\pm h) = F(0) = 0$, $F'(\pm\xi) = 0$ for some $\xi \in (0, h)$. But

$$F'(x) = y'(x) - y'(-x) - 2y'(0) - \frac{x^2}{h^2}\nabla^2 y'(h)$$
$$-\lambda \left[\binom{1 - x/h}{4} + \binom{1 + x/h}{4} \right]$$

vanishes also for $x = 0$ and $x = \pm h$. Thus, in view of the fact that $F'(x)$

has five distinct zeros in the closed interval $[-h, h]$, the fifth derivative $F^{\text{v}}(x)$ has at least one zero in $(-h, h)$. We easily find

$$F^{\text{v}}(x) = y^{\text{v}}(x) + y^{\text{v}}(-x) - 2\lambda h^{-4}$$

and thus have, for some $\xi_1 \in (-h, h)$,

$$\lambda = \tfrac{1}{2}h^4[y^{\text{v}}(\xi_1) + y^{\text{v}}(-\xi_1)]$$

By the continuity of $y^{\text{v}}(x)$ it follows as before that

$$\lambda = h^4 y^{\text{v}}(\xi)$$

for some ξ between ξ_1 and $-\xi_1$. Inserting this value in (5-44), (5-42) follows.

5.1-4. Methods based on numerical differentiation. The methods based on integration were derived by replacing the function $f(t, y(t))$ in the identity

$$y(x + k) - y(x) = \int_x^{x+k} f(t, y(t))\, dt$$

by an interpolating polynomial and integrating it. There is no a priori reason why it should not be feasible to replace the function $y(x)$ on the left side of

$$y'(x) = f(x, y(x)) \tag{5-45}$$

by an interpolating polynomial and differentiating it. As a matter of fact, methods based on differentiation are much less frequently met in practice than methods based on integration. As will become evident from the analysis below, the reasons are to be found in unpleasant properties of error propagation.

The interpolating polynomial of the function $y(x)$ using the interpolating points $x_p, x_{p-1}, \cdots, x_{p-q}$ is given by

$$P(x) = \sum_{m=0}^{q} (-1)^m \binom{-s}{m} \nabla^m y_p, \qquad s = \frac{x - x_p}{h}$$

The result of differentiating at the point $x = x_{p-r}$ is

$$P'(x_{p-r}) = \frac{1}{h} \sum_{m=0}^{q} \delta_{r,m} \nabla^m y_p \tag{5-46}$$

where

$$\delta_{r,m} = (-1)^m\, h \frac{d}{dx} \binom{-s}{m}_{x=x_{p-r}} = (-1)^m \frac{d}{ds} \binom{-s}{m}_{s=-r} \tag{5-47}$$

Numerical values and further properties of the coefficients $\delta_{r,m}$ will be discussed below. Equating (5-46) to $f(x_{p-r}, y_{p-r})$, we obtain the following discrete approximation to $y' = f(x, y)$:

$$\sum_{m=0}^{q} \delta_{r,m} \nabla^m y_p = h f_{p-r} \tag{5-48}$$

The left side is of the form $(\delta_{r,0} + \delta_{r,1} + \cdots + \delta_{r,m}) y_p$ + a linear combination of $y_{p-1}, \cdots, y_{p-q}$. If

$$\delta_{r,0} + \delta_{r,1} + \cdots + \delta_{r,m} \neq 0 \tag{5-49}$$

(5-48) thus represents an equation for y_p which can be solved when the values $y_{p-1}, \cdots, y_{p-q}$ are known. The equation is explicit if $r > 0$, and implicit if $r = 0$. In the latter case it can be solved by iteration. Thus, in principle, (5-48) could be used much as one of the Adams formulas. It turns out, however, that for reasons of numerical stability the formula is practicable only in the cases

$$r = 0, \qquad q \leq 6$$
$$r = 1, \qquad q \leq 2$$

Even in these cases the formula is less accurate than the corresponding Adams-Moulton formula, using the same number of points.

Since formula (5-48) may nevertheless be useful for purposes other than the step-by-step integration of differential equations, we shall study it in somewhat greater detail. For the coefficients $\delta_{r,m}$ we find, if $m > r$, using the definition of the binomial coefficient,

$$\begin{aligned} \delta_{r,m} &= \frac{d}{ds}\left\{\frac{s(s+1)\cdots(s+m-1)}{1\cdot 2\cdots m}\right\}_{s=-r} \\ &= \frac{1}{m!}\lim_{s\to -r} s(s+1)\cdots(s+r-1)(s+r+1)\cdots(s+m-1) \\ &= (-1)^r \frac{r!\,(m-r-1)!}{m!} \end{aligned} \tag{5-50}$$

For $m \leq r$ direct differentiation is less convenient. A relation which holds for all m can be established by using a generating function. Setting

$$D_r(t) = \sum_{m=0}^{\infty} \delta_{r,m} t^m$$

we have

$$\begin{aligned} D_r(t) &= \sum_{m=0}^{\infty} (-t)^m \frac{d}{ds}\binom{-s}{m}_{s=-r} = \frac{d}{ds}\left(\sum_{m=0}^{\infty}\binom{-s}{m}(-t)^m\right)_{s=-r} \\ &= \frac{d}{ds}(1-t)^{-s}\Big|_{s=-r} = \frac{d}{ds} e^{-s\log(1-t)}\Big|_{s=-r} \end{aligned}$$

and thus

$$D_r(t) = -\log(1-t)\cdot(1-t)^r \tag{5-51}$$

It follows that

$$\delta_{r,0} + \delta_{r,1}t + \delta_{r,2}t^2 + \cdots = (t + \tfrac{1}{2}t^2 + \tfrac{1}{3}t^3 + \cdots) \times \left[1 - \binom{r}{1}t + \binom{r}{2}t^2 - \cdots + (-1)^r t^r\right]$$

By comparing coefficients, we see that $\delta_{r,0} = 0$ ($r = 0, 1, 2, \cdots$); furthermore,

$$\delta_{r,m} = \frac{1}{m} - \frac{1}{m-1}\binom{r}{1} + \frac{1}{m-2}\binom{r}{2} - \cdots + (-1)^{m-1}\binom{r}{m-1} \tag{5-52}$$

if $m \le r$, and

$$\delta_{r,m} = \frac{1}{m} - \frac{1}{m-1}\binom{r}{1} + \frac{1}{m-2}\binom{r}{2} - \cdots + (-1)^r \frac{1}{m-r} \tag{5-53}$$

if $m > r$. [The fact that (5-50) and (5-53) are identical is not obvious.] From the identities

$$D_{r+1}(t) = (1-t)D_r(t), \qquad (1-t)^{-1}D_r(t) = D_{r-1}(t)$$

we find again by expanding in powers of t and comparing coefficients,

$$\delta_{r+1,m} = \delta_{r,m} - \delta_{r,m-1}, \qquad r \ge 0, m \ge 1 \tag{5-54}$$

$$\delta_{r-1,m} = \delta_{r,0} + \delta_{r,1} + \cdots + d_{r,m}, \qquad r \ge 1, m \ge 0 \tag{5-55}$$

We finally note that

$$\delta_{0,m} = 1/m, \qquad m \ge 1$$

Table 5.9 is now easily calculated.

Table 5.9

Coefficients $\delta_{r,m}$

r \ m	0	1	2	3	4	5	6
0	0	1	$\frac{1}{2}$	$\frac{1}{3}$	$\frac{1}{4}$	$\frac{1}{5}$	$\frac{1}{6}$
1	0	1	$-\frac{1}{2}$	$-\frac{1}{6}$	$-\frac{1}{12}$	$-\frac{1}{20}$	$-\frac{1}{30}$
2	0	1	$-\frac{3}{2}$	$\frac{1}{3}$	$\frac{1}{12}$	$\frac{1}{30}$	$\frac{1}{60}$
3	0	1	$-\frac{5}{2}$	$\frac{7}{6}$	$-\frac{17}{12}$	$-\frac{1}{20}$	$-\frac{1}{60}$

The accuracy of the differentiation operator (5-48) is measured by substituting into (5-48) a solution of the differential equation. We obtain

$$hy'(x_{p-r}) - \sum_{m=0}^{q} \delta_{r,m}\nabla^m y(x_p) = R_{r,q}^D \tag{5-56}$$

where, by Lemma 5.2, if $x_{p-q} \le x_r \le x_p$,

$$R_{r,q}^D = \frac{h}{(q+1)!} y^{(q+1)}(\xi)L'(x_{p-r}), \qquad x_{p-q} < \xi < x_p$$

with $L(x) = (x - x_p)(x - x_{p-1}) \cdots (x - x_{p-q})$. Since we may write

$$L(x) = (-1)^{q+1} h^{q+1} \binom{-s}{q+1}$$

we have

$$hL'(x_{p-r}) = h^{q+1}\delta_{r,q+1}$$

It follows that

$$R^D_{r,q} = h^{q+1}\delta_{r,q+1} y^{(q+1)}(\xi), \qquad x_{p-q} < \xi < x_p \tag{5-57}$$

and it follows that the error in (5-48) is approximately proportional to the first neglected term, provided that the difference is replaced by the corresponding derivative. It should be noted that R^D_q is proportional to h^{q+1}, whereas the remainders R^{AM}_q and R^{MS}_q of the implicit integration formulas using the same number of points are proportional to h^{q+2} or h^{q+3}.

5.2. General Discussion of Linear Multistep Methods

All the methods discussed in §5.1 can be considered as special cases of the formula

$$\begin{aligned} &\alpha_k y_{n+k} + \alpha_{k-1} y_{n+k-1} + \cdots + \alpha_0 y_n \\ &\quad = h\{\beta_k f_{n+k} + \beta_{k-1} f_{n+k-1} + \cdots + \beta_0 f_n\}, \qquad n = 0, 1, 2, \cdots \end{aligned} \tag{5-58}$$

where k is a fixed integer, $f_m = f(x_m, y_m)$ $(m = 0, 1, 2, \cdots)$, and where α_μ and β_μ $(\mu = 0, 1, \cdots, k)$ denote real constants which do not depend on n. We shall always assume that $\alpha_k \neq 0$, $|\alpha_0| + |\beta_0| > 0$. Equation (5-58) is said to define the *general linear k-step method.* The method is called linear because the values f_m enter linearly in (5-58); it is not assumed that f is a linear function of y.

For the study of accumulated discretization and round-off error to be undertaken in §§5.3 and 5.4, it is more efficient to deal with the general formula (5-58) than to consider each special case separately. At the same time this approach, which was first adopted by Dahlquist [1956], leads to a mathematically well-rounded theory. It also leads to the discovery of new integration formulas which could not be obtained by the heuristic methods of §5.1.

In the present section we shall deal with the following question: *What properties must formula* (5-58) *have in order to define a "good" method for integrating a differential equation?* Theorems 5.5 and 5.6 will indicate two necessary conditions for a method to be "good"; later, in §5.3, these necessary conditions will be shown to be sufficient also.

We begin by discussing some elementary facts on linear difference equations.

5.2-1. Linear difference equations. Let k be a positive integer, and denote by I a set of consecutive integers. Let the functions $F_n(\eta_0, \eta_1, \cdots, \eta_k)$ be defined for $n \in I$ and for all real values of the $k + 1$ variables $\eta_0, \eta_1, \cdots, \eta_k$. The equation

$$F_n(y_n, y_{n+1}, \cdots, y_{n+k}) = 0 \tag{5-59}$$

is called a *difference equation of order* k. It symbolizes the problem of finding a sequence of numbers $\{y_n\}$, called a *solution* of the difference equation, such that (5-59) is satisfied for $n \in I$. Examples of difference equations are

$$y_{n+1} - y_n = n, \qquad -\infty < n < \infty$$

$$n^2(y_{n+3} - y_{n+1})^2 = -1, \qquad n \geq 27$$

If the function $f(x, y)$ is defined for $x \in [a, b]$ and for all values of y, and if $x_n = a + nh$, then (5-58) represents a difference equation of order k, and I is given by $n \geq 0$, $n + k \leq (b - a)h^{-1}$.

A difference equation is called *linear* (and *homogeneous*) if for each $n \in I$ the function F_n is linear (and homogeneous) in the variables $\eta_0, \eta_1, \cdots, \eta_k$. A linear difference equation can be written in the form

$$\alpha_{k,n}y_{n+k} + \alpha_{k-1,n}y_{n+k-1} + \cdots + \alpha_{0,n}y_n = \gamma_{n+k} \tag{5-60}$$

where the coefficients $\alpha_{k,n}, \cdots, \alpha_{0,n}$ and γ_{n+k} are given functions of n defined for $n \in I$. No additional difficulties arise if we admit complex-valued functions. We shall now discuss some basic facts concerning such linear difference equations. Some of these facts show a marked similarity to the theory of linear differential equations of order k, but there are also some striking differences.

We begin by considering the homogeneous equation

$$\alpha_{k,n}y_{n+k} + \alpha_{k-1,n}y_{n+k-1} + \cdots + \alpha_{0,n}y_n = 0 \tag{5-61}$$

whose coefficients we assume to be defined for all $n \geq 0$. We stipulate that $\alpha_{k,n} \neq 0$, $n \geq 0$. It is then clear that a solution of (5-61) can be found for every choice of the initial values $y_0, y_1, \cdots, y_{k-1}$, and that this solution is unique. In particular, to the starting values $y_0 = y_1 = \cdots = y_{k-1} = 0$ there corresponds the *trivial solution* $y_n \equiv 0$. If the sequences $\{y_n\}$ and $\{z_n\}$ are solutions of (5-61), then obviously the sequence formed with the elements $Ay_n + Bz_n$, where A and B are arbitrary constants, is again a solution of (5-61). Two solutions $\{y_n\}$ and $\{z_n\}$ are called *linearly independent* at the point $n = n_0$ if there do not exist two constants A and B, not both zero, such that

$$Ay_n + Bz_n = 0, \qquad n - n_0 = 0, 1, \cdots, k - 1$$

More generally, m solutions $\{y_n^{(\mu)}\}$ ($\mu = 1, 2, \cdots, m$) of (5-61) are called linearly independent at $n = n_0$ if it is not possible to find m constants A_μ ($\mu = 1, 2, \cdots, m$), not all zero, such that the relation

$$A_1 y_n^{(1)} + A_2 y_n^{(2)} + \cdots + A_m y_n^{(m)} = 0$$

holds simultaneously for $n - n_0 = 0, 1, \cdots, k - 1$. By elementary matrix algebra this condition is equivalent to the condition that the matrix

$$\mathbf{M} = \begin{pmatrix} y_{n_0}^{(1)} & y_{n_0+1}^{(1)} & \cdots & y_{n_0+k-1}^{(1)} \\ y_{n_0}^{(2)} & y_{n_0+1}^{(2)} & \cdots & y_{n_0+k-1}^{(2)} \\ \cdots & \cdots & \cdots & \cdots \\ y_{n_0}^{(m)} & y_{n_0+1}^{(m)} & \cdots & y_{n_0+k-1}^{(m)} \end{pmatrix} \tag{5-62}$$

should have rank m. It follows that the maximum number of linearly independent solutions at any point is k.

A system of k solutions of (5-61) which are linearly independent at $n = n_0$ is called a *fundamental system* of (5-61) at $n = n_0$. Such a fundamental system is (in principle) easy to construct. It suffices, for instance, to determine the k solutions $\{y_n^{(\mu)}\}$ ($\mu = 1, \cdots, k$) for which the matrix $\mathbf{M}$ reduces to the identity matrix. In special cases other choices of the fundamental system may be more convenient.

If it is assumed that

$$\alpha_{0,n} \neq 0, \qquad n \geq n_0 \tag{5-63}$$

it can be proved (see Problem 10) that a system which is fundamental at $n = n_0$ is fundamental at all $n \geq n_0$. On the other hand, if condition (5-63) is violated, a fundamental system does not necessarily remain fundamental. The following theorem, however, holds even without condition (5-63).

THEOREM 5.1. *A sequence $\{y_n\}$ which for $n \geq n_0$ is a solution of* (5-61) *can for $n \geq n_0$ be expressed as a linear combination of the solutions of a fundamental system at $n = n_0$.*

Proof. Let $\{y_n^{(\mu)}\}$ ($\mu = 1, \cdots, k$) be the given fundamental system. Determine the constants $A_1, \cdots, A_k$ by the conditions

$$\sum_{\mu=1}^{k} A_\mu y_n^{(\mu)} = y_n, \qquad n - n_0 = 0, \cdots, k - 1 \tag{5-64}$$

The determinant of this system of k linear equations for the k unknowns A_μ is different from zero because the k solutions $\{y^{(\mu)}\}$ are linearly independent at $n = n_0$. Hence the system possesses a unique solution $A_1, \cdots, A_k$. If we now define

$$z_n = y_n - \sum_{\mu=1}^{k} A_\mu y_n^{(\mu)}, \qquad n - n_0 = 0, 1, \cdots$$

then the sequence $\{z_n\}$, being a linear combination of solutions of the difference equation (5-61), is likewise a solution of (5-61). In view of $z_{n_0} = z_{n_0+1} = \cdots = z_{n_0+k-1} = 0$, this solution vanishes identically for $n \geq n_0$. It follows that (5-64) holds for all $n \geq n_0$.

On the basis of Theorem 5.1 it may be said that the "general solution" of (5-61) for $n \geq n_0$ is of the form $\sum_{\mu=1}^{k} A_\mu y_n^{(\mu)}$, where the sequences $\{y_n^{(\mu)}\}$ form a fundamental system at $n = n_0$ and where the constants A_μ are arbitrary. It is obvious that the general solution of the corresponding nonhomogeneous equation

$$\alpha_{k,n} y_{n+k} + \alpha_{k-1,n} y_{n+k-1} + \cdots + \alpha_{0,n} y_n = \gamma_{n+k} \tag{5-65}$$

can be written in the form $\{y_n + z_n\}$, where $\{y_n\}$ denotes the general solution of the homogeneous equation (corresponding to $\gamma_{n+k} = 0$) and $\{z_n\}$ is some particular solution of the nonhomogeneous equation. A particular solution of the nonhomogeneous equation may be found by solving (5-65) with the starting values $z_0 = z_1 = \cdots = z_{k-1} = 0$. This solution can be represented in terms of certain solutions of the homogeneous equation by a device which may be called the *discrete analog of Duhamel's principle*. For $m = 0, 1, 2, \cdots$ let the sequence $\{y_{n,m}\}$ satisfy the conditions

$$\begin{aligned} y_{n,m} &= 0, \qquad n = 0, 1, \cdots, m-1 \\ y_{m,m} &= 1/\alpha_{k,m} \end{aligned} \tag{5-66}$$

and let it be a solution of (5-61) for $n > m$. We then have:

THEOREM 5.2. *The solution of* (5-65) *satisfying* $z_0 = z_1 = \cdots = z_{k-1} = 0$ *is for* $n \geq k$ *represented by*

$$z_n = \sum_{m=k}^{n} \gamma_m y_{n,m} \tag{5-67}$$

Proof. It follows from the definition of $y_{n,m}$ that

$$\alpha_{k,n} y_{n+k,m} + \alpha_{k-1,n} y_{n+k-1,m} + \cdots + \alpha_{0,n} y_{n,m} = \begin{cases} 0, & n \neq m-k \\ 1, & n = m-k \end{cases} \tag{5-68}$$

Since $y_{n,m} = 0$ for $m > n$, (5-67) may be replaced by

$$z_n = \sum_{m=k}^{\infty} \gamma_m y_{n,m}$$

without convergence difficulties. Using (5-68), we find

$$\begin{aligned} &\alpha_{k,n} z_{n+k} + \alpha_{k-1,n} z_{n+k-1} + \cdots + \alpha_{0,n} z_n \\ &\qquad = \sum_{m=k}^{\infty} \gamma_m \{\alpha_{k,n} y_{n+k,m} + \cdots + \alpha_{0,m} y_{n,m}\} = \gamma_{n+k} \end{aligned}$$

as desired.

We conclude this section by constructing explicitly a fundamental system at $n = 0$ for the linear difference equation with constant coefficients, viz.,

$$\alpha_k y_{n+k} + \alpha_{k-1} y_{n+k-1} + \cdots + \alpha_0 y_n = 0 \tag{5-69}$$

where $\alpha_k \neq 0$, $\alpha_0 \neq 0$. Pursuing the analogy to ordinary differential equations with constant coefficients, which for suitable choices of λ have solutions of the form $e^{\lambda x}$, we shall first find all solutions of (5-69) of the form $y_n = e^{\lambda n}$, where λ is to be determined. Actually, it is more convenient to write $e^{\lambda} = \zeta$, so that $y_n = \zeta^n$, and to determine ζ. Since we want nontrivial solutions, we must require $\zeta \neq 0$. Substituting $y_n = \zeta^n$ into (5-69), we find

$$\alpha_k \zeta^{n+k} + \alpha_{k-1} \zeta^{n+k-1} + \cdots + \alpha_0 \zeta^n = 0$$

or, upon canceling ζ^n,

$$\alpha_k \zeta^k + \alpha_{k-1} \zeta^{k-1} + \cdots + \alpha_0 = 0$$

It follows that if $y_n = \zeta^n$ is a solution of (5-69), then ζ must be a (real or complex) root of the *characteristic polynomial*

$$\rho(\zeta) = \alpha_k \zeta^k + \alpha_{k-1} \zeta^{k-1} + \cdots + \alpha_0$$

Conversely, by backtracing our steps we can easily verify that if ζ satisfies $\rho(\zeta) = 0$, $y_n = \zeta^n$ is a solution of the difference equation. We now distinguish two cases.

(i) The polynomial $\rho(\zeta)$ has k distinct roots. Let the roots be denoted by $\zeta_1, \zeta_2, \cdots, \zeta_k$. We have just shown that the k sequences formed with the numbers $y_n^{(\mu)} = \zeta_\mu^n$ are solutions of (5-69). In view of $\alpha_0 \neq 0$, $\zeta = 0$ is not a root, and none of the solutions $\{y_n^{(\mu)}\}$ is trivial. The linear independence of the solutions follows from the fact that the matrix (5-62) for $n_0 = 0$ reduces to

$$\begin{pmatrix} 1 & \zeta_1 \cdots \zeta_1^{k-1} \\ 1 & \zeta_2 \cdots \zeta_2^{k-1} \\ \cdots & \cdots \cdots \\ 1 & \zeta_k \cdots \zeta_k^{k-1} \end{pmatrix}$$

The determinant of this matrix* has the value $\Pi_{i<j}(\zeta_i - \zeta_j) \neq 0$.

(ii) The polynomial $\rho(\zeta)$ has repeated roots. If the distinct roots of $\rho(\zeta)$ are denoted by $\zeta_1, \zeta_2, \cdots, \zeta_m$, and if $m < k$, then the sequences $\{\zeta_\mu^n\}$ $(\mu = 1, \cdots, m)$ do not form a fundamental system, because there

* Known as a Vandermonde determinant. (See Birkhoff and MacLane [1953], p. 303.)

are not enough of them. Additional solutions can be found by a procedure which is familiar from the theory of ordinary differential equations. If ζ is a root of multiplicity 2, say, we may think of ζ as having originated from the confluence of two distinct simple roots ζ and $\zeta + \varepsilon$, where ε is small. For every ε the sequence formed with the elements $\varepsilon^{-1}[(\zeta + \varepsilon)^n - \zeta^n]$ is a solution, because it is a linear combination of two solutions. We may thus expect that the resulting limiting sequence as $\varepsilon \to 0$, whose elements are $y_n = n\zeta^{n-1}$, is a solution corresponding to $\varepsilon = 0$. That this is indeed so is shown by the following theorem, which contains the statement under (i) as a special case.

THEOREM 5.3. *Let the distinct roots of the characteristic polynomial* $\rho(\zeta)$ *be denoted by* $\zeta_1, \zeta_2, \cdots, \zeta_m$ $(m \le k)$, *and let the root* ζ_μ *have multiplicity* p_μ. *Then the* $k = \Sigma p_\mu$ *sequences with the elements*

$$
\begin{aligned}
&y_n = \zeta_\mu^n \\
(5\text{-}70)\qquad &y_n = n\zeta_\mu^n \\
&\cdots\cdots\cdots\cdots\cdots\cdots \\
&y_n = n(n-1)\cdots(n-p_\mu+2)\zeta_\mu^n, \qquad \mu = 1, 2, \cdots, m
\end{aligned}
$$

form a fundamental system of solutions of the difference equation (5-69) *at* $n = 0$.

Proof. We shall first show that all sequences defined by (5-70) are solutions of (5-69). If ζ_μ is a root of multiplicity p_μ of $\rho(\zeta)$, then ζ_μ is also a root of multiplicity p_μ of the function $f(\zeta) = \zeta^n\rho(\zeta)$, where n is an arbitrary nonnegative integer. Consequently, the first $p_\mu - 1$ derivatives of $f(\zeta)$ vanish at $\zeta = \zeta_\mu$, which yields the relations

$$
\begin{aligned}
\alpha_k\zeta_\mu^{n+k} + \alpha_{k-1}\zeta_\mu^{n+k-1} + \cdots + \alpha_0\zeta_\mu^n = 0 \\
\alpha_k(n+k)\zeta_\mu^{n+k-1} + \alpha_{k-1}(n+k-1)\zeta_\mu^{n+k-2} + \cdots + \alpha_0 n\zeta_\mu^{n-1} = 0 \\
\cdots\cdots\cdots\cdots\cdots\cdots\cdots\cdots\cdots \\
\alpha_k(n+k)(n+k-1)\cdots(n+k-p_\mu+2)\zeta^{n+k-p_\mu+1} + \cdots \\
+ \alpha_0 n(n-1)\cdots(n-p_\mu+2)\zeta^{n-p_\mu+1} = 0
\end{aligned}
$$

But these relations are equivalent to stating that the p_μ sequences defined by (5-70) solve the difference equation (5-69).

It remains to be shown that the solutions defined by (5-69) are linearly independent. By explicit computation (see Problem 12 at the end of the chapter) it can be shown that the determinant of the matrix (5-62) for $n_0 = 0$ has the nonzero value

$$
\prod_{1 \le \mu < \nu \le m} (\zeta_\mu - \zeta_\nu)^{p_\mu + p_\nu} \prod_{\mu=1}^{m} (p_\mu - 1)!!
$$

where $0!! = 1$, $k!! = k!\,(k-1)!\cdots 1!$ for $k = 1, 2, \cdots$. This completes the proof of Theorem 5.3.

Throughout this section no assumptions about the reality of the coefficients of the difference equation have been made. If the coefficients of the difference equation and the initial values are real, the solution will, of course, be real. However, the polynomial $\rho(\zeta)$ may still have complex roots. Some of the functions of the fundamental system will then appear in complex form. An equivalent system involving real functions only can then be obtained exactly as in the analogous situation in ordinary differential equations. If ζ is a complex root of the real polynomial $\rho(\zeta)$, then the conjugate complex number $\bar{\zeta}$ is a root with the same multiplicity. Thus the sequences whose elements are the conjugate complex values of (5-70) are also solutions, and it follows that the real and imaginary parts of the sequences (5-70), which are by definition real, are themselves solutions of the difference equation. For instance, for the difference equation $y_{n+2} + y_n = 0$ (5-70) yields the complex fundamental system $y_n = i^n$, $y_n = (-i)^n$. Instead, we may use the real fundamental system $y_n = \cos n(\pi/2)$, $y_n = \sin n(\pi/2)$.

5.2-2. Existence and uniqueness of the solution of the approximating difference equation. We shall now study the following question: Suppose that $f(x, y)$ satisfies the conditions of Theorem 1.1, and assume that $\alpha_k \neq 0$. Does the difference equation (5-58) have a unique solution $\{y_n\}$ ($x_n \in [a, b]$) for arbitrarily chosen initial values $y_0, y_1, \cdots, y_k$? Clearly, the answer is in the affirmative if we can show that relation (5-58), considered as an equation for y_{n+k}, has a unique solution for arbitrary values of $y_n, y_{n+1}, \cdots, y_{n+k-1}$. This is trivially the case if $\beta_k = 0$, because (5-58) then represents y_{n+k} explicitly as a function of $y_n, \cdots, y_{n+k-1}$ which is defined for all values of its arguments. If $\beta_k \neq 0$, however, y_{n+k} occurs [as an argument in $f_{n+k} = f(x_{n+k}, y_{n+k})$] also in the right-hand term, and (5-58) represents an (in general nonlinear) equation for y_{n+k} which conceivably might have several solutions or no solution at all. If a unique solution exists, there arises the practical problem of finding it.

In order to analyze this problem, we write (5-58) in the form

$$y = F(y) \tag{5-71}$$

where $y = y_{n+k}$,

$$F(y) = h\frac{\beta_k}{\alpha_k} f(x_{n+k}, y) + c \tag{5-72}$$

$$c = \frac{1}{\alpha_k}\{h[\beta_{k-1}f_{n+k-1} + \cdots + \beta_0 f_n] - \alpha_{k-1}y_{n+k-1} - \cdots - \alpha_0 y_n\}$$

The iteration procedure described in §5.1-2 can be set up in the general case and takes the form

$$y^{(\nu+1)} = F(y^{(\nu)}), \qquad \nu = 0, 1, 2, \cdots \tag{5-73}$$

where $y^{(0)}$ is a suitable first approximation. The following theorem enables us to prove (for sufficiently small values of h) not only the convergence of the sequence $\{y^{(\nu)}\}$ to a solution of (5-71) but also the uniqueness of the solution.

THEOREM 5.4. *Let the function $F(y)$ be defined for $-\infty < y < \infty$ and let there exist a constant K such that $0 \leq K < 1$, and*

$$|F(y^*) - F(y)| \leq K\,|y^* - y| \tag{5-74}$$

for arbitrary y^* and y. Then the following statements hold:

(i) *Equation* (5-71) *has a unique solution y.*
(ii) *For arbitrary $y^{(0)}$ the sequence defined by* (5-73) *converges to y.*
(iii) *For $\nu = 1, 2, \cdots$ there hold the estimates*

$$|y - y^{(\nu)}| \leq \frac{K}{1-K}\,|y^{(\nu)} - y^{(\nu-1)}| \leq \frac{K^\nu}{1-K}\,|y^{(1)} - y^{(0)}| \tag{5-75}$$

If $F(y)$ is defined by (5-72), and if $f(x, y)$ satisfies a Lipschitz condition with respect to y with Lipschitz constant L, then condition (5-74) is satisfied with

$$K = \left|\frac{h\beta_k}{\alpha_k}\right| L \tag{5-76}$$

and this is <1 for all sufficiently small values of h.

Proof of Theorem 5.4. If y and y^* are two solutions of (5-71), then $y = F(y)$, $y^* = F(y^*)$. Subtracting the first relation from the second and using (5-74), there follows

$$|y^* - y| \leq K\,|y^* - y|$$

which by virtue of $|K| < 1$ is impossible unless $y^* = y$. Thus (5-71) can have at most one solution. In order to prove the existence of a solution, subtract from (5-73) the relation $y^{(\nu)} = F(y^{(\nu-1)})$ and use (5-74). There follows

$$|y^{(\nu+1)} - y^{(\nu)}| \leq K\,|y^{(\nu)} - y^{(\nu-1)}|$$

Using this estimate repeatedly, there result the relations

$$|y^{(\nu+\mu)} - y^{(\nu+\mu-1)}| \leq K^\mu\,|y^{(\nu)} - y^{(\nu-1)}|$$

and

$$|y^{(\nu+1)} - y^{(\nu)}| \leq K^\nu\,|y^{(1)} - y^{(0)}| \qquad (\nu, \mu = 1, 2, \cdots)$$

As a consequence, for any positive integer μ,

$$(5\text{-}77)\quad \begin{aligned} |y^{(\nu+\mu)} - y^{(\nu)}| &\le |y^{(\nu+\mu)} - y^{(\nu+\mu-1)}| + \cdots + |y^{(\nu+1)} - y^{(\nu)}| \\ &\le (K^\mu + K^{\mu-1} + \cdots + K)\, |y^{(\nu)} - y^{(\nu-1)}| \\ &\le \frac{K^\nu}{1-K} |y^{(1)} - y^{(0)}| \end{aligned}$$

Given any $\varepsilon > 0$, there exists an integer ν_0 such that

$$\frac{K^\nu}{1-K} |y^{(1)} - y^{(0)}| < \varepsilon \quad \text{for all } \nu > \nu_0$$

The sequence $\{y^{(\nu)}\}$ is thus shown to satisfy the *Cauchy criterion* for convergence and thus has a finite limit y. Letting $\nu \to \infty$ in (5-73) and using the fact that $F(y)$ is continuous by virtue of its satisfying a Lipschitz condition, we get

$$y = \lim_{\nu\to\infty} y^{(\nu)} = \lim_{\nu\to\infty} F(y^{(\nu)}) = F(\lim_{\nu\to\infty} y^{(\nu)}) = F(y)$$

The limit y is thus recognized as being a solution, and, in view of the uniqueness already established, the only solution of (5-71). The estimate (5-75) follows by letting $\mu \to \infty$ in (5-77) while keeping ν fixed.

The theorem thus established is actually only a very special case of a classical theorem on functional equations in Banach spaces (see, e.g., Collatz [1960], p. 38). In the case considered here the conclusions can be rendered evident geometrically.

5.2-3. Convergence of a multistep method. Let the function $f(x, y)$ satisfy the conditions of the existence theorem 1.1. If h is sufficiently small, and if the initial values $y_0, y_1, \cdots, y_{k-1}$ are prescribed, the difference equation (5-58) determines uniquely a sequence of values $\{y_n\}$ as long as $a \le x_n \le b$. Although our notation stresses only the dependence on the discrete variable n, the elements y_n of this sequence depend also on h; first, because h figures as a parameter in the difference equation (5-58), and second, because the initial values of the sequence are normally functions of h:

$$(5\text{-}78)\qquad y_\mu = \eta_\mu(h), \qquad \mu = 0, 1, \cdots, k-1$$

For a "good" method we expect that the values y_n thus generated tend to the value of the desired exact solution at the point x as $h \to 0$ and $x_n = x$, if the starting values are properly chosen. The following definition formalizes this intuitive notion of convergence.

Definition. The linear multistep method defined by (5-58) is called *convergent*, if the following statement is true for all functions $f(x, y)$

satisfying the conditions of Theorem 1.1 and all values of η: If $y(x)$ denotes the solution of the initial value problem

$$y' = f(x, y), \quad y(a) = \eta$$

then

$$\lim_{\substack{h \to 0 \\ x_0 = x}} y_n = y(x) \tag{5-79}$$

holds for all $x \in [a, b]$ and all solutions $\{y_n\}$ of the difference equation (5-58) having starting values $y_\mu = \eta_\mu(h)$ satisfying

$$\lim_{h \to 0} \eta_\mu(h) = \eta, \qquad \mu = 0, 1, \cdots, k - 1 \tag{5-80}$$

It should be noted that this definition requires that condition (5-79) be satisfied not only for the sequence $\{y_n\}$ defined with the *exact* starting values—for these (5-80) is certainly satisfied—but also for all sequences whose starting values *tend* to the right value as $h \to 0$. This more stringent condition is imposed because in practice it is almost never possible to start a computation with mathematically exact values.

5.2-4. Convergence: condition of stability. In this and the next two sections we shall consider some implications of the definition of convergence. At this point it becomes convenient to associate with the difference equation (5-58) the polynomials

$$\begin{aligned} \rho(\zeta) &= \alpha_k \zeta^k + \alpha_{k-1}\zeta^{k-1} + \cdots + \alpha_0 \\ \sigma(\zeta) &= \beta_k \zeta^k + \beta_{k-1}\zeta^{k-1} + \cdots + \beta_0 \end{aligned} \tag{5-81}$$

Conversely, if two polynomials $\rho(\zeta)$ and $\sigma(\zeta)$ are given, we can associate with them a difference equation of the form (5-58).

THEOREM 5.5. *A necessary condition for convergence of the linear multistep method* (5-58) *is that the modulus of no root of the associated polynomial* $\rho(\zeta)$ *exceeds* 1, *and that the roots of modulus* 1 *be simple.*

The condition thus imposed on $\rho(\zeta)$ is called the *condition of stability.*

Proof. If the method is convergent, it is convergent for the initial value problem $y' = 0$, $y(0) = 0$, whose exact solution is $y(x) = 0$. For this problem (5-58) reduces to the difference equation with constant coefficients

$$\alpha_k y_{n+k} + \alpha_{k-1} y_{n+k-1} + \cdots + \alpha_0 y_n = 0 \tag{5-82}$$

If the method is convergent, then by (5-80), for any $x > 0$,

$$\lim_{n \to \infty} y_n = 0 \qquad (h = x/n) \tag{5-83}$$

for all solutions $\{y_n\}$ of (5-82) satisfying

$$\lim_{h \to 0} \eta_\mu(h) = 0, \qquad \mu = 0, 1, \cdots, k - 1 \tag{5-84}$$

where $y_\mu = \eta_\mu(h)$. Let $\zeta = re^{i\varphi}$ $(r \geq 0,\ 0 \leq \varphi < 2\pi)$ be a root of $\rho(\zeta)$. By Theorem 5.3, the numbers

$$y_n = hr^n \cos n\varphi \tag{5-85}$$

define a solution of (5-82), and they also satisfy (5-84). If the method is convergent, (5-83) must hold. If $\varphi = 0$ or $\varphi = \pi$, this immediately implies $r \leq 1$. If $\varphi \neq 0$, $\varphi \neq \pi$, we note that

$$\frac{y_n^2 - y_{n+1}y_{n-1}}{\sin^2 \varphi} = h^2 r^{2n}$$

Since the term on the left tends to zero as $n \to \infty$, $h = x/n$, the term on the right must do the same, which again implies $r \leq 1$. This proves the first part of the assertion of Theorem 5.5. In order to prove the second part, assume that $\zeta = re^{i\varphi}$ is a root of $\rho(\zeta)$ of multiplicity exceeding 1. Then, again by Theorem 5.3, the numbers

$$y_n = h^{1/2} n r^n \cos n\varphi \tag{5-86}$$

represent a solution of (5-82). They also satisfy (5-84), for $|\eta_\mu(h)| = |y_\mu| \leq h^{1/2}\mu r^\mu$ $(\mu = 0, 1, \cdots, k-1)$. Hence they must satisfy (5-84) for a convergent method. If $\varphi = 0$ or $\varphi = \pi$, we have for $h = x/n$ that $|y_n| = x^{1/2}n^{1/2}r^n$, and it follows immediately that $r < 1$. If $\varphi \neq 0$, $\varphi \neq \pi$, we can make use of the relation

$$\frac{z_n^2 - z_{n+1}z_{n-1}}{\sin^2 \varphi} = r^{2n}$$

where $z_n = n^{-1}h^{-1/2}y_n$. Since $z_n \to 0$ as $n \to \infty$ in view of (5-83), the term on the left tends to zero as $n \to \infty$, and we conclude that $r < 1$. Theorem 5.5 is thus proved.

It might perhaps be thought that the divergence of the sequences defined by (5-85) and (5-86) for $r > 1$ and $r \geq 1$ has to do with the rather special choice of the starting values. This is not true, however. If the polynomial $\rho(\zeta)$ violates the condition of stability, it may be shown that, in some sense, "almost all" solutions of the difference equation (5-82) diverge even if (5-84) is satisfied.

As a numerical example of an *unstable* method (i.e., a method violating the condition of stability), we consider the solution of the problem $y' = y$, $y(0) = 1$ by the differentiation method (5-48), choosing $q = 2$, $r = 2$. We note that to a term $\nabla^m y_{n+k}$ in a difference equation of the form (5-58) there corresponds a term $\zeta^k(1 - \zeta^{-1})^m$ in the polynomial $\rho(\zeta)$. Thus we find for the method in question

$$\begin{aligned}
\rho(\zeta) &= \zeta^2[\delta_{21}(1 - \zeta^{-1}) + \delta_{22}(1 - \zeta^{-1})^2] \\
&= \zeta(\zeta - 1) - \tfrac{3}{2}(\zeta - 1)^2 \\
&= -\tfrac{1}{2}\zeta^2 + 2\zeta - \tfrac{3}{2}
\end{aligned}$$

which has the roots $\zeta = 1$, $\zeta = 3$. Equation (5-58) now reads

$$-\tfrac{1}{2}y_{n+2} + 2y_{n+1} - \tfrac{3}{2}y_n = hy_n$$

Solving this difference equation numerically with $h = 0.1$ and using the starting values $y_0 = 1.00000$, $y_1 = 1.10517$ (which agree with the exact solution to the number of places given), we obtain the values given in Table 5.10. A violent type of divergence is evident; it would take place even sooner for smaller values of h.

Table 5.10

x_n	y_n	$y_n - e^{x_n}$
0	1.00000	0.00000
0.1	1.10517	0.00000
0.2	1.22068	−0.00072
0.3	1.34618	−0.00368
0.4	1.47854	−0.01329
0.5	1.60638	−0.04234
0.6	1.69419	−0.12793
0.7	1.63634	−0.37741
0.8	1.12395	−1.10159
0.9	−0.74049	−3.20009
1.0	−6.55860	−9.27688

As to the remaining methods discussed in §5.1, we note that all methods based on integration satisfy the condition of stability. This follows at once from the fact that for these methods the polynomial $\rho(\zeta)$ is of the form $\zeta^k - \zeta^{k-p}$ $(1 \le p \le k)$ and thus has p distinct roots on the unit circle and a root of multiplicity $k - p$ at $\zeta = 0$. For the methods based on numerical differentiation there is no such simple general result, because the location of the roots of $\rho(\zeta)$ is not so easily ascertained. Either by numerical calculation or by applying an algebraic criterion due to Hurwitz it can be shown that for $r = 0$ (differentiation at the foremost point) these methods are stable for $q = 1, 2, \cdots, 6$. For $q = 7$, and very probably for all larger q, the methods are unstable. For $r = 1$ the stability breaks down at $q = 3$, and for $r = 2$, as was seen above, at $q = 2$.

5.2-5. The associated difference operator; order and error constant. The condition of stability has the purpose of preventing a small initial error in the computation from growing at such a rate that convergence is jeopardized. It is clear, however, that stability alone does not guarantee convergence. A further condition must be added which ensures that the difference equation (5-58) is a good approximation to the differential equation $y' = f(x, y)$.

When considering the problem of measuring the accuracy of a *one-step* method, we looked at the expression

$$y(x + h) - y(x) - h\Phi(x, y(x);\ h)$$

where $y(x)$ is a solution of the given differential equation. The smaller this quantity as a function of h, the higher was the accuracy of the method. Similarly, if (5-58) is to define a good method, we expect the discrepancy between the two sides of (5-58) to be small if h is small and if the values y_m are replaced by $y(x_m)$, where $y(x)$ is an exact solution of the given differential equation. In order to measure this discrepancy, we associate with (5-58) the *difference operator*

$$\begin{aligned} \text{(5-87)} \quad L[y(x);\ h] = {} & \alpha_k y(x + kh) + \alpha_{k-1} y(x + (k-1)h) + \cdots + \alpha_0 y(x) \\ & - h\{\beta_k y'(x + kh) + \beta_{k-1} y'(x + (k-1)h) + \cdots + \beta_0 y'(x)\} \end{aligned}$$

This may be regarded as a linear operator that "acts" on any differentiable function $y(x)$. For the time being, however, we shall apply the operator L only to functions which have continuous derivatives of sufficiently high order. We then may expand $L[y(x);\ h]$ in powers of h, and the expansion can be pushed as far as we please. In view of the formulas

$$\begin{aligned} y(x + mh) &= y(x) + mhy'(x) + \tfrac{1}{2}m^2h^2y''(x) + \cdots \\ hy'(x + mh) &= hy'(x) + mh^2y''(x) + \cdots \end{aligned}$$

it turns out that

$$\text{(5-88)} \quad L[y(x);\ h] = C_0 y(x) + C_1 hy'(x) + \cdots + C_q h^q y^{(q)}(x) + \cdots$$

where the coefficients C_q ($q = 0, 1, \cdots$) are constants which do not depend on the choice of the function $y(x)$:

$$\begin{aligned} C_0 &= \alpha_0 + \alpha_1 + \alpha_2 + \cdots + \alpha_k \\ C_1 &= \alpha_1 + 2\alpha_2 + \cdots + k\alpha_k - (\beta_0 + \beta_1 + \beta_2 + \cdots + \beta_k) \\ C_q &= \frac{1}{q!}(\alpha_1 + 2^q\alpha_2 + \cdots + k^q\alpha_k) \\ &\quad - \frac{1}{(q-1)!}(\beta_1 + 2^{q-1}\beta_2 + \cdots + k^{q-1}\beta_k), \qquad q = 2, 3, \cdots \end{aligned}$$

A given difference operator of the form (5-87) is said to be of *order* p if $C_0 = C_1 = \cdots = C_p = 0$, but $C_{p+1} \neq 0$. As in the case of one-step methods, the order may be considered as a first crude measure of the accuracy of a method. In the few cases where the definition of a one-step method and a linear multistep method overlap (as for Euler's method and the trapezoidal rule considered in Problem 3, Chapter 2), the two definitions of order given are easily seen to be consistent.

From the practical point of view the difference equation (5-58) is completely equivalent to the equation

$$\text{(5-89)}\quad \alpha_k y_{n+t+k} + \cdots + \alpha_0 y_{n+t} = h(\beta_k f_{n+t+k} + \cdots + \beta_0 f_{n+t})$$

where t is any fixed (positive or negative) integer. Proceeding as above, we may associate with (5-89) the difference operator

$$\text{(5-90)}\quad \begin{aligned} L_t[y(x);\ h] = {} & \alpha_k y(x + (t+k)h) + \cdots + \alpha_0 y(x + th) \\ & - h\{\beta_k y'(x + (t+k)h) + \cdots + \beta_0 y'(x + th)\} \end{aligned}$$

and define its order as the order of the first nonvanishing term in its Taylor expansion in powers of h minus 1. It is an important fact that *the order p as well as the constant C_{p+1} do not depend on t.* For, expanding (5-90) in powers of h is equivalent to expanding (5-87) in powers of h, where $y(x)$ is replaced by $y(x + th)$. We thus find, if L is of the order p,

$$L_t[y(x);\ h] = L[y(x + th);\ h] = C_{p+1}h^{p+1}y^{(p+1)}(x + th) + O(h^{p+2})$$

Since $y(x)$ was assumed sufficiently differentiable,

$$y^{(p+1)}(x + th) = y^{(p+1)}(x) + O(h),$$

and thus

$$L_t[y(x);h] = C_{p+1}h^{p+1}y^{(p+1)}(x) + O(h^{p+2})$$

This proves the assertion, even without making use of the assumption that t was an integer. At the same time the argument shows that the constants $C_{p+2}, C_{p+3}, \cdots$ must in general be expected to *depend* on t.

As an example, consider the mid-point rule [(5-25) with $q = 0$], which in the standardized form (5-58) appears as follows:

$$y_{n+2} - y_n = h \cdot 2f_{n+1}$$

The corresponding operator (5-87) is given by

$$\begin{aligned} L[y(x);\ h] &= y(x + 2h) - y(x) - 2hy'(x + h) \\ &= 2hy' + \frac{(2h)^2}{2}y'' + \frac{(2h)^3}{6}y''' + \frac{(2h)^4}{24}y^{\text{iv}} + \cdots \\ &\quad - 2hy' - 2h^2y'' - 2\frac{h^3}{2}y''' - 2\frac{h^4}{6}y^{\text{iv}} - \cdots \end{aligned}$$

We readily find $C_0 = C_1 = C_2 = 0$, $C_3 = \frac{1}{3}$, $C_4 = \frac{1}{3}$ and thus $p = 2$. Alternatively, by choosing $t = -1$, we may consider the operator

$$\begin{aligned} L_{-1}[y(x);\ h] &= y(x + h) - y(x - h) - 2hy'(x) \\ &= 2hy' + 2\frac{h^3}{6}y''' + 2\frac{h^5}{120}y^{\text{v}} + \cdots - 2hy' \end{aligned}$$

Again $C_0 = C_1 = C_2 = 0$, $C_3 = \frac{1}{3}$, and hence $p = 2$, but now $C_4 = 0$. From the point of view of practical computation, the second method of calculating p and C_{p+1} is clearly preferable, since in the Taylor expansion only terms of odd order occur. This remark is generally applicable to difference equations such that $\alpha_\mu = -\alpha_{k-\mu}$, $\beta_\mu = \beta_{k-\mu}$; see Problem 32 at the end of this chapter.

In practice it is frequently not necessary to determine p and C_{p+1} by explicit calculation of all the constants $C_0, C_1, \cdots$. For instance, for most of the formulas considered in §5.1, it was shown that with a proper determination of t,

$$L_t[y(x);\, h] = Kh^{q+1}y^{(q+1)}(\xi) \tag{5-91}$$

where q and $K \neq 0$ are constants independent of $y(x)$, and ξ is a point in a specified interval. Whenever (5-91) holds, we may deduce that $p = q$, $C_{p+1} = K$. For we have by the above remark

$$L_t[y(x);\, h] = C_{p+1}h^{p+1}y^{(p+1)}(x) + \cdots \tag{5-92}$$

where the dots represent terms of the form $h^r y^{(r)}(x)$ with $r > p + 1$. Now choose $y(x) = x^{p+1}/(p + 1)!$ in both (5-91) and (5-92). The operators on the left are then identical. Thus $Kh^{q+1} = C_{p+1}h^{p+1}$, and, since this must be true for all values of h, it follows that $q = p$, $K = C_{p+1}$.

The order p of a difference operator was introduced as a first crude measure of the accuracy of the operator. Would it be wise to consider the value of C_{p+1} as a finer measure of the accuracy within the class of all difference operators of a given order? The answer is obviously no, since C_{p+1} can be made as small numerically as we please by multiplying (5-58) by a suitable constant, a process which does not change the values y_n defined by the formula and therefore does not improve the accuracy of the procedure. It will turn out, however, that the constant

$$C = \frac{C_{p+1}}{\beta_0 + \beta_1 + \cdots + \beta_k} \tag{5-93}$$

which is invariant under such trivial modifications of formula (5-58), is a good and in many ways convenient measure of accuracy. This constant will be called the *error constant* of the method defined by (5-58). It can be defined for all convergent methods, because, as is shown in §5.2-6, all these methods satisfy the condition $\beta_0 + \beta_1 + \cdots + \beta_k \neq 0$.

In Table 5.11 orders and error constants are listed for the special methods discussed in §5.1.

Table 5.11

Orders and Error Constants for Special Difference Operators. [q = order of highest difference employed; k = order of associated difference equation (5-58)]

Method	Order of Associated Difference Operator	Error Constant
Adams-Bashforth	$q + 1 = k$	γ_{q+1}
Adams-Moulton	$q + 1 = k + 1$	γ^*_{q+1}
Nyström $q \neq 0$	$q + 1 = k$	$\frac{1}{2}\kappa_{q+1}$
Nyström $q = 0$	2	$\frac{1}{6}$
Milne-Simpson $q \neq 2$	$q + 1 = k + 1$	$\frac{1}{2}\kappa^*_{q+1}$
Milne-Simpson $q = 2$	4	$-\frac{1}{180}$
Differentiation at x_{n+k-r}	$q = k$	$\delta_{r,q+1}$

5.2-6. Convergence: condition of consistency. The condition hinted at at the beginning of §5.2-5 can now be stated as follows:

THEOREM 5.6. *A necessary condition for convergence of the linear multistep method defined by* (5-58) *is that the order of the associated difference operator be at least* 1.

The condition that the order $p \geq 1$ is called the condition of *consistency*. In terms of the constants C_n introduced in §5.2-5 the condition is equivalent to $C_0 = 0$, $C_1 = 0$; in terms of the polynomials $\rho(\zeta)$ and $\sigma(\zeta)$ the condition of consistency is expressed by the relations

$$\rho(1) = 0, \qquad \rho'(1) = \sigma(1) \tag{5-94}$$

It can readily be verified that all the special methods discussed in §5.1 are consistent.

Proof of Theorem 5.6. We begin by showing that $C_0 = 0$. If the method is convergent, it is convergent for the initial value problem $y' = 0$, $y(0) = 1$, with the exact solution $y(x) = 1$. The difference equation (5-58) again reduces to

$$\alpha_k y_{n+k} + \alpha_{k-1} y_{n+k-1} + \cdots + \alpha_0 y_n = 0 \tag{5-95}$$

Assuming that the method is convergent, the solution $\{y_n\}$ of (5-95) assuming the exact starting values $y_\mu = 1$ ($\mu = 0, 1, \cdots, k-1$) must satisfy $y_n \to 1$ as $h \to 0$, $nh = x$. Since in this case y_n does not depend on h, this is the same as saying that $y_n \to 1$ as $n \to \infty$. Letting $n \to \infty$ in (5-95), we obtain $\alpha_k + \alpha_{k-1} + \cdots + \alpha_0 = 0$. This is equivalent to $C_0 = 0$. It follows that $p \geq 0$.

In order to show that $C_1 = 0$, consider the initial value problem $y' = 1$, $y(0) = 0$. The exact solution is $y(x) = x$. The difference equation (5-58) now reads

$$\alpha_k y_{n+k} + \alpha_{k-1} y_{n+k-1} + \cdots + \alpha_0 y_n = h(\beta_k + \beta_{k-1} + \cdots + \beta_0) \tag{5-96}$$

For a convergent method every solution of (5-96) satisfying

$$\lim_{h\to 0} \eta_\mu(h) = 0, \qquad \mu = 0, 1, \cdots, k-1 \tag{5-97}$$

where $y_\mu = \eta_\mu(h)$, must also satisfy

$$\lim_{h\to 0} y_n = x \qquad (x_n = x) \tag{5-98}$$

For a convergent method we may furthermore assume that $k\alpha_k + (k-1)\alpha_{k-1} + \cdots + \alpha_1 = \rho'(1) \neq 0$ in view of Theorem 5.5.

Let the sequence $\{y_n\}$ be defined by $y_n = nhK$, where

$$K = \frac{\beta_k + \beta_{k-1} + \cdots + \beta_0}{k\alpha_k + (k-1)\alpha_{k-1} + \cdots + \alpha_1}$$

This sequence obviously satisfies (5-97) and is easily shown to be a solution of (5-96). From $\lim_{h\to 0} nhK = xK$, $nh = x$, we conclude that $K = 1$. This is equivalent to $C_1 = 0$. This completes the proof of Theorem 5.6.

The conditions of stability and consistency, which thus have been demonstrated to be necessary for convergence, were derived from a consideration of almost trivial examples. In view of this fact it is all the more surprising that these conditions will turn out to be not only necessary but also sufficient for convergence. Furthermore, for sufficiently accurate starting values the accumulated discretization error of a method of order p is $O(h^p)$, as in the case of one-step methods. The proof of these results, which is by no means trivial, will be given in §5.3.

5.2-7. The construction of operators of maximum order. In this section we shall study the following problem: Given a polynomial $\rho(\zeta)$ of degree k and satisfying $\rho(1) = 0$, what is the order that can be achieved in the associated operator L by choosing a suitable polynomial $\sigma(\zeta)$? An answer is contained in Theorem 5.7. This answer will enable us to look at the special operators discussed in §5.1 from a more general point of view. Some complex variable theory is required for the discussion.

With any difference operator of the form (5-87) we associate the function of the complex variable ζ:

$$\varphi(\zeta) = (\log \zeta)^{-1}\rho(\zeta) - \sigma(\zeta) \tag{5-99}$$

The function $\log \zeta$ is made single-valued by cutting the complex plane along the negative real axis and setting $\log 1 = 0$. The function $(\log \zeta)^{-1}$ has a pole of order 1 at $\zeta = 1$. If the operator is consistent, this pole is canceled by a zero of $\rho(\zeta)$, and the function $\varphi(\zeta)$ is holomorphic at $\zeta = 1$. More generally, the following lemma holds.

LEMMA 5.3. *The difference operator* (5-87) *associated with the polynomials* $\rho(\zeta)$ *and* $\sigma(\zeta)$ *has the exact order* p *if and only if the function* $\varphi(\zeta)$ *has a zero of the exact order* p *at* $\zeta = 1$.

Proof. Suppose the operator has order p. Then expression (5-87) is $O(h^{p+1})$ for any sufficiently differentiable function $y(x)$. In particular, choosing $y(x) = e^x$, we have

$$L[e^x; h] = e^x\{\rho(e^h) - h\sigma(e^h)\} = e^x C_{p+1} h^{p+1} + O(h^{p+2}) \tag{5-100}$$

as $h \to 0$ where $C_{p+1} \neq 0$. This means that the function $f(h) = \rho(e^h) - h\sigma(e^h)$, which is holomorphic at $h = 0$, has a zero of order $p + 1$, and $h^{-1}f(h)$ a zero of order p, at $h = 0$. Since the transformation $\zeta = e^h$ maps a neighborhood of $h = 0$ in a one-to-one fashion onto a neighborhood of $\zeta = 1$, it follows by a classical theorem in the theory of complex variables* that the function $\varphi(\zeta) = (\log \zeta)^{-1} f(\log \zeta)$ has a zero of order p at $\zeta = 1$.

Conversely, assume that $\varphi(\zeta)$ has a zero of order p at $\zeta = 1$. It follows as above that the function $f(h) = h\varphi(e^h)$ has a zero of order $p + 1$ at $h = 0$. Hence (5-100) holds with some nonzero value of the constant C_{p+1}. Hence the order of $L[y(x); h]$ is p when $y(x) = e^x$ is substituted. Since the order depends only on the coefficients α_μ and β_μ, it follows that the order of L is p.

The following positive result is now an almost immediate consequence.

THEOREM 5.7. *Let* $\rho(\zeta)$ *be a polynomial of degree* k *satisfying* $\rho(1) = 0$, *and let* k' *be an integer*, $0 \leq k' \leq k$. *Then there exists a unique polynomial* $\sigma(\zeta)$ *of degree* $\leq k'$ *such that the order of the associated difference operator is at least* $k' + 1$.

Proof. The function $(\log \zeta)^{-1}\rho(\zeta)$ is holomorphic at $\zeta = 1$ and therefore can be expanded in nonnegative powers of $\zeta - 1$:

$$\frac{\rho(\zeta)}{\log \zeta} = c_0 + c_1(\zeta - 1) + c_2(\zeta - 1)^2 + \cdots \tag{5-101}$$

If we set

$$\sigma(\zeta) = c_0 + c_1(\zeta - 1) + \cdots + c_{k'}(\zeta - 1)^{k'} \tag{5-102}$$

the function of $\varphi(\zeta)$ has a zero of multiplicity $k' + 1$ at $\zeta = 1$, and the associated operator has order $\geq (k' + 1)$ by Lemma 5.3. (The order exceeds $k' + 1$ if $c_{k'+1} = 0$.) Conversely, if L has order $k' + 1$, then $\varphi(\zeta)$ has a zero of multiplicity $k' + 1$ at $\zeta = 1$. In view of the uniqueness of Taylor's expansion, $\sigma(\zeta)$ must then be identical with (5-102). This completes the proof of Theorem 5.7. Only the cases $k' = k - 1$ and $k' = k$ are of practical interest; the former leads to the best possible explicit operator, and the latter to the best possible implicit operator.

* See Ahlfors [1953], p. 107, Theorem 11.

The method of proof also furnishes a means to determine the error constant. If the difference operator associated with $\rho(\zeta)$ and $\sigma(\zeta)$ has order p, then by (5-100)

$$\rho(e^h) - h\sigma(e^h) = C_{p+1}h^{p+1} + O(h^{p+2}) \tag{5-103}$$

where $C_{p+1} \neq 0$. Since $\log \zeta = \zeta - 1 + O((\zeta - 1)^2)$ as $\zeta \to 1$, it follows that

$$\frac{\rho(\zeta)}{\log \zeta} - \sigma(\zeta) = C_{p+1}(\zeta - 1)^p + O((\zeta - 1)^{p+1}) \tag{5-104}$$

Consequently, C_{p+1} equals the coefficient of the first nonvanishing term in the expansion of $\rho(\zeta)/\log \zeta$ in powers of $\zeta - 1$ which has not been absorbed into $\sigma(\zeta)$. For the error constant C we thus find in view of $\sigma(1) = c_0$,

$$C = c_p/c_0 \tag{5-105}$$

As an example, let $\rho(\zeta) = (\zeta - 1)(\zeta - \lambda)$, where λ is real. This is the most general quadratic polynomial compatible with consistency. It satisfies the stability condition for $-1 \leq \lambda < 1$. We have

$$\frac{\rho(\zeta)}{\log \zeta} = \frac{(\zeta - 1)[1 - \lambda + (\zeta - 1)]}{\log [1 + (\zeta - 1)]} = 1 - \lambda + \frac{3 - \lambda}{2}(\zeta - 1)$$
$$+ \frac{5 + \lambda}{12}(\zeta - 1)^2 - \frac{1 + \lambda}{24}(\zeta - 1)^3 + O((\zeta - 1)^4)$$

The optimal implicit method associated with the above $\rho(\zeta)$ is thus given by

$$\sigma(\zeta) = 1 - \lambda + \frac{3 - \lambda}{2}(\zeta - 1) + \frac{5 + \lambda}{12}(\zeta - 1)^2$$
$$= -\frac{1 + 5\lambda}{12} + \frac{2 - 2\lambda}{3}\zeta + \frac{5 + \lambda}{12}\zeta^2$$

and the corresponding difference equation (5-58) reads

(5-106)
$$y_{n+2} - (1 + \lambda)y_{n+1} + \lambda y_n = h\left\{\frac{5 + \lambda}{12}f_{n+2} + \frac{2 - 2\lambda}{3}f_{n+1} - \frac{1 + 5\lambda}{12}f_n\right\}$$

For $\lambda \neq -1$ the order of the method is 3, and the error constant is given by

$$C = -\frac{1}{24}\frac{1 + \lambda}{1 - \lambda}$$

For $\lambda = -1$ the order is 4, in accordance with the fact that the method then becomes identical with the Milne method.

If a representation of the right-hand side of (5-58) in terms of backward differences is desired, it is convenient to introduce the new variable $t = 1 - \zeta^{-1}$. For the derivation of an implicit method of order $p \geq k + 1$ from a given $\rho(\zeta)$ we set

$$\rho(\zeta) = \zeta^k R(t), \qquad \sigma(\zeta) = \zeta^k S(t) \tag{5-107}$$

where R and S are polynomials of order k in t, $R(0) = 0$. From (5-104) it follows in view of $\log \zeta = -\log(1 - t)$, $\zeta - 1 = t(1 - t)^{-1} = t + O(t^2)$ that

$$-\frac{R(t)}{\log(1 - t)} - S(t) = C_{p+1}t^p + O(t^{p+1}) \tag{5-108}$$

The polynomial $S(t)$ is thus identical with the Taylor polynomial of order k of the function $-[\log(1 - t)]^{-1}R(t)$ at $t = 0$, and $C_{p+1}t^p$ equals the next nonvanishing term in the Taylor expansion of this function.

As an example, let $\rho(\zeta) = \zeta^k - \zeta^{k-1}$. We find $R(t) = t$, and from the expansion

$$-\frac{t}{\log(1 - t)} = \gamma_0^* + \gamma_1^* t + \cdots$$

(see §5.1-2) we identify the coefficients of the polynomial $S(t)$ as the coefficients occurring in the Adams-Moulton formula. Since the term $\zeta^k t^\mu$ in $S(t)$ contributes a term $\nabla^\mu f_{n+k}$ to the right-hand side of (5-58), the method thus generated is indeed identical with the Adams-Moulton method. Besides rediscovering this method from a more general point of view, we now have shown that this is the best method that can be obtained from the polynomial $\rho(\zeta) = \zeta^k - \zeta^{k-1}$. Similarly, it can be shown that the Milne-Simpson method is the best method obtainable from $\rho(\zeta) = \zeta^k - \zeta^{k-2}$.

If an explicit method of order k is desired, it is more convenient to set

$$\rho(\zeta) = \zeta^k R(t) = \zeta^{k-1}\frac{R(t)}{1 - t}, \qquad \sigma(\zeta) = \zeta^{k-1}S(t) \tag{5-109}$$

where $S(t)$ is now of degree $k - 1$. We now find that

$$-\frac{R(t)}{(1 - t)\log(1 - t)} - S(t) = C_{p+1}t^p + O(t^{p+1}) \tag{5-110}$$

and again $S(t)$ can be identified as a Taylor polynomial of a function holomorphic at $t = 0$. Letting $\rho(\zeta) = \zeta^k - \zeta^{k-1}$ and $\rho(\zeta) = \zeta^k - \zeta^{k-2}$, we rediscover the Adams-Bashforth and the Nyström methods, and at the same time show that these are the best explicit methods obtainable with the polynomials $\rho(\zeta)$ in question.

5.2-8. The maximum order of a stable operator. In the preceding section it was shown that given any polynomial $\rho(\zeta)$ of degree k and satisfying $\rho(1) = 0$ one can always find a polynomial $\sigma(\zeta)$ such that the associated operator has at least order $k + 1$. We have thus found a lower bound for the maximum order of an operator that can be associated with a given $\rho(\zeta)$. The question may now be raised as to how far beyond $k + 1$ the order can be pushed by choosing $\rho(\zeta)$ judiciously. A count of

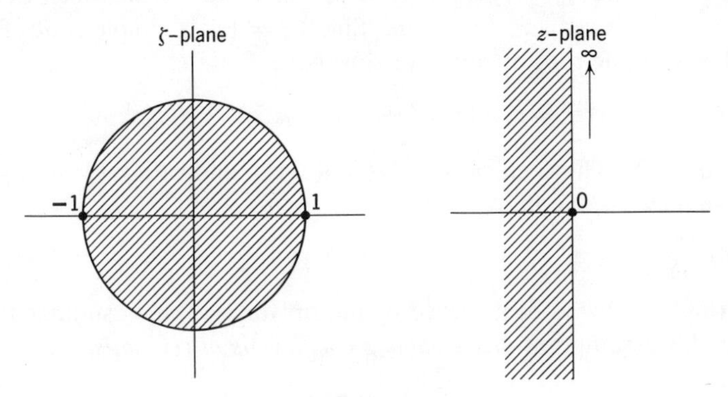

Fig. 5.2

the number of parameters at our disposal ($k + 1$ coefficients α_μ and β_μ each) and of the number of conditions to be satisfied suggests that $2k$ is the highest value of p that can be achieved. The truth of this conjecture has been established by Dahlquist [1956]. However, this result is of no great practical significance, for we now shall prove the important result, also due to Dahlquist, that any operator of order $>(k + 2)$ is necessarily unstable. *The maximum order that can be achieved with a $\rho(\zeta)$ satisfying the stability condition is $k + 2$, and even this is possible only with a rather restricted class of polynomials $\rho(\zeta)$.*

In order to prove this result, let $\rho(\zeta)$ be a polynomial compatible with consistency and satisfying the condition of stability.* In order to manage these hypotheses analytically, we introduce a new variable $z = x + iy$ by the linear fractional transformation

$$z = \frac{\zeta - 1}{\zeta + 1}, \qquad \zeta = \frac{1 + z}{1 - z} \tag{5-111}$$

This transformation maps the disk $|\zeta| < 1$ onto the half-plane $Re\, z < 0$, the circle $|\zeta| = 1$ onto the imaginary axis, the point $\zeta = 1$ onto $z = 0$, and the point $\zeta = -1$ onto $z = \infty$ (see Fig. 5.2).

* Only the following weaker conditions will actually be used: no root of $\rho(\zeta)$ exceeds 1 in modulus, and the root at $\zeta = 1$ has exact multiplicity 1.

In place of the polynomials $\rho(\zeta)$ and $\sigma(\zeta)$ we now consider the functions

$$r(z) = \left(\frac{1-z}{2}\right)^k \rho\left(\frac{1+z}{1-z}\right), \qquad s(z) = \left(\frac{1-z}{2}\right)^k \sigma\left(\frac{1+z}{1-z}\right) \tag{5-112}$$

which are again polynomials of degree $\leq k$. If $\rho(\zeta)$ has a root of multiplicity p at the point $\zeta = \zeta_0$, then $r(z)$ has a root of the same multiplicity at the point $z = (\zeta_0 - 1)/(\zeta_0 + 1)$, except if $\zeta_0 = -1$, in which case the degree of $r(z)$ is reduced to $k - p$. Since $\zeta = 1$ is a simple root of $\rho(\zeta)$, $z = 0$ is a simple root of $r(z)$. We thus have

$$r(z) = a_1 z + a_2 z^2 + \cdots + a_{k-1} z^{k-1} + a_k z^k \tag{5-113}$$

where $a_1 \neq 0$. The coefficients of $r(z)$ are real, and without loss of generality we may assume that

$$a_1 > 0 \tag{5-114}$$

since this can always be achieved by multiplying $\rho(\zeta)$ by a suitable factor. Under this assumption *the remaining coefficients of* $r(z)$ *satisfy*

$$a_\mu \geq 0, \qquad \mu = 1, \cdots, k \tag{5-115}$$

The proof is obtained by considering the product expansion of the polynomial $r(z)$. Denoting the roots of $r(z)$ by $x_\nu + iy_\nu$ and by a_l the nonvanishing coefficient with highest index, we have

$$r(z) = a_l z \prod_\lambda (z - x_\lambda) \prod_\mu ((z - x_\mu)^2 + y_\mu^2) \tag{5-116}$$

where λ ranges over the real roots and μ over the pairs of conjugate complex roots. In virtue of stability, $x_\nu \leq 0$ for all roots, so that upon expanding (5-116) the powers of z are multiplied by nonnegative constants only. It follows that all nonvanishing coefficients of $r(z)$ have the same sign as a_l. Since $a_1 > 0$, it follows that they are all positive.

We now consider the function

$$p(z) = \left(\frac{1-z}{2}\right)^k \varphi\left(\frac{1+z}{1-z}\right) = \frac{1}{\log\dfrac{1+z}{1-z}} r(z) - s(z) \tag{5-117}$$

The function $p(z)$ has a zero of order p at $z = 0$ if and only if $\varphi(\zeta)$ has a zero of order p at $\zeta = 1$ and thus, by Lemma 5.3, if and only if the operator associated with $\rho(\zeta)$ and $\sigma(\zeta)$ has order p. Consequently, if the operator has order p, then

$$s(z) = b_0 + b_1 z + b_2 z^2 + \cdots + b_{p-1} z^{p-1} \tag{5-118}$$

where

$$\frac{z}{\log \dfrac{1+z}{1-z}} \frac{r(z)}{z} = b_0 + b_1 z + b_2 z^2 + \cdots \tag{5-119}$$

Since the degree of $s(z)$ must not exceed k, the existence of a stable operator with $p > k + 1$ now hinges on the possibility that $b_{k+1} = \ldots = b_{p-1} = 0$. In order to investigate this possibility we determine the coefficients b_ν explicitly. Setting

$$\frac{z}{\log \dfrac{1+z}{1-z}} = c_0 + c_2 z^2 + c_4 z^4 + \cdots \tag{5-120}$$

we have, defining $a_\nu = 0$ for $\nu > k$,

$$\begin{aligned} b_0 &= c_0 a_1 \\ b_1 &= c_0 a_2 \\ b_{2\nu} &= c_0 a_{2\nu+1} + c_2 a_{2\nu-1} + \cdots + c_{2\nu} a_1 \\ b_{2\nu+1} &= c_0 a_{2\nu+2} + c_2 a_{2\nu} + \cdots + c_{2\nu} a_2, \qquad \nu = 1, 2, \cdots \end{aligned} \tag{5-121}$$

The further development depends on the fact *that the coefficients* $c_{2\nu}$ *defined by* (5-120) *satisfy*

$$c_{2\nu} < 0, \qquad \nu = 1, 2, \cdots \tag{5-122}$$

In order not to interrupt the main idea of the proof, we postpone the proof of this inequality and distinguish two cases.

(i) If k is odd, we have from (5-121)

$$b_{k+1} = c_2 a_k + c_4 a_{k-2} + \cdots + c_{k+1} a_1$$

Since $a_1 > 0$ and no a_ν is <0, it follows in view of (5-122) that $b_{k+1} < 0$. Thus we have proved

THEOREM 5.8. *The order of a stable operator whose stepnumber k is odd cannot exceed $k + 1$.*

(ii) If k is even, then

$$b_{k+1} = c_2 a_k + c_4 a_{k-2} + \cdots + c_k a_2$$

In view of what we know about the signs of $c_{2\nu}$ and a_ν, a necessary and sufficient condition for $b_{k+1} = 0$ is that $a_2 = a_4 = \cdots = a_k = 0$. This is the case if and only if the polynomial $r(z)$ satisfies the identity $r(-z) = -r(z)$, i.e., if $r(z)$ is odd. Since $r(z)$ cannot have any roots with positive real part, this identity prohibits $r(z)$ from having roots with negative real part. All roots of $r(z)$ thus must lie on the imaginary axis, which means that all roots of $\rho(\zeta)$ lie on $|\zeta| = 1$. Since $a_k = 0$, the degree of $r(z)$ is $k - 1$, and -1 is a root of $\rho(\zeta)$. From the fact that

$$b_{k+2} = c_4 a_{k-1} + c_6 a_{k-3} + \cdots + c_{k+2} a_1$$

is negative in view of (5-115), (5-122), and $a_1 > 0$, it follows that the order cannot exceed $k + 2$. We thus have proved

THEOREM 5.9. *The order p of a stable operator cannot exceed $k + 2$. A necessary and sufficient condition for $p = k + 2$ is that k be even, that all roots of $\rho(\zeta)$ have modulus* 1, *and that $\sigma(\zeta)$ be determined by* (5-113).

An operator satisfying the conditions of Theorem 5.9 will be called an *optimal* operator. The proof of Theorem 5.9 furnishes some interesting inequalities for the error constant associated with an optimal operator. From (5-104) we find in view of $\zeta - 1 = 2z + O(z^2)$ that $p(z) = 2^{p-k} C_{p+1} z^p + O(z^{p+1})$. If $p > k$ it thus follows from (5-119) that $C_{p+1} = 2^{k-p} b_p$. Using the fact that $\sigma(1) = 2^k s(0) = 2^k b_0$, we thus have

$$C = b_p / 2^p b_0 \tag{5-123}$$

For an optimal operator we have $p = k + 2$ and thus, using (5-121),

$$C = \frac{c_4 a_{k-1} + c_6 a_{k-3} + \cdots + c_{k+2} a_1}{2^{k+2} c_0 a_1}$$

Again making use of our knowledge of the signs of $c_{2\nu}$ and a_ν, and using the fact that $c_0 = \frac{1}{2}$, $c_4 = -\frac{2}{45}$, we find the inequalities

$$C \le -\frac{1}{2^k \cdot 45} \frac{a_{k-1}}{a_1} \tag{5-124}$$

$$C \le 2^{-k-1} c_{k+2} \tag{5-125}$$

Since $c_{k+2} < 0$, (5-125) shows that $|C|$ cannot be arbitrarily small for given k.

Numerical values of the coefficients $c_{2\nu}$ may be calculated from the recurrence relation $c_0 = \frac{1}{2}$,

$$c_{2\nu} + \tfrac{1}{3} c_{2\nu-2} + \tfrac{1}{5} c_{2\nu-4} + \cdots + \frac{1}{2\nu + 1} c_0 = 0, \qquad \nu = 1, 2, \cdots.$$

Some values are given in Table 5.12.

Table 5.12

The Coefficients $c_{2\nu}$ in (5-120)

ν	0	1	2	3	4
$c_{2\nu}$	$\frac{1}{2}$	$-\frac{1}{6}$	$-\frac{2}{45}$	$-\frac{22}{945}$	$-\frac{214}{14175}$

We still have to establish the inequality (5-122). The quickest way is by considering it a corollary of the following result, due to Kaluza [1928], which we state without proof.*

LEMMA 5.4. *Let* $f(t) = \sum_{\nu=0}^{\infty} A_\nu t^\nu$ *and* $g(t) = \sum_{\nu=0}^{\infty} B_\nu t^\nu$ *be two power series such that* $f(t)g(t) = 1$, $A_\nu > 0$ $(\nu = 0, 1, 2, \cdots)$, $A_{\nu+1}A_{\nu-1} - A_\nu^2 > 0$ $(\nu = 1, 2, \cdots)$. *Then* $B_\nu < 0$ $(\nu = 1, 2, \cdots)$.

For the function of $t = z^2$,

$$f(t) = \frac{1}{z} \log \frac{1+z}{1-z} = 2 + \tfrac{2}{3}z^2 + \tfrac{2}{5}z^4 + \cdots$$

the hypotheses of the lemma are easily shown to be satisfied, so that the conclusion holds for the coefficients $B_\nu = c_{2\nu}$.

5.2-9. The construction of optimal operators. The construction of the polynomial $\sigma(\zeta)$ from a given $\rho(\zeta)$ by the methods of §5.2-7 can be awkward when k is large. For optimal operators the following alternative procedure is recommended.

If the polynomial $\rho(\zeta)$ satisfies the conditions of Theorem 5.9, and if the roots different from ± 1 are located at the points $e^{\pm i\varphi}$, $e^{\pm i\psi}$, $\cdots$, it can be written in the form

$$\rho(\zeta) = \alpha_k \zeta^{\frac{1}{2}k}(\zeta - \zeta^{-1})(\zeta - 2\cos\varphi + \zeta^{-1})(\zeta - 2\cos\psi + \zeta^{-1}) \cdots$$

It follows that

$$\rho(\zeta) = \zeta^{\frac{1}{2}k}(\zeta - \zeta^{-1})P(\zeta + \zeta^{-1} - 2) \tag{5-126}$$

where P denotes a polynomial of degree $\frac{1}{2}k - 1$. If $r(z)$ is odd, the function $s(z)$ determined from (5-118) is a Taylor polynomial at $z = 0$ of the product of two odd functions and therefore even. From $s(-z) = s(z)$ it follows that $\sigma(\zeta) = \zeta^k \sigma(\zeta^{-1})$ or that

$$\beta_{k-\nu} = \beta_\nu, \qquad \nu = 0, 1, \cdots, k \tag{5-127}$$

It is now easily shown by induction that $\sigma(\zeta)$ can be represented in the form

$$\sigma(\zeta) = \zeta^{\frac{1}{2}k} Q(\zeta + \zeta^{-1} - 2) \tag{5-128}$$

where Q denotes a polynomial of degree $\frac{1}{2}k$. [We choose the variable $\zeta + \zeta^{-1} - 2$ rather than $\zeta + \zeta^{-1}$ in (5-126) and (5-128) because to a term $(\zeta + \zeta^{-1} - 2)^\mu$ in Q there corresponds a contribution $\nabla^{2\mu} f_{n+\frac{1}{2}k+\mu}$ in the right-hand member of (5-58).] We now set

$$\zeta + \zeta^{-1} - 2 = t^2 \tag{5-129}$$

* A simple proof may be found in *Amer. Math. Monthly*, **66**, 430 (1959).

To every t in a neighborhood of $t = 0$ there correspond two values of ζ in a neighborhood of $\zeta = 1$. The two values are reciprocal to each other and can be obtained by picking the two possible denominations of the square root in the formula $\zeta = (\sqrt{1 + \frac{1}{4}t^2} + \frac{1}{2}t)^2$. We make the function $\zeta(t)$ single-valued by choosing the branch of the square root that reduces to $+1$ as $t = 0$. We then have $\zeta - \zeta^{-1} = 2t\sqrt{1 + \frac{1}{4}t^2}$, $\zeta - 1 = t + O(t^2)$. Relation (5-104) thus is equivalent to

$$\text{(5-130)} \qquad \frac{t\sqrt{1 + \frac{1}{4}t^2}}{\log\left(\frac{1}{2}t + \sqrt{1 + \frac{1}{4}t^2}\right)} P(t^2) - Q(t^2) = C_{k+3}t^{k+2} + O(t^{k+3})$$

It follows that $Q(t^2)$ is identical with the Taylor polynomial of order k of the even function

$$\text{(5-131)} \qquad \frac{t\sqrt{1 + \frac{1}{4}t^2}}{\log\left(\frac{1}{2}t + \sqrt{1 + \frac{1}{4}t^2}\right)} P(t^2) = q_0 + q_2t^2 + q_4t^4 + \cdots$$

Since $\sigma(1) = Q(0) = q_0$, we find for the error constant the relation

$$\text{(5-132)} \qquad C = q_{k+2}/q_0$$

The coefficients $q_{2\nu}$ can be calculated by multiplying the polynomial

$$\text{(5-133)} \qquad P(t^2) = p_0 + p_2t^2 + \cdots + p_{k-2}t^{k-2}$$

by the Taylor series of the function

$$\text{(5-134)} \qquad \frac{t\sqrt{1 + \frac{1}{4}t^2}}{\log\left(\frac{1}{2}t + \sqrt{1 + \frac{1}{4}t^2}\right)} = k_0 + k_2t^2 + k_4t^4 + \cdots$$

By the recurrence relation established in Problem 23 the numerical values of Table 5.13 are readily obtained.

Table 5.13

The Coefficients $k_{2\nu}$ in (5-134)

ν	0	1	2	3	4
$k_{2\nu}$	2	$\frac{1}{3}$	$-\frac{1}{90}$	$\frac{1}{756}$	$-\frac{6013}{7257600}$

Examples. (i) ***The optimal method for*** $k = 2$. Since $\rho(\zeta) = \zeta^2 - 1$, we have $P(t^2) = 1$, and consequently $Q(t^2) = 2 + \frac{1}{3}t^2$. A term $t^{2\nu}$ corresponds to a difference $\nabla^{2\nu}f_{n+\frac{1}{2}k+\nu}$ on the right-hand side of (5-58); we are thus led once again to the Milne method

$$y_{n+2} - y_n = h(2f_{n+1} + \tfrac{1}{3}\nabla^2 f_{n+2})$$

(ii) ***The most general optimal method for*** $k = 4$. If the roots of $\rho(\zeta)$ are at $\zeta = \pm 1$, $\zeta = e^{\pm i\varphi}$, we have

$$\rho(\zeta) = \zeta^2\left(\zeta - \frac{1}{\zeta}\right)\left(\zeta - 2\cos\varphi + \frac{1}{\zeta}\right)$$

Hence $P(t^2) = 4\lambda + t^2$, where $\lambda = \sin^2 \frac{1}{2}\varphi$, $0 < \lambda < 1$. From

$$\frac{t\sqrt{1 + \frac{1}{4}t^2}}{\log\left(\frac{1}{2}t + \sqrt{1 + \frac{1}{4}t^2}\right)} P(t^2) = 8\lambda + \left(2 + \frac{4\lambda}{3}\right)t^2 + \left(\frac{1}{3} - \frac{2\lambda}{45}\right)t^4 + \left(-\frac{1}{90} + \frac{\lambda}{189}\right)t^6 + \cdots$$

we find

$$Q(t^2) = 8\lambda + \left(2 + \frac{4\lambda}{3}\right)t^2 + \left(\frac{1}{3} - \frac{2\lambda}{45}\right)t^4, \qquad C = \frac{1}{1512} - \frac{1}{720\lambda}$$

For $\lambda \to 1$, C approaches the upper bound $2^{-5}c_6$ given by (5-125).

(iii) ***A special optimal method with*** $k = 6$. The polynomial

$$\rho(\zeta) = \zeta^6 - \zeta^5 + \zeta^4 - \zeta^2 + \zeta - 1$$

has its roots at $\zeta = \pm 1$, $\zeta = e^{\pm i\pi/2}$, $\zeta = e^{\pm i\pi/3}$. We thus have

$$P(t^2) = \left(\zeta + \frac{1}{\zeta}\right)\left(\zeta + \frac{1}{\zeta} - 1\right) = (t^2 + 2)(t^2 + 1)$$

We thus find the polynomial

$$Q(t^2) = 4 + \frac{20}{3}t^2 + \frac{134}{45}t^4 + \frac{286}{675}t^6$$

and the resulting method can be written

$$y_{n+6} - y_{n+5} + y_{n+4} - y_{n+2} + y_{n+1} - y_n$$
$$= h\left(4f_{n+3} + \frac{20}{3}\nabla^2 f_{n+4} + \frac{134}{45}\nabla^4 f_{n+5} + \frac{286}{675}\nabla^6 f_{n+6}\right)$$

5.3. The Discretization Error of Linear Multistep Methods

5.3-1. The discretization error in a special case. As far as possible the treatment of discretization error to be given below will conform to the pattern established by the one-step methods. In view of the greater complexity of the problem it seems advisable, however, to begin the discussion with a thorough study of the special problem

(5-135) $$y' = Ay, \qquad y(0) = 1$$

where A is a (real or complex) constant, $A \neq 0$. The special result obtained here also will be needed for theoretical purposes.

For the special problem (5-135) the difference equation (5-58) assumes the form

$$\begin{aligned}\alpha_k y_{n+k} + \alpha_{k-1} y_{n+k-1} + \cdots + \alpha_0 y_n \\ = Ah(\beta_k y_{n+k} + \beta_{k-1} y_{n+k-1} + \cdots + \beta_0 y_n)\end{aligned} \tag{5-136}$$

This linear difference equation with constant coefficients can be solved by the methods of §5.2-1. The characteristic equation is $\rho(\zeta) = 0$, where

$$\tilde{\rho}(\zeta) = \rho(\zeta) - Ah\sigma(\zeta) \tag{5-137}$$

Here $\rho(\zeta)$ and $\sigma(\zeta)$ denote the polynomials defined by (5-81). If the roots $\tilde{\zeta}_1, \tilde{\zeta}_2, \cdots, \tilde{\zeta}_k$ of $\tilde{\rho}(\zeta)$ are all distinct, the solution of (5-136) can according to Theorem 5.3 be written in the form

$$y_n = \sum_{\mu=1}^{k} A_\mu \tilde{\zeta}_\mu^n \tag{5-138}$$

where the A_μ denote suitable constants. Before determining these from the initial conditions, we shall establish some properties of the roots $\tilde{\zeta}_\mu$, $\mu = 1, \cdots, k$.

We shall assume that the method defined by (5-58) is *stable* and *consistent*. We shall furthermore assume that the polynomials $\rho(\zeta)$ and $\sigma(\zeta)$ have *no common factors.** It is intuitively obvious that for small values of h the roots $\tilde{\zeta}_\mu$ approach certain of the roots $\zeta_1, \zeta_2, \cdots, \zeta_k$ of the polynomial $\rho(\zeta)$. More precisely, it may be deduced from a theorem in the theory of complex variables† that for each sufficiently small $\varepsilon > 0$ there exists a $\delta > 0$ such that for $0 \leq h < \delta$ the equation $\tilde{\rho}(\zeta) = 0$ has exactly as many roots in each of the disks $|\zeta - \zeta_\mu| < \varepsilon$ $(\mu = 1, 2, \cdots, k)$ as the equation $\rho(\zeta) = 0$. If ζ_μ is a root of multiplicity p, the p roots $\tilde{\zeta}_\nu$ approaching it can be shown to be the p different values of an analytic function of $h^{1/p}$ obtained by assigning to $h^{1/p}$ its p different values. It follows from $\sigma(\zeta_\mu) \neq 0$ that these p roots $\tilde{\zeta}_\nu$ are *distinct* for h sufficiently small, $h \neq 0$.

In order to simplify the following discussion we introduce a special labeling for the roots $\zeta_1, \cdots, \zeta_k$. By assumption of stability the polynomial $\rho(\zeta)$ has no roots of modulus exceeding 1. If there are m roots of modulus 1—necessarily simple again by the stability assumption—they are denoted by $\zeta_1, \zeta_2, \cdots, \zeta_m$. In particular, we put $\zeta_1 = 1$; this is always

* This hypothesis is satisfied by all special methods discussed in §5.1. It follows from the discussion of §5.2-7 that common factors of $\rho(\zeta)$ and $\sigma(\zeta)$ can be removed without changing either the order or the error constant of the method.

† See, e.g., Theorem 11, p. 107, of Ahlfors [1953].

a root by virtue of consistency. The roots of modulus 1 are called the *essential roots*; ζ_1 in particular will be termed the *principal root*. The remaining roots $\zeta_{m+1}, \cdots \zeta_k$, if there are any, will be called *nonessential roots*.

We shall require some quantitative statements about the roots $\tilde{\zeta}_\nu$. As to the roots approaching nonessential roots ζ_ν, the rough estimate,

$$|\tilde{\zeta}_\mu| \leq t, \qquad \mu = m + 1, \cdots, k \tag{5-139}$$

where

$$t = \frac{1}{2}\left(1 + \max_{m<\mu\leq k} |\zeta_\mu|\right), \qquad |t| < 1 \tag{5-140}$$

is valid for sufficiently small values of h. It states that for small enough h the roots approximating nonessential roots all lie within a circle contained in the unit circle.

The roots approaching essential roots are, according to the above, analytic functions of h. We wish to determine the linear term in their expansion in powers of h. By the method of undetermined coefficients we find

$$\tilde{\zeta}_\mu = \zeta_\mu(1 + \lambda_\mu Ah + O(h^2)), \qquad \mu = 1, \cdots, m \tag{5-141}$$

where

$$\lambda_\mu = \frac{\sigma(\zeta_\mu)}{\zeta_\mu \rho'(\zeta_\mu)} \tag{5-142}$$

The constants λ_μ depend only on the method and not on the differential equation to be integrated. They are called *growth parameters* for reasons which will become apparent later. We note that

$$\lambda_1 = \frac{\sigma(1)}{\rho'(1)} = 1 \tag{5-143}$$

by virtue of consistency.

We are interested in the asymptotic behavior of $\tilde{\zeta}_\mu^n$ when $h \to 0$ and $n \to \infty$ in such a manner that $nh = x$ remains constant. In view of

$$\begin{aligned}(1 + \lambda_\mu Ah + O(h^2))^n &= \exp\left[xh^{-1}\log(1 + \lambda_\mu Ah + O(h^2))\right] \\ &= \exp\left[xh^{-1}(\lambda_\mu Ah + O(h^2))\right] \\ &= \exp(\lambda_\mu Ax) + O(h)\end{aligned}$$

(see also Problem 13, Chapter 1), it follows that, writing $\zeta_\mu = e^{in\varphi_\mu}$,

$$\tilde{\zeta}_\mu^n = e^{in\varphi_\mu}[e^{\lambda_\mu Ax} + O(h)], \qquad h \to 0 \tag{5-144}$$

In particular, since $\zeta_1 = 1$, $\lambda_1 = 1$,

$$\tilde{\zeta}_1^n = e^{Ax} + O(h) \tag{5-145}$$

We wish to determine the $O(h)$ term in the last equation more accurately. We begin by calculating a better approximation for $\tilde{\zeta}_1$. By relation (5-141), we may write

$$\tilde{\zeta}_1 = e^{Ah} + \tau$$

where τ is an analytic function of h at $h = 0$, $\tau = O(h^2)$. Substituting into (5-137), we find

$$\rho(e^{Ah}) + \rho'(e^{Ah})\tau + O(\tau^2) - hA\sigma(e^{Ah}) + O(h\tau) = 0$$

We now make use of the relation

$$\rho(e^{Ah}) - hA\sigma(e^{Ah}) = C_{p+1}(hA)^{p+1} + O(h^{p+2}) \tag{5-146}$$

which follows by substituting $y(x) = e^{Ax}$ in (5-88). We then obtain

$$\rho'(e^{Ah})\tau = -C_{p+1}(hA)^{p+1} + O(h\tau) + O(h^{p+2})$$

and thus, using $\rho'(e^{Ah}) = \rho'(1) + O(h)$,

$$\tau = -C(hA)^{p+1} + O(h^{p+2})$$

where C is the error constant defined by (5-93). It follows that

$$\tilde{\zeta}_1 = e^{Ah}[1 - (Ah)^{p+1}C + O(h^{p+2})] \tag{5-147}$$

For $h \to 0$, $nh = x$ we find

$$\tilde{\zeta}_1^n = e^{Ax}[1 - xh^pA^{p+1}C + O(h^{p+1})] \tag{5-148}$$

This is the desired more accurate version of (5-145).

We now turn to the problem of determining the constants A_μ in (5-138) from the condition that $y_0, y_1, \cdots, y_{k-1}$ assume given values. For convenience we write the initial conditions in the form

$$y_\mu = \tilde{\zeta}_1^\mu + \delta_\mu, \qquad \mu = 0, \cdots, k-1 \tag{5-149}$$

The equations to be satisfied by the A_μ then read

$$A_1\tilde{\zeta}_1^\mu + A_2\tilde{\zeta}_2^\mu + \cdots + A_k\tilde{\zeta}_k^\mu = \tilde{\zeta}_1^\mu + \delta_\mu, \qquad \mu = 0, 1, \cdots, k-1 \tag{5-150}$$

This system of k linear equations for the k unknowns $A_1, \cdots, A_k$ may be solved as follows. Define for $\nu = 1, 2, \cdots, k$ the polynomials

$$\tilde{\rho}_\nu(\zeta) = \frac{\tilde{\rho}(\zeta)}{\zeta - \tilde{\zeta}_\nu} = \tilde{\alpha}_{\nu,0} + \tilde{\alpha}_{\nu,1}\zeta + \cdots + \tilde{\alpha}_{\nu,k-1}\zeta^{k-1} \tag{5-151}$$

Since $\tilde{\rho}(\tilde{\zeta}_\mu) = 0$, $\mu = 1, \cdots, k$, we have

$$\tilde{\rho}_\nu(\tilde{\zeta}_\mu) = \begin{cases} 0, & \nu \neq \mu \\ \tilde{\rho}'(\tilde{\zeta}_\mu), & \nu = \mu \end{cases} \tag{5-152}$$

Multiplying the μth equation (5-150) by $\tilde{\alpha}_{\nu,\mu}$, adding, and using (5-152), we obtain

$$A_\nu \tilde{\rho}'(\tilde{\zeta}_\nu) = \tilde{\rho}_\nu(\tilde{\zeta}_1) + \tilde{\Delta}_\nu \tag{5-153}$$

where

$$\tilde{\Delta}_\nu = \tilde{\alpha}_{\nu,0}\delta_0 + \tilde{\alpha}_{\nu,1}\delta_1 + \cdots + \tilde{\alpha}_{\nu,k-1}\delta_{k-1}$$

Since $\tilde{\rho}'(\tilde{\zeta}_\nu) \neq 0$ for $h \neq 0$ in view of the fact that the roots $\tilde{\zeta}_\nu$ are simple, it follows that

$$\begin{aligned} A_1 &= 1 + \frac{\tilde{\Delta}_1}{\tilde{\rho}'(\tilde{\zeta}_1)} \\ A_\nu &= \frac{\tilde{\Delta}_\nu}{\tilde{\rho}'(\tilde{\zeta}_\nu)}, \qquad \nu = 2, \cdots, k \end{aligned} \tag{5-154}$$

As $h \to 0$, we have $\tilde{\Delta}_\nu = \Delta_\nu + O(h\delta)$, where $\delta = \max |\delta_\mu|$, and

$$\Delta_\nu = \alpha_{\nu,0}\delta_0 + \alpha_{\nu,1}\delta_1 + \cdots + \alpha_{\nu,k-1}\delta_{k-1} \tag{5-155}$$

where

$$\alpha_{\nu,0} + \alpha_{\nu,1}\zeta + \cdots + \alpha_{\nu,k-1}\zeta^{k-1} = \frac{\rho(\zeta)}{\zeta - \zeta_\nu} \tag{5-156}$$

Since $\tilde{\rho}'(\tilde{\zeta}_\nu) = \rho'(\tilde{\zeta}_\nu) - Ah\sigma'(\tilde{\zeta}_\nu)$, if ζ_ν is a root of multiplicity p,

$$\tilde{\rho}'(\tilde{\zeta}_\nu) = \frac{1}{(p-1)!}\rho^{(p)}(\zeta_\nu)(\tilde{\zeta}_\nu - \zeta_\nu)^{p-1} + O(h)$$

It follows that $\tilde{\rho}'(\tilde{\zeta}_\nu) = \rho'(\zeta_\nu) + O(h)$ if $p = 1$, and $\tilde{\rho}'(\tilde{\zeta}_\nu) = O(h^{1-1/p})$, if $p > 1$.

Collecting our results, we find

$$\begin{aligned} y_n = {} & \left(1 + \frac{\Delta_1}{\rho'(1)} + O(h\delta)\right) e^{Ax}[1 - xA^{p+1}Ch^p + O(h^{p+1})] \\ & + \sum_{\mu=2}^{m} \left(\frac{\Delta_\mu}{\rho'(\zeta_\mu)} + O(h\delta)\right) e^{in\varphi_\mu}[e^{\lambda_\mu Ax} + O(h)] \\ & + \text{terms majorized by } h^{-1}t^n. \end{aligned}$$

Since $h^{-1}t^n \to 0$ more rapidly than any power of h, we finally obtain for the error $e_n = y_n - e^{Ax_n}$, retaining terms of leading order in h and δ only, the following asymptotic relationship:

$$e_n = -CA^{p+1}xe^{Ax}h^p + O(h^{p+1}) + \sum_{\mu=1}^{m} \frac{\Delta_\mu}{\rho'(\zeta_\mu)} e^{in\varphi_\mu}e^{\lambda_\mu Ax} + O(h\delta) \tag{5-157}$$

This formula shows the discretization error e_n to consist of essentially two parts. The first part, represented by the term involving C in (5-157), may be called the *genuine discretization error*. It corresponds exactly to the discretization error of a one-step method of order p with the principal error function $-Cy^{(p+1)}(x)$. This error is a smooth function of x; it is, among other things, amenable to deferred approach to the limit procedures. The second part, represented by the sum in (5-157), owes its existence to the fact that $\delta \neq 0$ and may thus be called the *starting error*. This error is of a more complicated nature than the genuine discretization error and calls for several comments.

(i) If $\delta \to 0$ as $h \to 0$, the starting error clearly tends to zero as $h \to 0$. Since the genuine discretization error also tends to zero as $h \to 0$, this shows that a consistent and stable multistep method is convergent (in the sense defined in §5.2-3) for the special differential equation under consideration.

(ii) If the starting values are *exact*, i.e., if $y_\mu = e^{Ax_\mu}$ for $\mu = 0, 1, \cdots, k-1$ then, by (5-147), $\delta_\mu = e^{Ax_\mu} - \zeta_1^\mu = \mu C(Ah)^{p+1} + O(h^{p+2})$. Thus, if $Ah \neq 0$ and $\mu > 0$, $\delta_\mu \neq 0$, and the starting error does not vanish. It is, however, of the order of h^{p+1}, i.e., one order of magnitude smaller than the genuine discretization error.

(iii) If $m = 1$, i.e., if the principal root is the only essential root of $\rho(\zeta)$, the sum in (5-157) reduces to its first term. The starting error is then proportional to the exact solution, and the result of using a multistep method is like using a one-step method with slightly wrong starting values.

(iv) If $m > 1$, and if $\delta \neq 0$, additional terms appear in the sum in (5-157). These terms can be characterized as oscillations (owing to the factors $e^{in\varphi_\mu}$ with $\varphi_\mu \neq 0$) with amptitudes proportional to $|e^{\lambda_\mu Ax}|$. If $\text{Re}\,\lambda_\mu A \leq \text{Re}\,A$, these oscillations are not conspicuous in comparison with the exact solution e^{Ax}. If $\text{Re}\,\lambda_\mu A > \text{Re}\,A$ for some μ, however, then these oscillations will, for a fixed value of h, become large in comparison with the desired solution e^{Ax} as x increases and thus render meaningless the calculated values y_n. The described situation occurs, for instance, if $A < 0$ and if one of the growth parameters is negative. All commonly used methods with several essential roots have some negative growth parameters.* As an example we quote the *mid-point rule* $y_{n+1} - y_{n-1} = 2hf_n$ [(5-25) with $q = 0$]. Here we have $\zeta_2 = -1$, $\lambda_2 = -1$. If the differential equation to be integrated is $y' = -y$, and if $\delta \neq 0$, it follows from (5-157) that in addition to the genuine discretization error, which is asymptotic to $-\frac{1}{6}h^2xe^{-x}$, there is a starting error consisting of a linear combination

* It will be shown in §5.4-6 that for optimal difference operators (see §5.2-8 for definition) the growth parameter corresponding to $\zeta = -1$ is always less than $-\frac{1}{3}$.

of the terms e^{-x} and $(-1)^n e^x$. The first term is not dangerous. The second term, however, grows relative to the exact solution like e^{2x} and will, for sufficiently large values of x, spoil the numerical values y_n.

In Table 5.14 we present numerical values calculated with the step $h = 0.1$. Although only five decimal places are reproduced, the values

Table 5.14

	Mid-point Rule		Adams-Bashforth	
x_n	y_n	e_n	y_n	e_n
0.0	1.00000	0.00000	1.00000	0.00000
0.1	0.90484	0.00000	0.90484	0.00000
0.2	0.81903	0.00030	0.81911	0.00038
0.3	0.74103	0.00021	0.74149	0.00070
0.4	0.67083	0.00051	0.67122	0.00090
.	.		.	
.	.		.	
.	.		.	
3.0	0.05152	0.00173	0.05043	0.00064
3.1	0.04363	−0.00142	0.04564	0.00060
3.2	0.04280	0.00204	0.04132	0.00056
3.3	0.03507	−0.00181	0.03740	0.00052
3.4	0.03578	0.00241	0.03386	0.00049
.	.		.	
.	.		.	
.	.		.	
5.0	0.01803	0.01129	0.00688	0.00015
5.1	−0.00775	−0.01385	0.00623	0.00014
5.2	0.01958	0.01406	0.00564	0.00012
5.3	−0.01166	−0.01665	0.00511	0.00011
5.4	0.02191	0.01739	0.00462	0.00011

were calculated to ten places, and the starting value $y_1 = e^{-h}$ was exact to ten places. An oscillatory behavior of the error is manifest from the start and soon takes on dimensions which render the numerical values useless. For the sake of contrast, approximate values obtained with the second order Adams-Bashforth method are also given. Since $m = 1$ for this method, we expect no oscillatory component in the error. This is confirmed by the smoothness of the numerical values.

For the Milne rule (5-33) we find $\lambda_1 = 1$, $\lambda_2 = -\frac{1}{3}$. In the above example the starting error thus involves e^{-x} and $(-1)^n e^{x/3}$. The second, "extraneous" component grows less strongly than for the mid-point rule, although still very strongly in comparison with the solution itself.

The presence of oscillatory components in e_n which grow relative to the exact solution is frequently referred to as *numerical instability*. This instability is not of the same kind as the instability discussed in §5.2-4. That instability had to do with roots of $\rho(\zeta)$ whose modulus exceeds 1. The methods with this property are divergent. The phenomenon discussed in the present section arises in a convergent method, if some growth parameter associated with an essential root $\neq 1$ is negative and if the desired solution is exponentially decreasing. We shall call this phenomenon *weak stability* or *conditional stability*. Convergent methods with $m = 1$, which cannot produce weak stability, are called *strongly stable*.

Although the above considerations have been carried out only for a special, simple differential equation, the conclusions will be shown to remain qualitatively true for the general initial value problem. The distinction between weakly and strongly stable methods will be fundamental also in the discussion of round-off error.

5.3-2. Two lemmas. We are ultimately interested in establishing formulas similar to (5-157) for the discretization error in the approximate solution of an arbitrary differential equation by a linear multistep method. This task is less simple than it was for one-step methods, and a considerable analytical apparatus is required. Of the two lemmas stated in this section, the second will be used repeatedly in later sections. The first lemma is required for formulation and proof of the second.

LEMMA 5.5. *Let the polynomial* $\rho(\zeta) = \alpha_k\zeta^k + \alpha_{k-1}\zeta^{k-1} + \cdots + \alpha_0$ *satisfy the condition of stability, and let the coefficients* γ_l $(l = 0, 1, 2, \cdots)$ *be defined by*

$$\frac{1}{\alpha_k + \alpha_{k-1}\zeta + \cdots + \alpha_0\zeta^k} = \gamma_0 + \gamma_1\zeta + \gamma_2\zeta^2 + \cdots \tag{5-158}$$

Then

$$\Gamma = \sup_{l=0,1,\cdots} |\gamma_l| < \infty$$

Proof. If $\hat{\rho}(\zeta) = \alpha_k + \alpha_{k-1}\zeta + \cdots + \alpha_0\zeta^k = \zeta^k\rho(\zeta^{-1})$, the roots of $\hat{\rho}(\zeta)$ are the reciprocals of the roots of $\rho(\zeta)$. Since $\rho(\zeta)$ has no roots outside $|\zeta| = 1$, it follows that $1/\hat{\rho}(\zeta)$ is holomorphic in $|\zeta| < 1$. Moreover, since the roots $\zeta_1, \zeta_2, \cdots, \zeta_m$ of $\rho(\zeta)$ on $\zeta = 1$ are simple, the poles of $[\hat{\rho}(\zeta)]^{-1}$ on $|\zeta| = 1$ are simple, and there exist constants $A_1, A_2, \cdots, A_m$ such that the function

$$f(\zeta) = \frac{1}{\hat{\rho}(\zeta)} - \frac{A_1}{\zeta - \zeta_1^{-1}} - \cdots - \frac{A_m}{\zeta - \zeta_m^{-1}} \tag{5-159}$$

is holomorphic for $|\zeta| \leq 1$. By Cauchy's estimate (see Ahlfors [1953], p. 98) the coefficients of the Taylor expansion of $f(\zeta)$ at $\zeta = 0$ are thus bounded.

Since the coefficients in the expansion of each of the terms $A_\mu/(\zeta - \zeta_\mu^{-1})$ are also bounded, the assertion of the lemma follows.

We shall require the identity

$$\text{(5-160)} \qquad \alpha_k\gamma_l + \alpha_{k-1}\gamma_{l-1} + \cdots + \alpha_0\gamma_{l-k} = \begin{cases} 1, & l = 0 \\ 0, & l > 0 \end{cases}$$

where it is assumed that $\gamma_l = 0$ for $l < 0$. This is proved by multiplying both sides of (5-158) by $\alpha_k + \alpha_{k-1}\zeta + \cdots + \alpha_0\zeta^k$ and comparing coefficients in the resulting expansion in powers of ζ.

The following lemma concerns the growth of solutions of the nonhomogeneous linear difference equation

$$\text{(5-161)} \qquad \alpha_k z_{m+k} + \alpha_{k-1}z_{m+k-1} + \cdots + \alpha_0 z_m = h\{\beta_{k,m}z_{m+k} + \beta_{k-1,m}z_{m+k-1} + \cdots + \beta_{0,m}z_m\} + \lambda_m$$

LEMMA 5.6. *Let the polynomial* $\rho(\zeta) = \alpha_k\zeta^k + \cdots + \alpha_0$ *satisfy the condition of stability, let* B^*, β, *and* Λ *be nonnegative constants such that*

$$\text{(5-162)} \qquad |\beta_{k,n}| + |\beta_{k-1,n}| + \cdots + |\beta_{0,n}| \le B^*, \quad |\beta_{k,n}| \le \beta, \quad |\lambda_n| \le \Lambda, \quad n = 0, 1, 2, \cdots, N$$

and let $0 \le h < |\alpha_k|\,\beta^{-1}$. *Then every solution of* (5-161) *for which*

$$\text{(5-163)} \qquad |z_\mu| \le Z, \qquad \mu = 0, 1, \cdots, k-1$$

satisfies

$$\text{(5-164)} \qquad |z_n| \le K^* e^{nhL^*}, \qquad n = 0, 1, \cdots, N$$

where

$$\text{(5-165)} \qquad L^* = \Gamma^* B^*, \qquad K^* = \Gamma^*(N\Lambda + AZk)$$

$$A = |\alpha_k| + |\alpha_{k-1}| + \cdots + |\alpha_0|, \qquad \Gamma^* = \frac{\Gamma}{1 - h\,|\alpha_k|^{-1}\beta}$$

Proof. For $l = 0, 1, \cdots, n-k$, multiply the equation (5-161) corresponding to $m = n - k - l$ by γ_l and add the resulting equations. If the sum is called S_n, we obtain by summing on the left

$$S_n = (\alpha_k z_n + \alpha_{k-1}z_{n-1} + \cdots + \alpha_0 z_{n-k})\gamma_0 + (\alpha_k z_{n-1} + \alpha_{k-1}z_{n-2} + \cdots + \alpha_0 z_{n-k-1})\gamma_1 + \cdots + (\alpha_k z_k + \alpha_{k-1}z_{k-1} + \cdots + \alpha_0 z_0)\gamma_{n-k}$$

Rearranging, we get

$$S_n = \alpha_k\gamma_0 z_n + (\alpha_k\gamma_1 + \alpha_{k-1}\gamma_0)z_{n-1} + \cdots + (\alpha_k\gamma_{n-k} + \alpha_{k-1}\gamma_{n-k-1} + \cdots + \alpha_0\gamma_{n-2k})z_k + (\alpha_{k-1}\gamma_{n-k} + \cdots + \alpha_0\gamma_{n-2k+1})z_{k-1} + \cdots + \alpha_0\gamma_{n-k}z_0$$

By (5-160) this reduces to

$$\text{(5-166)} \qquad S_n = z_n + (\alpha_{k-1}\gamma_{n-k} + \cdots + \alpha_0\gamma_{n-2k+1})z_{k-1} + \cdots + \alpha_0\gamma_{n-k}z_0$$

Summing on the right, we find

$$S_n = h\{\beta_{k,n-k}\gamma_0 z_n + (\beta_{k-1,n-k}\gamma_0 + \beta_{k,n-k-1}\gamma_1)z_{n-1} + \cdots + (\beta_{0,n-k}\gamma_0 + \cdots + \beta_{k,n-2k}\gamma_k)z_{n-k} + \cdots + \beta_{0,0}\gamma_{n-k}z_0\} + \lambda_{n-k}\gamma_0 + \lambda_{n-k-1}\gamma_1 + \cdots + \lambda_0\gamma_{n-k} \tag{5-167}$$

Equating (5-166) and (5-167) we find, using (5-162) and (5-163),

$$|z_n| \le h\beta\,|\alpha_k^{-1}|\,|z_n| + h\Gamma B^* \sum_{m=0}^{n-1} |z_m| + N\Gamma\Lambda + A\Gamma Zk$$

Solving for $|z_n|$, there follows

$$|z_n| \le hL^* \sum_{m=0}^{n-1} |z_m| + K^* \tag{5-168}$$

where L^* and K^* are given by (5-165). We now proceed by induction. Since $A\Gamma \ge 1$ and hence $K^* \ge Z$, the estimate

$$|z_m| \le K^*(1 + hL^*)^m \tag{5-169}$$

is true for $m = 0, 1, \cdots, k - 1$. Assuming its truth for $m = 0, 1, \cdots, n - 1$ and using it on the right of (5-168), we obtain

$$|z_n| \le hL^*K^* \frac{(1 + hL^*)^n - 1}{hL^*} + K^* = K^*(1 + hL^*)^n$$

Relation (5-169) is thus established for $m = n$ and this holds generally for $m = 0, 1, \cdots, N$. The statement of the lemma now follows by using $1 + hL^* \le e^{hL^*}$.

In subsequent applications of Lemma 5.6 the constants Λ, Z, and N will usually depend on h. It is important to remember that even then Γ, A, and in most cases also B^* and β are independent of h.

5.3-3. A sufficient condition for convergence. In this section we shall show that the conditions of stability and consistency, which in §§5.2-4 and 5.2-6 were found to be necessary for convergence of a linear multistep method, are also sufficient for convergence.

THEOREM 5.10. *A stable and consistent linear multistep method is convergent.*

Proof. Let the function $f(x, y)$ satisfy the conditions of the existence theorem 1.1, and let η be an arbitrary constant. We shall denote by $y(x)$ the solution of the initial value problem $y' = f(x, y)$, $y(a) = \eta$. Let y_n $(n = 0, 1, 2, \cdots)$ be the solution of the difference equation (5-58), defined by the starting values $y_\mu = \eta_\mu(h)$, $\mu = 0, 1, \cdots, k - 1$. We set

$$\delta = \delta(h) = \max_{\mu=0,1,\cdots,k-1} |\eta_\mu(h) - y(a + \mu h)| \tag{5-170}$$

and assume that

(5-171) $$\lim_{h\to 0} \delta(h) = 0$$

We then have to show that for any $x \in [a, b]$

$$\lim_{\substack{h\to 0\\ x_n=x}} y_n = y(x)$$

We begin by estimating the quantity $|L[y(x_m); h]|$, where L denotes the difference operator defined by (5-87). Since $f(x, y)$ is not assumed to be differentiable, the derivatives $y''(x)$, $y'''(x)$, $\cdots$ do not necessarily exist and the method of §5.2-5 cannot be used. However, we may proceed as follows.

The function $y'(x) = f(x, y(x))$ is continuous in the closed interval $[a, b]$. We define for $\varepsilon \geq 0$ the quantity

$$\chi(\varepsilon) = \max_{\substack{|x^*-x|\leq\varepsilon\\ x,x^*\in[a,b]}} |y'(x^*) - y'(x)|$$

For $\mu = 0, 1, 2, \cdots, k$ we can write

$$y'(x_{m+\mu}) = y'(x_m) + \theta_\mu \chi(\mu h)$$

where $|\theta_\mu| \leq 1$. Furthermore, since

$$y(x_{m+\mu}) = y(x_m) + \mu h y'(\xi_\mu)$$

where $x_n < \xi_\mu < x_{n+\mu}$, we have

$$y(x_{m+\mu}) = y(x_m) + \mu h[y'(x_m) + \theta'_\mu \chi(\mu h)]$$

where $|\theta'_\mu| \leq 1$. Hence it follows that

$$\begin{aligned} L[y(x_m); h] = {} & (\alpha_0 + \alpha_1 + \cdots \alpha_k)y(x_m) + (\alpha_1 + 2\alpha_2 + \cdots + k\alpha_k)y'(x_m)h \\ & + \theta'(|\alpha_1| + 2\,|\alpha_2| + \cdots + k\,|\alpha_k|)\chi(kh)h \\ & - (\beta_0 + \beta_1 + \cdots + \beta_k)y'(x_m)h - \theta(|\beta_0| + \cdots + |\beta_k|)\chi(kh)h \end{aligned}$$

where $|\theta| \leq 1$, $|\theta'| \leq 1$. Since L is assumed to be consistent, we have $\alpha_0 + \alpha_1 + \cdots + \alpha_k = 0$, $\alpha_1 + 2\alpha_2 + \cdots + k\alpha_k - \beta_0 - \beta_1 - \cdots - \beta_k = 0$. Hence

(5-172) $$|L[y(x_m); h]| \leq K\chi(kh)h$$

where

$$K = |\alpha_1| + 2\,|\alpha_2| + \cdots + k\,|\alpha_k| + |\beta_0| + \cdots + |\beta_k|$$

We now subtract $L[y(x_m); h]$ from the corresponding relation

$$\alpha_k y_{m+k} + \cdots + \alpha_0 y_m - h\{\beta_k f_{m+k} + \cdots + \beta_0 f_m\} = 0$$

satisfied by the values y_m. Writing $e_m = y_m - y(x_m)$, $m = 0, 1, \cdots$, and setting

$$g_m = \begin{cases} [f(x_m, y_m) - f(x_m, y(x_m))]e_m^{-1}, & e_m \neq 0 \\ 0 & e_m = 0 \end{cases}$$

we get

$$\alpha_k e_{m+k} + \cdots + \alpha_0 e_m - h\{\beta_k g_{m+k} e_{m+k} + \cdots + \beta_0 g_m e_m\} = \theta_m K\chi(kh)h$$

where $|\theta_m| \leq 1$. In view of the Lipschitz condition, $|g_m| \leq L$, $m = 0, 1, 2, \cdots$. We thus are in a position to apply Lemma 5.6 with $z_m = e_m$, $Z = \delta(h)$, $\Lambda = K\chi(kh)h$, $N = (x_n - a)/h$, and $B^* = BL$, where $B = |\beta_0| + |\beta_1| + \cdots + |\beta_k|$. It follows that

$$|e_n| \leq \Gamma^*[A\delta(h) + (x_n - a)K\chi(kh)] \exp [(x_n - a)L\Gamma^* B] \tag{5-173}$$

where

$$A = |\alpha_0| + |\alpha_1| + \cdots + |\alpha_k|$$

$$\Gamma^* = \frac{\Gamma}{1 - h\,|\alpha_k^{-1}\beta_k|\,L} \tag{5-174}$$

Since $y'(x)$ is uniformly continuous on $[a, b]$, $\chi(kh) \to 0$ as $h \to 0$. In view of this fact and of (5-170) the above bound for $|e_n|$ tends to zero as $h \to 0$ for every $x_n \in [a, b]$, establishing the desired result.

We remark that the above proof applies with little change to the case where y is complex and $f(x, y)$ is complex-valued, if convergence is defined in the appropriate way. In this case it is not necessary to assume that the coefficients of $\rho(\zeta)$ and $\sigma(\zeta)$ are real; indeed, Lemmas 5.5 and 5.6 are true without this hypothesis.

5.3-4. An improved a priori bound for the discretization error. The bound (5-173) is in general very poor and does not exhibit the true order of magnitude of the accumulated discretization error which can be expected if the true solution is sufficiently smooth. Denoting by p the order of the difference operator (5-87), we shall now assume that the exact solution $y(x)$ has a continuous derivative of order $p + 1$ for $x \in [a, b]$, and set

$$Y = \max_{x \in [a,b]} |y^{(p+1)}(x)| \tag{5-175}$$

It will be shown that for sufficiently accurate starting values the accumulated discretization error is of the order of h^p. The following fact will be required.

LEMMA 5.7. *Let $L[y(x); h]$ be a difference operator of order $p > 0$. There exists a constant $G > 0$, depending only on L, such that*

$$(5\text{-}176) \qquad |L[y(x);\, h]| \le h^{p+1} G Y, \qquad a \le x, \qquad x + kh \le b$$

for all functions $y(x)$ having a continuous derivative of order $p + 1$ in $[a, b]$.

Remark. It follows from the results of §5.1-3 and §5.1-4 that for a number of special difference operators—namely those associated with the Adams methods, the Milne method, and the methods based on differentiation—we can write

$$(5\text{-}177) \qquad L[y(x);\, h] = h^{p+1} C_{p+1} y^{(p+1)}(\xi)$$

where C_{p+1} is defined in §5.2-5, and where ξ is a suitable number in the interval $(x, x + kh)$. We shall refer to (5-177) as the *generalized mean value theorem.* If the generalized mean value theorem holds, then (5-176) follows immediately with $G = |C_{p+1}|$. However, the generalized mean value theorem has not been proved for arbitrary operators of the form (5-87). By the definition of C_{p+1} we merely have

$$L[y(x);\, h] = h^{p+1} C_{p+1} y^{(p+1)}(x) + O(h^{p+2})$$

and from this we cannot deduce (5-176), since this would not take into account the $O(h^{p+2})$ term.

Proof of Lemma 5.7. An expression for G can be found in the general case by using Taylor's theorem with the integral form of the remainder. If $y(x)$ has a continuous derivative of order $q + 1$ in (a, b), and if x and $\bar{x}$ are in (a, b), then we have (see Taylor [1955], p. 112)

$$y(\bar{x}) = y(x) + (\bar{x} - x)y'(x) + \cdots + \frac{(\bar{x} - x)^q}{q!} y^{(q)}(x) + \frac{1}{q!}\int_x^{\bar{x}} (\bar{x} - t)^q y^{(q+1)}(t)\, dt$$

Applying this with $\bar{x} = x + \mu h$ $(\mu = 1, 2, \cdots, k)$ to both $y(x)$ and $y'(x)$, we obtain after putting $t = x + hs$

$$y(x + \mu h) = \cdots + h^{p+1} \int_0^{\mu} \frac{(\mu - s)^p}{p!} y^{(p+1)}(x + hs)\, ds$$

$$y'(x + \mu h) = \cdots + h^{p+1} \int_0^{\mu} \frac{(\mu - s)^{p-1}}{(p - 1)!} y^{(p+1)}(x + hs)\, ds$$

where the dots represent terms involving derivatives of order $\le p$. If the operator $L[y(x); h]$ has order p, and if each term on the right of (5-87) is replaced by one of the above expressions, then the terms represented by dots cancel owing to the definition of p, and what remains may be written in the form

$$L[y(x);\ h] = h^{p+1}\int_0^k G(s)y^{(p+1)}(x + hs)\,ds \tag{5-178}$$

The function $G(s)$ is called the *kernel* of this integral representation of $L[y(x); h]$; it is represented by a different analytical expression in each of the intervals $[\mu, \mu + 1)$ $(\mu = 0, 1, \cdots, k - 1)$. For $s \in [k - 1, k]$ we have

$$G(s) = \alpha_k \frac{(k-s)^p}{p!} - \beta_k \frac{(k-s)^{p-1}}{(p-1)!}$$

and for $s \in [\mu, \mu + 1)$ $(\mu = 0, 1, \cdots, k - 2)$

$$G(s) = \alpha_k \frac{(k-s)^p}{p!} + \alpha_{k-1}\frac{(k-1-s)^p}{p!} + \cdots + \alpha_\mu \frac{(\mu+1-s)^p}{p!}$$
$$-\beta_k \frac{(k-s)^{p-1}}{(p-1)!} - \cdots - \beta_\mu \frac{(\mu+1-s)^{p-1}}{(p-1)!}$$

It now easily follows that

$$|L[y(x);\ h]| \le h^{p+1}\int_0^k |G(s)|\,|y^{(p+1)}(x + sh)|\,ds$$

and that (5-176) is true with

$$G = \int_0^k |G(s)|\,ds$$

We are now ready to tackle the problem stated at the beginning of the section. With a view toward later applications, we shall drop the assumption that the sequence $\{y_n\}$ is an exact solution of the difference equation (5-58). Instead, we shall merely assume that

$$\alpha_k y_{m+k} + \alpha_{k-1}y_{m+k-1} + \cdots + \alpha_0 y_m$$
$$-h\{\beta_k f_{m+k} + \cdots + \beta_0 f_m\} = \theta_m K_1 h^{q+1}, \qquad m = 0, 1, 2, \cdots \tag{5-179}$$

where K_1 and q are nonnegative constants, and $|\theta_m| \le 1$. We then can prove:

THEOREM 5.11. *Under the above hypotheses, if* $|h\beta_k\alpha_k^{-1}|\,L < 1$ *and* $x_n \in [a, b]$,

$$|e_n| \le \Gamma^*[A\delta k + (x_n - a)(K_1h^q + GYh^p)]\exp[(x_n - a)L\Gamma^*B] \tag{5-180}$$

where δ *is the maximum starting error defined by* (5-170), *and* A *and* Γ^* *are given by* (5-174).

The *proof* is a simplified version of the proof of Theorem 5.10. Subtracting from (5-179) the quantity $L[y(x_m); h]$, we now get

$$\alpha_k e_{m+k} + \cdots + \alpha_0 e_m - h\{\beta_k g_{m+k} e_{m+k} + \cdots + \beta_0 g_m e_m\}$$
$$= \bar{\theta}_m [K_1 h^{q+1} + GYh^{p+1}], \qquad m = 0, 1, 2, \cdots$$

where again $|\bar{\theta}_m| \leq 1$. From this estimate (5-180) follows immediately by another application of Lemma 5.6.

The estimate (5-180) very clearly exhibits the influence of the various sources of local error on the accumulated error. The term $A\delta k$ represents the influence of the starting error, which is essentially independent of h. The term involving K_1 and GY represent the errors resulting from local inaccuracies and discretization. These errors are both magnified by the same factor, which is of the order of h^{-1}.

It is suggested that the reader determine the numerical values of the constants A, B, G and Γ^* occurring in (5-180) for some special methods (see Problem 31). It is clear that for methods where the constants β_k are large and of alternating signs (such as the Adams methods) the constant B turns out to be large, which leads to a less favorable estimate. As to the value of Γ, it is easy to show that $\Gamma = 1$ for all methods based on numerical integration. For such methods, $\rho(\zeta) = \zeta^k - \zeta^{k-q}$, where $1 \leq q \leq k$, hence $\hat{\rho}(\zeta) = 1 - \zeta^q$, and

$$\frac{1}{\hat{\rho}(\zeta)} = \frac{1}{1 - \zeta^q} = 1 + \zeta^q + \zeta^{2q} + \cdots$$

It follows that $|\gamma_l| \leq 1$. For explicit methods we have $\beta_k = 0$, and hence $\Gamma^* = \Gamma$.

We note without proof that the error bound of Theorem 5.11 applies also to the case where y and $f(x, y)$ are complex. The polynomials $\rho(\zeta)$ and $\sigma(\zeta)$ need not have real coefficients in this case.

5.3-5. Asymptotic behavior of the discretization error. We now shall study the asymptotic behavior of the errors e_n, for a fixed value of x_n, as $h \to 0$. The result of the preceding section implies that, under liberal conditions, $e_n \to 0$ as $h \to 0$. In order to get something more interesting, we must look at e_n through the magnifying glass of a suitable negative power of h. Also, in order to get a well-determined result, it is necessary to agree on a definite starting procedure. We shall assume that the starting values, and thus the starting errors, are known, sufficiently differentiable functions of h. We write

$$e_\mu = \delta_\mu(h), \qquad \mu = 0, 1, \cdots, k - 1$$

and define q as the largest integer for which

$$\delta_\mu(h) = O(h^q), \qquad \mu = 0, 1, \cdots, k-1 \tag{5-181}$$

It is assumed that $q \geq 1$. (It may happen that $q = \infty$, e.g., if all the starting values are exact.) Furthermore, we assume that $y^{(p+2)}(x)$ is continuous and that the function $g(x) = f_y(x, y(x))$ is continuous and continuously differentiable for $x \in [a, b]$. Lastly, we assume that the difference equation (5-58) is satisfied exactly, so that $K_1 = 0$ in (5-180). It then follows from Theorem 5.11 that $e_n = O(h^r)$, where $r = \min(p, q)$, $r \geq 1$.

Once more we subtract from the difference equation (5-58) the corresponding equation satisfied by $y(x)$, now writing the remainder term in the form $R = C_{p+1}y^{(p+1)}(x_n)h^{p+1} + O(h^{p+2})$. Using

$$f(x_n, y_n) - f(x_n, y(x_n)) = g_n e_n + O(h^{2r})$$

where* $g_n = g(x_n)$ and $g(x) = f_y(x, y(x))$, we obtain

$$\begin{aligned}\alpha_k e_{n+k} + \alpha_{k-1}e_{n+k-1} + \cdots + \alpha_0 e_n &= h\{\beta_k g_{n+k}e_{n+k} + \cdots \\ &+ \beta_0 g_n e_n + O(h^{2r})\} - C_{p+2}y^{(p+1)}(x_n)h^{p+1} + O(h^{p+2})\end{aligned}$$

For the magnified error $\bar{e}_n = h^{-r}e_n$ we thus find the equation

$$\begin{aligned}\alpha_k \bar{e}_{n+k} + \alpha_{k-1}e_{n+k-1} + \cdots + \alpha_0 \bar{e}_n &= h\{\beta_k g_{n+k}\bar{e}_{n+k} + \cdots \\ &+ \beta_0 g_n \bar{e}_n\} - C_{p+1}y^{(p+1)}(x_n)h^{p+1-r} + O(h^2)\end{aligned} \tag{5-182}$$

(We have used $2 \leq r + 1$, $2 \leq p + 2 - r$.) This has to be solved under the initial conditions

$$\bar{e}_\mu = h^{-r}\delta_\mu(h) \equiv \bar{\delta}_\mu(h), \qquad \mu = 0, 1, \cdots, k-1 \tag{5-183}$$

We shall represent the solution in the form $\bar{e}_n^I + \bar{e}_n^H$, where $\bar{e}_n^I$ denotes the solution of the full equation (5-182) having zero initial values and $\bar{e}_n^H$ is the solution of the corresponding homogeneous equation

$$\alpha_k \bar{e}_{n+k} + \cdots + \alpha_0 \bar{e}_n = h\{\beta_k g_{n+k}\bar{e}_{n+k} + \cdots + \beta_0 g_n \bar{e}_n\} \tag{5-184}$$

satisfying the initial conditions (5-183).

We begin by determining $\bar{e}_n^I$. Two cases have to be distinguished. If $p > r$, we can apply Lemma 5.6 with $z_m = \bar{e}_m^I$, $Z = 0$, $N = (b-a)/h$, $\beta_{\mu,m} = \beta_\mu g_{m+\mu}$ and, using a suitable constant K, $\Lambda = Kh^2$. It then follows that $\bar{e}_n^I = O(h)$.

A more interesting result is obtained if $p = r$, i.e., if the starting error is not large compared with the truncation error. Since $y^{(p+1)}(x)$ has a continuous derivative in $[a, b]$, we may write

$$C_{p+1}y^{(p+1)}(x_n) = C\{\beta_k y^{(p+1)}(x_{n+k}) + \cdots + \beta_0 y^{(p+1)}(x_n)\} + O(h)$$

* This definition of g_n differs slightly from that used in §5.3-4.

where $C = C_{p+1}/(\beta_k + \cdots + \beta_0)$ is the error constant defined in §5.2-5. Thus (5-182) may be written in the form (omitting the superscript in $\bar{e}_n^I$)

$$\alpha_k \bar{e}_{n+k} + \cdots + \alpha_0 \bar{e}_n = h\{\beta_k[g_{n+k}\bar{e}_{n+k} - Cy^{(p+1)}(x_{n+k})] + \cdots + \beta_0[g_n e_n - Cy^{(p+1)}(x_n)]\} + O(h^2)$$

The same equation [without the $O(h^2)$ term] would result if the initial value problem

$$e'(x) = g(x)e(x) - Cy^{(p+1)}(x), \qquad e(a) = 0 \tag{5-185}$$

were solved by the multistep method under discussion. The starting values $\bar{e}_\mu^I = 0$ differ from the corresponding values $e(x_\mu)$ by $O(h)$ at most. It follows by Theorem 5.11, as applied to the present problem, that

$$\bar{e}_n^I = e(x_n) + O(h) \tag{5-186}$$

where $e(x)$ is defined by (5-185).

We now turn to the determination of $\bar{e}_n^H$. We shall represent this solution in the form

$$\bar{e}_n^H = \sum_{\mu=1}^{k} A_\mu e_n^{(\mu)} \tag{5-187}$$

where the sequences $\{e_n^{(\mu)}\}$ $(\mu = 1, \cdots, k)$ represent a fundamental system of solutions at $n = 0$ of the homogeneous equation (5-184) which is defined in the following special manner. Consider the differential equation $z'(x) = g_0 z(x)$. If it is solved by the multistep method under consideration, there results the linear difference equation with *constant* coefficients

$$\alpha_k z_{n+k} + \cdots + \alpha_0 z_n = hg_0\{\beta_k z_{n+k} + \cdots + \beta_0 z_n\} \tag{5-188}$$

If $g_0 \neq 0$ and if h is sufficiently small, but not zero, this difference equation possesses a fundamental system of solutions $z_n^{(\mu)}$ of the form $\tilde{\zeta}_\mu^n$, where the numbers $\tilde{\zeta}_\mu$ $(\mu = 1, \cdots, k)$ are the roots of the equation

$$\rho(\zeta) - hg_0\sigma(\zeta) = 0 \tag{5-189}$$

In this case we define the solutions $e_n^{(\mu)}$ by the condition

$$e_\nu^{(\mu)} = \tilde{\zeta}_\mu^\nu, \qquad \nu = 0, 1, \cdots, k-1; \qquad \mu = 1, \cdots, k \tag{5-190}$$

If $g_0 = 0$, let $s \leq k$ be the number of nonzero roots of the polynomial $\rho(\zeta)$. The difference equation (5-188) has now the stepnumber s. Let $z_n^{(\mu)}$ $(\mu = 1, 2, \cdots, s)$ be the fundamental system given by the construction of Theorem 5.3. (It may now involve solutions like $n\zeta^{n-1}$ if $\rho(\zeta)$ has multiple nonzero roots.) We supplement these s solutions by the sequences $\{z_n^{(\mu)}\}$ ($\mu = s + 1, \cdots, k$) defined by

$$z_n^{(\mu)} = \begin{cases} 1, & n = \mu - s - 1 \\ 0, & \text{otherwise} \end{cases}$$

Clearly, they are also solutions of (5-188) for $n \geq 0$, and the determinant of the matrix

$$\begin{pmatrix} z_0^{(1)} & z_1^{(1)} \cdots z_{k-1}^{(1)} \\ z_0^{(2)} & z_1^{(2)} \cdots z_{k-1}^{(2)} \\ \cdots & \cdots \\ z_0^{(k)} & z_1^{(k)} \cdots z_{k-1}^{(k)} \end{pmatrix}$$

is different from zero. We now define the solutions $e_n^{(\mu)}$ by

$$\text{(5-191)} \qquad e_\nu^{(\mu)} = z_\nu^{(\mu)}, \qquad \nu = 0, 1, \cdots, k - 1; \ \mu = 1, \cdots, k$$

Whether $g_0 = 0$ or not, the solutions $e_n^{(\mu)}$ thus defined form a fundamental system at $n = 0$, and although they may become dependent later (since we do not assume that $g_n \neq 0$ for $n > 0$), $\bar{e}_n^H$ on the strength of Theorem 5.1 can be represented in the form (5-187). Also, if we denote by $\tilde{\zeta}_1, \cdots, \tilde{\zeta}_m$ those roots of (5-189) which approach the *essential* roots $\zeta_1 = 1, \zeta_2, \cdots, \zeta_m$ of $\rho(\zeta)$, we always have

$$\text{(5-192)} \qquad e_\nu^{(\mu)} = z_\nu^{(\mu)} = \tilde{\zeta}_\mu^\nu, \qquad \nu = 0, 1, \cdots, k - 1; \ \mu = 1, \cdots m$$

It follows from §5.3-1 that there exist constants K and t, $|t| < 1$, such that for all sufficiently small values of h

$$\text{(5-193)} \qquad |z_n^{(\mu)}| \leq Kt^n, \qquad \mu = m + 1, \cdots, k; \ n = 1, 2, \cdots$$

We shall now establish asymptotic relations for the solutions $e_n^{(\mu)}$. It will turn out that for $\mu = 1, \cdots, m$ these solutions can be approximated by solutions of a simple differential equation, whereas for $\mu > m$ they show a rapid decay similar to (5-193).

For $\mu = 1, \cdots, m$, let $f_n^{(\mu)} = \zeta_\mu^{-n} e_n^{(\mu)}$. By (5-184) and (5-192) the quantities $f_n^{(\mu)}$ satisfy

$$\text{(5-194)} \qquad f_\nu^{(\mu)} = \zeta_\mu^{-\nu} \tilde{\zeta}_\mu^\nu = 1 + O(h), \quad \nu = 0, 1, \cdots, k - 1$$

$$\text{(5-195)} \quad \alpha_k^{(\mu)} f_{n+k}^{(\mu)} + \cdots + \alpha_0^{(\mu)} f_n^{(\mu)} = h\{\beta_k^{(\mu)} g_{n+k} f_{n+k}^{(\mu)} + \cdots + \beta_0^{(\mu)} g_n f_n^{(\mu)}\}, \quad n = 0, 1, \cdots$$

where

$$\text{(5-196)} \qquad \begin{aligned} \alpha_k^{(\mu)} \zeta^k + \cdots + \alpha_0^{(\mu)} &= \rho(\zeta_\mu \zeta) \equiv \rho^{[\mu]}(\zeta) \\ \beta_k^{(\mu)} \zeta^k + \cdots + \beta_0^{(\mu)} &= \sigma(\zeta_\mu \zeta) \equiv \sigma^{[\mu]}(\zeta) \end{aligned}$$

The linear multistep method defined by the polynomials $\rho^{[\mu]}(\zeta)$ and $\sigma^{[\mu]}(\zeta)$ is stable and can be made consistent if $\sigma^{[\mu]}(\zeta)$ is divided by the growth parameter λ_μ defined in §5.3-1 [equation (5-142)], for it follows from (5-196) that

$$\rho^{[\mu]}(1) = \rho(\zeta_\mu) = 0$$
$$\rho^{[\mu]\prime}(1) = \zeta_\mu \rho'(\zeta_\mu) = \lambda_\mu^{-1}\sigma(\zeta_\mu) = \lambda_\mu^{-1}\sigma^{[\mu]}(1)$$

If we thus divide the constants $\beta_k^{(\mu)}$ in (5-195) by λ_μ and multiply the quantities g_m by λ_μ, the difference equation (5-195) becomes a consistent and stable approximation to the differential equation

$$e'_\mu(x) = \lambda_\mu g(x) e_\mu(x) \tag{5-197a}$$

Also, the initial values (5-194) deviate by at most $O(h)$ from the corresponding values $e_\mu(x_\nu)$ of the solution of (5-197a) satisfying

$$e_\mu(a) = 1 \tag{5-197b}$$

We now invoke Theorem 5.11 [with $p = 1$, $K = 0$, $\delta = O(h)$] and the remark following its proof to obtain

$$f_n^{(\mu)} = e_\mu(x_n) + O(h), \qquad \mu = 1, \cdots, m; \; n = 0, 1, \cdots$$

where $e_\mu(x)$ is defined by (5-197). It follows that

$$e_n^{(\mu)} = e^{in\varphi_\mu} e_\mu(x_n) + O(h) \tag{5-198}$$

This relation settles the asymptotic behavior of the solutions $e_n^{(\mu)}$ belonging to the essential roots. In order to prove that the remaining solutions tend to zero as $h \to 0$ we note first of all that by Lemma 5.6 [applied with $z_m = e_m^{(\mu)}$, $Z = 1$, $N = (b - a)/h$, $\beta_{\mu,m} = \beta_\mu g_{m+\mu}$, $\Lambda = 0$]

$$e_n^{(\mu)} = O(1), \qquad h \to 0. \tag{5-199}$$

In order to improve this result, we set $e_n^{(\mu)} = z_n^{(\mu)} + \delta_n$ and observe that by (5-191)

$$\delta_\nu = 0, \qquad \nu = 0, 1, \cdots, k - 1 \tag{5-200}$$

We also put $g_n = g_0 + \varepsilon_n$. Inserting into (5-184), we obtain

$$\begin{aligned} \alpha_k(z_{n+k}^{(\mu)} + \delta_{n+k}) + \cdots + \alpha_0(z_n^{(\mu)} + \delta_n) \\ = hg_0\{\beta_k(z_{n+k}^{(\mu)} + \delta_{n+k}) + \cdots + \beta_0(z_n^{(\mu)} + \delta_n)\} \\ + h\{\beta_k \varepsilon_{n+k} e_{n+k}^{(\mu)} + \cdots + \beta_0 \varepsilon_n e_{(\mu)}^{n}\} \end{aligned}$$

The terms involving $z_m^{(\mu)}$ cancel in view of (5-198). The last line may be simplified in view of (5-199) and the fact that $g(x)$ is differentiable. We thus obtain

(5-201) $$\alpha_k \delta_{n+k} + \cdots + \alpha_0 \delta_n = h g_0 \{\beta_k \delta_{n+k} + \cdots + \beta_0 \delta_0\} + hC\theta_n \varepsilon_n$$

when C is a suitable constant independent of h and $|\theta_n| \leq 1$. We now apply Lemma 5.6, letting $z_m = \delta_m$, $Z = 0$ [which is possible by (5-200)], $\beta_{\mu,m} = g_0 \beta_\mu$,

(5-202) $$N = \frac{2 \log h}{|\log t|} = l$$

[where t is given by (5-140)], and, since $|\varepsilon_n| \leq nhG$ if $G = \max |g'(x)|$, $\Lambda = CGlh^2$. It follows that for $n \leq l + k$, $|\delta_n| \leq C_1(h \log h)^2$, where C_1 is a suitable constant. On the other hand, for $n \geq l$ it follows from (5-193) and (5-202) that $|z_n^{(\mu)}| \leq Kt^n \leq Kh^2$. We thus have for $l \leq n \leq l + k$

(5-203) $$|e_n^{(\mu)}| = |z_n^{(\mu)} + \delta_n| \leq Kh^2 + C_1(h \log h)^2 \leq K_1 h^{2-\varepsilon}$$

where ε is an arbitrary positive number. It has thus been shown that $e_n^{(\mu)}$ is $O(h^{2-\varepsilon})$ for k consecutive values of n which are such that $nh \to 0$ when $h \to 0$. It remains to apply Lemma 5.6 once more to the difference equation (5-184), started at $n = l$. It then follows that the last estimate (with a different value of the constant K_1) continues to hold for $n \geq l$ as long as $x_n \in [a, b]$.

There remains the problem of determining the constants A_μ in (5-187) as functions of the starting values. Since $e_\nu^{(\mu)} = z_\nu^{(\mu)}$ for $\nu = 0, 1, \cdots, k - 1$, the $e_n^{(\mu)}$ can be replaced by the $z_n^{(\mu)}$ for this purpose. For the $z_n^{(\mu)}$, however, a similar calculation has already been performed in §5.3-1. For $\mu > m$ it suffices to determine the order of magnitude of the A_μ. By the definition of the exponent q, the starting values themselves are $O(1)$. If $g_0 \neq 0$, it follows from (5-154) that $A_\mu = O(h^{-1+1/p_\mu})$, where p_μ is the multiplicity of the root ζ_μ approached by $\tilde{\zeta}_\mu$ as $h \to 0$. If $g_0 = 0$, all A_μ are $O(1)$. Thus in either case, since we may choose $\varepsilon > 1/p_\mu$ ($\mu = m + 1, \cdots, k$) in (5-203), we have for $x_n = x$, $a < x \leq b$,

$$A_\mu e_n^{(\mu)} = O(h), \qquad \mu = m + 1, \cdots, k$$

For $\mu = 1, 2, \cdots, m$ the coefficients A_μ can be determined exactly as in §5.3-1. Setting

(5-204) $$\Delta_\mu = \lim_{h \to 0} h^{-r} \{\alpha_{\mu,0} \delta_0(h) + \alpha_{\mu,1} \delta_1(h) + \cdots + \alpha_{\mu,k-1} \delta_{k-1}(h)\}$$

we find that

$$A_\mu = \frac{\Delta_\mu}{\rho'(\zeta_\mu)} + O(h), \qquad \mu = 1, \cdots, m$$

Thus finally,

$$\bar{e}_n^H = \sum_{\mu=1}^{m} \frac{\Delta_\mu}{\rho'(\zeta_\mu)} e^{in\varphi_\mu} e_\mu(x_n) + O(h)$$

The results of this section may now be summarized as follows:

THEOREM 5.12. *Under the hypotheses stated at the beginning of* §5.3-5 *the asymptotic behavior of the discretization error in the solution of* $y' = f(x, y)$ *by the linear multistep method* (5-58) *as* $h \to 0$, $nh = x - a$, $a < x \le b$ *is given by the formula*

$$e_n = h^p e(x) + h^r \sum_{\mu=1}^{m} \frac{\Delta_\mu}{\rho'(\zeta_\mu)} e^{in\varphi_\mu} e_\mu(x) + O(h^{r+1}) \tag{5-205}$$

where $r = \min(p, q)$, *and where* Δ_μ *is a function of the starting values described by* (5-204). *The functions* $e(x)$ *and* $e_\mu(x)$ *are defined as the solutions of the initial value problems* (5-185) *and* (5-197), *respectively.*

This result is a direct generalization of (5-157). Again the error is seen to consist of two parts. The *genuine discretization error* is represented by the term involving $e(x)$. It corresponds exactly to the discretization error of a one-step method with principal error function $-Cy^{(p+1)}(x)$. The *starting error* is represented by the sum involving the functions $e_\mu(x)$. If $m > 1$, even the leading term in the expression for the starting error is, in general, not a smooth function of x. Consequently, if $q \le p$ (i.e., if the starting values are relatively inaccurate), the leading term in the asymptotic expansion of e_n is not smooth, and extrapolation to the limit procedures cannot be rigorously applied.* Moreover, in the approximation of an exponentially decreasing solution where $g(x) < 0$, if $\lambda_\mu < 0$ for some μ, the corresponding function $e_\mu(x)$ will increase exponentially and will make itself conspicuous by rapid oscillations owing to the factor $e^{in\varphi_\mu}$. Such oscillations may become troublesome even if $q > p$ (i.e., if the starting errors are small compared to the truncation error). On the other hand, if $m = 1$ (as for all Adams methods), the leading term in (5-181) is always a smooth function of x, even if the starting values are not accurate. This *self-stabilizing* feature represents one of the main advantages of the strongly stable methods.

5.3-6. Milne's method for estimating the local discretization error. If the discretization error consists mainly of genuine discretization error [as in the case $q < p$ in (5-205)], a rough idea about the size of the error can be obtained by integrating the differential equation defining the function $e(x)$. This makes it necessary to know, approximately at least,

* See Wasow [1955] for a related example where extrapolation to the limit does not work.

the function $-Cy^{(p+1)}(x)$ which plays the role of the principal error function. Even if it is not desired to integrate (5-185) a knowledge of this function can be valuable, for instance, for the purpose of deciding whether the current step h is adequate. To determine $y^{(p+1)}(x)$ by p times totally differentiating $f(x, y)$ is usually not practical. A device due to W. E. Milne makes it possible to obtain an approximation to $y^{(p+1)}(x)$ in a much simpler way. It is based on comparing a "predicted" value of y_{n+k} with the finally accepted, "corrected" value.†

We shall first describe Milne's method under the simplifying assumption that the values $y_n, y_{n+1}, \cdots, y_{n+k-1}$ coincide with the values of the exact solution $y(x)$. Let the predictor formula be

$$\text{(5-206}p\text{)} \quad \alpha_k^* y_{n+k}^* + \alpha_{k-1}^* y_{n+k-1} + \cdots + \alpha_0^* y_n = h\{\beta_{k-1}^* f_{n+k-1} + \cdots + \beta_0^* f_n\}$$

and the corrector formula be

$$\text{(5-206}c\text{)} \quad \alpha_k y_{n+k} + \alpha_{k-1} y_{n+k-1} + \cdots + \alpha_0 y_n = h\{\beta_k f_{n+k} + \beta_{k-1} f_{n+k-1} + \cdots + \beta_0 f_n\}$$

It is not assumed here that $|\alpha_0| + |\beta_0| > 0$. It is essential, however, that the order of the predictor and the corrector formulas be the same. Thus, if $z(x)$ is any sufficiently differentiable function, we have

$$\text{(5-207}p\text{)} \quad \alpha_k^* z(x + kh) + \cdots + \alpha_0^* z(x) = h\{\beta_{k-1}^* z'(x + (k - 1)h) + \cdots + \beta_0^* z'(x)\} + C_{p+1}^* z^{(p+1)}(x)h^{p+1} + O(h^{p+2})$$

$$\text{(5-207}c\text{)} \quad \alpha_k(x + kh) + \cdots + \alpha_0 z(x) = h\{\beta_k z'(x + kh) + \cdots + \beta_0 z'(x)\} + C_{p+1} z^{(p+1)}(x)h^{p+1} + O(h^{p+2})$$

where C_{p+1}^* and C_{p+1} are two nonvanishing constants depending only on $\alpha_k^*, \cdots, \beta_0^*$ and $\alpha_k, \cdots, \beta_0$, respectively. If, in particular, we choose $x = x_n$, $z(x) = y(x)$ and subtract (5-207p) from (5-206p) and (5-207c), taking into account that $y_m = y(x_m)$ $(m = n, n + 1, \cdots, n + k - 1)$, we obtain

$$\text{(5-208}p\text{)} \quad \alpha_k^*(y_{n+k}^* - y(x_{n+k})) = -C_{p+1}^* y^{(p+1)}(x_n)h^{p+1} + O(h^{p+2})$$

$$\text{(5-208}c\text{)} \quad \alpha_k(y_{n+k} - y(x_{n+k})) = h\beta_k[f(x_{n+k}, y_{n+k}) - f(x_{n+k}, y(x_{n+k}))] - C_{p+1} y^{(p+1)}(x_n)h^{p+1} + O(h^{p+2})$$

The expressions in brackets can be written $f_y(x_{n+k}, \eta)(y_{n+k} - y(x_{n+k}))$. Solving for $y_{n+k} - y(x_{n+k})$, we may thus correctly write

$$\text{(5-208}c'\text{)} \quad \alpha_k(y_{n+k} - y(x_{n+k})) = -C_{p+1} y^{(p+1)}(x_n)h^{p+1} + O(h^{p+2})$$

† The reader is asked to turn back to §5.1-2(ii) for a description of the mechanism of predictor-corrector methods.

Eliminating $y(x_{n+k})$ from (5-208p) and (5-208c) and solving for $y^{(p+1)}(x_n)$, we obtain

$$\text{(5-209)} \quad y^{(p+1)}(x_n) = h^{-p-1}\left(\frac{C^*_{p+1}}{\alpha^*_k} - \frac{C_{p+1}}{\alpha_k}\right)^{-1}(y_{n+k} - y^*_{n+k}) + O(h)$$

This formula expresses the unknown derivative $y^{(p+1)}$ (approximately) in terms of the difference between the predicted value y_{n+k} and the corrected value y^*_{n+k}.

To justify a similar device for calculating $y^{(p+1)}$ in the general case where the values $y_n, y_{n+1}, \cdots, y_{n+k-1}$ are not assumed to be exact, it seems necessary to assume that the left-hand terms of the predictor and the corrector formulas should be the same, i.e., that $\alpha^*_\mu = \alpha_\mu$, $\mu = 0, \cdots, k$. Moreover, the corrector formula and the starting values must be such that

$$\text{(5-210)} \qquad e_n = y_n - y(x_n) = h^p e(x_n) + O(h^{p+1})$$

where $e(x)$ is a continuously differentiable function of x [not necessarily the one defined by (5-185)] and $p \geq 1$. This condition is satisfied if $q > p$ in (5-205), or if $q = p$ and $m = 1$. Under these hypotheses we obtain by putting $x = x_n$, $z(x) = y(x)$ in (5-207p) and subtracting the resulting relation from (5-206p)

$$\begin{aligned}\text{(5-211)} \quad & \alpha_k e^*_{n+k} + \alpha_{k-1}e_{n+k-1} + \cdots + \alpha_0 e_n \\ &= h\{\beta^*_{k-1}[f(x_{n+k-1}, y_{n+k-1}) - f(x_{n+k-1}, y(x_{n+k-1}))] + \cdots \\ &+ \beta^*_0[f(x_n, y_n) - f(x_n, y(x_n))]\} - h^{p+1}C^*_{p+1}y^{(p+1)}(x_n) + O(h^{p+2})\end{aligned}$$

where $e^*_{n+k} = y^*_{n+k} - y(x_{n+k})$. The expressions in brackets can by virtue of (5-210) be replaced by

$$h^p g(x_m)(e(x_m) + O(h)) \qquad (m = n, n+1, \cdots, n+k-1)$$

where $g(x) = f_y(x, y(x))$. Furthermore, since $e(x)$ was assumed to be continuously differentiable,

$$g(x_{n+\mu})e(x_{n+\mu}) = g(x_n)e(x_n) + O(h), \qquad \mu = 1, \cdots, k-1$$

so that (5-211) may finally be written in the form

$$\begin{aligned}\text{(5-212}p\text{)} \quad & \alpha_k e^*_{n+k} + \alpha_{k-1}e_{n+k-1} + \cdots + \alpha_0 e_n \\ &= h^{p+1}\{(\beta^*_{k-1} + \cdots + \beta^*_0)g(x_n)e(x_n) - C^*_{p+1}y^{(p+1)}(x_n)\} + O(h^{p+2})\end{aligned}$$

In an entirely analogous manner we obtain from (5-207c) and (5-208c)

$$\begin{aligned}\text{(5-212}c\text{)} \quad & \alpha_k e_{n+k} + \alpha_{k-1}e_{n+k-1} + \cdots + \alpha_0 e_n \\ &= h^{p+1}\{(\beta_k + \cdots + \beta_0)g(x_n)e(x_n) - C_{p+1}y^{(p+1)}(x_n)\} + O(h^{p+2})\end{aligned}$$

We now subtract (5-212*p*) from (5-212*c*) and note that, by virtue of consistency, $\beta_{k-1}^* + \cdots + \beta_0^* = k\alpha_k + (k-1)\alpha_{k-1} + \cdots + \alpha_1 = \beta_k + \cdots + \beta_0$. Thus since $e_{n+k} - e_{n+k}^* = y_{n+k} - y_{n+k}^*$, we obtain

$$\alpha_k(y_{n+k} - y_{n+k}^*) = h^{p+1}(C_{p+1}^* - C_{p+1})y^{(p+1)}(x_n) + O(h^{p+2})$$

It is now easy to solve for the value of the principal error function. We find

$$-Cy^{(p+1)}(x_n) = h^{-p-1}K(y_{n+k} - y_{n+k}^*) + O(h) \tag{5-213}$$

where the constant

$$K = \frac{C\alpha_k}{C_{p+1} - C_{p+1}^*}$$

depends only on the predictor and corrector formulas used. Introducing the error constant

$$C^* = \frac{C_{p+1}^*}{\beta_0 + \cdots + \beta_k}$$

associated with the predictor formula, we can also write

$$K = \frac{\alpha_k}{\beta_0 + \cdots + \beta_k}\frac{C}{C - C^*} \tag{5-214}$$

We thus have proved:

THEOREM 5.13. *If the approximate solution satisfies* (5-210), *if the predictor and the corrector formulas are of the same order, and if the polynomials* $\rho(\zeta)$ *associated with them are the same (possibly after multiplication by a suitable power of* ζ), *then* (5-213) *holds for any fixed* $x \in [a, b]$ *as* $h \to 0$, $x_n = x$.

Table 5.15

Values of the Constant K in (5.214)

	Order p	1	2	3	4	5	6
K	Adams-Bashforth-Moulton	$\frac{1}{2}$	$\frac{1}{6}$	$\frac{1}{10}$	$\frac{19}{270}$	$\frac{27}{502}$	$\frac{863}{19950}$
	Nyström-Milne-Simpson	$-\frac{1}{2}$	∞	0	$\frac{1}{60}$	$\frac{1}{58}$	$\frac{37}{2352}$

Suitable pairs of predictor-corrector formulas are, for instance, the Adams-Bashforth and the Adams-Moulton methods, or the Nyström and the Milne-Simpson methods. We list in Table 5.15 the values of the constant K for these pairs, for various values of the order p.

The irregular values in the last row are due to the fact that for $p = 2$ the Nyström and the Milne-Simpson formulas coincide and that for $p = 3$ the Milne-Simpson formula is automatically of order 4. It should be noted that for $p \geq 3$ the values of K are quite small (<0.1), so that the local discretization error $-Ch^{p+1}y^{(p+1)}(x)$ is only a small fraction of the difference between the corrected and the predicted value. For paper and pencil work, Milne recommends carrying the value of $K(y_{n+k} - y^*_{n+k})$ in a special "check" column to indicate the local discretization error and to guard against blunders.

5.3-7. Correcting only once. If the equation defining y_{n+k} was implicit (i.e., if $\beta_k \neq 0$), we have up to now always assumed that (5-58) was satisfied exactly. If the equation is solved by an iterative procedure, as described in §5.2-2, infinitely many iterations are required in general. In practice, of course, only a finite number of iterations are performed, and as a result (5-58) is *not* satisfied exactly. The resulting discrepancy may be considered as a local round-off error and treated by the methods to be discussed in §5.4. Another way of looking at the same problem consists in fixing once for all the number of iterations to be performed. Any such rule defines a new, in general no longer linear, multistep method for solving differential equations. We shall consider below in some detail the case where only *one* iteration is performed.

Let the predictor formula be given by

$$y^*_{n+k} + \alpha^*_{k-1}y_{n+k-1} + \cdots + \alpha^*_0 y_n = h\{\beta^*_{k-1}f_{n+k-1} + \cdots + \beta^*_0 f_n\} \tag{5-215}$$

and let its order be q, so that for a sufficiently differentiable function $z(x)$

$$\begin{aligned} z(x + kh) + \alpha^*_{k-1}z(x + (k-1)h) + \cdots + \alpha^*_0 z(x) &= h\{\beta^*_{k-1}z'(x + (k-1)h) + \cdots + \beta^*_0 z'(x)\} \\ &\quad + C^*_{q+1}z^{(q+1)}(x)h^{q+1} + O(h^{q+2}) \end{aligned} \tag{5-216}$$

We have assumed that $\alpha^*_k = 1$ for simplification of the later algebra. The corrector formula is assumed in the usual form

$$\alpha_k y_{n+k} + \cdots + \alpha_0 y_n = h\{\beta_k f_{n+k} + \cdots + \beta_0 f_n\}$$

Its order is assumed to be p, where not necessarily $p = q$, so that

$$\begin{aligned} \alpha_k z(x + kh) + \cdots + \alpha_0 z(x) &= h\{\beta_k z'(x + kh) + \cdots \\ &\quad + \beta_0 z'(x)\} + C_{p+1}z^{(p+1)}(x)h^{p+1} + O(h^{p+2}) \end{aligned} \tag{5-217}$$

The algorithm of predicting only once can then be stated in the following way. Assuming that the values $y_{n+k-1}, \cdots, y_n$ are known, one calculates y^*_{n+k} from (5-215) and y_{n+k} from

$$\text{(5-218)} \quad \alpha_k y_{n+k} + \cdots + \alpha_0 y_n = h\{\beta_k f(x_{n+k}, y^*_{n+k}) + \beta_{k-1} f(x_{n+k-1}, y_{n+k-1}) + \cdots + \beta_0 f(x_n, y_n)\}$$

which is now an explicit equation for y_{n+k}.

We shall obtain an estimate for $e_n = y_n - y(x_n)$ via an expression for $e^*_n = y^*_n - y(x_n)$. Defining for $m = 0, 1, 2, \cdots$ the quantities g_m and g^*_m by the relations

$$f(x_m, y_m) - f(x_m, y(x_m)) = g_m e_m \qquad (g_m = 0 \text{ if } e_m = 0)$$
$$f(x_m, y^*_m) - f(x_m, y(x_m)) = g^*_m e^*_m \qquad (g^*_m = 0 \text{ if } e^*_m = 0)$$

we obtain by letting $z(x) = y(x_n)$ in (5-216) and (5-217) and subtracting from the corresponding relations (5-215) and (5-218),

$$e^*_{n+k} + \alpha^*_{k-1} e_{n+k-1} + \cdots + \alpha^*_0 e_n = h\{\beta^*_{k-1} g_{n+k-1} e_{n+k-1} + \cdots + \beta^*_0 g_n e_n\} - C^*_{q+1} y^{(q+1)}(x_n) h^{q+1} + O(h^{q+2})$$

$$\alpha_k e_{n+k} + \alpha_{k-1} e_{n+k-1} + \cdots + \alpha_0 e_n = h\{\beta_k g^*_{n+k} e^*_{n+k} + \beta_{k-1} g_{n+k-1} e_{n+k-1} + \cdots + \beta_0 g_n e_n\} - C_{p+1} y^{(p+1)}(x_n) h^{p+1} + O(h^{p+2})$$

Eliminating e^*_{n+k}, we find, after some rearranging,

$$\text{(5-219)} \quad \begin{aligned} \alpha_k e_{n+k} &+ \alpha_{k-1} e_{n+k-1} + \cdots + \alpha_0 e_n \\ &= h\{[\beta_{k-1} g_{n+k-1} + \beta_k g^*_{n+k}(-\alpha^*_{k-1} \\ &\quad + h\beta^*_{k-1} g_{n+k-1})] e_{n+k-1} + \cdots \\ &\quad + [\beta_0 g_n + \beta_k g^*_{n+k}(-\alpha^*_0 + h\beta^*_0 g_n)] e_n\} \\ &\quad - C_{p+1} y^{(p+1)}(x_n) h^{p+1} - \beta_k g^*_{n+k} C^*_{q+1} y^{(q+1)}(x_n) h^{q+2} \\ &\quad + O(h^{\min(p+2, q+3)}) \end{aligned}$$

To this expression we apply Lemma 5.6, letting $z_m = e_m$, $N = (b - a)/h$, $\beta_{k,m} = 0$,

$$\beta_{\mu,m} = \beta_\mu g_{m+\mu} + \beta_k g^*_{m+k}(-\alpha^*_\mu + h\beta^*_\mu g_{m+\mu}), \qquad \mu = 0, 1, \cdots, k-1$$

$\Lambda = Kh^{r+1}$, where $r = \min(p, q+1)$, and assuming that the starting values are also $O(h^{r+1})$. From the Lipschitz condition it follows easily that $|g_m| \le L$, $|g^*_m| \le L$ and thus that the coefficients $\beta_{\mu,m}$ are bounded. We thus conclude from Lemma 5.6 that $e_m = O(h^r)$ as $h \to 0$, $x_n = x$, uniformly with respect to x for $a \le x \le b$. A quantitative estimate can be derived from (5-164), if desired. We prefer to take a quick look at the asymptotic behavior of e_n. In order to make things simple, we continue to assume that $e_\mu = O(h^{r+1})$, $\mu = 0, 1, \cdots, k-1$, so that the starting error is negligible if compared with the genuine discretization error. Three cases have to be distinguished, according to whether $q+1 > p$, $q+1 = p$, or $q+1 < p$.

(i) $q+1 > p = r$. We change the definition of g_m and g_m^* somewhat by now putting

$$g_m = g_m^* = f_y(x_m, y(x_m))$$

For the quantities $\bar{e}_n = h^{-r}e_n$ we now in place of (5-182) obtain the relation

$$\begin{aligned}(5\text{-}220)\quad \alpha_k\bar{e}_{n+k} + \cdots + \alpha_0\bar{e}_n &= h\{(\beta_{k-1} - \alpha_{k-1}^*\beta_k)g_{n+k-1}\bar{e}_{n+k-1} + \cdots + (\beta_0 - \alpha_0^*\beta_k)g_n\bar{e}_n\} \\ &\quad - hC_{p+1}y^{(p+1)}(x_n) + O(h^2)\end{aligned}$$

The multistep method implied in (5-220) is consistent, since

$$\begin{aligned}(\beta_{k-1} - \alpha_{k-1}^*\beta_k) + \cdots + (\beta_0 - \alpha_0^*\beta_k) &= \beta_{k-1} + \cdots + \beta_0 \\ &\quad - (\alpha_0^* + \alpha_1^* + \cdots + \alpha_{k-1}^*)\beta_k = \beta_{k-1} + \cdots + \beta_0 + \beta_k\end{aligned}$$

in view of $1 + \alpha_{k-1}^* + \cdots + \alpha_0^* = 0$.

In view of the differentiability of $y^{(p+1)}(x)$ and of the definition of the error constant

$$C = \frac{C_{p+1}}{\beta_0 + \beta_1 + \cdots + \beta_k}$$

the third line in (5-220) is

$$-h\{(\beta_{k-1} - \alpha_{k-1}^*\beta_k)Cy^{(p+1)}(x_{n+k-1}) + \cdots + (\beta_0 - \alpha_0^*\beta_k)Cy^{(p+1)}(x_n)\} + O(h^2)$$

An application of Theorem 5.11 now shows that

$$(5\text{-}221)\qquad \bar{e}_n = e(x_n) + O(h)$$

where

$$(5\text{-}222)\qquad \begin{aligned} e'(x) &= g(x)e(x) - Cy^{(p+1)}(x) \\ e(a) &= 0 \end{aligned}$$

Thus, *if the predictor formula has at least the order of the corrector formula, the asymptotic behavior of the discretization error for correcting once and for correcting infinitely many times is the same.*

(ii) $q+1 = p = r$. In place of (5-182) we now obtain for $\bar{e}_n = h^{-r}e_n$ the relation (5-220) in which the third line is replaced by

$$-h[C_{p+1}y^{(p+1)}(x_n) + \beta_k g(x_n)C_p^* y^{(p)}(x_n)] + O(h^2)$$

By Theorem 5.11 it follows that

$$\bar{e}_n = e(x_n) + O(h) \tag{5-223}$$

where

$$\begin{aligned} e'(x) &= g(x)e(x) - Cy^{(p+1)}(x) - \frac{\beta_k C_p^*}{\beta_k + \cdots + \beta_0} g(x)y^{(p)}(x) \\ e(a) &= 0 \end{aligned} \tag{5-224}$$

Thus, *if the order of the corrector formula exceeds the order of the predictor formula by* 1, *the order of the discretization error is not changed by correcting only once; however, in the definition of the magnified error function* $e(x)$ *there now appears an extra term.* In the special case

$$y_{n+1}^* - y_n = hf_n, \qquad y_{n+1} - y_n = \tfrac{1}{2}h(f_{n+1} + f_n)$$

we obtain a result relating to a special case of the simplified Runge-Kutta method [equation (2-10*b*)] which by a completely different method was obtained in §2.2-5 [cf. equation (2-35) for $\alpha = \frac{1}{2}$].

(iii) $r = q + 1 < p$. We again obtain for $\bar{e}_n = h^{-r}e_n$ the relation (5-220) but with the last line now replaced by

$$-h\beta_k C_{q+1}^* g(x_n)y^{(q+1)}(x_n) + O(h^2)$$

Again by applying Theorem 5.11 we find that $\bar{e}_n = e(x_n) + O(h)$, where

$$\begin{aligned} e'(x) &= g(x)e(x) - \frac{\beta_k C_{q+1}^*}{\beta_0 + \cdots + \beta_k} g(x)y^{(q+1)}(x) \\ e(a) &= 0 \end{aligned} \tag{5-225}$$

Thus, *if the order of the corrector formula exceeds the order of the predictor formula by more than* 1, *the order of the discretization error is decreased by correcting only once, and the form of the magnified error function changed.*

The results of this section to some extent run parallel to the results on the use of quadrature formulas in the one-step procedures discussed in §3.2-4. The algorithm of correcting only once is widely used in practice.

5.4. Round-off Error in the Integration by Multistep Methods

5.4-1. An a priori bound.

Instead of satisfying the difference equation (5-58) exactly, the quantities $\tilde{y}_n$ actually calculated satisfy an equation which we write in the form

$$\begin{aligned} \alpha_k\tilde{y}_{n+k} + \alpha_{k-1}\tilde{y}_{n+k-1} + \cdots + \alpha_0\tilde{y}_n &= h\{\beta_k f(x_{n+k}, \tilde{y}_{n+k}) + \cdots \\ &+ \beta_0 f(x_n, \tilde{y}_n)\} + \varepsilon_{n+k}, \qquad n = 0, 1, 2, \cdots \end{aligned} \tag{5-226}$$

The quantities ε_n will be called the *local round-off errors.* The problem to be studied in §5.4 is the influence of these local errors on the *accumulated round-off error* $r_n = \tilde{y}_n - y_n$.

Without making any speculations about the nature of the local round-off error, we shall derive in this section an a priori estimate for r_n under the sole assumption that $|\varepsilon_{n+k}| \le \varepsilon$ $(n = 0, 1, 2, \cdots)$, where ε is a constant. Subtracting from (5-226) the corresponding equation (5-58) and writing

$$g_m = r_m^{-1}[f(x_m, \tilde{y}_m) - f(x_m, y_m)], \qquad \text{if } r_m \ne 0$$
$$g_m = 0, \qquad \text{if } r_m = 0$$

so that in any case $|g_m| \le L$, the Lipschitz constant, we obtain

$$\text{(5-227)} \quad \alpha_k r_{n+k} + \alpha_{k-1} r_{n+k-1} + \cdots + \alpha_0 r_n = h\{\beta_k g_{n+k} r_{n+k} + \cdots + \beta_0 g_n r_n\} + \varepsilon_{n+k}$$

An application of Lemma 5.6 to this relation with $z_m = r_m$, $\Lambda = \varepsilon$, $N = (b - a)/h$, and (since $r_0 = r_1 = \cdots = r_{k-1} = 0$) $Z = 0$ yields

$$\text{(5-228)} \qquad r_n \le \varepsilon h^{-1}(x_n - a)\Gamma^* \exp[(x_n - a)\Gamma^* BL]$$

where the constants are defined as in Theorem 5.11.

Relation (5-228) in all likelihood overestimates the actual round-off error by a large margin. The essential, and for the later developments significant, result contained in it is that $r_n = O(\varepsilon h^{-1})$; in the light of a result to be derived later (§6.3-2) for equations of the second order this is not entirely trivial.

5.4-2. An a posteriori bound. We shall now assume that $Nh^{p+1} \le \varepsilon \le Kh^2$, where N and K are independent of h. According to (5-228) we then have $r_n = O(h)$. Assuming that $f_{yy}(x, y)$ exists and is continuous in a neighborhood of $y = y(x)$, we then can write, if h is sufficiently small,

$$f(x_m, \tilde{y}_m) - f(x_m, y_m) = g(x_m) r_m + \theta_m K_2 \varepsilon \qquad (|\theta_m| \le 1)$$

where $g(x) = f_y(x, y(x))$. Equation (5-227) can then be replaced by

$$\text{(5-229)} \quad \alpha_k r_{n+k} + \cdots + \alpha_0 r_n = h\{\beta_k g_{n+k} r_{n+k} + \cdots + \beta_0 g_n r_n\} + \varepsilon_{n+k} + \theta_n K_3 h\varepsilon$$

where now $g_m = g(x_m)$. We accordingly write

$$r_n = r_n^{(1)} + r_n^{(2)}$$

where $\{r_n^{(1)}\}$ is the solution of (5-229) for $\theta_n \equiv 0$ and $\{r_n^{(2)}\}$ is the solution for $\varepsilon_{n+k} \equiv 0$, both solutions satisfying $r_\mu^{(i)} = 0$, $\mu = 0, 1, \cdots, k - 1$. We may call $r_n^{(1)}$ the *primary error* and $r_n^{(2)}$ the *secondary error*. It will be noted that the secondary error is due exclusively to the nonlinearity of the differential equation. From Lemma 5.6 it follows that the secondary error is $O(\varepsilon)$, whereas the primary error must be expected to be $O(\varepsilon h^{-1})$. We may therefore assume that for $h \to 0$ the behavior r_n is governed by the behavior of $r_n^{(1)}$, and the following considerations will be directed exclusively toward the primary error $r_n^{(1)}$.

According to Theorem 5.2, the primary error can be represented in the form

$$r_n^{(1)} = \sum_{l=k}^{n} \varepsilon_l d_{nl} \tag{5-230}$$

where for $l = k, k+1, \cdots \{d_{nl}\}$ is the solution of the difference equation

$$\begin{aligned} \alpha_k d_{n+k,l} + \alpha_{k-1} d_{n+k-1,l} + \cdots + \alpha_0 d_{n,l} \\ = h\{\beta_k g_{n+k} d_{n+k,l} + \cdots + \beta_0 g_n d_{n,l}\} \end{aligned} \tag{5-231}$$

assuming the initial values

$$d_{nl} = \begin{cases} 0, & n < l \\ \dfrac{1}{\alpha_k - h\beta_k g_l}, & n = l \end{cases} \tag{5-232}$$

By comparing (5-231) and (5-232) to (5-184) and (5-183) it is seen that the initial value problem for the $d_{n,l}$ becomes identical with the initial value problem for the quantities $\bar{e}_n^H$ considered in §5.3-5. If x_0 is replaced by x_{l-k+1} and if for the initial value functions the special functions

$$\begin{aligned} \delta_\mu(h) &= 0, & \mu &= 0, 1, \cdots, k-2 \\ \delta_\mu(h) &= \frac{h^r}{\alpha_k - h\beta_k g_l}, & \mu &= k-1 \end{aligned}$$

are chosen, we may apply (5-213). Noting that $\rho_\mu(\zeta) = \rho(\zeta)/(\zeta - \zeta_\mu) = \alpha_k \zeta^{k-1} + \cdots$, we have $\alpha_{\mu,k-1} = \alpha_k$ and thus $\Delta_\mu = 1$. The result is thus given by

$$d_{n,l} = \sum_{\mu=1}^{m} \frac{1}{\rho'(\zeta_\mu)} \exp\,[i(n - l + k - 1)\varphi_\mu]\, d_{l,\mu}(x_n) + O(h) \tag{5-233}$$

where the functions $d_{l,\mu}(x)$ are defined as the solutions of the initial value problems

$$\begin{aligned} d'_{l,\mu}(x) &= \lambda_\mu g(x) d_{l,\mu}(x) \\ d_{l,\mu}(x_l) &= 1, \qquad \mu = 1, 2, \cdots, m \end{aligned} \tag{5-234}$$

analogous to (5-197). (At this point the growth parameters λ_μ are seen to influence decisively the propagation of round-off error.) It is easily verified that for $x \geq x_l$

$$d_{l,\mu}(x) = \frac{e_\mu(x)}{e_\mu(x_l)} \qquad (\mu = 1, \cdots, m)$$

where the functions $e_\mu(x)$ are defined by (5-197). Substituting in (5-230), we find

$$r_n^{(1)} = \sum_{l=k}^{n} \varepsilon_l \left\{ \sum_{\mu=1}^{m} \frac{\exp\,[i(n-l+k-1)\varphi_\mu]}{\rho'(\zeta_\mu)} \frac{e_\mu(x_n)}{e_\mu(x_l)} + O(h) \right\} \tag{5-235}$$

As a first application of this formula we shall obtain an asymptotic bound for $r_n^{(1)}$ under the assumption that

$$|\varepsilon_l| \le p(x_l)\varepsilon \tag{5-236}$$

where ε is independent of x and $p(x)$ is a known continuous (or piecewise continuous) function of x. Rearranging the order of the summations in (5-235), we obtain

$$|r_n^{(1)}| \le \frac{\varepsilon}{h} \sum_{\mu=1}^{m} \frac{1}{|\rho'(\zeta_\mu)|} m_{\mu,n}$$

where

$$m_{\mu,n} = |e_\mu(x_n)|\, h \sum_{l=k}^{n} \{p(x_l)\, |e_\mu(x_l)|^{-1} + O(h)\} \tag{5-237}$$

At this point it is convenient to interpolate the following simple lemma, which will be used repeatedly.

LEMMA 5.8. *Let $f(x)$ be continuous and continuously differentiable for $x \in [a, b]$, and let, for $x_n \in [a, b]$,*

$$\Sigma_n = h \sum_{p=0}^{n} \{f(x_p) + h\theta_p K\} \tag{5-238}$$

where $|\theta_p| \le 1$ and K is a constant independent of h or n. Then there exists a constant K_1 such that

$$\left| \Sigma_n - \int_a^{x_n} f(t)\, dt \right| \le hK_1 \tag{5-239}$$

This is a special case $[f(x, y)$ independent of $y]$ of Theorem 1.4. Intuitively the sum Σ_n may be regarded a Riemann sum approximating a definite integral.

The function $f(x) = p(x)\, |e_\mu(x)|^{-1}$ satisfies the conditions of the lemma, because $p(x)$ and $e_\mu(x)$ are differentiable and $e_\mu(x) \ne 0$ for $x \in [a, b]$. We may therefore apply the lemma to the sum appearing in (5-237) and find that

$$m_{\mu,n} = m_\mu(x_n) + O(h)$$

where

$$m_\mu(x) = |e_\mu(x)| \int_a^x p(t)\, |e_\mu(t)|^{-1}\, dt \tag{5-240}$$

We thus have obtained

THEOREM 5.14. *If the local round-off errors* ε_l *satisfy* (5-236), *where* $\varepsilon = O(h^2)$, *then*

$$|r_n^{(1)}| \leq \frac{\varepsilon}{h} \sum_{\mu=1}^{m} \{|\rho'(\zeta_\mu)|^{-1} m_\mu(x_n) + O(h)\} \tag{5-241}$$

where the functions $m_\mu(x)$ *are defined by* (5-240).

In line with the one-step case, we shall characterize the functions $m_\mu(x)$ as solutions of certain initial value problems [cf. (2-77)]. Temporarily setting $e_\mu(x) = u(x) + iv(x)$ and $\lambda_\mu = \xi + i\eta$, we have from (5-197), separating real and imaginary parts,

$$u' = (\xi u - \eta v)g, \qquad v' = (\xi v + \eta u)g$$

Thus, if $|e_\mu| = (u^2 + v^2)^{1/2} \equiv r$, we find

$$rr' = uu' + vv' = (\xi u^2 - \eta uv + \xi v^2 + \eta uv)g = \xi r^2 g$$

hence

$$|e_\mu|' = \mathrm{Re}\lambda_\mu\, g\, |e_\mu|$$

We now obtain easily by differentiating (5-240)

$$\begin{aligned} m_\mu'(x) &= \mathrm{Re}\lambda_\mu\, g(x) m_\mu(x) + p(x) \\ m_\mu(a) &= 0 \end{aligned} \tag{5-242}$$

Thus, as far as the growth of the functions $m_\mu(x)$ is concerned, only the real parts of the growth parameters λ_μ are significant.

5.4-3. Statistical estimates. It has been pointed out previously that a truly satisfactory theory of round-off can only be achieved on a statistical basis. This is especially true for multistep methods. In particular, the statistical approach brings to light the real nature of conditional stability. It will be shown that conditional stability owing to growth parameters with negative real part is not confined to starting error. It will also appear as a consequence of round-off error in a quantitatively predictable way.

As in §2.3-4, we shall assume that the local round-off errors ε_l are independent random variables whose mean and variance satisfy

$$|E(\varepsilon_l)| \leq \mu p(x_l) \tag{5-243}$$

$$\mathrm{var}\,(\varepsilon_l) = \sigma^2 q(x_l) \tag{5-244}$$

where $p(x)$ and $q(x)$ are known nonnegative, piecewise smooth functions in $[a, b]$. The quantities μ and σ^2 are assumed to be independent of x. Appropriate assumptions on these functions and quantities for various arithmetic systems are discussed in §5.4-4. It may suffice at the moment to point out that the assumption $\mu = 0$ is in many cases not realistic for implicit linear multistep methods.

From (5-235) we have, using (1-94),

$$E(r_n^{(1)}) = \sum_{l=k}^{n} E(\varepsilon_l) \left\{ \sum_{\mu=1}^{m} \frac{\exp[i(n-l+k-1)\varphi_\mu]}{\rho'(\zeta_\mu)} \frac{e_\mu(x_n)}{e_\mu(x_l)} + O(h) \right\}$$

and thus, using (5-243) and proceeding as in the derivation of (5-241),

$$|E(r_n^{(1)})| \le \frac{\mu}{h} \left\{ \sum_{\mu=1}^{m} |\rho'(\zeta_\mu)|^{-1} m_\mu(x_n) + O(h) \right\} \tag{5-245}$$

where the functions $m_\mu(x)$ are defined by (5-242).

The calculation of var $(r_n^{(1)})$ is somewhat less painless. In order to illustrate the procedure we first consider the case of the simple equation $y' = Ay$ (A a real constant), assuming $q(x) = 1$. From §5.3-2 we recall that in this case $e_\mu(x) = e^{\lambda_\mu A x}$. Equation (5-235) then becomes

$$r_n^{(1)} = \sum_{l=k}^{n} \varepsilon_l d_{nl}$$

where

$$d_{nl} = \sum_{\mu=1}^{m} \frac{1}{\rho'(\zeta_\mu)} \exp[i(n-l+k-1)\varphi_\mu] \exp[\lambda_\mu A(x_n - x_l)] + O(h) \tag{5-246}$$

Using (1-95), we have

$$\text{var}\,(r_n^{(1)}) = \frac{\sigma^2}{h} v_n, \qquad v_n = h \sum_{l=k}^{m} d_{nl}^2$$

By their nature the coefficients d_{nl} are real. We thus may put $d_{nl}^2 = d_{nl}\bar{d}_{nl}$, where the bar indicates the conjugate complex number. From (5-246) we find

$$d_{nl}\bar{d}_{nl} = \left(\sum_{\mu=1}^{m} \frac{1}{\rho'(\zeta_\mu)} \exp[i(n-l+k-1)\varphi_\mu] \exp[\lambda_\mu A(x_n - x_l)] \right)$$
$$\times \left(\sum_{\nu=1}^{m} \frac{1}{\overline{\rho'(\zeta_\nu)}} \exp[-i(n-l+k-1)\varphi_\nu] \exp[\bar{\lambda}_\nu A(x_n - x_l)] \right) + O(h)$$

Singling out the cross terms with $\mu = \nu$, this can be written in the form

$$d_{nl}\bar{d}_{nl} = \sum_{\mu=1}^{m} |\rho'(\zeta_\mu)|^{-2} \exp[2\,\text{Re}\,\lambda_\mu A(x_n - x_l)]$$
$$+ \sum_{\substack{\mu,\nu=1 \\ \mu \ne \nu}}^{m} C_{\mu\nu} \exp[i(n-l+k-1)\delta_{\mu\nu}] \exp[(\lambda_\mu + \bar{\lambda}_\nu) A(x_n - x_l)] + O(h)$$

where $C_{\mu\nu} = [\rho'(\zeta_\mu)\, \overline{\rho'(\zeta_\nu)}]^{-1}$, $\delta_{\mu\nu} = \varphi_\mu - \varphi_\nu$.

The sums required for forming v_n can now be formed by evaluating geometric series. The results are (assuming $A \neq 0$)

(5-247)

$$h \sum_{l=k}^{n} \exp [2 \operatorname{Re}\lambda_\mu A(x_n - x_l)] = h \frac{\exp [2 \operatorname{Re}\lambda_\mu Ah(n - k + 1)] - 1}{\exp (2 \operatorname{Re}\lambda_\mu Ah) - 1}$$
$$= \frac{\exp (2 \operatorname{Re}\lambda_\mu Ax_n) - 1}{2 \operatorname{Re}\lambda_\mu A} + O(h)$$

and if $\mu \neq \nu$,

$$\text{(5-248)} \quad h \sum_{l=k}^{n} \exp [i(n - l + k - 1)\delta_{\mu\nu}] \exp [\lambda_\mu + \bar{\lambda}_\nu)A(x_n - x_l)]$$
$$= h \exp [i(k - 1)\delta_{\mu\nu}] \sum_{p=0}^{n-k} \exp \{[(\lambda_\mu + \bar{\lambda}_\nu)Ah + i\delta_{\mu\nu}]p\}$$
$$= h \exp [i(k - 1)\delta_{\mu\nu}] \frac{\exp \{[(\lambda_\mu + \bar{\lambda}_\nu)Ah + i\delta_{\mu\nu}](n - k + 1)\} - 1}{\exp [(\lambda_\mu + \lambda_\mu)Ah + i\delta_{\mu\nu}] - 1}$$

Since $\delta_{\mu\nu} \not\equiv 0 \pmod{2\pi}$ (the roots $\zeta_1, \zeta_2, \cdots, \zeta_m$ are simple for a stable method), the denominator in the last function approaches $\exp (i\delta_{\mu\nu}) - 1 \neq 0$ as $h \to 0$. The last expression can thus be written

$$h \frac{\exp (in\delta_{\mu\nu}) \exp [(\lambda_\mu + \bar{\lambda}_\nu)Ax_n] - \exp [i(k - 1)\delta_{\mu\nu}]}{\exp (i\delta_{\mu\nu}) - 1} + O(h^2)$$

and it is seen that the contribution to v_n arising from cross products with $\mu \neq \nu$ is $O(h)$. Thus,

$$\text{(5-249)} \qquad v_n = \sum_{\mu=1}^{m} |\rho'(\zeta_\mu)|^{-2} \frac{\exp (2 \operatorname{Re}\lambda_\mu Ax_n) - 1}{2 \operatorname{Re}\lambda_\mu A} + O(h)$$

The functions

$$v_\mu(x) = \frac{\exp (2 \operatorname{Re}\lambda_\mu Ax) - 1}{2 \operatorname{Re}\lambda_\mu A}$$

appearing in (5-249) can be characterized as the solutions of

$$\text{(5-250)} \qquad \begin{aligned} &v_\mu'(x) = 2 \operatorname{Re}\lambda_\mu Av_\mu(x) + 1 \\ &v_\mu(0) = 0, \qquad \mu = 1, 2, \cdots, m \end{aligned}$$

We now turn to the general case, where the trick of summing a geometric series has to be replaced by a procedure of broader applicability. Equation (5-230) now holds with

$$d_{nl} = \sum_{\mu=1}^{m} \frac{1}{\rho'(\zeta_\mu)} \exp [i(n - l + k - 1)\varphi_\mu] e_\mu(x_n)[e_\mu(x_l)]^{-1} + O(h)$$

and we now have

$$\operatorname{var}(r_n^{(1)}) = \frac{\sigma^2}{h} v_n, \qquad v_n = h \sum_{l=k}^{n} q_l d_{nl}^2$$

where $q_l = q(x_l)$. Proceeding as in the above special case, we find

$$d_{nl}\bar{d}_{nl} = \sum_{\mu=1}^{m} C_{\mu\mu} d_{nl\mu\mu} + \sum_{\substack{\mu,\nu=1 \\ \mu\neq\nu}}^{m} C_{\mu\nu} \exp\,[i(n - l + k - 1)\delta_{\mu\nu}] d_{nl\mu\nu} + O(h) \tag{5-251}$$

where for $\mu, \nu = 1, \cdots, m$

$$\begin{aligned} C_{\mu\nu} &= [\rho'(\zeta_\mu)\, \overline{\rho'(\zeta_\nu)}]^{-1} \\ \delta_{\mu\nu} &= \varphi_\mu - \varphi_\nu \\ d_{nl\mu\nu} &= e_\mu(x_n)[e_\mu(x_l)\, \overline{e_\nu(x_l)}]^{-1}\, \overline{e_\nu(x_n)} \end{aligned} \tag{5-252}$$

Clearly,

$$v_n = \sum_{\mu=1}^{m} C_{\mu\mu} v_{n\mu\mu} + \sum_{\substack{\mu,\nu=1 \\ \mu\neq\nu}}^{n} C_{\mu\nu} v_{n\mu\nu} + O(h)$$

where

$$v_{n\mu\nu} = h \sum_{l=k}^{n} q_l d_{nl\mu\nu} \exp\,[i(n - l + k - 1)\delta_{\mu\nu}] \tag{5-253}$$

We shall first evaluate $v_{n\mu\mu}$. Considering the sum (5-253) as an approximation to a Riemann integral, we have by Lemma 5.8

$$v_{n\mu\mu} = v_\mu(x_n) + O(h)$$

where

$$v_\mu(x) = e_\mu(x)\overline{e_\mu(x)} \int_a^x q(t)[e_\mu(t)\overline{e_\mu(t)}]^{-1}\, dt$$

By using (5-197) we can easily verify that the functions $v_\mu(x)$ satisfy the differential equations

$$\begin{aligned} v_\mu'(x) &= 2\,\mathrm{Re}\lambda_\mu\, g(x) v_\mu(x) + q(x) \\ v_\mu(a) &= 0, \qquad \mu = 1, 2, \cdots, m \end{aligned} \tag{5-254}$$

which are a generalization of (5-250).

We now wish to show that $v_{n\mu\nu} = O(h)$ for $\mu \neq \nu$. This is accomplished by an appeal to the following lemma.

LEMMA 5.9. *Let the function $f(x)$ be continuous and continuously differentiable for $a \leq x \leq b$, and let $\delta \not\equiv 0 \bmod 2\pi$. Then there exists a constant C such that*

$$\left| \sum_{p=0}^{n} e^{ip\delta} f(x_p) \right| \leq C \tag{5-255}$$

for all $h > 0$ and all $x_n \in [a, b]$.

The proof is by summation by parts. Setting

$$S_p = \sum_{l=0}^{p} e^{il\delta} = \frac{e^{i(p+1)\delta} - 1}{e^{i\delta} - 1}$$

we have

$$\begin{aligned}\sum_{p=0}^{n} e^{ip\delta} f(x_p) &= f(x_0) + \sum_{p=1}^{n} (S_p - S_{p-1}) f(x_p) \\ &= \sum_{p=0}^{n-1} S_p[f(x_p) - f(x_{p+1})] + S_n f(x_n)\end{aligned}$$

By the mean value theorem,

$$f(x_p) - f(x_{p+1}) = -hf'(\xi)$$

for some $\xi \in [x_p, x_{p+1}]$. Using

$$|S_p| \leq \frac{2}{|e^{i\delta} - 1|}$$

and the fact that $n \leq (b - a)/h$, we find that (5-255) is true with

$$C = \frac{2}{|e^{i\delta} - 1|}\left\{(b - a) \max_{x \in [a,b]} |f'(x)| + \max_{x \in [a,b]} |f(x)|\right\} \tag{5-256}$$

An application of Lemma 5.8 to the functions

$$f(t) = q(t)\, |e_\mu(t)\, \overline{e_\nu(t)}|^{-1}$$

—which satisfy the conditions of continuity and differentiability since $e_\mu(t) \neq 0$—putting $\delta = \delta_{\mu\nu}$, shows that $v_{n\mu\nu} = O(h)$ for $\mu \neq \nu$. We thus can state the following main result:

THEOREM 5.15. *If the local round-off errors are independent random variables satisfying* (5-243) *and* (5-244), *then the primary component of the accumulated round-off error is a random variable whose mean satisfies* (5-245) *and whose variance is given by*

$$\operatorname{var}(r_n^{(1)}) = \frac{\sigma^2}{h}\left\{\sum_{\mu=1}^{m} |\rho'(\zeta_\mu)|^{-2} v_\mu(x_n) + O(h)\right\}, \qquad x_n \in [a, b] \tag{5-257}$$

the function $v_\mu(x)$ being defined by (5-254).

If $q(x) > 0$ for $a \leq x \leq b$, it is also easy to show that condition (1-96) is satisfied, so that the distribution of $r_n^{(1)}$ as $n \to \infty$ is governed by the central limit theorem.

It should be noted that the relative strength with which the functions $v_\mu(x)$ participate in expression (5-257) depends only on the polynomial $\rho(\zeta)$ and thus on the underlying integration formula. It is not possible to eliminate undesirable terms from (5-257) by special arithmetical devices.

5.4-4. The local round-off error. We shall now assume that $\alpha_k = 1$, so that the new value y_{n+k} can be calculated from

$$(5\text{-}258)\quad y_{n+k} = -\alpha_{k-1}y_{n+k-1} - \cdots - \alpha_0 y_n + h\{\beta_k f_{n+k} + \cdots + \beta_0 f_n\}$$

(Since it was assumed that $\alpha_k \neq 0$, this assumption could have been made from the start, but it seemed preferable to preserve the symmetric appearance of the formulas.) The right-hand side is a function F of the variables $x_n, y_n, y_{n+1}, \cdots, y_{n+k-1}$, and h. This is obvious for $\beta_k = 0$ and is also true for $\beta_k \neq 0$ if h is sufficiently small, by virtue of Theorem 5.4. Assuming that x_n and h are exact numbers, the local round-off error is the difference between the calculated value and the mathematical value of this function F, both times evaluated at the points of $\tilde{y}_n, \cdots, \tilde{y}_{n+k-1}$. The study of the local round-off error is complex, and each special method and arithmetic procedure must be judged on its own merits. We shall confine ourselves to some general remarks of a qualitative nature.

(i) Fixed point arithmetic. We first discuss the case $\beta_k = 0$. If single precision arithmetic is used, and if the expression $\Phi = \beta_{k-1}f_{n+k-1} + \cdots + \beta_0 f_n$ is evaluated first, then the *inherent* error of Φ is of the order of hu (u = basic unit) and is seen to be negligible in comparison with the error *induced* by forming $h\Phi$, which is of the order of u. If some of the coefficients α_μ are not integers, additional errors arise from forming the products $\alpha_\mu y_{n+\mu}$. These errors are of the same order of magnitude and vary between the same limits as the induced error. If $\beta_k \neq 0$, the expression

$$C = -\alpha_{k-1}y_{n+k-1} - \cdots - \alpha_0 y_n + h\{\beta_{k-1}f_{n+k-1} + \cdots + \beta_0 f_n\}$$

is still usually evaluated only once, with an error described as above. An additional error is introduced by the iteration procedure which involves repeated evaluations of the function $h\beta_k f(x_{n+k}, y_{n+k})$. The errors arising in successive iterations do not accumulate, but even so this error amounts to one additional error of the size of an induced error. This error may be called the *iteration* error. Experience shows that the iteration error is frequently biased, owing to the fact that the limit y_{n+k} of the iteration procedure is repeatedly approached from the same side. If partial double precision is used, the induced error is zero. The local error then consists of inherent error, resulting from truncating the arguments $y_{n+\mu}$ in f, and possibly iteration error, all of the order of hu. If some of the coefficients α_μ are not integers, the products $\alpha_\mu y_{n+\mu}$ have to be formed in double precision in order to avoid errors of the size of u.

(ii) Floating point operation. Inherent and produced errors are now both of the order of hfu, where u denotes the basic unit of the mantissa. The error committed in forming C, on the other hand, is of the order of uy.

This error may be called the *adduced* error. Another error of the order of magnitude of the adduced error arises in implicit methods when the last correction $h\beta_k f_{n+k}$ is added to C.

As to statistical assumptions in the error, the safest hypothesis consists in assuming that the local error is uniformly distributed between its extreme values. Theoretically better results can frequently be obtained by breaking down the local error into its various components and by assuming that each component is an independent random variable which is uniformly distributed between its largest and smallest possible value.

5.4-5. Numerical examples. We have tested the statistical results obtained above by integrating the initial value problem

$$\begin{aligned} y' &= -16xy \\ y(-0.75) &= y_{0,q} = y_{0,0}(1 + q\Delta), \qquad q = 0, \cdots, Q-1 \end{aligned} \tag{5-259}$$

(already considered in §2.3-7) by the following three multistep methods:

(I) The second order Adams-Bashforth method

$$y_{n+2} - y_{n+1} = h\{\tfrac{3}{2}f_{n+1} - \tfrac{1}{2}f_n\}$$

(II) The second order Nyström method ("mid-point rule")

$$y_{n+2} - y_n = 2hf_{n+1}$$

(III) The fourth order Milne method

$$y_{n+2} - y_n = h\{\tfrac{1}{3}f_{n+2} + \tfrac{4}{3}f_{n+1} + \tfrac{1}{3}f_n\}$$

The following data were employed for all three methods:

$$\begin{aligned} h &= 2^{-6} \\ y_{0,0} &= 0.0022159242 \cdot 2^{-15} \\ y_{-1,0} &= 0.0018334831 \cdot 2^{-15} \text{ (an accurate value)} \\ \Delta &= \tfrac{1}{3} \cdot 2^{-8} \\ Q &= 500 \\ u &= 2^{-36} \end{aligned}$$

The exact values $y_{n,0}$ were obtained by numerical calculation using higher precision. A discussion of the results in the three cases follows.

(I) Since $m = 1$, $\rho'(1) = 1$, the pattern indicated by (5-257) is identical with the pattern established for one-step methods. Assuming $q(x) = 1$ as appropriate to fixed point operation, $v_1(x)$ satisfies

$$v_1' = 1 - 32xv_1, \qquad v_1(-0.75) = 0 \tag{5-260}$$

The values of $v_1(x)$ given in Table 5.16-I were found by numerical integration.

Assuming $\mu = 0$, $\sigma^2 = \frac{1}{12}u^2$, the predicted and experimental values of mean and variance were found as shown in Table 5.16-I. The experimental

Table 5.16-I

x		−0.50	−0.25	0	0.25	0.50	0.75
$v_1(x)$		6.5	132.7	361.1	133.0	6.7	0.1
$u^{-1}E(r_n)$	exp.	1.1	5.1	8.8	4.7	0.8	0.0
	pred.	0	0	0	0	0	0
u^{-2} var (r_n)	exp.	47.7	972.1	2586.5	931.0	44.1	0.5
	pred.	34.9	707.7	1925.9	709.3	35.7	0.5

values of the variance exceed the predicted values by an approximately constant percentage. This discrepancy appears to be due to the neglect of secondary and inherent error.

(II) The growth parameters λ_1 and λ_2 associated with the roots $\zeta_1 = 1$ and $\zeta_2 = -1$ of the mid-point rule have in §5.3-1 been determined as $\lambda_1 = 1$, $\lambda_2 = -1$. Since $\rho'(1) = 2$, $\rho'(-1) = -2$, we now expect that approximately

$$\text{var}\,(r_n^{(1)}) = \frac{\sigma^2}{4h}\,(v_1(x_n) + v_2(x_n))$$

where $v_1(x)$ is given by (5-260), and $v_2(x)$ satisfies

$$v_2' = 1 + 32xv_2, \qquad v_2(-0.75) = 0 \tag{5-261}$$

Numerical integration of (5-261) yields the values shown in Table 5.16-II.

Table 5.16-II

x		−0.50	−0.25	0	0.25	0.50	0.75
$v_2(x)$		0.1	0.1	0.2	1.1	24.1	3590.5
$u^{-1}E(r_n)$	exp.	0.4	2.6	4.3	2.3	0.5	2.7
	pred.	0	0	0	0	0	0
u^{-2} var (r_n)	exp.	21.6	395.5	981.5	330.2	55.3	6335.5
	pred.	16.6	357.0	963.4	357.6	82.2	9574.0

The predicted values given in Table 5.16-II were calculated with $\mu = 0$, $\sigma^2 = \frac{1}{6}u^2$. Numerical results are also shown. The steep increase of var (r_n) for $x \geq 0.50$ because of the increase in $v_2(x)$ is in sharp contrast to the exponential decrease of the solution.

(III) The growth parameters for the Milne-Simpson procedure were

Table 5.16-III

x		−0.50	−0.25	0	0.25	0.50	0.75
$v_2(x)$		0.1	0.2	0.4	0.8	2.7	15.2
$u^{-1}E(r_n)$	exp.	18.3	87.0	141.6	84.7	20.1	2.6
	pred.	—	—	—	—	—	—
$u^{-2}\operatorname{var}(r_n)$	exp.	86.9	1518.9	3751.4	1270.7	79.1	117.2
	pred.	70.4	1417.6	3856.0	1427.2	100.2	163.6

determined as $\lambda_1 = 1$, $\lambda_2 = -\frac{1}{3}$. Since again $\rho'(1) = -\rho'(-1) = 2$, we expect that

$$\operatorname{var}(r_n^{(1)}) = \frac{\sigma^2}{4h}(v_1(x) + v_2(x))$$

where $v_1(x)$ is defined by (5-260) and $v_2(x)$ by

$$v_2' = 1 + \tfrac{1}{3}\cdot 32xv_2, \qquad v_2(-0.75) = 0 \tag{5-262}$$

Numerical values of $v_2(x)$ are given in Table 5.16-III.

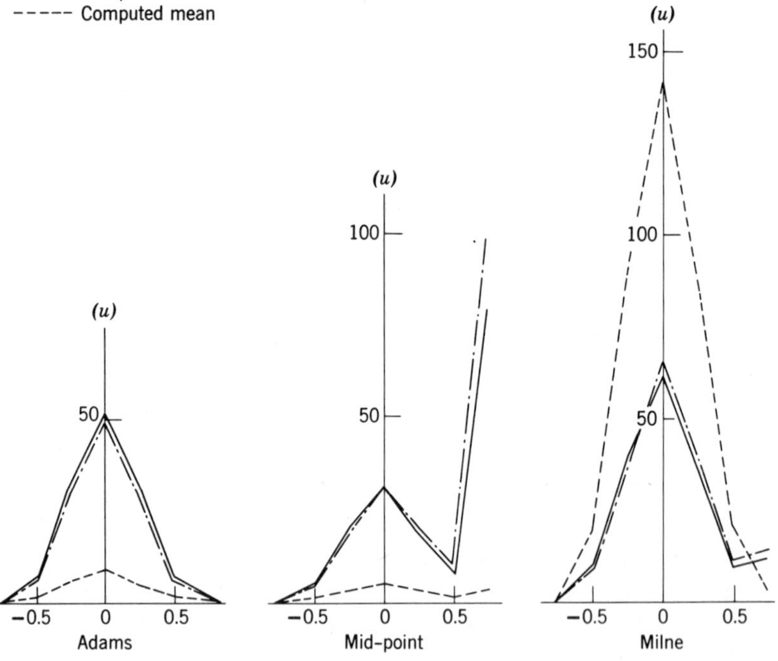

Fig. 5.3

There is again an increase for $x \geq 0$, although not nearly as pronounced as for the mid-point rule. No attempt has been made to predict $E(r_n)$ in Table 5.16-III; to assume $\mu = 0$ would obviously not be realistic. For the variance of the local error we assume $\sigma^2 = \frac{2}{3}u^2$ by virtue of the fact that an iteration error is now present. (The mid-point rule was used to obtain a first predicted value.) The disquieting final upswing in the values of var (r_n) should again be noted.

Predicted values for the standard deviation and experimental values of both mean and standard deviation are shown in Fig. 5.3 for all three cases. While the numerical agreement between predicted and experimental values is not always very pronounced, qualitative agreement obtains throughout. It thus appears that the statistical theory can produce qualitatively correct results even in relatively complicated situations.

5.4-6. Inequalities for the growth parameters. In view of the predominant role which is played by the growth parameters λ_μ in the statistical theory of round-off error as well as in the nonstatistical asymptotic theory of discretization error, the question arises naturally whether it is not possible to choose the polynomials $\rho(\zeta)$ and $\sigma(\zeta)$ in such a way that these parameters all have some harmless value such as $+1$. In the present section we shall establish some results showing that this cannot be done without decreasing the order of the method.

If ζ is an essential root of $\rho(\zeta)$, the corresponding growth parameter has been defined by the relation

$$\lambda = \frac{\sigma(\zeta)}{\zeta\rho'(\zeta)} \tag{5-263}$$

(We omit subscripts for the sake of simplicity.) From the relations

$$\rho(\zeta) = (\zeta + 1)^k r(z)$$
$$\sigma(\zeta) = (\zeta + 1)^k s(z)$$

where $z = (\zeta - 1)/(\zeta + 1)$, we find

$$\rho'(\zeta) = k(\zeta + 1)^{k-1} r(z) + 2(\zeta + 1)^{k-2} r'(z) \tag{5-264}$$

If $\zeta \neq -1$, we thus can express λ in terms of $r(z)$ and $s(z)$ as follows:

$$\lambda = \frac{2s(z)}{(1 - z^2)r'(z)} \tag{5-265}$$

If the root $\zeta = -1$, then $r(z)$ has the exact degree $k - 1$. By letting $\zeta \to -1$ (or equivalently, $z \to \infty$) in (5-264) and substituting in (5-263), we obtain

$$\lambda = 2b_k/a_{k-1} \tag{5-266}$$

where a_{k-1} and b_k are the leading coefficients of the polynomials $r(z)$ and $s(z)$, respectively.

If the roots of $r(z)$ all lie on the imaginary axis, and if $s(z)$ is determined by (5-118), then $r'(z)$ and $s(z)$ are either both odd or both even. It follows that the expression on the right of (5-265) is real. The same is trivially true of (5-226), and we thus find that under the assumptions stated *the growth parameters are real* (Dahlquist [1959], p. 39).

If the roots of $r(z)$ are all purely imaginary, $\zeta = -1$ is a root of $\rho(\zeta)$ if and only if k is even. If the method is optimal, we can determine b_k from (5-121) and find for the corresponding growth parameter the value

$$\lambda = \frac{2(c_2 a_{k-1} + c_4 a_{k-3} + \cdots + c_k a_1)}{a_{k-1}} \tag{5-267}$$

By virtue of (5-115) and (5-122) there follow the inequalities*

$$\lambda \leq 2c_2 = -\tfrac{1}{3} \tag{5-268}$$

$$\lambda \leq 2c_k(a_1/a_{k-1}) \tag{5-269}$$

If viewed in conjunction with the inequality (5-124) for the error constant, the last result shows that C and λ cannot simultaneously attain their upper bounds (5-124) and (5-269) (see Problem 37).

It still might appear from the above results that negative values of the growth parameter could be avoided by making k odd. This hope is futile, as is shown by the following theorem, which makes no assumption on the parity of k.

THEOREM 5.16. *If $m = k \geq 2$ and if $\sigma(\zeta)$ is chosen so as to maximize the order, then*

$$\sum_{\mu=2}^{k} \lambda_\mu < 0 \tag{5-270}$$

The proof is accomplished by calculating the product of the roots ζ_μ of the polynomial (5-137) in two different ways. From Vieta's rule we have

(5-271)

$$\prod_{\mu=1}^{k} \tilde{\zeta}_\mu = (-1)^k \frac{\alpha_0 - hA\beta_0}{\alpha_k - hA\beta_k} = (-1)^k \frac{\alpha_0}{\alpha_k}\left\{1 + hA\left(\frac{\beta_k}{\alpha_k} - \frac{\beta_0}{\alpha_0}\right) + O(h^2)\right\}$$

From (5-141) we find

$$\prod_{\mu=1}^{k} \tilde{\zeta}_\mu = \prod_{\mu=1}^{k} \zeta_\mu \prod_{\mu=1}^{k} [1 + \lambda_\mu Ah + O(h^2)] \tag{5-272}$$

$$= (-1)^k \frac{\alpha_0}{\alpha_k}\left\{1 + Ah \sum_{\mu=1}^{k} \lambda_\mu + O(h^2)\right\}$$

* For (5-268) see Dahlquist [1959], p. 40.

Comparing the $O(h)$ terms in the expression to the right of (5-271) and (5-272), we get

$$\sum_{\mu=1}^{k} \lambda_\mu = \frac{\beta_k}{\alpha_k} - \frac{\beta_0}{\alpha_0} \tag{5-273}$$

We now observe that $\Pi_{\mu=1}^{k} \zeta_\mu$ is -1 or $+1$ according to whether $\zeta = -1$ is a root or not, or, in view of an above remark, according to whether k is even or odd. Thus, using Vieta again, $\alpha_0 = -\alpha_k$ in every case. Hence

$$\sum_{\mu=1}^{k} \lambda_\mu = \frac{\beta_k + \beta_0}{\alpha_k} \tag{5-274}$$

From (5-112) we find

$$\beta_k = \beta_0 = b_0 + b_2 + \cdots + b_i$$

$$\alpha_k = a_1 + a_3 + \cdots + a_j$$

where $i = k$, $j = k - 1$, if k is even, and $i = k - 1$, $j = k$, if k is odd. Hence

$$\begin{aligned} \sum_{\mu=1}^{k} \lambda_\mu &= 2\,\frac{a_1c_0 + (a_3c_0 + a_1c_2) + \cdots + (a_{i+1}c_0 + \cdots + a_1c_i)}{a_1 + a_3 + \cdots + a_j} \\ &\leq 2c_0 + \frac{c_2(a_1 + \cdots + a_{i-1}) + \cdots + c_ia_1}{a_1 + a_3 + \cdots + a_j} \end{aligned} \tag{5-275}$$

The result (5-270) now follows in virtue of $\lambda_1 = 2c_0 = 1$, using (5-115) in conjunction with $c_{2\nu} < 0$, $\nu = 1, 2, \cdots$.

5.4-7. Stable operators with integral coefficients. It follows from the discussion in §5.4-4 that for fixed point operations the local round-off error is minimized if the constants α_μ/α_k ($\mu = 0, \cdots, k-1$) are integers. This assertion is also valid for partial double precision and double precision operations, and, in a qualitative way, even for floating point arithmetic. In the present section we shall study and solve the problem of determining all polynomials $\rho(\zeta)$ of a given degree k which satisfy the condition of stability, are compatible with the condition of consistency, and which have the above property.

Without loss of generality we may assume that $\alpha_k = 1$. The desired polynomials are then of the form

$$\rho(\zeta) = \zeta^k + \alpha_{k-1}\zeta^{k-1} + \cdots + \alpha_0 \tag{5-276}$$

where the coefficients $\alpha_{k-1}, \cdots, \alpha_0$ are integers. A polynomial with leading coefficient 1 and integral coefficients will be called a *monic* polynomial.*

* The complete algebraic description would be "monic polynomial over the domain of all integers"; see Birkhoff and MacLane [1953], Chapter III.

For brevity we shall call a polynomial satisfying the stability conditions of Theorem 5.5 a *stable* polynomial. We then have

THEOREM 5.17. *Let $\rho(\zeta)$ be a monic, stable polynomial. There exists an integer M such that all nonzero roots of $\rho(\zeta)$ satisfy $\zeta^M = 1$.*

*Proof.** If $\rho(\zeta)$ has only zero roots, the theorem is trivial. If $\rho(\zeta)$ has nonzero roots, we may assume that the zero roots have been divided out. If

$$\rho(\zeta) = \zeta^k + \alpha_{k-1}\zeta^{k-1} + \cdots + \alpha_0$$

we then have $\alpha_0 \neq 0$. Let the roots of $\rho(\zeta)$ be denoted by $\zeta_1, \zeta_2, \cdots, \zeta_k$. By hypothesis, all $|\zeta_\mu| \leq 1$; but since $\alpha_0 = (-1)^k \Pi \zeta_\mu$ is a nonvanishing integer, $|\zeta_\mu| = 1$, $\mu = 1, 2, \cdots, k$. Now consider the function

$$f(\zeta) = (-1)^k \rho(\zeta)\rho(-\zeta)$$

This is an even function of ζ, and also a monic polynomial of degree $2k$. It therefore can be written as a monic polynomial of degree k in the variable $\zeta^2 : f(\zeta) = \rho^{(1)}(\zeta^2)$. From the product expansion of $\rho(\zeta)$ we see that

$$f(\zeta) = (\zeta^2 - \zeta_1^2)(\zeta^2 - \zeta_2^2) \cdots (\zeta^2 - \zeta_k^2)$$

It follows that the roots of the polynomial $\rho^{(1)}(\zeta)$ are the numbers $\zeta_1^2, \zeta_2^2, \cdots, \zeta_k^2$. Using the recurrence relation

$$\rho^{(m+1)}(\zeta) = (-1)^k \rho^{(m)}(\sqrt{\zeta})\rho^{(m)}(-\sqrt{\zeta}), \qquad m = 0, 1, 2, \cdots$$

we can similarly construct an infinite sequence of monic polynomials $\rho^{(m)}(\zeta)$, $m = 1, 2, \cdots$, having the roots $\zeta_\mu^{2^m}$ $(\mu = 1, 2, \cdots, k)$. However, only a finite number of different monic polynomials of a fixed degree k exist whose roots have unit modulus. This follows easily from the fact that the coefficients of all such polynomials, being the first k elementary symmetric functions† of k variables of modulus 1, are bounded. The finitely many different ones among the polynomials $\rho^{(m)}(\zeta)$ thus have only a finite number of different roots. Hence, in each of the k sequences $\{\zeta_\mu^{2^m}\}$ $(\mu = 1, \cdots, k)$ there must be repeated elements. Thus, for suitable m and $n > m$ (which still may depend on μ)

$$\zeta_\mu^{2^n} = \zeta_\mu^{2^m} \qquad \text{or} \qquad \zeta_\mu^{2^n - 2^m} = 1$$

Thus ζ_μ satisfies $\zeta_\mu^{M_\mu} = 1$ for $M_\mu = 2^n - 2^m$. If we let $M = \Pi M_\mu$, then every root ζ_μ satisfies $\zeta_\mu^M = 1$. This completes the proof of Theorem 5.17.

* This proof was suggested to the author by E. G. Straus, who attributed the theorem to Kronecker.

The proof uses the stability hypothesis only in the weakened form that no root of $\rho(\zeta)$ exceeds 1 in modulus.

† See Birkoff and MacLane [1953], p. 146.

It follows from Theorem 5.17 that $\rho(\zeta)$ is a factor of the polynomial $\zeta^M - 1$. In other words, $\zeta^M - 1 = \rho(\zeta)\pi(\zeta)$, where $\pi(\zeta)$ is a certain polynomial of degree $M - k$, which, as is readily seen by carrying out the usual long division process, again has integral coefficients. We now appeal to the unique factorization theorem for monic polynomials.* According to this theorem, every monic polynomial can be factored into a product of irreducible monic polynomials (i.e., monic polynomials which cannot be factored further into monic polynomials), and this factorization is unique up to the order in which the factors appear. The factorization of the special monic polynominal $\zeta^M - 1$ is well known. It is given by

$$\zeta^M - 1 = \prod_{d \mid M} \Phi_d(\zeta) \tag{5-277}$$

Here the symbol $d \mid M$ indicates that the index d runs through the divisors of M, and $\Phi_d(\zeta)$ denotes the dth *cyclotomic polynomial.* Such is called the monic polynomial whose roots are those dth roots of unity which have the property that $\zeta^m \neq 1$ for $0 < m < d$ (i.e., the so-called *primitive* dth roots of unity). It is known that the degree of Φ_d is given by Euler's function $\varphi(d)$, which is defined in the following manner. If the decomposition of d in prime factors is given by

$$d = p_1^{\nu_1} p_2^{\nu_2} \cdots p_m^{\nu_m}$$

then

$$\varphi(d) = p_1^{\nu_1 - 1}(p_1 - 1)p_2^{\nu_2 - 1}(p_2 - 1) \cdots p_m^{\nu_m - 1}(p_m - 1) \tag{5-278}$$

For instance, $\varphi(2) = 1$, $\varphi(6) = 2$, $\varphi(13) = 12$. Confronting the factorization (5-277) with the earlier factorization of $\zeta^M - 1$, we find in view of the uniqueness of the decomposition into monic factors that $\rho(\zeta)$ must be of the form

$$\rho(\zeta) = \prod \Phi_{d_i}(\zeta) \tag{5-279}$$

where $\Sigma\varphi(d_i) = k$. If $\rho(\zeta)$ satisfies the condition of stability, then any Φ_{d_i} may appear at most once in the product (5-279), because otherwise there would be multiple roots on the unit circle. If $\rho(\zeta)$ is compatible with consistency, then it must contain the factor $\zeta - 1$, which happens to be the first cyclotomic polynomial $\Phi_1(\zeta)$. We thus have proved:

THEOREM 5.18. *Every monic stable polynomial compatible with consistency can be written in the form $\zeta^m\rho(\zeta)$, where $\rho(\zeta)$ is the product of different cyclotomic polynomials including $\Phi_1(\zeta)$.*

* See van der Waerden [1950], p. 75, or Birkhoff and MacLane [1953], p. 76.

In order to actually construct all polynomials $\rho(\zeta)$ of a given degree k, we need a table of all cyclotomic polynomials $\Phi_d(\zeta)$ of degree $\leq k$. The construction of this table is facilitated by the fact that

$$\varphi(k) \geq k^{1/2} \text{ for } k > 6 \tag{5-280}$$

Thus, if the degree of $\rho(\zeta)$ is 7, only cyclotomic polynomials $\Phi_d(\zeta)$ with $d \leq 36$ can be used [since $\rho(\zeta)$ also contains $\zeta - 1$]. The proof of (5-280)

Table 5.17

Cyclotomic Polynomials

$$\varphi(d) = 1 \begin{cases} \Phi_1(\zeta) = \zeta - 1 \\ \Phi_2(\zeta) = \zeta + 1 \end{cases}$$

$$\varphi(d) = 2 \begin{cases} \Phi_3(\zeta) = \zeta^2 + \zeta + 1 \\ \Phi_4(\zeta) = \zeta^2 + 1 \\ \Phi_6(\zeta) = \zeta^2 - \zeta + 1 \end{cases}$$

$$\varphi(d) = 4 \begin{cases} \Phi_5(\zeta) = \zeta^4 + \zeta^3 + \zeta^2 + \zeta + 1 \\ \Phi_8(\zeta) = \zeta^4 + 1 \\ \Phi_{10}(\zeta) = \zeta^4 - \zeta^3 + \zeta^2 - \zeta + 1 \\ \Phi_{12}(\zeta) = \zeta^4 - \zeta^2 + 1 \end{cases}$$

Table 5.18

Stable, Monic, and Consistent Polynomials $\rho(\zeta)$ of Even Degree ≤ 6

$$k = 2 \quad \rho(\zeta) = \zeta^2 - 1$$

$$k = 4 \begin{cases} \rho(\zeta) = \zeta^4 + \zeta^3 - \zeta - 1 \\ \rho(\zeta) = \zeta^4 - 1 \\ \rho(\zeta) = \zeta^4 - \zeta^3 + \zeta - 1 \end{cases}$$

$$k = 6 \begin{cases} \rho(\zeta) = \zeta^6 + \zeta^5 - \zeta - 1 \\ \rho(\zeta) = \zeta^6 - \zeta^4 + \zeta^2 - 1 \\ \rho(\zeta) = \zeta^6 - \zeta^5 + \zeta - 1 \\ \rho(\zeta) = \zeta^6 - 2\zeta^4 + 2\zeta^2 - 1 \\ \rho(\zeta) = \zeta^6 + \zeta^5 + \zeta^4 - \zeta^2 - \zeta - 1 \\ \rho(\zeta) = \zeta^6 - 1 \\ \rho(\zeta) = \zeta^6 - \zeta^5 + \zeta^4 - \zeta^2 + \zeta - 1 \end{cases}$$

easily follows from (5-278). Explicit formulas for $\Phi_d(\zeta)$ can be given (see van der Waerden [1950], p. 120) for arbitrary d. For our purposes it is sufficient to tabulate all $\Phi_d(\zeta)$ with $\varphi(d) \leq 4$.

From Table 5.17 we can now easily obtain a list of all $\rho(\zeta)$ of degree ≤ 6. [$\varphi(d)$ is even for $d > 2$ and hence there is no d such that $\varphi(d) = 5$.] In Table 5.18 we list the polynomials of even degree.

5.5. Problems and Supplementary Remarks

Section 5.1

1. (a) Using the fact that Newton's representation of the interpolating polynomial holds for arbitrary values of $z_p, z_{p-1}, \cdots, z_{p-q}$, deduce that

$$\sum_{i=m}^{q}(-1)^{m+i}\binom{q}{i}\binom{i}{m} = \begin{cases}1, & m = q\\ 0, & m < q\end{cases}$$

(b) Express $\nabla^2 \cos(x + h)$ and $\nabla^4 \cos(x + 2h)$ in the forms $\varphi(h)\cos x$ and $\psi(h)\cos x$, respectively, and show that

$$\lim_{h\to 0} h^{-2}\varphi(h) = -1, \qquad \lim_{h\to 0} h^{-4}\psi(h) = 1$$

2. Using a cubic interpolating polynomial, derive Simpson's $\frac{3}{8}$ rule

$$y_p - y_{p-3} = \tfrac{3}{8}h(f_p + 3f_{p-1} + 3f_{p-2} + f_{p-3})$$

and show that the error is of the same order as in Simpson's $\frac{1}{3}$ rule.

3. Using generating functions, show that

$$\kappa_m = 2\gamma_m - \gamma_{m-1}, \qquad \kappa_m^* = 2\gamma_m^* - \gamma_{m-1}^*, \qquad m = 0, 1, 2, \cdots$$

4. Solve the initial value problem $y' = 1 - xy^2$, $y(0) = 0$ numerically, using the step $h = 0.1$, by

(a) determining the initial values y_{-1} and y_{+1} by the Runge-Kutta method, and

(b) continuing the computation by the Adams-Moulton formula with $q = 3$, using suitable Adams-Bashforth formulas as predictors.

5. Solve the initial value problem $y' = x - y^2$, $y(0) = 0$ using the step $h = 0.1$ by determining the initial values $y(-0.1)$ and $y(0.1)$ by a power series method and continuing the computation by the Milne-Simpson formula, using as predictors

(a) the Adams-Bashforth formula with $q = 2$,

(b) Milne's predictor formula (5-34).

[Ans. $y_{-1} = 0.00500\,05004$, $y_1 = 0.00499\,94997$, $y_{10} = 0.45554\,47054$]

6. Let $P_{2q}(x)$ be the interpolating polynomial of degree $\leq 2q$ using the interpolating points $x_{-q}, x_{-q+1}, \cdots, x_q$. Show that the even part of $P_{2q}(x)$ can be represented in the form

$$\tfrac{1}{2}(P_{2q}(x) + P_{2q}(-x)) = \sum_{m=0}^{q}\binom{s+m}{2m}\nabla^{2m}f_m \qquad \left(s = \frac{x}{h}\right)$$

From this deduce the following generalization* of Simpson's rule:

$$y_1 - y_{-1} = h\sum_{m=0}^{q}\sigma_m\nabla^{2m}y_m' + R \qquad \text{(5-281)}$$

* Not applicable to the solution of initial value problems for $q > 1$. Why?

where

$$\sigma_m = \int_{-1}^{1} \binom{s+m}{2m} ds, \qquad m = 0, 1, 2, \cdots$$

$$R = \sigma_{q+1} y^{(2q+3)}(\xi) h^{2q+3}, \qquad x_{-q} < \xi < x_q$$

[Apply the technique used in deriving (5-42).]

7*. Show that the coefficients σ_m defined in Problem 6 satisfy

$$\sum_{m=0}^{\infty} \sigma_m t^{2m} = \frac{t\sqrt{1 + \frac{1}{4}t^2}}{\log\left(\sqrt{1 + \frac{1}{4}t^2} + \frac{1}{2}t\right)} \tag{5-282}$$

and deduce the recurrence relation $\sigma_0 = 2$,

$$\sigma_m = 2^{1-2m}\binom{\frac{1}{2}}{m} - \tfrac{1}{3}\binom{-\frac{1}{2}}{1} 2^{-2}\sigma_{m-1} - \tfrac{1}{5}\binom{-\frac{1}{2}}{2} 2^{-4}\sigma_{m-2}$$
$$- \cdots - \frac{1}{2m+1}\binom{-\frac{1}{2}}{m} 2^{-2m}\sigma_0, \qquad m = 1, 2, \cdots$$

[Use the expansion

$$\sum_{m=0}^{\infty} \binom{s+m}{2m} t^{2m} = \frac{1}{2\sqrt{1 + \frac{1}{4}t^2}} \left\{\left(\sqrt{1 + \tfrac{1}{4}t^2} + \tfrac{1}{2}t\right)^{2s+1} + \left(\sqrt{1 + \tfrac{1}{4}t^2} - \tfrac{1}{2}t\right)^{2s+1}\right\}$$

and integrate term by term with respect to s.]

Section 5.2

8. Find the general solutions of the following difference equations ($p = 1, 2, \cdots$): (a) $y_{n+2} = y_{n+1} + y_n$; (b) $y_n - y_{n-p} = 0$; (c) $\nabla^p y_n = 0$. Also find the solution of (a) satisfying the conditions $y_0 = 0$, $y_1 = 1$. [The sequence thus generated is known as the Fibonacci sequence.]

9. Find a particular solution of the difference equation

$$\alpha_k y_{n+k} + \alpha_{k-1} y_{n+k-1} + \cdots + \alpha_0 y_n = \zeta^n$$

where ζ is a root of multiplicity p ($p \geq 0$) of the characteristic polynomial $\rho(\zeta) = \alpha_k \zeta^k + \alpha_{k-1}\zeta^{k-1} + \cdots + \alpha_0$.

10. Assuming that $\alpha_{0,n} \neq 0$, $\alpha_{k,n} \neq 0$, show that a system of m solutions of the linear difference equation

$$\alpha_{k,n} y_{n+k} + \alpha_{k-1,n} y_{n+k-1} + \cdots + \alpha_{0,n} y_n = 0$$

which is independent at $n = n_0$ cannot suddenly become dependent at some $n > n_0$.

11. Let $f(x)$ be q times continuously differentiable in the interval $[x_{n-q}, x_n]$. Show that for some ξ in that interval,

$$f^{(q)}(\xi) = h^{-q} \nabla^q f_n$$

[Apply Rolle's theorem q times to the function $f(x) - P(x)$, where $P(x)$ is given by (5-12).] As an application, show that

$$\lim_{h \to 0} h^{-k} \nabla^k \zeta^n = n(n-1) \cdots (n-k+1)\zeta^{n-k} \tag{5-283}$$

12. Evaluate the determinant of the matrix (5-62) for the system (5-70) at $n = 0$, proceeding as follows:

(a) Replace the system (5-70) by the system defined by

$$y_n = h^{-k} \nabla^k \zeta_\mu^n \qquad (\mu = 1, 2, \cdots, m;\ k = 0, 1, \cdots, p_\mu - 1)$$

where $h \neq 0$.

(b) By elementary transformations on rows, reduce the determinant to the Vandermonde determinant of the numbers $h^{-k}(\zeta_\mu - kh)$ $(\mu = 1, 2, \cdots, m;$ $k = 0, 1, \cdots, p_\mu - 1)$.

(c) Using (5-283), let $h \to 0$.

13*. Let the numbers a_μ, φ_μ, δ_μ $(\mu = 1, 2, \cdots, k)$ be real, $0 < \varphi_\mu \le \pi$, let the numbers φ_μ be all different from each other, and let at least one a_μ be different from zero. Show that the sequence $\{s_n\}$ where

$$s_n = a_1 \cos(n\varphi_1 + \delta_1) + \cdots + a_k \cos(n\varphi_k + \delta_k)$$

cannot have a limit as $n \to \infty$.

14. Solve by iteration Kepler's equation

$$x = y - \varepsilon \sin y$$

for y, where $x = 0.8$, $\varepsilon = 0.2$, starting with $y^{(0)} = x$. Estimate the error after five steps by applying Theorem 5.4. Compare the solution with the analytical solution given by

$$y = x + \sum_{n=1}^{\infty} \frac{2}{n} J_n(n\varepsilon) \sin nx \qquad (J_n = \text{Bessel function of order } n)$$

15. Show that the operators associated with the difference equations

$$y_{n+4} - y_n = \tfrac{4}{3}h(2f_{n+3} - f_{n+2} + 2f_{n+1}) \tag{5-284}$$

$$y_{n+3} - y_n = \tfrac{3}{8}h(f_{n+3} + 3f_{n+2} + 3f_{n+1} + f_n) \tag{5-285}$$

are both of order 4 and that the error constants have the respective values $\frac{7}{90}$ and $-\frac{1}{80}$ {Expand $L[y(x); h]$ around the points x_{n+2} and $x_{n+3/2}$, respectively.}

16. Determine the constants α and β in such a way that the operator associated with

$$y_{n+3} - y_n + \alpha(y_{n+2} - y_{n+1}) = h\beta(f_{n+2} + f_{n+1})$$

is of order 4, and determine the error constant. Verify that the resulting operator is unstable.

17. Show by direct numerical computation that the unstable operator obtained in the preceding problem cannot be used to solve the initial value problem stated in Problem 5, but that as a predictor in conjunction with the Milne formula it is superior* to either of the predictor formulas suggested in Problem 5.

18. Determine the constants α, β, and γ such that the operator associated with the explicit difference equation

$$y_{n+2} - y_{n-2} + \alpha(y_{n+1} - y_{n-1}) = h[\beta(f_{n+1} + f_{n-1}) + \gamma f_n]$$

is of order 6. Is the resulting operator stable? [Ans. $\alpha = 28, \beta = 12, \gamma = 36$; no!]

19. *Sufficient condition for stability.* Let $\rho(1) = 0$, and let the coefficients of the polynomial

$$\rho_1(\zeta) = \frac{\rho(\zeta)}{\zeta - 1} = \gamma_{k-1}\zeta^{k-1} + \gamma_{k-2}\zeta^{k-2} + \cdots + \gamma_0$$

satisfy $\gamma_{k-1} > \gamma_{k-2} > \cdots > \gamma_0$. Prove that $\rho(\zeta)$ satisfies the condition of stability. Apply the result to show that the methods (5-48) based on numerical differentiation are stable for $r = 0$ and $m = 1, 2, 3, 4$.

20*. The methods (5-48) based on numerical differentiation are of order q. Make them of order $q + 1$ by adding a suitable term to the polynomial $\sigma(\zeta)$. [Ans. $\sigma(\zeta) = \zeta^{q-r} - \delta_{r,q+1}(\zeta - 1)^q$].

21*. Taking for granted that the coefficients c_{2n} defined by (5-120) satisfy $\sum_{n=0}^{\infty} c_{2n} = 0$, prove Dahlquist's theorem (see Dahlquist [1956], p. 52) that $p \leq k$ for an explicit operator. [Show that $p > k$ implies $\beta_k = s(1) > 0$.]

22. Determine the optimal operators corresponding to the polynomials

(a) $\rho(\zeta) = \zeta^6 - 2\zeta^4 + 2\zeta^2 - 1$; (b) $\rho(\zeta) = \zeta^6 - \zeta^4 + \zeta^2 - 1$.

23. Show that the coefficients k_{2n} defined by (5-134) satisfy the recurrence relation

$$k_{2n} + \frac{1}{3}\binom{-\frac{1}{2}}{1}2^{-2}k_{2n-2} + \cdots + \frac{1}{2n+1}\binom{-\frac{1}{2}}{n}2^{-2n}k_0 = \binom{\frac{1}{2}}{n}2^{-2n+1},$$

$$n = 0, 1, 2, \cdots$$

Section 5.3

24. Calculate the growth parameters λ_μ for Simpson's $\frac{3}{8}$ rule (5-285) and for Milne's predictor formula (5-34).

25*. Show that the unstable component $(-1)^n e^{-Ax_n/3}$ in the solution of $y' = Ay$, $y(0) = 1$ by Milne's method can be annihilated by averaging every fixed number of times (i.e., for $n = mp$, p an integer, $m = 1, 2, \cdots$) the value obtained by Milne's method with the value obtained from Simpson's $\frac{3}{8}$ rule (5-285). (See Milne and Reynolds [1959].)

26. Determine the constant Γ defined in Lemma 5.5 for the polynomials

(a) $\rho(\zeta) = -\frac{3}{2}\zeta^2 + 2\zeta - \frac{1}{2}$; (b) $\rho(\zeta) = (\zeta^2 - 1)(\zeta^2 - 2\zeta\cos\alpha + 1)$.

Show that $\Gamma = \infty$ for $\rho(\zeta) = (\zeta - 1)^2$.

* Numerical experiments performed by J. Titus indicate that unstable operators may safely be used as predictor formulas.

27. Show that the kernel $G(s)$ in the integral representation (5-178) of a difference operator $L[y(x); h]$ satisfies

$$\int_0^k G(s)\, ds = C_{p+1}$$

where C_{p+1} is the constant defined in §5.2-5.

28. An operator $L[y(x); h]$ is called *definite* if the kernel $G(s)$ does not change sign in the interval $[0, k]$. Show that the generalized mean value theorem (5-177) holds for all functions $y(x)$ with continuous $y^{(p+1)}(x)$ if and only if the operator $L[y(x); h]$ is definite.

29. Show that the operators considered in Problem 15 are definite.

30. *Necessary conditions for definiteness.* If the operator $L[y(x); h]$ is definite, and if $\beta_k = 0$, show that sign α_k = sign C_{p+1}. If the operator is definite and if $\beta_k \neq 0$, show that $p\beta_k/\alpha_k$ lies outside the interval $[0, 1]$, and that

$$\text{sign } \beta_k = -\text{sign } C_{p+1}.$$

31. Work out the estimate of Theorem 5.11 in the following special cases: (a) Milne-Simpson method with $q = 2$; (b) Adams-Bashforth method with $q = 3$; (c) Adams-Moulton method with $q = 4$.

32. *Improved extrapolation to the limit.* In this and the following two problems we consider symmetric (but not necessarily optimal) difference equations of even order $k = 2s$. Let

$$\text{(5-286)} \qquad \begin{aligned} &\alpha_s y_{n+s} + \alpha_{s-1} y_{n+s-1} + \cdots + \alpha_{-s} y_{n-s} \\ &= h\{\beta_s f_{n+s} + \beta_{s-1} f_{n+s-1} + \cdots + \beta_{-s} f_{n-s}\} \end{aligned}$$

where $\alpha_\mu = -\alpha_{-\mu}$, $\beta_\mu = \beta_{-\mu}$, $\mu = 0, 1, \cdots, s$, be such a difference equation. Show that the order p of the difference operator $L[y(x); h]$ associated with (5-286) is even, and that

$$L[y(x); h] = C_{p+1} h^{p+1} y^{(p+1)}(x) + O(h^{p+3})$$

33*. Let the initial value problem $y' = f(x, y)$, $y(a) = \eta$ be solved by a stable and consistent operator of the form (5-286), and let the given starting values be $y_{-s}, y_{-s+1}, \cdots, y_{s-1}$, so that (5-286) is applied the first time with $n = 0$. If the starting errors e_μ are $O(h^{p+1})$, $\mu = -s, -s+1, \cdots, s-1$, then it follows from Theorem 5.12 that

$$\text{(5-287)} \qquad e_n = h^p e(x_n) + O(h^{p+1})$$

where

$$\text{(5-288)} \qquad e'(x) = g(x)e(x) - C y^{(p+1)}(x), \qquad e(a) = 0$$

$$C = \frac{C_{p+1}}{\beta_{-s} + \beta_{-s+1} + \cdots + \beta_s}$$

Show that if the starting errors satisfy

$$e_\mu = h^p e(x_\mu) + O(h^{p+2}) \tag{5-289}$$

where $e(x)$ is defined by (5-288), relation (5-287) can be improved to

$$e_n = h^p e(x_n) + O(h^{p+2}) \tag{5-290}$$

[*Hint:* Using the result of Problem 32, derive a recurrence relation for the quantities $z_n = e_n - h^p e(x_n)$ to which Lemma 5.6 can be applied with $Z = O(h^{p+2})$ and $\Lambda = O(h^{p+3})$.] The result (5-290) means that the application of Richardson's deferred approach to the limit to a properly started solution will improve the order of the approximation by two units.* (In practice, this statement is true only if round-off error is negligible.)

34. Show that condition (5-289) holds for the starting values

$$y_\mu = y(x_\mu) - Ch^{p+1}\mu y^{(p+1)}(a), \qquad \mu = -s, -s+1, \cdots, s-1 \tag{5-291}$$

35*. If the trapezoidal rule is used with the mid-point rule as a predictor formula, and if only one correction is performed, there results the formula

$$y_{n+1} - y_n = \tfrac{1}{2}h\{f_n + f^*_{n+1}\} \tag{5-292}$$

where $f_n = f(x_n, y_n)$, $f^*_{n+1} = f(x_{n+1}, y_{n-1} + 2hf_n)$. According to the results of §5.3-7 the asymptotic error of this method is the same as for the trapezoidal rule with infinite correction. Is it true that without detriment to the accuracy (5-292) can for $n \geq 1$ be replaced by the formula

$$y_{n+1} - y_n = \tfrac{1}{2}h\{f^*_n + f^*_{n+1}\} \tag{5-293}$$

which requires only one evaluation of $f(x, y)$ per step? [No; consider the example $y' = Ay$, $y(0) = 1$.]

Section 5.4

36. Determine the functions $v_\mu(x)$ for the integration of the problem

$$y' = -\frac{1}{x}y, \qquad y(1) = 1$$

by the Milne-Simpson procedure, when

(a) $q = 1$ (fixed point arithmetic)
(b) $q = y^2$ (floating point arithmetic)

37*. It has been proved that for a maximal operator the growth parameter λ belonging to $\zeta = -1$ and the error constant C satisfy the inequalities

$$\text{(5-268)}\quad \lambda \leq -\tfrac{1}{3}, \qquad \text{(5-125)}\quad C \leq 2^{-k-1}c_{k+2}$$

* For special and somewhat incomplete results in this direction, see Gaunt [1927] and de Vogelaere [1957].

If $A = \lambda + \frac{1}{3}$, $B = C - 2^{-k-1}c_{k+2}$, show that for $k \geq 4$

$$AB \geq 2^{-k-1}c_4 c_k \tag{5-294}$$

[The result shows that the bounds for λ and C, although individually sharp, cannot be approached simultaneously.]

38. Although the hypotheses under which (5-268) was derived are not satisfied for $q \geq 3$, show that the inequality still holds for the generalized Milne-Simpson method. [The coefficients of the polynomial $r(z)$ are still positive.]

39. For an *explicit* method of order $p = k \geq 2$, $\zeta = -1$ is a root of $\rho(\zeta)$, and the corresponding growth parameter is ≤ -1. [Dahlquist, oral communication.]

40. Determine all monic stable polynomials of order 5 that are compatible with consistency.

Notes

§§5.1-2 and 5.1-3. Other linear multistep formulas have been proposed by Capra [1956], Keitel [1956], Löwdin [1952], Sconzo [1954]. For linear multistep methods with variable steps see Urabe and Mise [1955]. Related methods, not strictly of the linear multistep type considered here, are given by Urabe and Tsushima [1953], Wilf [1957], Wall [1956]. The computational aspects of the Adams methods are discussed by Collatz [1960], p. 78, and Ralston [1960]. All methods discussed in Chapter 5 are essentially based on the idea that the solution is best approximated by polynomials. For some differential equations, especially those whose solutions have highly oscillatory components, this approach may be unsuitable, and approximation by some other simple class of functions, such as exponential functions, may be preferable. This approach is pursued by Brock and Murray [1952]; see also Certaine [1960]. The idea definitely deserves further study.

§5.2-1. A more complete treatment of difference equations is given by Goldberg [1958], Fort [1948], Milne-Thomson [1933], Nörlund [1924].

§5.2-3. The definition of convergence given here is not identical with the concept of "stable convergence" introduced by Dahlquist [1956].

§5.2-4. The stability condition introduced here was discussed heuristically by Todd [1950]. The precise definition used here is taken from Dahlquist [1956]. Unstable methods for the numerical integration of ordinary differential equations are proposed by Muhin [1952a], Salzer [1956], Quade [1957], Kopal [1958]. Many different definitions of stability have been suggested in the literature; cf. Carr [1957], Liniger [1957], Hamming [1959], Wilf [1959], Budak and Gorbunov [1959], Ralston [1960], Hull and Luxemburg [1960]. Some of these definitions are closely related to our definition of convergence, others to our concept of weak or conditional stability introduced in §5.3-1. Frequently the definitions apply rigorously only to the integration of $y' = Ay$, where A is a constant. The stability of the differentiation methods with $r = 0$ and $q = 1, 2, 3, 4, 5, 6$ was shown by Mitchell and Craggs [1953].

§5.2-7. For other methods of constructing difference operators see Frei [1954], Hamming [1959], Hull and Newbery [1959], Robertson [1960].

§5.2-8. For a related result see Ceschino and Kuntzmann [1958].

§5.3-1. A similar, but more special, calculation is performed by Loud [1949]. The phenomenon of weak stability was reported in a special case by Dahlquist [1951], and a comprehensive discussion for equations with constant coefficients was given by Rutishauser [1952].

§5.3-3. The fact that stability and consistency together are necessary and sufficient for convergence is—for *linear* differential equations—proved in a more general context by Lax and Richtmyer [1956] (see also Richtmyer [1957]).

§5.3-4. There is an extensive literature on error bounds for the Adams methods; see von Mises [1930], Tollmien [1938, 1953], Fricke [1949], Hamel [1949], Weissinger [1950, 1952], Matthieu [1951], Mohr [1951], Vietoris [1953a, 1953b], Bahvalov [1955a, 1955b], Gaier [1956], Zondek and Sheldon [1959]. The error of Milne's method is estimated by Richter [1951], and that of quadrature methods in general by Shura-Bura [1952] and Hildebrand [1956], p. 219. Eltermann [1955], Uhlmann [1957a], Richter [1952], and Dahlquist [1959] give error bounds for the integration of systems of first order equations by various linear multistep methods. See also Hull and Newbery [1959].

§5.3-5. Asymptotic formulas for the discretization error are given by Brodskii [1953], Lotkin [1954], Gray [1955].

§5.3-6. Estimates of the local discretization error by comparing predicted and corrected values are informally discussed by Lotkin [1955].

§5.3-7. The algorithm of predicting only once is proposed by Wall [1956]. See also the review of this paper by Young [1957].

§5.4-1. Bounds for the accumulated round-off error are given by Bahvalov [1955a, 1955b], Dahlquist [1956, 1959], Lotkin [1954].

§5.4-3. A statistical model of round-off propagation in multistep methods is discussed by Sterne [1953]. The results of this section were announced by Henrici [1960].

chapter

6 Multistep Methods for Special Equations of the Second Order

Although they have been explicitly formulated only for single differential equations, the methods and results discussed in Chapter 5 can be generalized to systems of differential equations, much as the results of Chapter 2 were extended to systems in Chapter 3. It is thus possible, for instance, to integrate a differential equation of the second order,

$$y'' = f(x, y, y') \tag{6-1}$$

by reducing it to the system $y' = z$, $z' = f(x, y, z)$, and applying one of the methods described in Chapter 5. As in the case of one-step methods, this procedure is a perfectly legitimate one. No accuracy is lost, nor is there any unnecessary outlay of computational effort.

The situation is slightly different if the equation to be integrated is of the form

$$y'' = f(x, y) \tag{6-2}$$

i.e., if no derivatives appear in the right-hand member of the differential equation. Equations of this type will be called *special* differential equations, and so will those of the form $y^{(n)} = f(x, y)$ where n is an integer > 2. Special differential equations of the second order, and in particular systems of such equations, occur frequently, e.g., in mechanical problems without dissipation. If one is not particularly interested in the values of the first derivatives, it seems unnatural to introduce them artificially in order to produce systems of first order equations. And indeed, astronomers have for over a century used methods for integrating (6-2) which are of the multistep type and which work without first derivatives.

The present chapter is devoted to a study of these methods. Although the theory resembles in outline the theory of multistep methods for first order equations, there are sufficiently many new and unexpected features, notably in the field of error propagation, to warrant their discussion in a separate chapter. We shall emphasize in the following discussion these new aspects and pass over briefly the developments which are more or less analogous to the first order case. A similar theory could be established for special equations of any order. However, we shall confine ourselves to the second order case in view of the relatively rare occurrence of special equations of order higher than 2 in practical applications.

6.1. Local Study of Linear Multistep Methods

6.1-1. Some special methods. Our first task consists in deriving a formula similar to (5-13) for a function satisfying

$$y''(x) = f(x, y(x)) \tag{6-3}$$

By integrating twice, we obtain the formula

$$y(x + k) - y(x) = ky'(x) + \int_x^{x+k} (x + k - t)f(t, y(t))\, dt$$

which also may be regarded as a form of Taylor's formula with a remainder term. This result is not yet satisfactory for our purposes, because we do not wish to use the first derivative $y'(x)$. However, by writing down the same result with k replaced by $-k$ and adding, the first derivative drops out, and the sum of the two integrals, writing $f(t) = f(t, y(t))$, may be transformed as follows:

$$\int_x^{x+k} (x + k - t)f(t)\, dt + \int_x^{x-k} (x - k - t)f(t)\, dt$$
$$= \int_x^{x+k} (x + k - t)[f(t) + f(2x - t)]\, dt$$

There results the identity

$$y(x + k) - 2y(x) + y(x - k) = \int_x^{x+k} (x + k - t)[f(t) + f(2x - t)]\, dt \tag{6-4}$$

which forms the basis of many integration formulas. Replacing $f(t)$ by the interpolating polynomial of degree q, using the points $x_p, \cdots, x_{p-q},$

Table 6.1

Method	x	$x + k$	q
Störmer	x_p	x_{p+1}	≥ 0
Cowell	x_{p-1}	x_p	≥ 2
(6-15)	x_{p-1}	x_{p+1}	≥ 2
(6-19)	x_{p-2}	x_p	≥ 4

we can obtain a number of special methods according to our choice of x, k, and q. Some of them are listed in Table 6.1. A detailed discussion of these methods follows.

(i) Störmer's method (Störmer [1907, 1921]). The program outlined above yields

$$y_{p+1} - 2y_p + y_{p-1} = h^2 \sum_{m=0}^{q} \sigma_m \nabla^m f_p \tag{6-5}$$

where

$$\sigma_m = \frac{(-1)^m}{h^2} \int_{x_p}^{x_{p+1}} (x_{p+1} - x) \left[\binom{-s}{m} + \binom{s}{m} \right] dx \quad \left(s = \frac{x - x_p}{h} \right) \tag{6-6}$$

$$= (-1)^m \int_0^1 (1 - s) \left[\binom{-s}{m} + \binom{s}{m} \right] ds$$

By interchanging summation and integration, we find as in §5.1-2

$$S(t) = \sum_{m=0}^{\infty} \sigma_m t^m = \left[\frac{t}{\log(1 - t)} \right]^2 \frac{1}{1 - t} \tag{6-7}$$

From

$$\frac{d}{dt} [\log(1 - t)]^2 = -\frac{2}{1 - t} \log(1 - t)$$

$$= 2(1 + t + t^2 + \cdots)(t + \tfrac{1}{2}t^2 + \tfrac{1}{3}t^3 + \cdots)$$

$$= 2(h_1 t + h_2 t^2 + h_3 t^3 + \cdots)$$

where $h_m = 1 + \frac{1}{2} + \cdots + 1/m$ denotes the mth partial sum of the harmonic series, there follows

$$\left[\frac{\log(1 - t)}{t} \right]^2 = 1 + \tfrac{2}{3} h_2 t + \tfrac{2}{4} h_3 t^2 + \cdots \tag{6-8}$$

(A similar expansion is needed in a different context in §6.1-7.) Multiplying both sides of (6-8) by (6-7), we thus find

$$(1 + \tfrac{2}{3} h_2 t + \tfrac{2}{4} h_3 t^2 + \cdots)(\sigma_0 + \sigma_1 t + \sigma_2 t^2 + \cdots) = 1 + t + t^2 + \cdots$$

There follows the recurrence relation $\sigma_0 = 1$,

$$\sigma_m = 1 - \tfrac{2}{3}h_2\sigma_{m-1} - \tfrac{2}{4}h_3\sigma_{m-2} - \cdots - \frac{2}{m+2}h_{m+1}\sigma_0, \qquad m = 1, 2, \cdots$$

from which the numerical values given in Table 6.2 are readily found.

Table 6.2

Values of σ_m

m	0	1	2	3	4	5	6
σ_m	1	0	$\frac{1}{12}$	$\frac{1}{12}$	$\frac{19}{240}$	$\frac{3}{40}$	$\frac{863}{12096}$

Formula (6-5) is used in much the same manner as the Adams-Bashforth formula. Once the values $y_p, \cdots, y_{p-q}$ are known, y_{p+1} can be calculated explicitly, without iteration. There remains the difficulty of determining the starting values. Either a power series or one of the methods described in Chapter 4 can be used to obtain these. In order to save labor and increase the accuracy, it is recommended that the starting points be placed *symmetrically* about x_0. For instance, if five starting values are needed, they should be located at x_{-2}, x_{-1}, x_0, x_1, x_2. For $q = 0$ and $q = 1$, Störmer's formula reduces to the simple rule

(6-9) $$y_{p+1} - 2y_p + y_{p-1} = h^2 f_p$$

(ii) Cowell's method (Cowell and Crommelin [1910]). In accordance with Table 6.2, we now let

(6-10) $$y_p - 2y_{p-1} + y_{p-2} = h^2 \sum_{m=0}^{q} \sigma_m^* \nabla^m f_p$$

where

(6-11) $$\sigma_m^* = \frac{(-1)^m}{h^2}\int_{x_{p-1}}^{x_p} (x_p - x)\left[\binom{-s}{m} + \binom{s+2}{m}\right]$$

$$= (-1)^m \int_{-1}^{0} (-s)\left[\binom{-s}{m} + \binom{s+2}{m}\right] ds$$

The generating function of the coefficients σ_m^* is determined as above; the result is

(6-12) $$S^*(t) = \sum_{m=0}^{\infty} \sigma_m^* t^m = \left[\frac{t}{\log(1-t)}\right]^2$$

From the relation $S^*(t) = (1 - t)S(t)$ it follows that for $m = 1, 2, \cdots$

(6-13) $$\sigma_m^* = \sigma_m - \sigma_{m-1}$$

Multiplying both sides of (6-12) by expansion (6-8), we obtain

$$(1 + \tfrac{2}{3}h_2 t + \tfrac{2}{4}h_3 t^2 + \cdots)(\sigma_0^* + \sigma_1^* t + \sigma_2^* t^2 + \cdots) = 1$$

By comparing coefficients, we get the recurrence relation $\sigma_0^* = 1$,

$$\sigma_m^* = -\tfrac{2}{3}h_2\sigma_{m-1}^* - \tfrac{2}{4}h_3\sigma_{m-2}^* - \cdots - \frac{2}{m+2}\,h_{m+1}\sigma_0^*, \qquad m = 1, 2, \cdots$$

The numerical values presented in Table 6.3 are readily found.

Table 6.3

Values of σ_m^*

m	0	1	2	3	4	5	6
σ_m^*	1	-1	$\frac{1}{12}$	0	$-\frac{1}{240}$	$-\frac{1}{240}$	$-\frac{221}{60480}$

The Cowell formula is not used for $q = 0$. For $q = 1$ it reduces to (6-9). For $q = 2$ and $q = 3$ we obtain the frequently used formula

(6-14)

$$y_p - 2y_{p-1} + y_{p-2} = h^2\{f_{p-1} + \tfrac{1}{12}\nabla^2 f_p\} = \tfrac{1}{12}h^2\{f_p + 10f_{p-1} + f_{p-2}\}$$

The Cowell formulas are implicit for $q \geq 2$; i.e., the unknown value y_p occurs not only on the left but also, as an argument in f_p, on the right. The resulting equation for y_p, which is nonlinear except when the differential equation is linear, can be solved by iteration, as described in §5.1-2(ii). The convergence of the iteration process is guaranteed by Theorem 5.4, and the error estimates given there can be applied (substitute h^2 for h). As a predictor formula, Störmer's formula (6-5) is recommended, because it uses the same combination of values of y_n on the left.

(iii) Some further special methods. Further useful formulas can be derived by doubling the step k, analogously to the doubling of the step in the derivation of the Nyström and Milne-Simpson methods for first order equations. By taking $k = 2h$ and $x = x_{p-1}$, we obtain the formula

$$y_{p+1} - 2y_{p-1} + y_{p-3} = h^2 \sum_{m=0}^{q} \tau_m \nabla^m f_p \tag{6-15}$$

where the coefficients τ_m are given by

$$\tau_m = (-1)^m \int_{-1}^{1} (1 - s)\left[\binom{-s}{m} + \binom{s+2}{m}\right] ds \tag{6-16}$$

The generating function is

$$T(t) = \sum_{m=0}^{\infty} \tau_m t^m = \left[\frac{t}{\log(1-t)}\right]^2 \frac{4 - 4t + t^2}{1 - t} \tag{6-17}$$

and leads to the recurrence relations $\tau_0 = 4$, $\tau_1 = -4$,

$$\tau_m = 1 - \tfrac{2}{3}h_2\tau_{m-1} - \tfrac{2}{4}h_3\tau_{m-2} - \cdots - \frac{2}{m+2}h_{m+1}\tau_0, \qquad m = 2, 3, \cdots$$

Numerical values are given in Table 6.4.

Table 6.4

Values of τ_m

m	0	1	2	3	4	5	6
τ_m	4	−4	$\frac{4}{3}$	0	$\frac{1}{15}$	$\frac{1}{15}$	$\frac{61}{945}$

Formula (6-15) is explicit. It is not recommended for $q < 2$. For $q = 2$ there results the simple and relatively accurate formula

$$y_{p+1} - 2y_{p-1} + y_{p-3} = \tfrac{4}{3}h^2\{f_p + f_{p-1} + f_{p-2}\} \tag{6-18}$$

Since $\tau_3 = 0$, the same formula results also from picking the value $q = 3$.

Another set of formulas is derived by choosing $k = 2h$, $x = x_{p-2}$ in (6-4). We obtain

$$y_p - 2y_{p-2} + y_{p-4} = h^2 \sum_{m=0}^{q} \tau_m^* \nabla^m f_p \tag{6-19}$$

where

$$\tau_m^* = (-1)^m \int_{-2}^{0} (-s)\left[\binom{-s}{m} + \binom{s+4}{m}\right] ds \tag{6-20}$$

From the generating function

$$T^*(t) = \sum_{m=0}^{\infty} \tau_m^* t^m = \left[\frac{t}{\log(1-t)}\right]^2 (4 - 4t + t^2) \tag{6-21}$$

the recurrence relations $\tau_0^* = 4$, $\tau_1^* = -8$, $\tau_2^* = 4 + \frac{4}{3}$,

$$\tau_m^* = -\tfrac{2}{3}h_2\tau_{m-1}^* - \tfrac{2}{4}h_3\tau_{m-2}^* - \cdots - \frac{2}{m+2}h_{m+1}\tau_0^*, \qquad m = 3, 4, \cdots$$

are derived in the usual manner. By comparing (6-17) and (6-21) we also find that $\tau_m^* = \tau_m - \tau_{m-1}$, $m = 1, 2, \cdots$. The numerical values of Table 6.5 are readily found.

For $q = 0, 1, 2$, (6-19) has an irregular appearance, and its use for practical purposes is not recommended. For $q = 3$ the formula reads

$$y_p - 2y_{p-2} + y_{p-4} = \tfrac{4}{3}h^2\{f_{p-1} + f_{p-2} + f_{p-3}\}$$

and thus becomes equivalent to (6-18). For $q = 4$ and, since $\tau_5^* = 0$, also for $q = 5$, (6-19) reduces to

(6-22) $$y_p - 2y_{p-2} + y_{p-4} = \tfrac{1}{15}h^2\{f_p + 16f_{p-1} + 26f_{p-2} + 16f_{p-3} + f_{p-4}\}$$

Further formulas can be derived by forming linear combinations of some of the above formulas; see also §6.1-6.

Table 6.5

Values of τ_m^*

m	0	1	2	3	4	5	6
τ_m^*	4	-8	$\frac{16}{3}$	$-\frac{4}{3}$	$\frac{1}{15}$	0	$-\frac{2}{945}$

6.1-2. General operators for special second order equations. All methods considered in the preceding section are special cases of the difference equation

(6-23) $$\alpha_k y_{n+k} + \alpha_{k-1} y_{n+k-1} + \cdots + \alpha_0 y_n = h^2\{\beta_k f_{n+k} + \beta_{k-1} f_{n+k-1} + \cdots + \beta_0 f_n\}$$

where $f_m = f(x_m, y_m)$. This expression is identical with (5-58), with the important exception that h is now replaced by h^2. We shall concentrate in the remaining part of this chapter on the general expression (6-23) rather than on any of the special methods considered in §6.1-1. Again we shall try to discover necessary and sufficient conditions in order for (6-23) to define a "good" method for integrating $y'' = f(x, y)$. In the course of this investigation an insight will be obtained into the properties of the concrete methods studied at the beginning of the chapter. We shall assume throughout that $\alpha_k \neq 0$, $|\alpha_0| + |\beta_0| > 0$. The order k of the difference equation (6-23) is then uniquely determined. We note that for the special methods considered in §6.1-1 the stepnumber k is at least 2.

With the difference equation (6-23) we associate the difference operator

(6-24) $$L[y(x); h] = \alpha_k y(x + kh) + \alpha_{k-1} y(x + (k-1)h) + \cdots + \alpha_0 y(x) - h^2\{\beta_k y''(x + kh) + \beta_{k-1} y''(x + (k-1)h) + \cdots + \beta_0 y''(x)\}$$

This operator can act on any function $y(x)$ which has a second derivative, but will at present only be applied to functions which have continuous derivatives of sufficiently high order. We may then expand (6-24) in powers of h and obtain

(6-25) $$L[y(x);\ h] = C_0y(x) + C_1y'(x)h + \cdots + C_qy^{(q)}(x)h^q + \cdots$$

where the coefficients C_q ($q = 0, 1, 2, \cdots$) are independent of $y(x)$. In particular,

(6-26)
$$\begin{aligned} C_0 &= \alpha_0 + \alpha_1 + \cdots + \alpha_k \\ C_1 &= \alpha_1 + 2\alpha_2 + \cdots + k\alpha_k \\ C_q &= \frac{1}{q!}\{\alpha_1 + 2^q\alpha_2 + \cdots + k^q\alpha_k\} \\ &\quad - \frac{1}{(q-2)!}\{0^{q-2}\alpha_0 + 1^{q-2}\alpha_1 + \cdots + k^{q-2}\alpha_k\}, \qquad q = 2, 3, \cdots \end{aligned}$$

The *order* p of the difference operator (6-24) is defined as the (unique) integer such that $C_q = 0$, $q = 0, 1, \cdots, p+1$; $C_{p+2} \neq 0$. As in §5.2-5 it can easily be shown that p and C_{p+2} do not depend on t if (6-23) is replaced by

$$\alpha_k y_{n+t+k} + \cdots + \alpha_0 y_{n+t} = h^2\{\beta_k f_{n+t+k} + \cdots + \beta_0 f_{n+t}\}$$

By definition,

(6-27) $$L[y(x);\ h] = h^{p+2}C_{p+2}y^{(p+2)}(x) + O(h^{p+3})$$

The order of an operator may thus be regarded as a measure for the accuracy with which (6-23) is satisfied if the values of an exact solution of $y'' = f(x, y)$ are substituted for y_m. Within the class of all difference operators of a given order, the constant

(6-28) $$C = \frac{C_{p+2}}{\beta_0 + \beta_1 + \cdots + \beta_k}$$

called the *error constant*, represents a further measure for this accuracy. It will be noted that p as well as C is invariant with respect to the trivial operation of multiplying $L[y(x);\ h]$ by a nonzero constant.

The order and error constant of an operator can frequently be determined without calculating the constants $C_0, C_1, \cdots$ explicitly. If it is known that $L[y(x);\ h] = 0$ whenever $y(x)$ is a polynomial of degree not exceeding $p + 1$, but $\neq 0$ for some polynomial of degree $p + 2$, then the order is clearly p. Let us now look at the special formulas of §6.1-1. With a suitable choice of t and s they can all be brought into the form

(6-29) $$y_{n+t} - 2y_n + y_{n-t} = h^2 \sum_{m=0}^{q} \gamma_m \nabla^m f_{n+s}$$

By construction, these formulas are exact whenever $f(x, y(x)) = y''(x)$ is a polynomial of degree q or less. In order to make them exact for a polynomial of degree $q + 1$, the term $\gamma_{q+1}\nabla^{q+1}f_{n+s}$ has to be added on the right. It follows that the order of (6-29) is p, where p is the first integer $>q$ such that $\gamma_p \neq 0$. In order to determine the error constant, let us assume that (6-29) is of order $p = q + 1$, so that $\gamma_{q+1} \neq 0$. If $y(x) = x^{q+3}$, we then have

$$y_{n+t} - 2y_n + y_{n-t} - h^2 \sum_{m=0}^{q} \gamma_q \nabla^m y''_{n+s}$$
$$= h^2\gamma_{q+1}\nabla^{q+1}y''_{n+s} = h^{q+3}\gamma_{q+1}(q + 3)!$$

in view of $\nabla^m x^m = h^m m!$. By comparing the last expression with (6-27), it follows that $C_{p+2} = \gamma_{q+1}$. The value of $\beta_0 + \beta_1 + \cdots + \beta_k$ is identical with $\Sigma\gamma_q\nabla^q f_{n+s}$ when $f(x, y) = 1$. It follows that $\beta_0 + \beta_1 + \cdots + \beta_k = \gamma_0$, and we thus obtain, if $\gamma_{q+1} \neq 0$,

$$C = \gamma_{q+1}/\gamma_0 \tag{6-30}$$

for all methods of the form (6-29).

Table 6.6 gives the values of p and C for the special methods discussed in §6.1-1.

Table 6.6

Order and Error Constants for Special Multistep Methods for Special Second Order Equations*

Method		Order of Associated Difference Operator	Error Constant
Störmer	$q = 0$	2	$\frac{1}{12}$
	$q > 0$	$k = q + 1$	σ_{q+1}
Cowell	$q = 2$	4	$-\frac{1}{240}$
	$q > 2$	$k + 1 = q + 1$	σ^*_{q+1}
(6-15)	$q = 2$	4	$\frac{1}{60}$
	$q > 2$	$k = q + 1$	$\frac{1}{4}\tau_{q+1}$
(6-19)	$q = 4$	6	$-\frac{1}{1890}$
	$q > 4$	$k + 1 = q + 1$	$\frac{1}{4}\tau^*_{q+1}$

q = order of highest difference employed.
k = order of difference equation.

6.1-3. Bounds for the remainders. If $L[y(x);\ h]$ is an operator of order p, we shall require bounds for $|L[y(x);\ h]|$ in terms of a bound Y for $|y^{(p+2)}(x)|$. If a generalized mean value theorem such as

$$L[y(x);\ h] = h^{p+2}C_{p+2}y^{(p+2)}(\xi) \tag{6-31}$$

holds, where ξ is a point in some specified interval, then the bound

$$|L[y(x);\ h]| = h^{p+2}\,|C_{p+2}|\ Y \tag{6-32}$$

follows immediately. It seems difficult to establish (6-31) for any large class of special operators. For a number of the most interesting formulas a generalized mean value theorem can, however, be proved by special devices. Throughout the following discussion we shall assume that $y^{(p+2)}(x)$ is continuous.

The operator associated with Störmer's method is for $q = 0$ given by

$$L[y(x);\ h] = y(x + h) - 2y(x) + y(x - h) - h^2y''(x)$$

Expanding $y(x + h)$ and $y(x - h)$ in powers of h through the term in h^4 and employing the derivative form of the remainder, we find

$$L[y(x);\ h] = \tfrac{1}{24}h^4\{y^{\text{iv}}(\xi_1) + y^{\text{iv}}(\xi_2)\} \tag{6-33}$$

where (if $h > 0$) $x < \xi_1 < x + h$ and $x - h < \xi_2 < x$. This formula is in itself sufficient to establish (6-32). In order to deduce from (6-33) a generalized mean value theorem, we note that $y^{\text{iv}}(x)$ as a continuous function assumes all values between $y^{\text{iv}}(\xi_1)$ and $y^{\text{iv}}(\xi_2)$ in the interval $\xi_2 < x < \xi_1$. Thus in particular, for some ξ satisfying $\xi_2 < \xi < \xi_1$,

$$y^{\text{iv}}(\xi) = \tfrac{1}{2}[y^{\text{iv}}(\xi_1) + y^{\text{iv}}(\xi_2)]$$

The relation

$$L[y(x);\ h] = \tfrac{1}{12}h^4y^{\text{iv}}(\xi) \tag{6-34}$$

now follows immediately.

For the case $q = 2$ of the Cowell method one can show by an argument similar to that used in the proof of (5-42) that

$$y(x + h) - 2y(x) + y(x - h) - \tfrac{1}{12}h^2\{y''(x + h) + 10y''(x) + y''(x - h)\} = -\tfrac{1}{240}h^6y^{\text{vi}}(\xi) \tag{6-35}$$

for some ξ in $(x - h, x + h)$.

As was the case for operators approximating differential equations of the first order, every operator (6-24) of order p can be represented in the form

$$L[y(x);\ h] = h^{p+2}\int_0^k G(s)y^{(p+2)}(x + sh)\, ds \tag{6-36}$$

by using Taylor's formula with the integral representation of the remainder. If the kernel $G(s)$ does not change sign in $[0, k]$, we can deduce the generalized mean value theorem as in Problem 28, Chapter 5. If $G(s)$ does change sign, we still have

$$|L[y(x);\ h]| \leq h^{p+2} G Y \tag{6-37}$$

where

$$G = \int_0^k |G(s)|\, ds \tag{6-38}$$

We shall apply representation (6-36) in order to derive the generalized mean value theorem for the operator defined by (6-18). Setting $x_{p-1} = 0$ without loss of generality and expanding around the origin, we find

$$y(2h) - 2y(0) + y(-2h) - \tfrac{4}{3}h^2(y''(h) + y(0) + y''(-h)) = h^6 \int_{-2}^{2} G(s) y^{\mathrm{vi}}(sh)\, ds$$

where

$$G(s) = \begin{cases} \dfrac{1}{5!}(2-s)^5 - \dfrac{4}{3\cdot 3!}(1-s)^3, & 0 \leq s \leq 1 \\[2ex] \dfrac{1}{5!}(2-s)^5, & 1 \leq s \leq 2 \end{cases}$$

and $G(-s) = G(s)$. It is evident that $G(0) > 0$, $G(s) > 0$ for $1 \leq s < 2$. By Descartes' rule of signs it is easily shown that $G(s)$ has no zero for $0 < s < 1$. It thus follows that $G(s) \geq 0$ for $-2 \leq s \leq 2$, and that the generalized mean value theorem holds.

In like manner it can be shown that the operator associated with (6-22) satisfies $G(s) \leq 0$ and thus the generalized mean value theorem.

6.1-4. Convergence: definition. Although (6-23) represents an implicit equation for y_{n+k} if $\beta_k \neq 0$, it follows by Theorem 5.4 that if the function $f(x, y)$ satisfies a Lipschitz condition with Lipschitz constant L, (6-23) has a unique solution y_{n+k} for all values of h satisfying

$$|h| < \left|\frac{\alpha_k}{\beta_k L}\right|^{1/2} \tag{6-39}$$

This solution can be found by the method of successive substitutions described in §5.2-2. For all h satisfying (6-39), the values y_m ($m = k, k+1, \cdots$) may thus be regarded as uniquely determined functions of the starting values $y_0, y_1, \cdots, y_{k-1}$, which in turn are functions of h:

$$y_\mu = \eta_\mu(h), \qquad \mu = 0, \cdots, k-1$$

We expect that for a "good" method the values y_n thus generated tend to the value of the exact solution at the point x as $h \to 0$ and $x_n = x$, provided that the starting values are properly chosen. This intuitive notion is formalized by the following definition of convergence.

Definition. The linear multistep method defined by (6-23) is called *convergent* if the following statement is true for all functions $f(x, y)$ satisfying the conditions of Theorem 1.1 and all constants η and η': If $y(x)$ denotes the solution of the initial value problem

$$y'' = f(x, y), \qquad y(a) = \eta, \qquad y'(a) = \eta' \tag{6-40}$$

then

$$\lim_{\substack{h \to 0 \\ x_n = x}} y_n = y(x) \tag{6-41}$$

holds for all $x \in [a, b]$, and for all sequences $\{y_n\}$ defined by (6-23) with starting values $y_\mu = \eta_\mu(h)$ satisfying the two conditions

$$\lim_{h \to 0} \eta_\mu(h) = \eta, \qquad \mu = 0, 1, \cdots, k - 1 \tag{6-42a}$$

$$\lim_{h \to 0} \frac{\eta_\mu(h) - \eta_0(h)}{\mu h} = \eta', \qquad \mu = 1, \cdots, k - 1 \tag{6-42b}$$

This definition differs from the corresponding definition for first order equations only in the addition of condition (6-42*b*), which guarantees that the starting values approximate correctly the initial slope of the desired solution. That the starting values are exact [i.e., that $y_\mu = y(x_\mu)$] is clearly sufficient, but not necessary, in order that (6-42*a*) and (6-42*b*) are satisfied.

6.1-5. Convergence: condition of stability. It is convenient to associate with the difference equation (6-23) or with the difference operator (6-24) the polynomials

$$\begin{aligned} \rho(\zeta) &= \alpha_k \zeta^k + \alpha_{k-1}\zeta^{k-1} + \cdots + \alpha_0 \\ \sigma(\zeta) &= \beta_k \zeta^k + \beta_{k-1}\zeta^{k-1} + \cdots + \beta_0 \end{aligned} \tag{6-43}$$

Conversely, with any two polynomials (6-43) with real coefficients and satisfying $\alpha_k \neq 0$, $|\alpha_0| + |\beta_0| \neq 0$ we may associate a difference operator (6-24).

THEOREM 6.1. *A necessary condition for the convergence of the linear multistep method defined by* (6-23) *is that the modulus of no root of the polynomial* $\rho(\zeta)$ *exceed* 1, *and that the multiplicity of the roots of modulus* 1 *be at most* 2.

The condition thus imposed on the roots of $\rho(\zeta)$ is again called the *condition of stability*.

The *proof* is similar to that of Theorem 5.3. We consider the initial value problem $y'' = 0$, $y(0) = y'(0) = 0$, whose exact solution vanishes identically. If ζ is a root of $\rho(\zeta)$, then $y_n = h^2 \operatorname{Re} \zeta^n$ defines a solution of the corresponding difference equation which also satisfies (6-42). If the method is convergent, we thus have $\lim_{n\to\infty} y_n = 0$. For ζ real it follows immediately that $|\zeta| \leq 1$. If ζ is complex, we eliminate $\bar{\zeta}^n$ from the two equations

$$\tfrac{1}{2}h^2(\zeta^n + \bar{\zeta}^n) = y_n \qquad \text{and} \qquad \tfrac{1}{2}h^2(\zeta^{n+1} + \bar{\zeta}^{n+1}) = y_{n+1}$$

finding

$$\tfrac{1}{2}h^2\zeta^n = \frac{y_n\bar{\zeta} - y_{n+1}}{\bar{\zeta} - \zeta}$$

The term on the right tends to zero for a convergent method. Therefore the term on the left must do the same, which is impossible unless $|\zeta| \leq 1$.

If ζ is a root on $\rho(\zeta)$ whose multiplicity exceeds 2, then $y_n = h^{3/2} \operatorname{Re} n^2\zeta^n$ defines a solution of the difference equation which again satisfies (6-42) for our problem. We thus have $\lim_{n\to\infty} y_n = 0$ if $nh = x > 0$ for a convergent method. If ζ is real, it immediately follows that $|\zeta| < 1$. If it is complex, we eliminate $\bar{\zeta}^n$ from the equations

$$\tfrac{1}{2}(n^2h^{3/2})(\zeta^n + \bar{\zeta}^n) = y_n \qquad \text{and} \qquad \tfrac{1}{2}[(n+1)^2h^{3/2}](\zeta^{n+1} + \bar{\zeta}^{n+1}) = y_{n+1}$$

finding

$$\zeta^n = 2\frac{z_n\bar{\zeta} - z_{n+1}}{\bar{\zeta} - \zeta}$$

where $z_n = n^{-2}h^{-3/2}y_n$ $(n = 1, 2, \cdots)$. Since $z_n = h^{1/2}\, x^{-2}y_n \to 0$ as $h \to 0$, $x_n = x > 0$, it follows that $|\zeta| < 1$, as desired.

It will be noted that the condition of stability is satisfied by all special methods based on integration which were considered in §6.1-1.

6.1-6. Convergence: condition of consistency. While the condition of stability guarantees that small initial disturbances are not unduly amplified, it is clearly not strong enough to insure convergence. A further condition must be added to the effect that the difference operator (6-24) is, locally at least, a good approximation to the differential equation.

THEOREM 6.2. *The order of a convergent linear multistep method* (6–23) *is at least* 1.

This *condition* of *consistency* demands that the constants C_i introduced in §6.1-2 satisfy $C_0 = C_1 = C_2 = 0$; in terms of the polynomials (6-43) this may be expressed in the form

$$\rho(1) = 0, \qquad \rho'(1) = 0, \qquad \rho''(1) = 2\sigma(1) \tag{6-44}$$

It will be noted that, by virtue of the condition of stability, $\rho''(1)$ and thus $\sigma(1)$ are different from zero for a convergent method.

Proof of Theorem 6.2. We shall show successively that the constants C_0, C_1, C_2 defined by (6-26) are zero for a convergent method.

We first consider the initial value problem $y'' = 0$, $y(0) = 1$, $y'(0) = 0$ with the exact solution $y(x) = 1$. The difference equation (6-23) now reads

(6-45) $$\alpha_k y_{n+k} + \alpha_{k-1} y_{n+k-1} + \cdots + \alpha_0 y_n = 0, \qquad n = 0, 1, 2, \cdots$$

Assuming exact starting values, (6-42) is satisfied, and we must have

$$\lim_{\substack{h \to 0 \\ nh = x}} y_n = 1$$

for every $x > 0$. Since the y_n defined by (6-45) do not depend on h, this is equivalent to

(6-46) $$\lim_{n \to \infty} y_n = 1$$

Letting $n \to \infty$ in (6-45), we obtain

(6-47) $$\alpha_k + \alpha_{k-1} + \cdots + \alpha_0 = 0$$

which is the same as $C_0 = 0$.

Next consider the problem $y'' = 0$, $y(0) = 0$, $y'(0) = 1$, whose exact solution is $y(x) = x$. Again the values y_n satisfy (6-45). Assuming the exact starting values $y_\mu = \mu h$ ($\mu = 0, 1, \cdots, k - 1$), it is easy to see that $y_n = hz_n$, where the sequence $\{z_n\}$ is the solution of (6-45) with the starting values $z_\mu = \mu$. This solution is independent of h. In view of the convergence of the method,

$$\lim_{\substack{h \to 0 \\ nh = x}} \frac{y_n}{x} = 1$$

for every $x > 0$. It follows that

(6-48) $$\lim_{n \to \infty} \frac{z_n}{n} = 1$$

Summing the equations

$$\alpha_k z_{m+k} + \alpha_{k-1} z_{m+k-1} + \cdots + \alpha_0 z_m = 0$$

for $m = 0, 1, \cdots, n$, we obtain in view of (6-47) the relation

(6-49)
$$\alpha_k z_{n+k} + (\alpha_k + \alpha_{k-1}) z_{n+k-1} + \cdots + (\alpha_k + \alpha_{k-1} + \cdots + \alpha_1) z_{n+1} = -D$$

where

$$D = z_0\alpha_0 + z_1(\alpha_1 + \alpha_0) + \cdots + z_{k-1}(\alpha_{k-1} + \alpha_{k-2} + \cdots + \alpha_0)$$

is independent of n. Dividing (6-49) by n and letting $n \to \infty$, we get, using (6-48),

$$\alpha_k + (\alpha_k + \alpha_{k-1}) + \cdots + (\alpha_k + \alpha_{k-1} + \cdots + \alpha_1) = 0$$

or

$$k\alpha_k + (k-1)\alpha_{k-1} + \cdots + 1\alpha_1 = 0$$

which is equivalent to $C_1 = 0$.

Consider finally the problem $y'' = 2$, $y(0) = y'(0) = 0$, with the exact solution $y(x) = x^2$. The difference equation (6-23) now reads

$$\alpha_k y_{n+k} + \alpha_{k-1} y_{n+k-1} + \cdots + \alpha_0 y_n = 2h^2(\beta_k + \cdots + \beta_0) \tag{6-50}$$

We have proved that, for a convergent method, $\rho(1) = \rho'(1) = 0$ and, in view of stability, $\rho''(1) \neq 0$. Hence,

$$k^2\alpha_k + (k-1)^2\alpha_{k-1} + \cdots + 1^2\alpha_1 = \rho''(1) + \rho'(1) \neq 0$$

It can be verified that

$$y_n = Kh^2n^2 \tag{6-51}$$

where

$$K = \frac{2(\beta_k + \beta_{k-1} + \cdots + \beta_0)}{k^2\alpha_k + (k-1)^2\alpha_{k-1} + \cdots + 1^2\alpha_1}$$

defines a solution of (6-50) that also satisfies (6-42) for $\eta = \eta' = 0$. Thus, in view of the convergence of the method,

$$Kx^2 = \lim_{\substack{h\to 0\\ nh = x}} y_n = x^2$$

for every $x > 0$. It follows that $K = 1$, which implies that $C_2 = 0$. This completes the proof of Theorem 6.2.

It is easily verified that all special methods described in §6.1-1 are consistent in the sense of Theorem 6.2.

6.1-7. The construction of operators of maximum order. In this section we shall show how to adjoin to a given polynomial $\rho(\zeta)$ satisfying $\rho(1) = \rho'(1) = 0$ a polynomial $\sigma(\zeta)$ such that the associated operator L defined by (6-24) has the greatest possible order. To this purpose we define the function

$$\varphi(\zeta) = (\log \zeta)^{-2}\rho(\zeta) - \sigma(\zeta) \tag{6-52}$$

made single-valued in the cut ζ-plane by giving $\log \zeta$ its principal value. Exactly as in §5.2-7, we may prove the following lemma:

LEMMA 6.1. *The difference operator* (6-24) *associated with the polynomials* $\rho(\zeta)$ *and* $\sigma(\zeta)$ *has the exact order* p *if and only if the function* $\varphi(\zeta)$ *has a zero of exact order* p *at* $\zeta = 1$.

The following analog of Theorem 5.7 now follows easily:

THEOREM 6.3. *Let* $\rho(\zeta)$ *be a polynomial of degree* k *satisfying* $\rho(1) = \rho'(1) = 0$. *For every integer* k' *such that* $0 \leq k' \leq k$ *there exists a unique polynomial* $\sigma(\zeta)$ *of degree* $\leq k'$ *such that the order of the associated operator* L *is at least* $k' + 1$.

Proof. In view of the double root of $\rho(\zeta)$ at $\zeta = 1$, the function $(\log \zeta)^{-2}\rho(\zeta)$ is holomorphic at $\zeta = 1$ and can therefore be expanded in a Taylor series. Let

$$\frac{\rho(\zeta)}{(\log \zeta)^2} = c_0 + c_1(\zeta - 1) + c_2(\zeta - 1)^2 + \cdots \tag{6-53}$$

By defining

$$\sigma(\zeta) = c_0 + c_1(\zeta - 1) + \cdots + c_{k'}(\zeta - 1)^{k'} \tag{6-54}$$

the function $\varphi(\zeta)$ achieves a zero of multiplicity $\geq(k' + 1)$ at $\zeta = 1$, and the associated operator has order $\geq(k' + 1)$ by Lemma 6.1. (The order *equals* $k' + 1$ unless $c_{k'+1} = 0$.) Conversely, if L has order $k' + 1$, then $\varphi(\zeta)$ has a zero of multiplicity $k' + 1$ at $\zeta = 1$, and since Taylor's expansion is unique this can be the case only if $\sigma(\zeta)$ is defined by (6-54).

Only the cases $k' = k - 1$ and $k' = k$ of Theorem 6.3 are of practical interest, the former leading to the best explicit, and the latter to the best implicit, method (unless it should happen that $c_k = 0$). The method of proof of Theorem 6.3 also leads to a simple formula for the error constant. If the difference operator (6-24) has order p, then by letting $y(x) = e^x$ in (6-27) we find that

$$\rho(e^h) - h^2\sigma(e^h) = C_{p+2}h^{p+2} + O(h^{p+3})$$

where $C_{p+2} \neq 0$. Replacing e^h by ζ and noting that $h = \log \zeta = \zeta - 1 + O((\zeta - 1)^2)$ as $h \to 0$, we find that

$$(\log \zeta)^{-2}\rho(\zeta) - \sigma(\zeta) = C_{p+2}(\zeta - 1)^p + O((\zeta - 1)^{p+1}) \tag{6-55}$$

as $\zeta \to 1$. Consequently, $C_{p+2} = c_p$, where c_p is the coefficient of the first nonvanishing term in the expansion (6-53) that has not been absorbed into $\sigma(\zeta)$. Since $\sigma(1) = c_0$, it thus follows that

$$C = c_p/c_0 \tag{6-56}$$

If a representation of the right-hand side of (6-23) in terms of backward differences of f_{n+k} is wanted, it is convenient to introduce the variable $t = (1 - \zeta^{-1})$. A difference $\nabla^q f_{n+k}$ then contributes the term $\zeta^k t^q$ to the polynomial $\sigma(\zeta)$. Defining the polynomials $R(t)$ and $S(t)$ by writing

$$\rho(\zeta) = \zeta^k R(t), \qquad \sigma(\zeta) = \zeta^k S(t) \tag{6-57}$$

we obtain from (6-55) in view of $\log \zeta = -\log(1 - t)$ and $\zeta - 1 = t(1 - t)^{-1} = t + O(t^2)$ the relation

$$\frac{R(t)}{[\log(1 - t)]^2} - S(t) = C_{p+2} t^p + O(t^{p+1}) \tag{6-58}$$

If $p > k$, the polynomial $S(t)$ is thus shown to be identical with the Taylor polynomial of order k of the function $[\log(1 - t)]^{-2} R(t)$ at $t = 0$, and C_{p+2} is the coefficient of the next nonvanishing term in this expansion.

For the sake of illustration, let $\rho(\zeta) = \zeta^k - 2\zeta^{k-1} + \zeta^{k-2}$. We find $R(t) = t^2$, and from the expansion

$$\frac{t^2}{[\log(1 - t)]^2} = \sigma_0^* + \sigma_1^* t + \sigma_2^* t^2 + \cdots$$

(see §6.1-1) recognize the coefficients of the polynomial $S(t)$ as the coefficients of the Cowell formula (6-10). In a similar manner the other methods discussed in §6.1-1 can be derived systematically from our present more general point of view.

6.1-8. The maximum order of a stable operator. The construction of the preceding section can be applied to any polynomial $\rho(\zeta)$ which is compatible with the condition of consistency, i.e., which satisfies $\rho(1) = \rho'(1) = 0$. By a judicious choice of $\rho(\zeta)$ one may hope, by counting the number of parameters involved, to push the order of the resulting operator as high as $2k$. This possibility does not exist, however, when $\rho(\zeta)$ is assumed to be stable. For such polynomials the maximum order of any associated operator is $k + 2$, if k is even, and $k + 1$, if k is odd, as we shall now show.

Let $\rho(\zeta)$ be a polynomial compatible with consistency and satisfying the condition of stability.† We again introduce the variable z defined by (5-111) and define the polynomials $r(z)$ and $s(z)$ by (5-112). Since $\zeta = 1$ is a root of $\rho(\zeta)$ of exact multiplicity 2, $z = 0$ is a root of exact multiplicity 2 of $r(z)$. We thus have

$$r(z) = a_2 z^2 + a_3 z^3 + \cdots + a_k z^k$$

† We actually need only the following slightly weaker form of the stability condition: No root of $\rho(\zeta)$ exceeds 1 in modulus, and the root at $\zeta = 1$ has exact multiplicity 2.

where $a_2 \neq 0$. We may assume that

(6-59a) $$a_2 > 0$$

without loss of generality. By considering the product expansion of $r(z)$ and using the fact that no root of $r(z)$ has a positive real part, it follows as in §5.2-8 that

(6-59b) $$a_\mu \geq 0, \qquad \mu = 3, \cdots, k$$

We now consider the function

(6-60) $$p(z) = \left(\frac{1-z}{2}\right)^k \varphi\left(\frac{1+z}{1-z}\right) = \left[\log\frac{1+z}{1-z}\right]^{-2} r(z) - s(z)$$

In view of $\zeta - 1 = 2z + O(z^2)$, $p(z)$ has a zero of order p at $z = 0$ if and only if $\varphi(\zeta)$ has a zero of the same order at $\zeta = 1$ and thus, by Lemma 6.1, if and only if the associated operator has order p. Letting

(6-61) $$\left[\frac{z}{\log\dfrac{1+z}{1-z}}\right]^2 \frac{r(z)}{z^2} = b_0 + b_1 z + b_2 z^2 + \cdots$$

it follows that if the operator has order p, then

(6-62) $$s(z) = b_0 + b_1 z + \cdots + b_{p-1} z^{p-1}$$

On the other hand, the degree of $s(z)$ is at most k. The order of the associated operator thus exceeds $k + 1$ if and only if $b_{k+1} = \cdots = b_{p-1} = 0$. Setting

(6-63) $$\left[\frac{z}{\log\dfrac{1+z}{1-z}}\right]^2 = d_0 + d_2 z^2 + d_4 z^4 + \cdots$$

we find from (6-61), defining $a_\nu = 0$ for $\nu > k$,

(6-64) $$\begin{aligned} b_0 &= d_0 a_2 \\ b_1 &= d_0 a_3 \\ b_{2\nu} &= d_0 a_{2\nu+2} + d_2 a_2 + \cdots + d_{2\nu} a_2 \\ b_{2\nu+1} &= d_0 a_{2\nu+3} + d_2 a_{2\nu+1} + \cdots + d_{2\nu} a_3, \qquad \nu = 1, 2, \cdots \end{aligned}$$

We shall prove below that

(6-65) $$d_{2\nu} < 0, \qquad \nu = 1, 2, \cdots$$

Assuming the truth of this inequality, we distinguish two cases.

(i) If k is odd, then it follows from (6-64), using (6-65) and (6-59), that

$$b_{k+1} = d_2 a_{k+1} + d_4 a_{k-1} + \cdots + d_{k+1} a_2 < 0$$

and $p > k + 1$ is impossible. Thus:

THEOREM 6.4. *The order of a stable operator whose stepnumber k is odd cannot exceed* $k + 1$.

The *existence* of an operator of order $k + 1$ is already implied by Theorem 6.3.

(ii) If k is even, then

$$b_{k+1} = d_4 a_{k-1} + d_6 a_{k-3} + \cdots + d_k a_3$$

In view of (6-59) and (6-65) a necessary and sufficient condition for $b_{k+1} = 0$ is $a_3 = a_5 = \cdots = a_{k-1} = 0$. This is the case if and only if $r(z)$ is even, $r(z) = r(-z)$. Since $r(z)$ cannot have any roots with a positive real part in view of the stability assumption, there cannot be any root with a negative real part. It follows that all roots of $r(z)$ are purely imaginary, and that all roots of $\rho(\zeta)$ have modulus 1. In view of (6-59) and (6-65),

$$b_{k+2} = d_4 a_k + d_6 a_{k-2} + \cdots + d_{k+2} a_2 < 0$$

so that the order cannot exceed $k + 2$. We thus have proved:

THEOREM 6.5. *The order p of a stable operator cannot exceed* $k + 2$. *A necessary and sufficient condition for* $p = k + 2$ *is that k be even, that all roots of* $\rho(\zeta)$ *have modulus* 1, *and that* $\sigma(\zeta)$ *be determined by* (6-54).

Again we may call an operator satisfying the condition of Theorem 6.5 an *optimal* operator. The method of proof of Theorem 6.5 also yields some representations and inequalities for the error constant of an optimal operator. From (6-55) we find in view of $\zeta - 1 = 2z + O(z^2)$

$$p(z) = 2^{p-k} C_{p+2} z^p + O(z^{p+1})$$

If $p > k$, it thus follows that $C_{p+1} = 2^{k-p} b_p$. Using $\sigma(1) = 2^k s(0) = 2^k b_0$, we get

$$C = b_p / 2^p b_0 \tag{6-66}$$

For an optimal operator, $p = k + 2$, and thus, using (6-64),

$$C = \frac{d_4 a_k + d_6 a_{k-2} + \cdots + d_{2k+2} a_2}{2^{k+2} d_0 a_2} \tag{6-67}$$

In view of (6-59) and (6-65), and using the numerical values $d_0 = \frac{1}{4}$, $d_4 = -\frac{1}{60}$, we easily find the following inequalities for C:

$$C \le -\frac{1}{2^k \cdot 60} \frac{a_k}{a_2} \tag{6-68}$$

$$C \le 2^{-k} d_{k+2} \tag{6-69}$$

Equality is attained in both cases for Cowell's method (6-14).

In order to find a recurrence relation for the coefficients $d_{2\nu}$, and also for the proof of (6-65), we shall calculate the coefficients A_ν in the expansion of

$$\left[\frac{1}{2}\log\frac{1+z}{1-z}\right]^2 = A_0 z^2 + A_1 z^4 + A_2 z^6 + \cdots$$

By differentiating both sides, we find

$$2A_0 z + 4A_1 z^3 + 6A_2 z^5 + \cdots = \frac{1}{1-z^2}\log\frac{1+z}{1-z}$$
$$= 2(1 + z^2 + z^4 + \cdots)(z + \tfrac{1}{3}z^3 + \tfrac{1}{5}z^5 + \cdots)$$

Comparing the coefficients of $z^{2\nu+1}$ on both sides, we readily find

(6-70) $$A_\nu = \frac{1}{\nu+1}k_\nu, \qquad \nu = 0, 1, 2, \cdots$$

where

$$k_\nu = 1 + \tfrac{1}{3} + \cdots + \frac{1}{2\nu+1}$$

From

$$\left(d_0 + d_2 z^2 + \cdots\right) z^{-2}\left[\log\frac{1+z}{1-z}\right]^2 = 1$$

we have the recurrence relation $d_0 = \frac{1}{4}$,

$$d_{2\nu} = -\tfrac{1}{2}k_1 d_{2\nu-2} - \tfrac{1}{3}k_2 d_{2\nu-4} - \cdots - \frac{1}{\nu+1}k_\nu d_0, \qquad \nu = 1, 2, \cdots$$

from which the numerical values in Table 6.7 are readily calculated.

Table 6.7

Values of the Coefficients $d_{2\nu}$ in (6-63)

ν	0	1	2	3	4
$d_{2\nu}$	$\frac{1}{4}$	$-\frac{1}{6}$	$-\frac{1}{60}$	$-\frac{8}{945}$	$-\frac{76}{14175}$

For the proof of (6-65) we again apply Kaluza's Lemma 5.4, this time to the function

$$f(t) = \left[\frac{1}{2z}\log\frac{1+z}{1-z}\right]^2 = A_0 + A_1 t + A_2 t^2 + \cdots$$

where $t = z^2$. It is clear that $A_\nu > 0$. In order to show that $\Delta_\nu = A_{\nu+1}A_{\nu-1} - A_\nu^2 > 0$ we note that by (6-70), for $\nu \geq 2$,

$$\nu^2(\nu^2 - 1)\Delta_{\nu-1}^2 = k_{\nu-1}^2 - \frac{\nu^2 - 1}{4\nu^2 - 1}(2k_{\nu-1} - 1)$$

$$\geq k_{\nu-1}^2 - \tfrac{1}{2}k_{\nu-1} + \tfrac{4}{15}$$

The last expression is positive for $k_\nu \geq 1$, so that $\Delta_{\nu-1} > 0$. The hypotheses of Kaluza's lemma are thus satisfied, and the conclusion holds for $C_\nu = 4d_{2\nu}$, $\nu = 1, 2, \cdots$.

It is interesting to note that the analog of (6-65) for exponents >2 is not true, so that the results of this section cannot immediately be generalized to operators for differential equations of the form $y^{(\mu)} = f(x, y)$ with $\mu > 2$.

6.1-9. The construction of optimal operators. It has been stated that the polynomial $r(z)$ associated with an optimal method is even. It follows from (6-61) that $s(z)$ is also an even polynomial. By (5-112) we conclude that $\rho(\zeta) = \zeta^k\rho(\zeta^{-1})$, $\sigma(\zeta) = \zeta^k\sigma(\zeta^{-1})$, or that

$$\alpha_{k-\nu} = \alpha_\nu, \qquad \beta_{k-\nu} = \beta_\nu, \qquad \nu = 0, 1, \cdots, k$$

It follows that both $\rho(\zeta)$ and $\sigma(\zeta)$ can be written as $\zeta^{\frac{1}{2}k}$ times certain polynomials of degree $\frac{1}{2}k$ in the variable $\zeta + \zeta^{-1} - 2$. Furthermore, since $\zeta = 1$ is a double root of $\rho(\zeta)$, the polynomial representing $\rho(\zeta)$ contains $\zeta + \zeta^{-1} - 2$ as a factor. Letting

$$\zeta + \zeta^{-1} - 2 = t^2$$

we thus can find two polynomials P and Q of degrees $\frac{1}{2}k - 1$ and $\frac{1}{2}k$, respectively, such that

$$\rho(\zeta) = \zeta^{\frac{1}{2}k}t^2P(t^2), \qquad \sigma(\zeta) = \zeta^{\frac{1}{2}k}Q(t^2)$$

In view of

$$\zeta = (\tfrac{1}{2}t + \sqrt{1 + \tfrac{1}{4}t^2})^2$$

where the square root is defined as in §5.2-9, we have $\zeta - 1 = t + O(t^2)$ and

$$\log \zeta = 2 \log (\tfrac{1}{2}t + \sqrt{1 + \tfrac{1}{4}t^2})$$

Relation (6-55) is thus equivalent to

$$\left\{\frac{\frac{1}{2}t}{\log (\frac{1}{2}t + \sqrt{1 + \frac{1}{4}t^2})}\right\}^2 P(t^2) - Q(t^2) = C_{k+4}t^{k+2} + O(t^{k+4})$$

The first term on the left represents an even function holomorphic at $t = 0$. Setting

$$\left\{\frac{\frac{1}{2}t}{\log\left(\frac{1}{2}t + \sqrt{1 + \frac{1}{4}t^2}\right)}\right\}^2 [P(t^2) = q_0 + q_2 t^2 + q_4 t^4 + \cdots] \tag{6-71}$$

it follows from the uniqueness of Taylor's expansion that

$$Q(t^2) = q_0 + q_2 t^2 + \cdots + q_k t^k \tag{6-72}$$

Furthermore, in view of $\sigma(1) = Q(0) = q_0$,

$$C = q_{k+2}/q_0 \tag{6-73}$$

The coefficients $q_{2\nu}$ can be calculated by multiplying the polynomial

$$P(t^2) = p_0 + p_2 t^2 + \cdots + p_{k-2} t^{k-2}$$

by the Taylor series

$$\left\{\frac{\frac{1}{2}t}{\log\left(\frac{1}{2}t + \sqrt{1 + \frac{1}{4}t^2}\right)}\right\}^2 = l_0 + l_2 t^2 + l_4 t^4 + \cdots \tag{6-74}$$

It follows from

$$\frac{\log\left(\frac{1}{2}t + \sqrt{1 + \frac{1}{4}t^2}\right)}{\frac{1}{2}t} = 1 + \frac{1}{3}\binom{-\frac{1}{2}}{1}\left(\frac{t}{2}\right)^2 + \frac{1}{5}\binom{-\frac{1}{2}}{2}\left(\frac{t}{2}\right)^4 + \cdots$$

that the coefficients $l_{2\nu}$ satisfy the recurrence relation $l_0 = 1$,

$$nl_{2n} = -\frac{n+1}{3 \cdot 2^2}\binom{-\frac{1}{2}}{1} l_{2n-2} - \frac{n+2}{5 \cdot 2^4}\binom{-\frac{1}{2}}{2} l_{2n-4} - \cdots - \frac{2n}{(2n+3)2^{2n}}\binom{-\frac{1}{2}}{n} l_0, \qquad n = 1, 2, \cdots$$

The values in Table 6.8 are now readily obtained.

Table 6.8

The Coefficients $l_{2\nu}$ in (6-74)

ν	0	1	2	3	4
$l_{2\nu}$	1	$\frac{1}{12}$	$-\frac{1}{240}$	$\frac{31}{60480}$	$\frac{289}{3628800}$

Examples. (i) ***Optimal method for*** $k = 2$. We have $P(t^2) = 1$ and consequently $q_{2\nu} = 2^{-2\nu} l_{2\nu}$, $Q(t^2) = 1 + \frac{1}{12}t^2$. There results the Cowell formula (6-14), and the error constant has the value $C = l_4 = -\frac{1}{240}$, in agreement with previous calculations.

(ii) ***The general optimal method for*** $k = 4$. Assuming the nontrivial roots of $\rho(\zeta)$ at the points $\zeta = e^{\pm i\varphi}$, we have

$$\rho(\zeta) = \zeta^2\left(\zeta + \frac{1}{\zeta} - 2\right)\left(\zeta + \frac{1}{\zeta} - 2\cos\varphi\right) \tag{6-75}$$

and hence $P(t^2) = 4\lambda + t^2$, where $\lambda = \sin^2 \frac{1}{2}\varphi$. We thus obtain

$$\left\{\frac{\frac{1}{2}t}{\log\left(\frac{1}{2}t + \sqrt{1 + \frac{1}{4}t^2}\right)}\right\}^2 P(t^2) = 4\lambda + \left(1 + \frac{\lambda}{3}\right)t^2 + \left(\frac{1}{12} - \frac{\lambda}{60}\right)t^4 + \left(-\frac{1}{240} + \frac{31\lambda}{15120}\right)t^6 + \cdots$$

Consequently,

$$Q(t^2) = 4\lambda + \left(1 + \frac{\lambda}{3}\right)t^2 + \left(\frac{1}{12} - \frac{\lambda}{60}\right)t^4$$

$$C = \frac{31}{15120} - \frac{1}{960\lambda}$$

For $\lambda = 1$, $\varphi = \pi$, we again get (6-22). For $\lambda = \frac{3}{4}$ and $\lambda = \frac{1}{2}$ we obtain the new formulas

$$y_{n+4} - y_{n+3} - y_{n+1} + y_n = h^2\{3f_{n+2} + \tfrac{5}{4}\nabla^2 f_{n+3} + \tfrac{17}{240}\nabla^4 f_{n+4}\} \tag{6-76}$$

$$y_{n+4} - 2y_{n+3} + 2y_{n+2} - 2y_{n+1} + y_n = h^2\{2f_{n+2} + \tfrac{7}{6}\nabla^2 f_{n+3} + \tfrac{3}{40}\nabla^4 f_{n+4}\} \tag{6-77}$$

For $\lambda = 1$ the error constant achieves the upper bound (6-69).

(iii) ***A special method with*** $k = 6$. The polynomial $\rho(\zeta) = \zeta^6 - \zeta^4 - \zeta^2 + 1$ has its nontrivial roots at $-1, -1, +i, -i$ and thus satisfies the condition of Theorem 6.5. We have

$$\rho(\zeta) = \zeta^3(\zeta + \zeta^{-1} - 2)(\zeta + \zeta^{-1})(\zeta + \zeta^{-1} + 2)$$

and thus

$$P(t^2) = 8 + 6t^2 + t^4$$

Our algorithm yields

$$Q(t^2) = 8 + \tfrac{20}{3}t^2 + \tfrac{22}{15}t^4 + \tfrac{59}{945}t^6$$

thus giving rise to the formula of order 8:

$$y_{n+6} - y_{n+4} - y_{n+2} + y_n = h^2\{8f_{n+3} + \tfrac{20}{3}\nabla^2 f_{n+4} + \tfrac{22}{15}\nabla^4 f_{n+5} + \tfrac{59}{945}\nabla^6 f_{n+6}\} \tag{6-78}$$

6.2. The Discretization Error

6.2-1. Two lemmas. We shall state and prove in this section two lemmas which are analogous to Lemmas 5.5 and 5.6. The first lemma will be needed in the proof of the second, and the second lemma is the main tool for obtaining the a priori bound for the discretization error to be derived in the next section.

LEMMA 6.2. *Let the polynomial* $\rho(\zeta) = \alpha_k\zeta^k + \cdots + \alpha_0$ *satisfy the condition of stability for the integration of second order equations, and let the coefficients* γ_l $(l = 0, 1, 2, \cdots)$ *be defined by*

$$\frac{1}{\alpha_k + \alpha_{k-1}\zeta + \cdots + \alpha_0\zeta^k} = \gamma_0 + \gamma_1\zeta + \gamma_2\zeta^2 + \cdots \tag{6-79}$$

Then there exist two constants Γ *and* γ *such that*

$$|\gamma_l| \le l\Gamma + \gamma, \qquad l = 0, 1, 2, \cdots \tag{6-80}$$

Proof. Let $\alpha_k + \alpha_{k-1}\zeta + \cdots + \alpha_0\zeta^k = \hat{\rho}(\zeta)$. We have $\hat{\rho}(\zeta) = \zeta^k\rho(\zeta^{-1})$. Since $\rho(\zeta)$ has no roots outside $|\zeta| = 1$, and since the roots on $|\zeta| = 1$ have at most multiplicity 2, the polynomial $\hat{\rho}(\zeta)$ has no roots inside $|\zeta| = 1$, and its roots on $|\zeta| = 1$ have at most multiplicity 2.

If the double roots of $\rho(\zeta)$ on $|\zeta| = 1$ are denoted by $\zeta_2, \zeta_4, \cdots, \zeta_{2d}$, then for a suitable choice of the constants $A_1, \cdots, A_d$ the function

$$f(\zeta) = \frac{1}{\hat{\rho}(\zeta)} - \frac{A_1}{(\zeta - \zeta_2^{-1})^2} - \cdots - \frac{A_d}{(\zeta - \zeta_d^{-1})^2} \tag{6-81}$$

is holomorphic for $|\zeta| < 1$ and has at most a finite number of simple poles on $|\zeta| = 1$. By Lemma 5.5 the coefficients of its Taylor expansion at $\zeta = 0$ are bounded. In view of

$$\frac{1}{(\zeta - \zeta_\mu^{-1})^2} = \frac{\zeta_\mu^2}{(1 - \zeta_\mu\zeta)^2} = \zeta_\mu^2\{1 + 2\zeta_\mu\zeta + 3(\zeta_\mu\zeta)^2 + \cdots\} \tag{6-82}$$

the Taylor coefficients at $\zeta = 0$ of each of the terms on the right of (6-81) satisfy an inequality of the form (6-80). It follows that the same is true for the function

$$\frac{1}{\hat{\rho}(\zeta)} = f(\zeta) + \sum_{\mu=1}^{d} \frac{A_\mu}{(\zeta - \zeta_{2\mu}^{-1})^2} \tag{6-83}$$

The following examples show that values of Γ and γ are easy to find in concrete cases. For the Störmer and the Cowell method we have $\rho(\zeta) = \zeta^k(1 - \zeta^{-1})^2$, hence $\hat{\rho}(\zeta) = (1 - \zeta)^2$. It follows by (6-82) that

(6-80) is true with $\Gamma = 1$, $\gamma = 1$. For the "double step" methods (6-15) and (6-19) we have $\rho(\zeta) = \zeta^k(1 - \zeta^{-2})^2$ and $\hat{\rho}(\zeta) = (1 - \zeta^2)^2$. It follows that (6-80) holds with $\Gamma = \frac{1}{2}$, $\gamma = 1$.

The following lemma has to do with the growth of the solutions of the difference equation

$$\text{(6-84)} \quad \alpha_k z_{m+k} + \alpha_{k-1} z_{m+k-1} + \cdots + \alpha_0 z_m = h^2\{\beta_{k,m} z_{m+k} + \cdots + \beta_{0,m} z_m\} + \lambda_m$$

LEMMA 6.3. *Let the polynomial* $\rho(\zeta) = \alpha_k\zeta^k + \cdots + \alpha_0$ *satisfy the condition of stability (for the integration of second order equations), let* B^*, β, *and* Λ *be constants such that*

$$\text{(6-85)} \quad |\beta_{k,m}| + |\beta_{k-1,m}| + \cdots + |\beta_{0,m}| \le B^*, \qquad |\beta_{k,m}| \le \beta, \qquad |\lambda_m| \le \Lambda, \qquad 0 \le m \le N$$

and let $0 \le h^2 < |\alpha_k|\beta^{-1}$. *Then every solution of* (6-84) *for which*

$$\text{(6-86)} \qquad |z_\mu| \le Z, \qquad \mu = 0, 1, \cdots, k-1$$

satisfies

$$\text{(6-87)} \qquad |z_n| \le K^* e^{nh^2L^*}, \qquad 0 \le n \le N$$

Here

$$\text{(6-88)} \quad L^* = \frac{(N\Gamma + \gamma)B^*}{1 - h^2\,|\alpha_k^{-1}|\,\beta}, \qquad K^* = \frac{(\frac{1}{2}N^2\Gamma + N\gamma)\Lambda + (N\Gamma + \gamma)kAZ}{1 - h^2\,|\alpha_k^{-1}|\,\beta}$$

where $A = |\alpha_k| + |\alpha_{k-1}| + \cdots + |\alpha_0|$, *and* Γ *and* γ *are defined by* (6-80).

The *proof* is analogous to the proof of Lemma 5.6. For $l = 0, 1, \cdots, n-k$, multiply the equation (6-84) corresponding to $m = n - k - l$ by γ_l and add the resulting equations. We obtain on the left z_n plus some leftover terms from $m = 0, 1, \cdots, k-1$ which by (6-86) are bounded by $kAZ(N\Gamma + \gamma)$. On the right we have a contribution $\beta_{k,n-k}\gamma_0 z_n$ from $l = 0$, a sum bounded by

$$(N\Gamma + \gamma)B\sum_{m=0}^{n-1}|z_m|$$

and a weighted sum of values of λ_m for which by (6-80) and (6-85) we have the estimate

$$\left|\sum_{l=0}^{n-k}\gamma_l\lambda_{n-k-l}\right| \le \Lambda\sum_{l=0}^{n-k}(l\Gamma + \gamma) \le \Lambda(\tfrac{1}{2}N^2\Gamma + N\gamma)$$

Solving the resulting inequality for $|z_n|$, we find

$$|z_n| \le h^2L^*\sum_{m=0}^{n-1}|z_m| + K^*$$

By induction we now find as in §5.3-2 the estimate

$$|z_n| \leq K^*(1 + h^2 L^*)^n$$

of which (6-87) is an easy consequence.

6.2-2. Sufficient conditions for convergence; a priori bounds. As in the case of multistep methods for first order equations, the conditions of stability and consistency, which earlier were seen to be necessary for convergence, are also sufficient. We thus have:

THEOREM 6.6. *The linear multistep method defined by* (6-23) *is convergent if and only if it satisfies the conditions of stability and consistency.*

The "only if" part of the assertion was the content of Theorems 6.1 and 6.2 and has already been proved. The proof of the "if" part can be based on Lemma 6.3 exactly as the proof of Theorem 5.10 was based on Lemma 5.6. Details are omitted.

We now shall discuss a bound for the discretization error under the assumption that the exact solution $y(x)$ has a continuous derivative of order $p + 2$ for $x \in [a, b]$. We shall put

$$Y = \max_{a \leq x \leq b} |y^{(p+2)}(x)| \tag{6-89}$$

The numerical values y_n are only assumed to satisfy the relation

$$\alpha_k y_{m+k} + \cdots + \alpha_0 y_m = h^2(\beta_k f_{m+k} + \cdots + \beta_0 f_m) + \theta_m K h^{q+2} \tag{6-90}$$

where $|\theta_m| \leq 1$, and K and q are constants, $q > 0$. The term involving K is introduced to allow for small local errors, and also because it is needed later in the study of the asymptotic behavior of the numerical values. As to the starting errors, we assume that

$$|y_\mu - y(x_\mu)| \leq h\delta, \qquad \mu = 0, 1, \cdots, k - 1 \tag{6-91}$$

We also set

$$A = |\alpha_k| + \cdots + |\alpha_0|, \qquad B = |\beta_k| + \cdots + |\beta_0|$$

and denoting by Γ and γ the constants introduced in §6.2-1,

$$\Gamma^* = \frac{\Gamma}{1 - h^2 L\, |\alpha_k^{-1}\beta_k|}, \qquad a^* = a - \frac{h\gamma}{\Gamma}$$

Finally, we shall denote by G the constant defined by (6-38). We then can prove:

THEOREM 6.7. *Under the above conditions, if* $h^2 < L^{-1}|\alpha_k \beta_k^{-1}|$, *the discretization error* $e_n = y_n - y(x_n)$ *satisfies for* $a \leq x_n \leq b$

$$|e_n| \leq \Gamma^* \left[(x_n - a^*)kA\delta + \frac{(x_n - a^*)^2}{2}(Kh^q + GYh^p) \right] \exp\left[(x_n - a^*)^2 \Gamma^* LB\right] \tag{6-92}$$

Proof. Substituting the exact solution $y(x)$ into the difference operator (6-24), we find

$$\text{(6-93)}\quad \alpha_k y(x_m + kh) + \cdots + \alpha_0 y(x_m) = h^2\{\beta_k y''(x_m + kh) + \cdots + \beta_0 y''(x_m)\} + R_m$$

where, by the results of §6.1-2,

$$|R_m| \le GYh^{p+2}$$

Defining the numbers g_m $(m = 0, 1, 2, \cdots)$ by

$$g_m = \begin{cases} \dfrac{f(x_m, y_m) - f(x_m, y(x_m))}{y_m - y(x_m)}, & \text{if } y_m \ne y(x_m) \\ 0, & \text{if } y_m = y(x_m) \end{cases}$$

we obtain by subtracting (6-93) from (6-90)

$$\alpha_k e_{m+k} + \cdots + \alpha_0 e_m = h^2\{\beta_k g_{m+k} e_{m+k} + \cdots + \beta_0 g_m e_m\} + \theta'_m(Kh^{q+2} + GYh^{p+2})$$

To this difference equation we apply Lemma 6.3, setting $z_m = e_m$, $Z = h\delta$, $N = (x_n - a)/h$, $\Lambda = Kh^{q+2} + GYh^{p+2}$, $\beta_{\mu,m} = g_{m+\mu}\beta_\mu$. Since $|g_m| \le L$ in view of the Lipschitz condition, we have

$$B^* = LB, \qquad \beta = L\,|\beta_k|$$

The inequality of the theorem now follows by virtue of (6-87) and the relations

$$N\Gamma + \gamma = (x_n - a^*)\Gamma h^{-1}$$

$$\tfrac{1}{2}N^2\Gamma + N\gamma = \tfrac{1}{2}(x_n - a)^2\Gamma h^{-2} + (x_n - a)\gamma h^{-1} \le \tfrac{1}{2}(x_n - a^*)^2\Gamma h^{-2}$$

As in the first order case [cf. (5-180)], the estimate (6-92) exhibits very clearly the influence of starting error, local inaccuracies, and local discretization error on the accumulated error. In each case this influence is by a factor $1/h$ stronger than in the first order case.

The reader is urged to write down the estimate (6-92) for some concrete special methods, rendering numerically evident its dependence on Y and L.

6.2-3. Asymptotic formulas for the discretization error. For the study of the asymptotic behavior of the error as $h \to 0$ we shall assume that the following conditions are satisfied in addition to those stated at the beginning of §6.2-2:

(i) $y^{(p+3)}(x)$ is continuous in $[a, b]$;
(ii) the function $g(x) = f_y(x, y(x))$ is continuously differentiable in $[a, b]$.

In order to fix the numerical solution for each value of h, the starting values have to be specified functions of h. We shall write

$$e_\mu = \delta_\mu(h), \qquad \mu = 0, 1, \cdots, k-1 \tag{6-94}$$

and shall assume that the functions $\delta_\mu(h)$ are

(iii) p times differentiable at $h = 0$, and
(iv) $O(h^{q+1})$, where q is a positive integer.

Under the above hypotheses we shall study the behavior of the exact solution of the difference equation (6-23), so that $K = 0$ in (6-90). Some notations have to be introduced in order to formulate the result.

Let $\zeta_1, \zeta_2, \cdots, \zeta_k$ denote the roots of the polynomial $\rho(\zeta)$. Assuming that there are $2d$ double roots of modulus 1 (each counted twice), we number these roots so that $\zeta_1 = \zeta_2 = 1, \zeta_3 = \zeta_4, \cdots, \zeta_{2d-1} = \zeta_{2d}$. These roots are called *essential roots*. The roots ζ_ν with $2d < \nu \le k$ (if any) are either simple or have modulus <1. For $\nu = 1, 2, \cdots, 2d$ we define the polynomials

$$\rho_\nu(\zeta) = \frac{\rho(\zeta)}{\zeta - \zeta_\nu} = \alpha_{\nu,0} + \alpha_{\nu,1}\zeta + \cdots + \alpha_{\nu,k-1}\zeta^{k-1} \tag{6-95}$$

and set, if $r = \min(p, q)$,

$$\Delta_\nu = \lim_{h\to 0} h^{-r-1}\{\alpha_{\nu,0}\,\delta_0(h) + \alpha_{\nu,1}\,\delta(h) + \cdots + \alpha_{\nu,k-1}\,\delta_{k-1}(h)\} \tag{6-96}$$

Clearly, $\Delta_\nu = 0$ if $q > p$.

The function $e(x)$ is defined as the solution of the initial value problems

$$\begin{aligned} e''(x) &= g(x)e(x) - Cy^{(p+2)}(x) \\ e(a) &= e'(a) = 0 \end{aligned} \tag{6-97}$$

where C is the error constant of the method as defined by (6-28). With each essential root we associate the growth parameter

$$\mu_\nu^2 = \frac{2\sigma(\zeta_\nu)}{\zeta_\nu^2\rho''(\zeta_\nu)}, \qquad \nu = 1, 2, \cdots, 2d \tag{6-98}$$

The functions $s_k(x)$ are for $\kappa = 1, 2, \cdots, d$ defined as solutions of the initial value problems

$$\begin{aligned} s_\kappa''(x) &= \mu_{2\kappa}^2 g(x)s_\kappa(x) \\ s_\kappa(a) &= 0, \qquad s_\kappa'(a) = 1 \end{aligned} \tag{6-99}$$

With these notations and definitions we have the following theorem:

THEOREM 6.8. *Under the conditions stated at the beginning of* §6.2-3, *the asymptotic behavior of the discretization error in the solution of the initial value problem* (6-40) *by the method* (6-23) *as* $h \to 0$, $nh = x - a$, $a < x \leq b$ *is described by the formula*

$$e_n = h^p e(x) + h^r \sum_{\kappa=1}^{d} \frac{2\Delta_{2\kappa}}{\rho''(\zeta_{2\kappa})} \zeta_{2\kappa}^n s_\kappa(x) + O(h^{r+1}) \tag{6-100}$$

We omit the proof, which is similar to the proof of Theorem 5.12.

If $p \neq q$, only one of the terms on the right of (6-100) is significant. If $p = q = r$, we may distinguish between *genuine discretization error* [represented by the term involving $e(x)$] and *starting error* [represented by the terms involving the functions $s_\kappa(x)$]. If $g(x) < 0$, the function $e(x)$ will usually grow at most like x as $x \to \infty$. On the other hand, if $\mu_{2\kappa}^2 < 0$ for some κ, the corresponding function $s_\kappa(x)$ grows exponentially. This is another example of that irregular error growth which has been called *conditional stability*. Since $\mu_2^2 = 1$ by virtue of consistency, conditional stability in this sense can be caused only by double essential roots different from 1. A glance at §6.1-7 shows that now, contrary to the first order case, there exist methods of maximal order $k + 2$ for which $\zeta = 1$ is the only essential root and which are therefore free of conditional stability.

The essential root $\zeta = 1$ may give rise, however, to a different kind of instability. For the sake of concreteness, consider the initial value problem $y'' = y$, $y(0) = 1$, $y'(0) = -1$. This has the exponentially decreasing solution $y(x) = e^{-x}$. On the other hand, since $s_1(x) = \sinh x$, the starting error always contains an exponentially increasing component. This type of instability, however, is not due to an imperfection of the method, but is already inherent in the mathematical problem. If for the same differential equation the initial conditions are changed to $y(0) = 1 + 2\delta$, $y'(0) = -1$, where δ is arbitrarily small, the solution is changed into $y(x) = (1 + \delta)e^{-x} + \delta e^x$ which tends to infinity as $x \to \infty$. The character of the exact solution has thus been radically changed by an arbitrarily small change in the initial condition, and the numerical method in this case does nothing but correctly reflect a property of the mathematical problem. The phenomenon thus described may be called *mathematical instability.**

On the basis of Theorem 6.8 one now may go on to derive conclusions which are entirely analogous to those of §§5.3-7 and 5.3-8. No new points of view occur in this analysis, and the details are left to the reader.

* The question of how to deal effectively with mathematically unstable problems is considered by Fox and Mitchell [1957].

6.3. Propagation of Round-off Error

6.3-1. The a priori bound. According to our established procedure, we write the equation satisfied by the numerical approximation $\{\tilde{y}_n\}$ in the form

$$(6\text{-}101)\quad \alpha_k\tilde{y}_{n+k} + \alpha_{k-1}\tilde{y}_{n+k-1} + \cdots + \alpha_0\tilde{y}_n = h^2\{\beta_k f(x_{n+k}, \tilde{y}_{n+k}) + \cdots + \beta_0 f(x_n, \tilde{y}_n)\} + \varepsilon_{n+k}, \quad n = 0, 1, 2\cdots$$

The *local round-off errors* ε_n depend on the computational procedure and on the organization of the arithmetic unit of the computing equipment. We shall try in this section to derive statements on the (accumulated) round-off error $r_n = \tilde{y}_n - y_n$ under gradually tightening assumptions on ε_n. We begin by assuming only $|\varepsilon_n| \le \varepsilon$, where ε is independent of n. Subtracting from (6-101) the corresponding equation (6-23) and letting

$$g_m = r_m^{-1}[f(x_m, \tilde{y}_m) - f(x_m, y_m)],\ r_m \ne 0; \qquad g_m = 0,\ r_m = 0$$

we find

$$(6\text{-}102)\quad \alpha_k r_{n+k} + \cdots + \alpha_0 r_n = h^2\{\beta_k g_{n+k} r_{n+k} + \cdots + \beta_0 g_n r_n\} + \varepsilon_{n+k}$$

To this relation we apply Lemma 6.3, setting $z_m = r_m$, $Z = 0$ (no initial round-off error), $\Lambda = \varepsilon$, $N = (x_n - a)/h$, $B^* = L\Sigma\,|\beta_\mu|$, $\beta = L\,|\beta_k|$. Assuming that $h^2 < L^{-1}\,|\beta_k^{-1}\alpha_k|$, we find that for $a \le x_n \le b$,

$$(6\text{-}103)\qquad |r_n| \le \varepsilon h^{-2}\Gamma^*(x_n - a^*)^2 \exp\,[(x_n - a^*)^2\Gamma^* LB]$$

where the constants Γ^*, a^*, and B are defined as in Theorem 6.7.

The conspicuous feature of (6-103) consists in the fact that the bound for $|r_n|$ is now $O(\varepsilon/h^2)$ in place of the $O(\varepsilon/h)$ which has been obtained so far for all methods integrating *first* order equations. Although (6-103) represents merely a very pessimistic *bound*, with no certainty of ever being attained, the subsequent investigations will show that the method (6-23) is indeed more sensitive to round-off than the corresponding methods for equations of the first order.

6.3-2. An a posteriori bound. We shall now derive a bound for $|r_n|$ which is (asymptotically) sharp at least in the case where all $|\varepsilon_m|$ have a specified maximum value. A passing to the limit $h \to 0$ is now implied, and we have to assume that $Nh^{p+2} \le \varepsilon \le Kh^3$ (note the change in the exponent). By Theorem 6.7 we then can assert that $r_m = O(h)$. We also assume that $f_{yy}(x, y)$ exists and is continuous in a neighborhood of the exact solution $y = y(x)$. If h is sufficiently small, we then have

$$f(x_m, \tilde{y}_m) - f(x_m, y_m) = g(x_m)r_m + \theta_m K_1 \varepsilon h^{-1}$$

where $g(x) = f_y(x, y(x))$, and we can thus replace (6-102) by

$$\text{(6-104)} \quad \begin{aligned} &\alpha_k r_{n+k} + \cdots + \alpha_0 r_n \\ &\quad = h^2\{\beta_k g_{n+k} r_{n+k} + \cdots + \beta_0 g_n r_n\} + \varepsilon_{n+k} + \theta_n K_2 h\varepsilon \end{aligned}$$

where the meaning of g_m is now different: $g_m = g(x_m)$. Accordingly, we split r_n into the sum $r_n^{(1)} + r_n^{(2)}$, where $r_n^{(1)}$ is called the *primary error* and defined as the solution of (6-104) for $\theta_n \equiv 0$, and where $r_n^{(2)}$, the *secondary error*, is defined as the solution for $\varepsilon_n \equiv 0$, $\theta_n \neq 0$. By an application of Lemma 6.3 we can derive a bound for $r_n^{(2)}$ which shows that $r_n^{(2)} = O(\varepsilon h^{-1})$. The primary error, on the other hand, must be expected to be $O(\varepsilon h^{-2})$. The behavior of r_n as $h \to 0$ is thus determined by the behavior of the primary error. For this reason, the remainder of this section will be concerned exclusively with the primary error.

The primary error is defined as the solution of a linear, nonhomogeneous difference equation with zero initial values. According to Theorem 5.2 it can thus be represented in the form

$$\text{(6-105)} \qquad r_n^{(1)} = \sum_{l=k}^{n} \varepsilon_l d_{n,l}$$

where for $l = k, k+1, \cdots$, $\{d_{n,l}\}$ is the solution of the difference equation

$$\text{(6-106)} \quad \begin{aligned} &\alpha_k d_{n+k,l} + \alpha_{k-1} d_{n+k-1,l} + \cdots + \alpha_0 d_{n,l} \\ &\qquad = h^2\{\beta_k g_{n+k} d_{n+k,l} + \cdots + \beta_0 g_n d_{n,l}\} \end{aligned}$$

assuming the initial values

$$\text{(6-107)} \qquad d_{n,l} = \begin{cases} 0, & n < l \\ \dfrac{1}{\alpha_k - h^2\beta_k g_l} & n = l \end{cases}$$

An examination of the difference equation satisfied by the discretization errors e_n (see §6.2-3) shows that the initial value problem for $d_{n,l}$ is identical with the initial value problem for e_n if $C = 0$, if x_0 is replaced by x_{l-k+1}, and if the initial values are chosen in the following special manner:

$$\delta_\mu(h) = 0, \qquad \mu = 0, 1, \cdots, k-2$$

$$\delta_\mu(h) = \frac{h^r}{\alpha_k - h^2\beta_k g_l}, \qquad \mu = k-1$$

The asymptotic behavior of $d_{n,l}$ can thus be obtained from (6-100) after a suitable change of variable. Since $\rho_\nu(\zeta) = \rho(\zeta)/(\zeta - \zeta_\nu)$, it follows that $\alpha_{\nu,k-1} = \alpha_k$. Hence

$$h^{-r-1}\{\alpha_{\nu,k-1}\delta_{k-1}(h) + \cdots + \alpha_{\nu,0}\delta_0(h)\} = \frac{1}{h} + O(1)$$

and (6-100) yields

$$d_{n,l} = \sum_{\kappa=1}^{d} \frac{2}{h\rho''(\zeta_{2\kappa})} \exp\left[i(n - l + k - 1)\varphi_{2\kappa}\right] d_{l,\kappa}(x_n) + O(1) \tag{6-108}$$

Here the functions $d_{l,\kappa}(x)$ are defined as the solutions of the initial value problems

$$\begin{aligned} d''_{l,\kappa}(x) &= \mu_{2\kappa}^2 g(x) d_{l,\kappa}(x) \\ d_{l,\kappa}(x_l) = 0, &\qquad d'_{l,\kappa}(x_l) = 1 \end{aligned} \tag{6-109}$$

We shall next express the functions $d_{l,\kappa}(x)$ in terms of the functions $s_\kappa(x)$ defined by (6-99) and of similar functions $c_\kappa(x)$ defined as solutions of the initial value problem

$$\begin{aligned} &c''_\kappa(x) = \mu_{2\kappa}^2 g(x) c_\kappa(x) \\ &c_\kappa(a) = 1, \qquad c'_\kappa(a) = 0, \qquad \kappa = 1, 2, \cdots, d \end{aligned} \tag{6-110}$$

We assert that

$$d_{l,\kappa}(x) = f_\kappa(x_l, x) \tag{6-111}$$

where for $a \le t \le b$, $a \le x \le b$

$$f_\kappa(t, x) = c_\kappa(t) s_\kappa(x) - s_\kappa(t) c_\kappa(x) \tag{6-112}$$

We shall show that if the functions $d_{l,\kappa}(x)$ are defined by (6-111) they satisfy conditions (6-109). For every fixed value of t, $f_\kappa(t, x)$ is a linear combination of solutions of the differential equation (6-109) and thus itself a solution. The first of the two initial conditions in (6-109) is obviously satisfied since $f_\kappa(x, x) = 0$. In order to show that the second condition is satisfied, we note that

$$\left.\frac{d}{dx} f_\kappa(t, x)\right|_{x=t} = W(t)$$

where

$$W(t) = c_\kappa(t) s'_\kappa(t) - s_\kappa(t) c'_\kappa(t)$$

We have $W(a) = 1$ and, since

$$\begin{aligned} W'_\kappa(t) &= c_\kappa(t) s''_\kappa(t) - s_\kappa(t) c''_\kappa(t) \\ &= \mu_{2\kappa}^2 g(t)[c_\kappa(t) s_\kappa(t) - s_\kappa(t) c_\kappa(t)] = 0 \end{aligned}$$

$W(t) = 1$ for $t \geq a$. This establishes the desired result. Using (6-105), (6-108), and (6-111), we thus obtain the basic formula

$$r_n^{(1)} = \frac{1}{h} \sum_{l=k}^{n} \varepsilon_l \left\{ \sum_{\kappa=1}^{d} \frac{2 \exp [i(n-l+k-1)\varphi_{2\kappa}]}{\rho''(\zeta_{2\kappa})} f_\kappa(x_l, x_n) + O(h) \right\} \tag{6-113}$$

As a first application we shall obtain an asymptotic bound for $|r_n^{(1)}|$. We assume that

$$|\varepsilon_l| \leq p(x_l)\varepsilon \tag{6-114}$$

where $\varepsilon \geq 0$ is independent of x and l and $p(x)$ is a known piecewise continuous function which is independent of h. Rearranging the order of the two summations, we obtain from (6-113)

$$|r_n^{(1)}| \leq \frac{2\varepsilon}{h^2} \sum_{\kappa=1}^{d} |\rho''(\zeta_{2\kappa})|^{-1} m_{\kappa,n}$$

where

$$m_{\kappa,n} = h \sum_{l=k}^{n} \{p(x_l) \, |f_\kappa(x_l, x_n)| + O(h)\}$$

By Lemma 5.8 (on approximating sums by integrals) we have

$$m_{\kappa,n} = m_\kappa(x_n) + O(h)$$

where

$$m_\kappa(x) = \int_a^x p(t) \, |f_\kappa(t, x)| \, dt \tag{6-115}$$

The following result has now been obtained:

THEOREM 6.9. *If the local round-off errors satisfy* (6-114), *where* $\varepsilon = O(h^2)$, *then*

$$|r_n^{(1)}| \leq \frac{2\varepsilon}{h^2} \sum_{\kappa=1}^{d} \{|\rho''(\zeta_{2\kappa})|^{-1} m_k(x_n) + O(h)\} \tag{6-116}$$

where the functions $m_\kappa(x)$ *are defined by* (6-115).

Inequalities for the functions $m_\kappa(x)$ in terms of solutions of differential equations can be derived; see Problems 17 and 18.

6.3-3. Statistical theory. We now turn to our statistical model of round-off, assuming that the local round-off errors are independent random variables satisfying

$$|E(\varepsilon_l)| \leq \mu p(x_l) \tag{6-117}$$

$$\operatorname{var}(\varepsilon_l) = \sigma^2 q(x_l) \tag{6-118}$$

Here $p(x)$ and $q(x)$ are nonnegative, piecewise smooth functions on $[a, b]$, and μ and σ^2 are nonnegative constants depending only on h. We note—and we might have made a similar remark in earlier chapters—that this model contains the deterministic approach to round-off exploited in §6.3-2 as the special case $\sigma^2 = 0$, $\mu = \varepsilon$. The theory below thus permits a continuous transition from the statistical to the nonstatistical approach. We shall discuss later how μ, σ^2, $p(x)$, and $q(x)$ should or might be chosen in a concrete case.

A bound for $E(r_n^{(1)})$ is easily calculated. From (6-113) we find, using (1-94),

(6-119)

$$E(r_n^{(1)}) = \frac{1}{h} \sum_{l=k}^{n} E(\varepsilon_l) \left\{ \sum_{\kappa=1}^{d} \frac{2 \exp [i(n - l + k - 1)\varphi_{2\kappa}]}{\rho''(\zeta_{2\kappa})} f_\kappa(x_l, x_n) + O(h) \right\}$$

Applying (6-118) and using the definition (6-115) of the functions $m_\kappa(x)$, we obtain

$$|E(r_n^{(1)})| \le \frac{2\mu}{h^2} \left\{ \sum_{\kappa=1}^{d} |\rho''(\zeta_{2\kappa})|^{-1} m_\kappa(x_n) + O(h) \right\} \tag{6-120}$$

In order to calculate the variance, we start again from (6-105), obtaining

$$\text{var}\,(r_n^{(1)}) = \frac{\sigma^2}{h^3} v_n, \qquad v_n = h^3 \sum_{l=k}^{n} q_l d_{nl}^2 \tag{6-121}$$

where $q_l = q(x_l)$ and, by (6-108),

$$hd_{nl} = \sum_{\kappa=1}^{d} \frac{2 \exp [i(n - l + k - 1)\varphi_{2\kappa}]}{\rho''(\zeta_{2\kappa})} f_\kappa(x_l, x_n) + O(h)$$

Since d_{nl} is real, we have

$$\begin{aligned} h^2 d_{nl}^2 &= h^2 d_{nl} \bar{d}_{nl} \\ &= \sum_{\kappa=1}^{d} c_{\kappa\kappa} d_{nl\kappa\kappa} + \sum_{\substack{\kappa,\lambda=1 \\ \kappa \ne \lambda}}^{d} c_{\kappa\lambda} \exp [i(n - l + k - 1)\delta_{\kappa\lambda}]\, d_{nl\kappa\lambda} + O(h) \end{aligned}$$

where for $\kappa, \lambda = 1, 2, \cdots, d$,

$$\begin{aligned} c_{\kappa\lambda} &= [\rho''(\zeta_{2\kappa})\, \overline{\rho''(\zeta_{2\lambda})}]^{-1} \\ \delta_{\kappa\lambda} &= \varphi_{2\kappa} - \varphi_{2\lambda} \\ d_{nl\kappa\lambda} &= f_\kappa(x_l, x_n)\, \overline{f_\lambda(x_l, x_n)} \end{aligned} \tag{6-122}$$

Reversing the order of the summations, we get

$$v_n = \sum_{\kappa=1}^{d} c_{\kappa\kappa} v_{n\kappa\kappa} + \sum_{\substack{\kappa,\lambda=1 \\ \kappa \ne \lambda}}^{d} c_{\kappa\lambda} v_{n\kappa\lambda} \exp [i(n - l + k - 1)\delta_{\kappa\lambda}] + O(h) \tag{6-123}$$

where

$$v_{n\kappa\lambda} = h\sum_{l=k}^{n} q_l d_{nl\kappa\lambda}$$

If $\kappa = \lambda$, we find by Lemma 5.8 (on replacing sums by integrals)

$$v_{n\kappa\kappa} = v_\kappa(x_n) + O(h) \tag{6-124}$$

where

$$v_\kappa(x) = \int_a^x q(t)\,|f_\kappa(t, x)|^2\,dt \tag{6-125}$$

For $\kappa \neq \lambda$ we have by Lemma 5.9 (discrete analog of Riemann-Lebesgue lemma) $v_{n\kappa\lambda} = O(h)$. Thus the only dominant contributions to v_n arise from the first sum, and we can state the following result:

THEOREM 6.10. *If the local round-off errors are independent random variables satisfying* (6-117) *and* (6-118), *then the primary component of the accumulated round-off error is a random variable whose mean satisfies* (6-120) *and whose variance is given by*

$$\text{var}\,(r_n^{(1)}) = \frac{4\sigma^2}{h^3}\sum_{\kappa=1}^{d}\{|\rho''(\zeta_{2\kappa})|^{-2}v_\kappa(x_n) + O(h)\} \tag{6-126}$$

where $v_\kappa(x)$ is defined by (6-125).

If $q(x) > 0$ for $a \le x \le b$, it can also be shown that condition (1-96) for the validity of the central limit theorem is satisfied.

The result shows that in the statistical model the standard derivation is now $O(h^{-3/2})$ as compared to $O(h^{-1/2})$ for the first order case and, incidentally, for ordinary integration. The result also shows that the behavior of the round-off error as $h \to 0$, as of the starting error, depends primarily on the double essential roots of $\rho(\zeta)$ and the associated growth parameters.

The question may be raised how the differential equations (5-254) for the functions $v_\kappa(x)$ look in the second order case. We shall answer this under the assumption that $\mu_{2\kappa}^2$ is *real*. The functions $s_\kappa(x)$ and $c_\kappa(x)$ are then real, too, and $\overline{f_\kappa(t, x)} = f_\kappa(t, x)$. Omitting the subscript κ for simplicity and denoting differentiation with respect to x by primes, we have

$$\begin{aligned} v'(x) &= \int_a^x q(t)[f(t, x)^2]'\,dt \\ v''(x) &= \int_a^x q(t)[f(t, x)^2]''\,dt \end{aligned} \tag{6-127}$$

and, since $f'(x, x) = 1$,

$$v'''(x) = q(x) + \int_a^x q(t)[f(t, x)^2]'''\,dt$$

By virtue of $(f^2)' = 2ff'$, $(f^2)'' = 2(f')^2 + 2ff'' = 2f'^2 + 2\mu^2 g f^2$ we have

$$(f^2)''' = 4\mu^2 g (f^2)' + 2\mu^2 g' f^2$$

From the last relation and from the fact that the integrals in (6-127) reduce to zero when $x = a$, it follows that $v_\kappa(x)$ may be regarded as the solution of the initial value problem

$$\begin{aligned} v_\kappa'''(x) &= 4\mu_{2\kappa}^2 g(x) v_\kappa'(x) + 2\mu_{2\kappa}^2 g'(x) v_\kappa(x) + 2q(x) \\ v_\kappa(a) &= v_\kappa'(a) = v_\kappa''(a) = 0 \end{aligned} \tag{6-128}$$

As an application of Theorem 6.10 we consider the solution of the equation

$$y'' = -\omega^2 y \tag{6-129}$$

(ω real) by the two methods

(a) $$y_{n+1} - 2y_n + y_{n-1} = h^2 f_n$$

(b) $$y_{n+2} - 2y_n + y_{n-2} = \frac{4h^2}{3}(f_{n+1} + f_n + f_{n-1})$$

[see (6-9) and (6-18)]. We assume that the integration is begun at $x = 0$; the exact form of the initial condition is irrelevant.

For method (a) we have $\rho(\zeta) = (\zeta - 1)^2$, $\sigma(\zeta) = \zeta$, $d = 1$, $\zeta_1 = \zeta_2 = 1$, $\mu_2^2 = 1$. Hence the sum in (6-126) reduces to the term involving the function $v_1(x)$ defined as solution of

$$v_1''' = -4\omega^2 v_1' + 2q, \qquad v(0) = v'(0) = v''(0) = 0$$

If $q = 1$ (fixed decimal point operations), the solution is easily verified to be

$$v_1(x) = \frac{1}{4\omega^3}(2\omega x - \sin 2\omega x) \tag{6-130}$$

Thus, since $\rho''(1) = 2$,

$$\operatorname{var}(r_n^{(1)}) = \frac{\sigma^2}{4\omega^3 h^3}(2\omega x - \sin 2\omega x + O(h))$$

for method (a). The round-off error grows roughly linearly with x [as in the integration of (6-129) by reducing to a system of first order equations and applying one-step methods].

For method (b) we find $\rho(\zeta) = \zeta^4 - 2\zeta^2 + 1$, $\sigma(\zeta) = \frac{4}{3}(\zeta^3 + \zeta^2 + \zeta)$, $d = 2$, $\zeta_1 = \zeta_2 = 1$, $\zeta_3 = \zeta_4 = -1$, $\mu_2^2 = 1$, $\mu_4^2 = -\frac{1}{3}$. The sum in (6-126) now consists of two terms. The first involves the function $v_1(x)$ defined above; the second the function $v_2(x)$ defined by

$$v_2''' = \tfrac{4}{3}\omega^2 v_2' + 2q, \qquad v(0) = v'(0) = v''(0) = 0$$

For $q = 1$ the solution is found to be

$$v_2(x) = \frac{3^{3/2}}{4\omega^3}\left(\sinh \frac{2\omega}{\sqrt{3}}\, x - \frac{2\omega}{\sqrt{3}}\, x\right) \tag{6-131}$$

Using $\rho''(1) = \rho''(-1) = 8$, we find

$$\operatorname{var}(r_n^{(1)}) = \frac{\sigma^2}{64\omega^3 h^3}\left[\left(2 - \frac{2}{\sqrt{3}}\right)\omega x - \sin 2\omega x + \sinh \frac{2\omega x}{\sqrt{3}} + O(h)\right]$$

for method (b).

For equal values of σ, the variance first grows more slowly for method (b) than for method (a). Later the error grows exponentially in method (b) owing to the presence of the sinh function. The two curves intersect where

$$\frac{1}{16}\left[\left(2 - \frac{2}{\sqrt{3}}\right)\omega x - \sin 2\omega x + \sinh \frac{2\omega x}{\sqrt{3}}\right] = 2\omega x - \sin 2\omega x$$

i.e., somewhere in the interval $5.0 \leq \omega x \leq 5.1$.

In the above examples, and moreover for any *linear* differential equation of the form $y'' = f(x, y)$, the function $g(x) = f_y(x, y(x))$ is independent of the particular solution $y(x)$ under consideration and hence of the initial values $y(a)$, $y'(a)$. A drastic illustration is the mathematically unstable problem $y'' = y$, $y(0) = 1$, $y'(0) = -1$. The true solution $y(x) = e^{-x}$ decreases exponentially. Yet for the integration of the equation by method (a) we find, assuming $q(x) = 1$,

$$\operatorname{var}(r_n^{(1)}) = \frac{\sigma^2}{4h^3}(\sinh 2x - 2x + O(h))$$

Assuming $q(x) = e^{-2x}$ (which is reasonable for floating operations), we have

$$\operatorname{var}(r_n^{(1)}) = \frac{\sigma^2}{16h^3}[e^{2x} + (3 + 4x)e^{-2x} - 4 + O(h)]$$

The exponentially growing component is still present.

6.3-4. Some properties of the growth parameters. If ζ is a double essential root of $\rho(\zeta)$, the corresponding growth parameter was defined by

$$\mu^2 = \frac{2\sigma(\zeta)}{\zeta^2 \rho''(\zeta)}$$

We now shall express the growth parameter in terms of the polynomials $r(z)$ and $s(z)$ and derive some of its properties, in particular for optimal methods.

From (5-112) we find by differentiation

$$\rho''(\zeta) = k(k-1)(\zeta+1)^{k-2}r(z) + (4k-2)(\zeta+1)^{k-3}r'(z) + 4(\zeta+1)^{k-4}r''(z) \tag{6-132}$$

where $z = (\zeta - 1)/(\zeta + 1)$. If the root ζ is different from -1, then $r(z) = r'(z) = 0$, and we find

$$\mu^2 = \frac{8s(z)}{(1-z^2)^2 r''(z)} \tag{6-133}$$

If $\zeta = -1$ is a double root, the polynomial $r(z)$ is of the form

$$r(z) = a_2 z^2 + \cdots + a_{k-2} z^{k-2}$$

where we may assume that $a_2 > 0$, $a_{k-2} > 0$. If $s(z)$ is determined by (6-62), we obtain by letting $\zeta \to -1$ in (6-132)

$$\mu^2 = 4b_k/a_{k-2} \tag{6-134}$$

If all roots of $r(z)$ lie on the imaginary axis, then $r(z)$ is even or odd according to whether k is even or odd. The polynomials $r''(z)$ and, if determined from (6-62), $s(z)$ have the same parity as $r(z)$. It then follows from (6-133) that the numbers μ^2 are all real, and the same is trivially true of (6-134).

If k is even, if the method is optimal, and if $\zeta = -1$ is a double root, an interesting inequality for the corresponding growth parameter can be derived from the explicit representation of the coefficients b_k given in §6.1-8. Using (6-64) we find

$$\mu^2 = \frac{4(d_4 a_{k-2} + d_6 a_{k-4} + \cdots + d_k a_2)}{a_{k-2}}$$

By virtue of the fact that $d_{2\nu} < 0$ $(\nu = 1, 2, \cdots)$, $a_\nu \geq 0$ $(\nu = 2, \cdots, k-2)$ we have, using $d_4 = -\frac{1}{60}$,

$$\mu^2 \leq -\tfrac{1}{15} \tag{6-135}$$

There also results the further inequality

$$\mu^2 \leq 4d_k \frac{a_2}{a_{k-2}} \tag{6-136}$$

which, if considered together with the inequality (6-68) for the error constant C, shows that for $k \geq 4$ the quantities μ^2 and C cannot simultaneously achieve their upper bounds (6-135) and (6-69).

It does not seem possible to generalize Theorem 5.16 to the second order case.

6.3-5. The local round-off error. The local round-off error was defined as the quantity ε_{n+k} in the relation

$$\alpha_k \tilde{y}_{n+k} + \cdots + \alpha_0 \tilde{y}_n = h^2\{\beta_k f(x_{n+k}, \tilde{y}_{n+k}) + \cdots + \beta_0 f(x_n, \tilde{y}_n)\} + \varepsilon_{n+k}$$

in which the quantities $\tilde{y}_m$ ($m = 0, 1, \cdots$) denote the actually calculated values of y_m. Any study of ε_{n+k}, either statistical or nonstatistical, must start from this definition. We shall not attempt to present such a study for the many variations of computational procedure that are possible in practice. However, we do wish to emphasize one important consequence of the preceding theory. If fixed point arithmetic with basic unit u is used, and if all quantities are carried to the same precision, then ε_{n+k} is of the order of u owing to the final multiplication of the expression in braces by h^2. (This consideration neglects the inherent error and the error from iteration, which can only make matters worse.) It follows from (6-126) that the standard deviation of the accumulated error is of the order of $uh^{-3/2}$. As already emphasized, this is worse by a factor h^{-1} than what we get in the integration of a first order equation (or of a system of such equations). On the other hand, if partial double precision is used, then the local round-off error is of the order of h^2u, because the multiplication $h^2\{\ \}$ is now performed exactly.* An additional error may arise from iteration, but this can also be kept down to h^2u by proper precautions. It follows that the standard deviation of the local error is now of the order of h^2u, and that of the accumulated error of the order of $uh^{1/2}$, which is what we get in the integration of a first order equation by partial double precision. Similar considerations apply to floating operations. It thus appears that the device of partial double precision is relatively more effective for second order equations than for first order equations.

Still another, more radical device which increases the numerical accuracy of certain methods of type (6-23) is discussed now.

6.4. Summed Form of the Difference Equations

In this section we discuss a new way to calculate the values y_n defined by (6-23). If no round-off error is present, and if the starting values are identical, the values y_n produced by this new algorithm are identical with those produced in the conventional manner. However, with respect to propagation of round-off error, the performance of the new algorithm is quite different, at least if single-precision arithmetic is used.

* We suppose that $h^2 > u$.

6.4-1. Statement of the algorithm. We write down (6-23) for $n = 0, 1, 2, \cdots, N$, and call S_N the sum of the resulting equations. Summing on the left, we have

$$S_N = \alpha_k(y_k + y_{k+1} + \cdots + y_{N+k}) + \alpha_{k-1}(y_{k-1} + y_k + \cdots + y_{N+k-1}) + \cdots + \alpha_0(y_0 + y_1 + \cdots + y_k)$$

or, upon rearranging,

$$\begin{aligned} S_N = {} & \alpha_0 y_0 + (\alpha_1 + \alpha_0)y_1 + \cdots + (\alpha_{k-1} + \alpha_{k-2} + \cdots + \alpha_0)y_{k-1} \\ & + (\alpha_k + \alpha_{k-1} + \cdots + \alpha_0)(y_k + y_{k+1} + \cdots + y_N) \\ & + (\alpha_k + \alpha_{k-1} + \cdots + \alpha_1)y_{N+1} + (\alpha_k + \cdots + \alpha_2)y_{N+2} \\ & + \cdots + \alpha_k y_{N+k} \end{aligned} \tag{6-137}$$

Assuming the method to be consistent, we have $\rho(1) = 0$ and hence

$$\alpha_k + \alpha_{k-1} + \cdots + \alpha_0 = 0 \tag{6-138}$$

Also, defining the polynomial $\rho_1(\zeta)$ and the coefficients $\alpha'_{k-1}, \alpha'_{k-2}, \cdots, \alpha'_0$ by setting

$$\rho_1(\zeta) = \frac{\rho(\zeta)}{\zeta - 1} = \alpha'_{k-1}\zeta^{k-1} + \alpha'_{k-2}\zeta^{k-2} + \cdots + \alpha'_0 \tag{6-139}$$

we find upon multiplying (6-139) by $\zeta - 1$, comparing coefficients, and using (6-138) that

$$\begin{aligned} \alpha'_{k-1} &= \alpha_k = -\alpha_{k-1} - \alpha_{k-2} - \cdots - \alpha_0 \\ \alpha'_{k-2} &= \alpha_k + \alpha_{k-1} = -\alpha_{k-2} - \cdots - \alpha_0 \\ & \cdots\cdots\cdots\cdots\cdots\cdots\cdots\cdots \\ \alpha'_0 &= \alpha_k + \alpha_{k-1} + \cdots + \alpha_1 = -\alpha_0 \end{aligned}$$

Thus (6-137) may be written

$$S_N = \alpha'_{k-1}y_{N+k} + \alpha'_{k-2}y_{N+k-1} + \cdots + \alpha'_0 y_{N+1} - \alpha'_{k-1}y_{k-1} - \alpha'_{k-2}y_{k-2} - \cdots - \alpha'_0 y_0 \tag{6-140}$$

Summing the terms on the right of (6-23), we obtain

$$\begin{aligned} S_N = h^2\{ & \beta_k(f_k + f_{k+1} + \cdots + f_{N+k}) \\ & + \beta_{k-1}(f_{k-1} + f_k + \cdots + f_{N+k-1}) \\ & + \cdots + \beta_0(f_0 + f_1 + \cdots + f_N)\} \end{aligned} \tag{6-141}$$

No simplification similar to the above occurs, since $\beta_k + \beta_{k-1} + \cdots + \beta_0 \neq 0$ for a stable method. However, we may put for abbreviation

$$\frac{h}{\alpha}\sum_{\mu=0}^{n} f_\mu = F_n - H, \qquad n = 0, 1, 2, \cdots \tag{6-142}$$

Here $\alpha \neq 0$ is a constant parameter which will play the role of a scale factor. The constant H is chosen in such a manner that the identity

$$\text{(6-143)} \quad \alpha'_{k-1}y_{N+k} + \alpha'_{k-2}y_{N+k-1} + \cdots + \alpha'_0 y_{N+1} = \alpha h\{\beta_k F_{N+k} + \beta_{k-1}F_{N+k-1} + \cdots + \beta_0 F_N\}$$

holds. By comparing (6-140) and (6-141), this relation is seen to imply that

$$\begin{aligned} \text{(6-144)} \quad & \alpha'_{k-1}y_{k-1} + \cdots + \alpha'_0 y_0 \\ &= \alpha h\{\beta_k F_{k-1} + \beta_{k-1}F_{k-2} + \cdots + \beta_1 F_0 + \beta_0 H\} \\ &= h^2\{\beta_k f_{k-1} + (\beta_k + \beta_{k-1})f_{k-2} + \cdots + (\beta_k + \beta_{k-1} + \cdots + \beta_1)f_0\} \\ &\quad + \alpha h(\beta_k + \beta_{k-1} + \cdots + \beta_0)H \end{aligned}$$

Defining the polynomial $\sigma_1(\zeta)$ and the coefficients $\beta'_{k-1}, \beta'_{k-2}, \cdots, \beta'_0$ by setting

$$\sigma_1(\zeta) = \frac{\sigma(\zeta) - \sigma(1)}{\zeta - 1} = \beta'_{k-1}\zeta^{k-1} + \beta'_{k-2}\zeta^{k-2} + \cdots + \beta'_0$$

we find

$$\begin{aligned} \beta'_{k-1} &= \beta_k \\ \beta'_{k-2} &= \beta_k + \beta_{k-1} \\ &\cdots\cdots\cdots\cdots\cdots \\ \beta'_0 &= \beta_k + \beta_{k-1} + \cdots + \beta_1 \end{aligned}$$

We can now write (6-144) in the form

$$\text{(6-145)} \quad \alpha h\sigma(1)H = \alpha'_{k-1}y_{k-1} + \cdots + \alpha'_0 y_0 - h^2\{\beta'_{k-1}f_{k-1} + \cdots + \beta'_0 f_0\}$$

Here the expression on the right is a known function of the starting values, and we may solve for H without difficulty in view of the fact that $\sigma(1) \neq 0$.

The equations

(6-146*a*)

$$\alpha'_{k-1}y_{n+k} + \alpha'_{k-2}y_{n+k-1} + \cdots + \alpha'_0 y_{n+1} = \alpha h\{\beta_k F_{n+k} + \cdots + \beta_0 F_n\}$$

$$\text{(6-146}b\text{)} \qquad F_{-1} = H, \qquad F_{n+k} - F_{n+k-1} = \frac{h}{\alpha}f_{n+k}$$

define the *summed form* of the difference equation (6-23). By virtue of their derivation, if H is determined by (6-145), they are satisfied by every solution of (6-23). Conversely, every solution of (6-146) satisfies (6-23), as can be seen by forming the first backward difference of (6-146*a*) and then using (6-146*b*). The process of solving (6-23) is thus equivalent to the process of solving (6-146), and, if started with the same values, both methods will produce identical mathematical approximations y_n.

The *discretization error* of the two methods is thus identical. In particular, the method described by (6-23) is convergent if and only if the method described by (6-146) is convergent, and the asymptotic behavior of the discretization error is described by Theorem 6.8 in both cases. However, from a computational point of view the two methods are obviously different. In §6.4-2 and 6.4-3 the question of how these differences affect the propagation of round-off error is discussed.

Formulas (6-146*a*) and (6-145) have a particularly simple appearance for the Störmer and the Cowell methods. In both cases, $\rho(\zeta) = \zeta^k - 2\zeta^{k-1} + \zeta^{k-2}$, so that $\rho_1(\zeta) = \zeta^{k-1} - \zeta^{k-2}$, $\alpha'_{k-1} = 1$, $\alpha'_{k-2} = -1$, $\alpha'_{k-3} = \cdots = \alpha'_0 = 0$. For the Strömer method we have

$$\sigma(\zeta) = \sigma_0\zeta^{k-1} + \sigma_1\zeta^{k-2}(\zeta - 1) + \cdots + \sigma_{k-1}(\zeta - 1)^{k-1}$$

where the coefficients σ_i are defined by (6-7). We have $\sigma(1) = \sigma_0 = 1$ and

$$\frac{\sigma(\zeta) - \sigma(1)}{\zeta - 1} = \sigma_1\zeta^{k-2} + \sigma_2\zeta^{k-3}(\zeta - 1) + \cdots + \sigma_{k-1}(\zeta - 1)^{k-2}$$

Replacing $n + k - 1$ by m for convenience, we can thus summarize the "summed" form of Störmer's method in the formulas

$$\text{(6-147}a\text{)} \quad \alpha h H = \nabla y_{k-1} - h^2\{\sigma_1 f_{k-2} + \sigma_2 \nabla f_{k-2} + \cdots + \sigma_{k-1}\nabla^{k-2} f_{k-2}\}$$

$$\text{(6-147}b\text{)} \quad F_{-1} = H, \qquad \nabla F_m = \frac{h}{\alpha} f_m$$

$$\text{(6-147}c\text{)} \quad \nabla y_{m+1} = \alpha h\{\sigma_0 F_m + \sigma_1 \nabla F_m + \cdots + \sigma_{k-1}\nabla^{k-1} F_m\}$$

The practical application is as follows: we calculate $f_0, f_1, \cdots, f_{k-1}$ from the starting values and then determine H from (147*a*). Then $F_0, F_1, \cdots, F_{k-1}$ are calculated from (6-147*b*) and differences are formed. Now y_k is calculated by (6-147*c*). This enables us to find $f_k = f(x_k, y_k)$ and F_k. The new differences $\nabla^q F_k$ can now be found, and y_{k+1} can be determined from (6-147*c*). The computation proceeds cyclically from here onward. The computational procedure for the summed form of other methods based on explicit formulas is similar.

For Cowell's method,

$$\sigma(\zeta) = \sigma_0^*\zeta^k + \sigma_1^*\zeta^{k-1}(\zeta - 1) + \cdots + \sigma_k^*(\zeta - 1)^k$$

where the coefficients σ_i^* are given by (6-12). Hence

$$\frac{\sigma(\zeta) - \sigma(1)}{\zeta - 1} = \sigma_1^*\zeta^{k-1} + \sigma_2^*\zeta^{k-2}(\zeta - 1) + \cdots + \sigma_k^*(\zeta - 1)^k$$

The counterpart of formulas (6-147) now reads, writing m for $n + k$:

(6-148a) $\alpha h H = \nabla y_{k-1} - h^2\{\sigma_1^* f_{k-1} + \sigma_2^* \nabla f_{k-1} + \cdots + \sigma_k^* \nabla^{k-1} f_{k-1}\}$

(6-148b) $F_{-1} = H, \qquad \nabla F_m = \dfrac{h}{\alpha} f_m$

(6-148c) $\nabla y_m = \alpha h\{\sigma_0^* F_m + \sigma_1^* \nabla F_m + \cdots + \sigma_k^* \nabla^k F_m\}$

These formulas are applied in a manner similar to that for (6-147) with one important difference. Equation (6-148c) is an implicit equation in the unknown quantity y_m, since F_m also depends on y_m. (The implicitness of a method is not removed by the process of summation.) It is thus necessary to predict a value of y_m by employing a formula using backward differences of F_{m-1} only. Formula (6-147c) (with m moved back one unit) recommends itself naturally for this purpose. A tentative value of F_m can then be found from (6-148b) and a corrected value of y_m calculated from (6-148c). This process is repeated, if necessary, much as in the Adams-Moulton method (see §5.1-2). In both (6-147) and (6-148) the constant α may be chosen arbitrarily by considerations of computational convenience.

6.4-2. Bounds for the round-off error. The numerical values $\tilde{y}_n$ and $\tilde{F}_n$ actually calculated from (6-146) satisfy the relations

(6-149)
$$\alpha'_{k-1}\tilde{y}_{n+k} + \cdots + \alpha'_0\tilde{y}_{n+1} = \alpha h\{\beta_k \tilde{F}_{n+k} + \cdots + \beta_0 \tilde{F}_n\} + \varepsilon_{n+k}$$
$$\tilde{F}_{n+k} - \tilde{F}_{n+k-1} = \frac{h}{\alpha} f(x_{n+k}, \tilde{y}_{n+k}) + \eta_{n+k}$$

where ε_{n+k} and η_{n+k} are the local round-off errors in y_{n+k} and F_{n+k}, respectively. If fixed point operations are used, both ε_{n+k} and η_{n+k} are of the order of magnitude of the basic unit u. Letting $r_n = \tilde{y}_n - y_n$, $R_n = \tilde{F}_n - F_n$, we find by subtracting from (6-149) the corresponding relations (6-146)

(6-150) $\alpha'_{k-1} r_{n+k} + \cdots + \alpha'_0 r_{n+1} = \alpha h\{\beta_k R_{n+k} + \cdots + \beta_0 R_n\} + \varepsilon_{n+k}$

(6-151) $$R_{n+k} - R_{n+k-1} = \frac{h}{\alpha} g_{n+k} r_{n+k} + \eta_{n+k}$$

where

$$g_m = \begin{cases} \dfrac{f(x_m, \tilde{y}_m) - f(x_m, y_m)}{\tilde{y}_m - y_m}, & \tilde{y}_m \neq y_m, \\ 0, & \tilde{y}_m = y_m \end{cases} \qquad m = 0, 1, 2, \cdots$$

By the Lipschitz condition, $|g_m| \le L$, $m = 0, 1, 2, \cdots$.

We now shall establish a bound for $|r_n|$ under the hypothesis that

$$|\varepsilon_n| \le \varepsilon, \qquad |\eta_n| \le \eta, \qquad n = 1, 2, \cdots \tag{6-152}$$

It will be shown that *if* $d = 1$, i.e., *if* $\zeta = 1$ *is the only double root of* $\rho(\zeta)$ *on the unit circle*, this bound is of the order of h^{-1} max (ε, η). This represents an improvement by a factor h over the bound (6-103) for the standard form of the algorithm.

We shall require the following modification of Lemma 6.2:

LEMMA 6.4. *If the polynomials* $\rho(\zeta)$ *and* $\sigma(\zeta)$ *define a consistent and stable method, and if* $\zeta = 1$ *is the only double root of* $\rho(\zeta)$ *satisfying* $|\zeta| = 1$, *then there exists a constant* $\gamma > 0$ *such that the coefficients* γ_n $(n = 0, 1, 2, \cdots)$ *defined by*

$$\frac{1}{\alpha_k + \alpha_{k-1}\zeta + \cdots + \alpha_0\zeta^k} = \gamma_0 + \gamma_1\zeta + \gamma_2\zeta^2 + \cdots$$

are of the form

$$\gamma_n = \frac{n}{\sigma(1)} + \gamma_n' \tag{6-153}$$

where $|\gamma_n'| \le \gamma$.

Proof. Let $\alpha_k + \alpha_{k-1}\zeta + \cdots + \alpha_0\zeta^k = \hat{\rho}(\zeta)$. From the consistency of $\rho(\zeta)$ it readily follows that $\hat{\rho}(1) = \hat{\rho}'(1) = 0$, $\hat{\rho}''(1) = 2\sigma(1)$. Since $\sigma(1) \neq 0$, the expansion of $1/\hat{\rho}(\zeta)$ in partial fractions is of the form

$$\frac{1}{\hat{\rho}(\zeta)} = \frac{1}{(\zeta - 1)^2\sigma(1)} + f(\zeta) \tag{6-154}$$

where $f(\zeta)$ is a rational function whose poles ζ_i are simple and satisfy $|\zeta_i| \ge 1$. The Taylor expansion of $f(\zeta)$ at $\zeta = 0$ thus has bounded coefficients by the reasoning employed in the proof of Lemma 5.6. The coefficient of ζ^n in the Taylor expansion of the first term on the right of (6-154) is $(n + 1)/\sigma(1)$. Relation (6-153) is now obvious.

Example. For the Störmer and the Cowell methods, $\rho(\zeta) = \zeta^k - 2\zeta^{k-1} + \zeta^{k-2}$, $\sigma(1) = 1$. Thus the function $f(\zeta)$ in (6-154) is zero, and (6-153) holds with $\gamma = 1$.

We now state the main result of this section.

THEOREM 6.11. *If* $d = 1$ *and if the local round-off errors satisfy* (6-152), *then the accumulated round-off error in the numerical solution of* $y'' = f(x, y)$ *by the algorithm* (6-146) *satisfies for* $h^2 \le L^{-1} |\alpha_k\beta_k^{-1}|$

$$r_q \le \left\{\alpha B\Gamma^*(x_q - a^*)^2\frac{\eta}{2h} + (\Gamma^* + 2\gamma^*)(x_q - a)\frac{\varepsilon}{h}\right\} \tag{6-155}$$

$$\times \exp\,[(x_q - a^*)^2\Gamma^*LB], \qquad a \le x_q \le b$$

where B and a^* are defined as in Theorem 6.7, and where

$$\Gamma^* = \frac{\sigma(1)^{-1}}{1 - h^2L\,|\alpha_k^{-1}\beta_k|}, \qquad \gamma^* = \frac{\gamma}{1 - h^2L\,|\alpha_k^{-1}\beta_k|}$$

Proof. Applying the operator ∇ to relation (6-150) and expressing ∇R_{n+k} by means of (6-151), we find

$$\text{(6-156)}\quad \alpha_k r_{m+k} + \alpha_{k-1} r_{m+k-1} + \cdots + \alpha_0 r_m = h^2\{\beta_k g_{m+k} r_{m+k} + \cdots + \beta_0 g_m r_m\} + \alpha h\{\beta_k \eta_{m+k} + \cdots + \beta_0 \eta_m\} + \varepsilon_{m+k} - \varepsilon_{m+k-1}, \qquad m = 0, 1, 2, \cdots$$

To this relation we apply the technique used in the proof of Lemma 6.3. We multiply (6-156) for $m = 0, 1, \cdots, n-k$ by γ_{n-k-m} and add. In view of $r_0 = r_1 = \cdots = r_{k-1} = 0$ there results

$$\begin{aligned} r_n = h^2\{&\beta_k\gamma_0 g_n r_n + (\beta_k\gamma_1 + \beta_{k-1}\gamma_0)g_{n-1}r_{n-1} + \cdots \\ &+ (\beta_k\gamma_{k-1} + \beta_{k-1}\gamma_{k-2} + \cdots + \beta_1\gamma_0)g_{n-k+1}r_{n-k+1} \\ &+ \sum_{m=k}^{n-k}(\beta_k\gamma_m + \beta_{k-1}\gamma_{m-1} + \cdots + \beta_0\gamma_{m-k})g_{n-m}r_{n-m}\} \\ &+ \alpha h\{\beta_k\gamma_0\eta_n + (\beta_k\gamma_1 + \beta_{k-1}\gamma_0)\eta_{n-1} + \cdots \\ &+ (\beta_k\gamma_{k-1} + \cdots + \beta_1\gamma_0)\eta_{n-k+1} + \\ &+ \sum_{m=k}^{n-k}(\beta_k\gamma_m + \cdots + \beta_0\gamma_{m-k})\eta_{n-m}\} \\ &+ \gamma_0\varepsilon_n + \sum_{m=1}^{n-k}(\gamma_m - \gamma_{m-1})\varepsilon_{n-m} \end{aligned}$$

Using (6-151), (6-152), and (6-153), the three main terms on the right can be bounded by

$$h^2\,|\beta_k\alpha_k^{-1}|\,L\,|r_n| + h^2LB(n\,|\sigma(1)|^{-1} + \gamma)\sum_{m=1}^{n-k}|r_{n-m}|,$$

$$\alpha hB(\tfrac{1}{2}n^2\,|\sigma(1)|^{-1} + n\gamma)\eta, \qquad n(2\gamma + |\sigma(1)|^{-1})\varepsilon$$

respectively. If $a \le x_n \le x_q$, then $n \le (x_q - a)h^{-1}$, and we have in the notation of the theorem

$$|r_n| \le h(x_n - a^*)\Gamma^*LB\sum_{m=0}^{n-1}|r_m| + K$$

where K denotes the expression in braces in (6-155). From this (6-155) follows by induction and by applying Bernoulli's inequality as in the proof of Lemma 6.3.

6.4-3. Statistical theory of round-off. If $\varepsilon < Ch^2$, $\delta < Dh^2$, we may conclude from Theorem 6.11 that $|r_n| < Kh$. Hence, for some constant K_1,

$$R_{n+k} - R_{n+k-1} = \frac{h}{\alpha} g_{n+k} r_{n+k} + \eta_{n+k} + \theta_n K_1 h^3 \tag{6-157}$$

where the definition of g_m has now been changed to $g_m = f_y(x_m, y(x_m))$. Differencing (6-150) and using (6-157), we get

$$\begin{aligned}(6\text{-}158)\quad \alpha_k r_{n+k} + \alpha_{k-1} r_{n+k-1} + \cdots + \alpha_0 r_n &= h^2\{\beta_k g_{n+k} r_{n+k} + \cdots \\ + \beta_0 g_n r_n\} + \alpha h\{\beta_k \eta_{n+k} + \cdots + \beta_0 \eta_n\} &+ \varepsilon_{n+k} - \varepsilon_{n+k-1} + \theta_n K_2 h^4\end{aligned}$$

where K_2 is another suitable constant. Accordingly, we may write r_n as a sum of a *primary* and a *secondary round-off error*,

$$r_n = r_n^{(1)} + r_n^{(2)}$$

where $r_n^{(1)}$ is defined as the solution of (6-158) corresponding to $\theta_n = 0$ and $r_n^{(2)}$ as the solution corresponding to $\varepsilon_n = \eta_n = 0$. By Lemma 6.4 the secondary error $r_n^{(2)}$ is bounded by a constant times h^2 and is therefore of the order of magnitude of one single local round-off error. The primary round-off error, on the other hand, must be expected to be of the order of h. Henceforth we shall be concerned only with the primary error.

Since (6-158) (with $\theta_n = 0$) is identical with (6-102) except for a more complicated form of the nonhomogeneous term, and since $r_0 = r_1 = \cdots = r_{k-1} = 0$ in both cases, we have

$$r_n^{(1)} = \sum_{l=k}^{n} [\alpha h(\beta_k \eta_l + \cdots + \beta_0 \eta_{l-k}) + \varepsilon_l - \varepsilon_{l-1}] d_{n,l}$$

where $d_{n,l}$ is defined by (6-106). Rearranging, we have

$$\begin{aligned} r_n^{(1)} &= \sum_{l=k}^{n-1} (d_{n,l} - d_{n,l+1})\varepsilon_l + d_{n,n}\varepsilon_n \\ &\quad + \alpha h \sum_{l=k}^{n} (\beta_0 d_{n,l+k} + \beta_1 d_{n,l+k-1} + \cdots + \beta_k d_{n,l})\eta_l \end{aligned}$$

We continue to assume that there is only the obligatory essential double root at $\zeta = 1$. Hence $d = 1$, and from (6-108) et seq. we find

$$d_{n,l} = \frac{2}{h\rho''(1)} f(x_l, x_n) + O(1) \tag{6-159}$$

where

$$f(t, x) = c(t)s(x) - s(t)c(x)$$

$c(x)$ and $s(x)$ being respectively defined* as the solutions of the initial value problems

$$c'' = g(x)c, \qquad c(a) = 1, \qquad c'(a) = 0,$$
$$g(x) = f_y(x, y(x))$$
$$s'' = g(x)s, \qquad s(a) = 0, \qquad s'(a) = 1,$$

If, in addition to $d = 1$, the method is such that $m = 2$ (i.e., $\zeta = 1$ is the only essential root, as is the case, e.g., for the Störmer and the Cowell methods) it can be shown by a rather lengthy computation, which we shall not reproduce here, that the term denoted by $O(1)$ in (6-159) is of the form $k(x_l, x_n) + O(h)$, where k is a continuously differentiable function of its two arguments. We then have (using $\frac{1}{2}\rho''(1) = \sigma(1) = \beta_0 + \cdots + \beta_k$):

$$d_{n,l} - d_{n,l+1} = -\frac{2}{\rho''(1)}\frac{\partial f}{\partial t}(x_l, x_n) + O(h)$$

$$\alpha h(\beta_0 d_{n,l+k} + \cdots + \beta_k d_{n,l}) = \alpha f(x_l, x_n) + O(h)$$

We thus get

(6-160)
$$r_n^{(1)} = -\frac{2}{\rho''(1)}\sum_{l=k}^{n}\left(\frac{\partial f}{\partial t}(x_l, x_n) + O(h)\right)\varepsilon_l + \alpha\sum_{l=k}^{n}(f(x_l, x_n) + O(h))\eta_l$$

Assume now that ε_n and η_n are mutually independent random variables with variances

(6-161) $$\operatorname{var}(\varepsilon_n) = q(x_n)\sigma^2, \qquad \operatorname{var}(\eta_n) = Q(x_n)\tau^2$$

where $q(x)$ and $Q(x)$ are fixed, piecewise continuously differentiable functions (not depending on h), and σ and τ are independent of x [but depending on h in a manner consistent with the hypothesis that $\varepsilon_n = O(h^2)$, $\eta_n = O(h^2)$]. Then the variance of $r_n^{(1)}$ can be calculated in the usual fashion so as to yield

$$\operatorname{var}(r_n^{(1)}) = \frac{1}{h}\left(\frac{2}{\rho''(1)}\right)^2(t_n + O(h))\sigma^2 + \frac{\alpha^2}{h}(v_n + O(h))\tau^2$$

where

$$t_n = h\sum_{l=k}^{n}\left[\frac{\partial f}{\partial t}(x_l, x_n)\right]^2 q(x_l), \qquad v_n = h\sum_{l=k}^{n}[f(x_l, x_n)]^2\, Q(x_l)$$

Approximating sums by integrals, we also have

$$t_n = t(x_n) + O(h), \qquad v_n = v(x_n) + O(h)$$

where

(6-162) $$t(x) = \int_a^x\left[\frac{\partial f}{\partial t}(t, x)\right]^2 q(t)\,dt$$

* These functions were denoted by f_1, c_1, and s_1 in §6.3.

and

(6-163) $$v(x) = \int_a^x [f(t, x)]^2 Q(t)\,dt$$

We thus have obtained the following theorem:

THEOREM 6.12. If $\zeta = 1$ *is the only essential root of the polynomial* $\rho(\zeta)$ *defining the linear multistep method* (6-23), *and if the local round-off errors in the summed form* (6-146) *of the algorithm are mutually independent random variables satisfying* (6-161), *then the primary component of the accumulated round-off error satisfies*

(6-164) $$\mathrm{var}\,(r^{(1)}) = \frac{1}{h}\left(\frac{2\sigma}{\rho''(1)}\right)^2 \{t(x_n) + O(h)\} + \frac{(\alpha\tau)^2}{h}\{v(x_n) + O(h)\}$$

where the functions $t(x)$ *and* $v(x)$ *are defined by* (6-162) *and* (6-163).

A differential equation for $v(x)$ is obtained by setting $\kappa = 1$ and replacing q by Q in (6-128). For $t(x)$ we find in a similar manner, if $q(x)$ is sufficiently differentiable,

(6-165) $$t''' - 4gt' - 2g't = q'' - 2gq$$
$$t(a) = 0, \qquad t'(a) = q(a), \qquad t''(a) = q'(a)$$

6.4-4. A numerical example. We have integrated our standard problem $y'' = -y$, $y(0) = 1$, $y'(0) = 0$ by the simplest Störmer method [(6-5) with $q = 0$], using both the orthodox and the summed version. The orthodox version leads to

(6-166) $$y_{n+1} - 2y_n + y_{n-1} = -h^2 y_n$$

From the initial condition, $y_0 = 1$, and, since the desired solution is symmetric about $x = 0$ (this information can frequently be obtained directly from the differential equation without knowledge of the explicit solution), $y_1 = y_{-1}$ and thus, in conjunction with (6-166), $y_1 = y_0(1 - \frac{1}{2}h^2)$. The summed form of the method (with $\alpha = 1$) is equivalent to the pair of equations

(6-167*a*) $$y_{n+1} - y_n = hF_n$$

(6-167*b*) $$F_{n+1} - F_n = -hy_{n+1}$$

Using the above value for y_1, we find from (6-167*a*) $F_0 = \frac{1}{2}hy_0$, which is sufficient to start the computation.

The numerical calculations were performed* on an IBM 709 computer in *floating arithmetic*, using symmetric rounding (the FORTRAN system was not used). For both the orthodox and the summed version 500

* The author is indebted to the Lockheed Aircraft Corporation, Palo Alto, California, for sponsoring these computations, and to Mr. Dale M. Thoe for carrying them out.

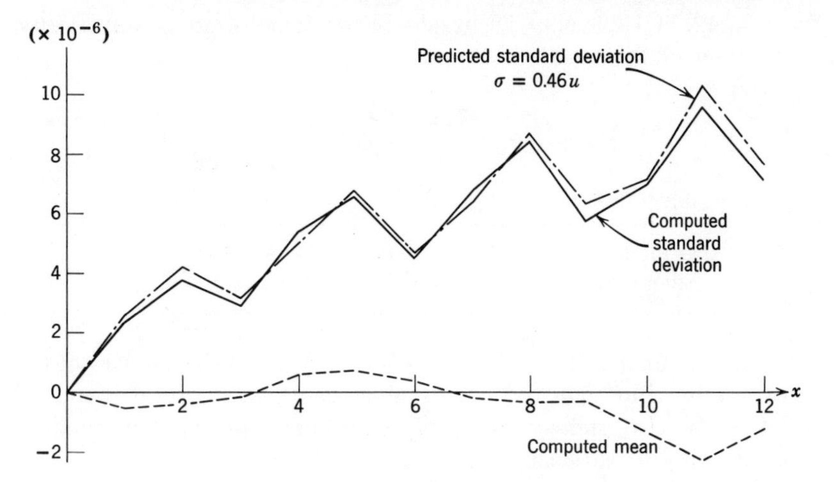

Fig. 6.1. Orthodox form.

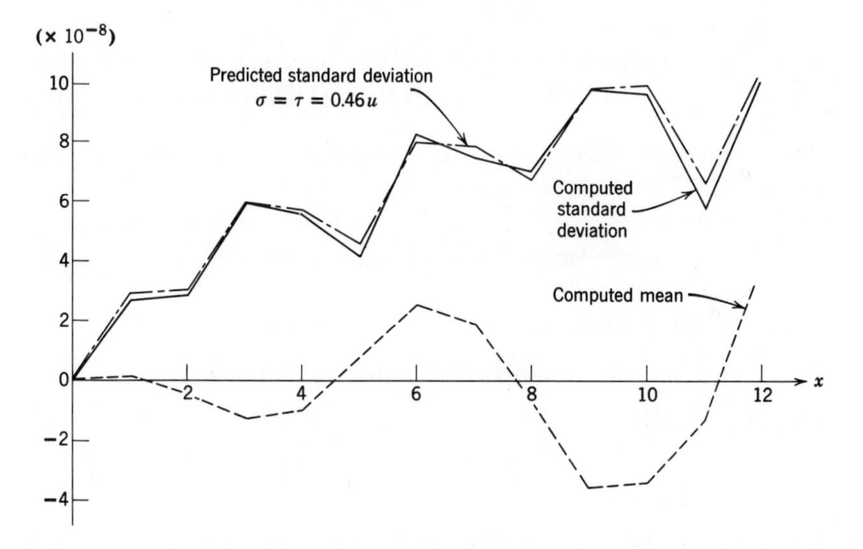

Fig. 6.2. Summed form.

numerical solutions $\{\tilde{y}_{n,q}\}$ were calculated corresponding to the starting values

$$\tilde{y}_{0,q} = 1 + q\Delta, \qquad q = 1, 2, \cdots, 500$$

(Δ a small constant), using the step $h = 2^{-7}$. The exact numerical solutions $\{y_{n,q}\}$ were obtained by working with double precision. The round-off errors $r_{n,q} = \tilde{y}_{n,q} - y_{n,q}$ were sampled for $x_n = 1, 2, \cdots, 12$. The means and standard deviations are shown in Figs. 6.1 and 6.2, and, for $x_n = 12$, in Table 6.9. The results clearly demonstrate the superiority of the summed form in this example.

Table 6.9

	Mean	Standard Deviation
Orthodox form	-1×10^{-6}	8×10^{-6}
Summed form	4×10^{-8}	10×10^{-8}

In order to analyze these results theoretically, we avail ourselves of the theory outlined in §3.4-8(iv). We thus assume that the local round-off errors ε_m and n_m are symmetrically distributed, independent random variables with the variances

$$\text{(6-168)} \qquad \begin{aligned} \operatorname{var}(\varepsilon_m) &= \tfrac{1}{3}\theta^2 u^2 y_m^2 \\ \operatorname{var}(\eta_m) &= \tfrac{1}{3}\theta^2 u^2 F_m^2 \end{aligned}$$

where θ is a constant. Since the IBM 709 works internally with the base $b = 2$, we have from (3-167) that $\frac{1}{2} \leq \theta \leq 1$. It is furthermore known that $u = 2^{-27}$.

Since the quantities y_m approximate $y(x_m) = \cos x_m$, Theorem 6.10 predicts that for the orthodox form

$$\operatorname{var}(r_n) = \frac{\theta^2 u^2}{3h^2}\{v(x_n) + O(h)\}$$

where $v(x)$ is the solution of the initial value problem

$$v''' = -4v' + 2(\cos x)^2, \qquad v(0) = v'(0) = v''(0) = 0$$

Integration yields

$$v(x) = \frac{4x - 2x\cos 2x - \sin 2x}{16}$$

The resulting values are also shown in Fig. 6.1. The constant θ was chosen a posteriori so as to produce the best possible agreement between experimental and theoretical values. It turned out that $\theta = 0.79$ produced the best agreement.

For the summed version we expect from Theorem 6.12, since F_m approximates $y'(x_m) = -\sin x_m$, that

$$\operatorname{var}(r_n) = \frac{\theta^2 u^2}{3h^2}\{v(x_n) + t(x_n) + O(h)\}$$

where $v(x)$ and $t(x)$ denote the solutions of the initial value problems

$$v''' = -4v' + 2(\sin x)^2, \qquad v(0) = v'(0) = v''(0) = 0$$
$$t''' = -4t' + 2(\sin x)^2, \; t(0) = 0, \; t'(0) = 1, \; t''(0) = 0$$

Integration yields

$$v(x) + t(x) = \frac{4x + 2x \cos 2x + \sin 2x}{8}$$

Resulting values are shown in Fig. 6.2. Again $\theta = 0.79$ was selected for best agreement. With this value of θ, agreement between theoretical and experimental values is excellent indeed throughout the considered range of x_n.

6.5. Problems and Supplementary Remarks

Section 6.1

1. *Starting values.* To begin a computation with $x_0 = a$ is satisfactory from the theoretical point of view. In practice, it is frequently convenient to use "virtual" points x_n with $n < 0$. Show, for instance, that if a solution $y(x)$ satisfying $y'(0) = 0$ of the equation $y'' = f(x)y$ is sought by Cowell's method, where $f(x) = f(-x)$, then it follows from $y_{-1} = y_1$ and from the condition that (6-14) be satisfied for $p = +1$ that

$$y_1 = y_0 \frac{1 + \frac{5}{12}h^2 f_0}{1 - \frac{1}{12}h^2 f_1}$$

Verify that for this value of y_1, $y_1 - y(x_1) = O(h^6)$, $h \to 0$.

2. Solve by Cowell's method (6-14) the initial value problem

$$y'' + \frac{4x^2}{1 + x^2} y = 0, \qquad y(0) = 1, \qquad y'(0) = 0$$

using the step $h = 1$. For large values of x, $y(x) \sim A \cos (2x + \delta)$, where A and δ are constants. Determine them.

3. *The Riccati transformation.* If $y'' = f(x)y$ and if the functions $u = y'/y$ and $v = y/y'$ are differentiable, then they satisfy, respectively, the first order equations

$$u' = f(x) - u^2, \qquad v' = 1 - f(x)v^2$$

4. Construct an explicit, unstable predictor formula with $p = 6$, $k = 4$.

[*Hint*: Determine the constants α, β, γ, and δ in such a manner that the operator

$$y_{n+2} + y_{n-2} + \alpha(y_{n+1} + y_{n-1}) + 2\beta y_n = h^2\{\gamma(f_{n+1} + f_{n-1}) + 2\delta f_n\}$$

has order 6. The unique solution is $\alpha = 16$, $2\beta = -14$, $\gamma = \frac{8}{3}$, and $2\delta = \frac{44}{3}$.]

5. Determine the constants α and β in such a way that the operator

$$y_{n+3} - y_{n+2} - y_{n+1} + y_n = h^2\{\alpha(f_{n+1} + f_{n+3}) + \beta(f_{n+1} + f_{n+2})\}$$

has order 4, and calculate the value of the error constant.

6. (a) Show that the operators associated with the Störmer formulas are definite for $q = 2$ and $q = 3$. (b) Show that the operator of order 6 associated with the polynomial (6-75) is definite for $0 < \varphi \leq \pi$. [Expand around x_{n+2}.]

7. *Generalization to equations of higher order.* It is clear that much of the theory of §6.1 can be extended to differential equations of the form

$$y^{(q)} = f(x, y)$$

where q is an arbitrary positive integer. The proofs of Theorems 6.4 and 6.5, however, depend essentially on the fact that the coefficients $c_{2l}^{(q)}$ defined by

$$\left\{\frac{z}{\log \dfrac{1+z}{1-z}}\right\}^q = c_0^{(q)} + c_2^{(q)}z^2 + c_4^{(q)}z^4 + \cdots$$

satisfy $c_{2l}^{(q)} < 0$ for $q \leq 2$ and $l = 1, 2, \cdots$. Show that $c_4^{(q)} > 0$ for $q > 2.6$ and that therefore these theorems cannot be generalized to the case $q \geq 3$.

8. (Continuation) Using the expansion

$$\left\{\frac{2z}{\log \dfrac{1+z}{1-z}}\right\}^3 = 1 - z^2 + \frac{1}{15}z^4 + \frac{1}{945}z^6 - \frac{16}{4725}z^8 + \cdots$$

construct an operator for the integration of

$$y''' = f(x, y)$$

which is stable (in the appropriate sense) and has $k = 6, p = 10$. [The associated polynomial $r(z)$ is, up to a multiplicative factor, unique: $r(z) = z^3 + \frac{5}{16}z^5$; see Dahlquist [1959], p. 30.]

9. Determine the optimal operator associated with the polynomial

$$\rho(\zeta) = \zeta^6 - 2\zeta^3 + 1$$

Section 6.2

10. Determine the growth parameters of the optimal methods associated with the polynomials

(a) $$\rho(\zeta) = \zeta^4 - 2\zeta^2 + 1$$

(b) $$\rho(\zeta) = \zeta^6 - \zeta^4 - \zeta^2 + 1$$

11. Calculate the polynomials $\rho_\nu(\zeta)$ and the quantities Δ_ν associated with the polynomials of the preceding problem.

12. For the polynomials considered in Problem 10, determine constants Γ and γ satisfying the conclusion of Lemma 6.2.

13. Discuss the asymptotic behavior of the discretization error in the numerical solution of the initial value problem

$$y'' = \frac{8y^2}{1 + 2x}, \qquad y(0) = 1, \qquad y'(0) = -2$$

[true solution: $y(x) = (1 + 2x)^{-1}$] by a method of order p with error constant C, when

(a) $d = 1$, and the starting values are exact;

(b) $d = 2$, $\mu_3^2 = \mu_4^2 = -1$, and the starting values are obtained from the Taylor polynomial of order p.

14*. Verify the result of Theorem 6.8 for the special initial value problem $y'' = A^2y$, $y(0) = 1$, $y'(0) = A$ (A = const.) by direct calculation in the manner of §5.3-1.

15*. Repeat Problem 14 for the initial value problem $y'' = (p + 1)(p + 2)x^{-2}y$, $y(1) = 1$, $y'(1) = p + 2$, where p is the order of the method.

16*. Derive results similar to those of Problems 32, 33, and 34 of Chapter 5 for a symmetric difference equation (6-23), written in the form

$$\alpha_s y_{n+s} + \alpha_{s-1} y_{n+s-1} + \cdots + \alpha_{-s} y_{n-s} = h^2\{\beta_s f_{n+s} + \cdots + \beta_{-s} f_{n-s}\}$$

where $\alpha_\mu = \alpha_{-\mu}$, $\beta_\mu = \beta_{-\mu}$, $\mu = 0, 1, \cdots, s$.

Section 6.3

17. *A lower bound for* $m_\kappa(x)$. Show that $m_\kappa(x) \geq |n_\kappa(x)|$, where $n_\kappa(x)$ is defined by

$$n_\kappa''(x) = \mu_{2\kappa}^2 g(x) n_\kappa(x) + p(x), \qquad n_\kappa(a) = n_\kappa'(a) = 0$$

[Use

$$\int_a^x p(t)\,|f(x, t)|\,dt \geq \left| \int_a^x p(t) f(x, t)\,dt \right|\Bigg]$$

18. *An upper bound for* $m_\kappa(x)$. Using Schwarz's inequality, show that

$$m_\kappa^2(x) \leq v_\kappa(x) \int_a^x p^2(t) q^{-1}(t)\,dt$$

19. If ζ_μ is a double root of a stable polynomial $\rho(\zeta)$, show that

$$|\rho''(\zeta_\mu)| \leq |\alpha_k|\, 2^{k-1}$$

[Use the representation

$$\rho(\zeta) = \alpha_k(\zeta - \zeta_\mu)^2 \prod_{\alpha \neq \mu} (\zeta - \zeta_\alpha)\Bigg]$$

20. Determine the function $v_1(x)$ for the numerical integration of

$$y'' = \frac{8y^2}{1 + 2x}, \qquad y(0) = 1, \qquad y'(0) = -2$$

for (a) $q = 1$, (b) $q = y^2$.

[*Ans.* With $\xi = 1 + 2x$, we find

(a) $$v(x) = \frac{1}{136}\left\{-\frac{17}{3}\xi^3 + \xi - \frac{\xi^{1+\sqrt{17}}}{2-\sqrt{17}} - \frac{\xi^{1-\sqrt{17}}}{2+\sqrt{17}}\right\}$$

(b) $$v(x) = \frac{1}{136}\left\{\frac{\xi^{1+\sqrt{17}}}{\sqrt{17}} - \frac{\xi^{1-\sqrt{17}}}{\sqrt{17}} - 2\xi \log \xi\right\}$$]

21. Repeat Problem 20 for the equation

$$y'' = \frac{8y}{(1+2x)^2}, \qquad y(0) = 1, \qquad y'(0) = -2$$

22. (a) Prove the following analog of Theorem 5.17: Every stable, monic polynomial $\rho(\zeta)$ associated with a consistent operator (6-23) is of the form $\zeta^m\Pi$, where Π is a product of cyclotomic polynomials in which $\Phi_1(\zeta)$ occurs exactly twice and every other factor at most twice. (b) Enumerate all stable, monic polynomials compatible with consistency and having $m = 0$, $k = 6$.

Section 6.4

23. Repeat Problem 2, using the "summed" version of Cowell's method.

24. Determine the function $t(x)$ for the case discussed in Problem 20.

25*. If we denote by $B_0, B_1, B_2, \cdots$ the *Bernoulli numbers* as defined in Hildebrand [1956], p. 150, show that the starting constant H defined by (6-145) satisfies

$$\text{(6-169)} \quad \alpha H = B_0 y'(a) + \frac{B_1}{1!}y''(a)h + \cdots + \frac{B_{p-1}}{(p-1)!}y^{(p)}(a)h^{p-1} + \left(\frac{B_p}{p!} + C\right)y^{(p+1)}(a)h^p + O(h^{p+1})$$

where p and C denote, respectively, the order and error constant of the method.

26*. Let the constants γ_ν^* ($\nu = 0, 1, 2, \cdots$) be defined by (5-20). If $p \geq k + 1$, show that

$$\text{(6-170)} \quad \alpha H = y'(a) + h\{\gamma_1^* f_0 - \gamma_2^* \nabla f_1 + \cdots + (-1)^{k-1}\gamma_k^* \nabla^{k-1} f_{k-1}\} + O(h^{k+1})$$

Notes

§6.1-1. Cowell and Commelin [1910] used (6-10) with $q = 4$ in their computation of the orbit of Halley's comet. Values of coefficients identical to our σ_m and σ_m^* are given by Bennett, Milne, and Bateman [1956], pp. 82, 83; however, the values corresponding to our σ_6 and σ_5^* are in error. Other special methods for the direct integration of second order equations are given by Zadiraka [1951], Jacobsen [1952], Mikelazde [1953a], Urabe and Yanagihara [1954], Sconzo [1954], de Vogelaere [1955], Salzer [1957].

§6.1-8. The results contained in Theorems 6.4 and 6.5 are due to Dahlquist [1959]. Some of them were obtained independently by Conte [1958].

§6.2-2. A priori bounds for special methods are stated by Muhin [1952a], Weissinger [1952], Matthieu [1953], Serrais [1956], Sheldon, Zondek, and Friedman [1957], Uhlmann [1957a]. The general linear multistep method is treated by Dahlquist [1959].

§6.2-3. A graphical discussion of weak stability is given by Collatz [1960], p. 136.

§6.4. The summed form of the Cowell method is closely related to an integration procedure which has been known to astronomers since at least 1800 and which is called by them the "second sum method" or the "Σ^2 procedure" (see von Oppolzer [1880], Jackson [1924]). The procedure was called to the attention of the general computing public by Herrick [1951]. It was apparently not recognized, however, that the method is algebraically equivalent to what we call here the Cowell method, that its principal merit lies in the reduction of round-off error, and that the beneficial effect of summing is not restricted to the Cowell method. The advantages of the summed form of Cowell's method in a nonastronomical problem are pointed out by Dettmar and Schlüter [1958]. The theoretical results of this section were briefly announced by Henrici [1960], and additional numerical experiments are reported in Henrici [1961].

part

III BOUNDARY VALUE PROBLEMS

THE METHODS discussed so far are all primarily designed for the solution of *initial value problems*. In these problems all subsidiary conditions which (in addition to the differential equation) are necessary in order to fix the solution are imposed at the same point. In many practical problems, however, the subsidiary conditions are imposed at several distinct points. It may, for instance, be desired to determine the trajectory of a ballistic missile which travels from one specified point on the surface of the earth to another in a given amount of time. Such problems are known as *boundary value problems.** Since for a single differential equation of the first order a single subsidiary condition usually suffices to fix the solution, genuine boundary value problems involve differential equations at least of the second order (or systems of at least two equations of the first order).

Perhaps the simplest boundary value problem is represented by the conditions

$$y'' = f(x, y), \qquad y(a) = A, \qquad y(b) = B$$

where $b > a$ and A and B are given constants. It is theoretically always possible to reduce the solution of a boundary value problem to the solution of a sequence of initial value problems. Let $y(x, \alpha)$ denote the solution of the initial value problem resulting from the above problem by replacing the condition for $y(b)$ by the condition $y'(a) = \alpha$, where α is a parameter. The above boundary value problem is then equivalent to solving the (in general nonlinear) equation $y(b, \alpha) = B$ for α. This can be effected by one

* For the case where the subsidiary conditions are imposed at exactly two distinct points, the designation *two-point boundary problem* is also common.

of the standard methods such as regula falsi or Newton's method. The latter is applicable by virtue of the fact that the function $\eta(x) = y_\alpha(x, \alpha)$ is a solution of the initial value problem (see §7.1-1) $\eta'' = f_y(x, y(x, \alpha))\eta$, $\eta(a) = 0$, $\eta'(a) = 1$.

Each evaluation of the function $y(x, \alpha)$ [and, if Newton's method is used, of $\eta(x, \alpha)$] requires the solution of an initial value problem. The above "shooting" technique (in the U.S.S.R. also called the "drive-through" method) nevertheless may represent a feasible procedure. In other cases, especially if systems of differential equations are involved or if the initial value problems show signs of mathematical instability, other procedures, which attack the given problem more directly, may be preferable. In Chapter 7 we discuss a method that is frequently used in practice, which consists in replacing the given problem by a set of implicit difference equations.

Even for the simple boundary value problem considered above it may happen that there are infinitely many solutions—as in the problem $y'' + \pi^2 y = 0$, $y(0) = 0$, $y(1) = 0$, for which $y(x) = C \sin \pi x$ is a solution for arbitrary C—or that there is no solution, as in the problem $y'' + \pi^2 y = 0$, $y(0) = 0$, $y(1) = 1$.

These simple examples point to the fact that the mathematical theory of boundary value problems is much more complicated than the theory of initial value problems. The same must therefore be expected of the theory of the numerical treatment of such problems. In order to present a mathematically consistent and at the same time not too voluminous treatment, we restrict our attention in Chapter 7 to a class of somewhat special, although still nonlinear, problems. This class exhibits most of the features that are typical also for more complicated problems; in addition, most of the methods which are discussed are easily adapted to more general situations. For an extensive discussion of techniques for the numerical solution of boundary value problems the reader is referred to Collatz [1949], Collatz [1960], Fox [1957].

chapter

7 Direct Methods for a Class of Nonlinear Boundary Value Problems of the Second Order

A boundary value problem will be said to be of class M if it is of the form

$$y'' = f(x, y), \qquad y(a) = A, \qquad y(b) = B \tag{7-1}$$

where $-\infty < a < b < \infty$, A and B are arbitrary constants, and the function $f(x, y)$, in addition to satisfying the conditions of the existence theorem 1.1, is such that $f_y(x, y)$ is continuous and satisfies

$$f_y(x, y) \geq 0, \qquad a \leq x \leq b, -\infty < y < \infty \tag{7-2}$$

The theoretical considerations of this chapter deal only with problems of class M. Some extensions are dealt with in the problems at the end of the chapter.

7.1. Method of Solution

7.1-1. Existence of a unique solution. The main motivation for concentrating on problems of class M is the following theorem:

THEOREM 7.1. *A boundary value problem of class M has a unique solution.*

Proof. Let $y(x, \alpha)$ denote the solution of the initial value problem $y'' = f(x, y)$, $y(a) = A$, $y'(a) = \alpha$. By virtue of a standard theorem in the theory of differential equations,* $y(x, \alpha)$ is a continuous function of x and α, and $y_\alpha(x, \alpha)$ also exists and is continuous for $x \in [a, b]$ and all

* See, e.g., Coddington and Levinson [1955], Chapter 1, Theorem 7.5.

values of α. In order to show that the equation for α, $y(b, \alpha) = B$, has exactly one solution, we shall prove that

$$y_\alpha(b, \alpha) \geq b - a \tag{7-3}$$

The desired result then follows from the fact that a monotone function defined for all values of α whose derivative is bounded away from zero assumes every value exactly once.

In order to establish (7-3), we differentiate the identity $y''(x, \alpha) = f(x, y(x, \alpha))$ with respect to α. Writing $\eta(x) = y_\alpha(x, \alpha)$, we obtain

$$\eta''(x) = f_y(x, y(x, \alpha))\eta(x), \quad a \leq x \leq b \tag{7-4}$$

From the definition of $y(x, \alpha)$, we have

$$\eta(a) = 0, \qquad \eta'(a) = 1 \tag{7-5}$$

We shall show that $\eta(x) \geq x - a$ for $a \leq x \leq b$. Assume that $\eta(\xi) < \xi - a$ for some $\xi \in (a, b]$. Since $\eta(x) > 0$ for small positive values of $x - a$, we may assume without loss of generality that

$$\eta(x) > 0 \text{ for } a < x \leq \xi \tag{7-6}$$

By the mean value theorem, $\eta(\xi) = (\xi - a)\eta'(\xi_1)$ for some $\xi_1 \in (a, \xi)$. It follows that $\eta'(\xi_1) < 1$. Applying the mean value theorem to the function $\eta'(x)$, we have $\eta'(\xi_1) - \eta'(a) = (\xi_1 - a)\eta''(\xi_2)$ for some $\xi_2 \in (a, \xi_1)$. In view of $\eta'(a) = 1$, we then have $\eta''(\xi_2) < 0$. This contradicts the differential equation (7-4) in view of $f_y \geq 0$ and (7-6). It thus follows that $\eta(x) \geq x - a$. The desired relation (7-3) is the special case $x = b$.

7.1-2. Setting up a finite difference scheme. For the direct numerical solution of a boundary value problem of class M we introduce the points $x_n = a + nh$ $(n = 0, 1, \cdots, N)$, where $h = (b - a)N^{-1}$ and N is an appropriate integer. A scheme is then designed for the determination of numbers y_n which, it is hoped, approximate the values $y(x_n)$ of the true solution at the points x_n. The natural way to obtain such a scheme is to demand that the y_n satisfy at each interior mesh point x_n a difference equation

$$\alpha_k y_{n+k} + \alpha_{k-1} y_{n+k-1} + \cdots + \alpha_0 y_n - h^2\{\beta_k f_{n+k} + \cdots + \beta_0 f_n\} = 0 \tag{7-7}$$

similar to (6-23). Again the coefficients are chosen in such a way that the associated difference operator (6-24) is small for a solution of $y'' = f(x, y)$. It is now no longer necessary to assume that $\alpha_k \neq 0$, since the resulting system of equations for the y_n is implicit in any case, and (7-7) no longer needs to be solvable for y_{n+k}. However, there is now another difficulty. As in any algebraic problem, we need as many equations for the determination of the unknowns as there are unknowns. Since y_0 and y_N are determined by the boundary condition, the unknowns in our case are $y_1, \cdots, y_{N-1}$. If the stepnumber of the difference expression >2, new unknown values such as y_{-1} or y_{N+1} are introduced for which there is no equation.

This difficulty can be circumvented by suitably modifying the difference equations near the boundary points; it does not arise at all if $k = 2$, the smallest possible value. It follows from §6.1 that if $k = 2$ in (7-7) and if the associated difference operator has positive order p, the difference equation is necessarily proportional to an equation of the form*

$$(7\text{-}8) \quad -y_{n-1} + 2y_n - y_{n+1} + h^2\{\beta_0 f_{n-1} + \beta_1 f_n + \beta_2 f_{n+1}\} = 0$$

where $\beta_0 + \beta_1 + \beta_2 = 1$. The difference equations most frequently used for boundary value problems are

$$(7\text{-}9) \qquad -y_{n-1} + 2y_n - y_{n+1} + h^2 f_n = 0 \qquad (p = 2)$$

and

$$(7\text{-}10) \quad -y_{n+1} + 2y_n - y_{n+1} + \tfrac{1}{12}h^2(f_{n-1} + 10f_n + f_{n+1}) = 0 \ \ (p = 4)$$

For most practical problems a scheme based on (7-8) seems adequate. We shall deal therefore in the present chapter only with difference equations of the type (7-8). For the construction of more elaborate schemes the reader is referred to Fox [1957].

The remarks in the two following subsections will not be subject to later theoretical scrutiny.

Boundary conditions of the second and third kind. In addition to the boundary conditions in (7-1), which are called conditions of the first kind, there occur in practice also conditions of the form

$$(7\text{-}11) \qquad \alpha y(a) + \beta y'(a) = A, \qquad \gamma y(b) + \delta y'(b) = B$$

where α, β, γ, and δ are constants, $\beta^2 + \delta^2 > 0$. Conditions (7-11) are said to be of the second or third kind according to whether $\alpha^2 + \gamma^2 = 0$ or >0. It is easy to extend the above scheme to these cases. An additional equation for y_0 is obtained by writing down (7-8) for $n = 0$ and eliminating y_{-1} by means of the first equation (7-11), where $y(a) = y_0$ and $y'(a) = (y_1 - y_{-1})(2h)^{-1}$. A similar procedure can be adopted in order to obtain an equation for y_N. For the existence of a solution of certain problems of the third kind, see Problem 8.

Variable meshsize. In many problems, especially those where the solution $y(x)$ must be expected to vary rapidly in some parts of the interval $[a, b]$ and only little in other parts, it is unreasonable to work with one and the same step h throughout the interval. A more natural procedure seems to be to split up the interval $[a, b]$ into a number of subintervals

* Sign and index in this expression have been chosen for later convenience.

and to work with a different h in each subinterval according to the expected behavior of $y(x)$. The above difference scheme is easily adapted to this case, provided that in any two adjoint intervals the stepsize in one interval

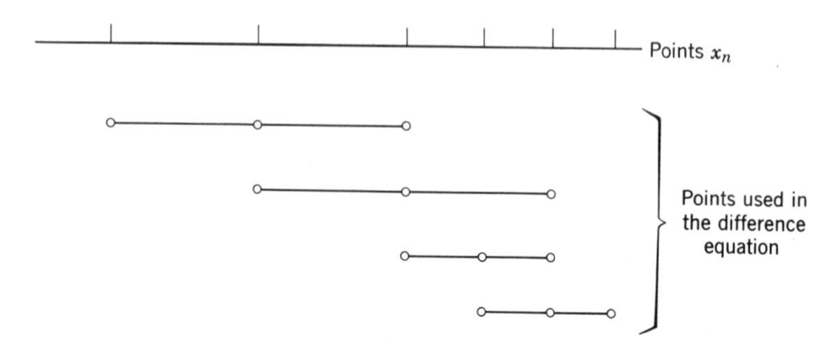

Fig. 7.1. Variable meshsize.

is an integral multiple of the stepsize in the other interval. The procedure is illustrated in Fig. 7.1 for the case where the two steps have the ratio 2 : 1.

7.1-3. Solution of the difference scheme: linear case. As in Chapter 3, the further discussion is simplified by the use of matrix and vector notation. Introducing the vectors*

$$\mathbf{y} = \begin{pmatrix} y_1 \\ y_2 \\ \cdot \\ \cdot \\ \cdot \\ y_{N-1} \end{pmatrix}, \qquad \mathbf{f}(\mathbf{y}) = \begin{pmatrix} f(x_1, y_1) \\ f(x_2, y_2) \\ \cdot \\ \cdot \\ \cdot \\ f(x_{N-1}, y_{N-1}) \end{pmatrix}, \qquad \mathbf{a} = \begin{pmatrix} A - \beta_0 h^2 f(x_0, A) \\ 0 \\ \cdot \\ \cdot \\ \cdot \\ 0 \\ B - \beta_2 h^2 f(x_N, B) \end{pmatrix}$$

and the matrices

$$\mathbf{J} = \begin{pmatrix} 2 & -1 & & & \\ -1 & 2 & -1 & & \\ & \cdot & & & \\ & & \cdot & & \\ & & & \cdot & \\ & & -1 & 2 & -1 \\ & & & -1 & 2 \end{pmatrix}, \qquad \mathbf{B} = \begin{pmatrix} \beta_1 & \beta_2 & & & \\ \beta_0 & \beta_1 & \beta_2 & & \\ & \cdot & & & \\ & & \cdot & & \\ & & & \cdot & \\ & & \beta_0 & \beta_1 & \beta_2 \\ & & & \beta_0 & \beta_1 \end{pmatrix}$$

* It should be noted that now the elements of the vector **y** are numbers approximating the same function at different points, whereas in Chapter 3 the elements of **y** were numbers approximating different functions at the same point.

(in which all elements not on the main diagonal or on the diagonals adjacent to it are zero), the system of equations arising from demanding that (7-7) hold for $n = 1, \cdots, N-1$ can be written compactly in the form

$$\mathbf{J}\mathbf{y} + h^2\mathbf{B}\mathbf{f}(\mathbf{y}) = \mathbf{a} \tag{7-12}$$

Methods for solving this system in the general case are discussed in §7.1-4. In this section we are concerned with the case where the given differential equation is *linear*, i.e., $f(x, y)$ is of the form

$$f(x, y) = g(x)y + k(x)$$

where $g(x)$ and $h(x)$ are given functions. Defining the diagonal matrix

$$\mathbf{G} = \begin{pmatrix} g(x_1) & & & & \\ & g(x_2) & & & \\ & & \cdot & & \\ & & & \cdot & \\ & & & & \cdot \\ & & & & & g(x_{N-1}) \end{pmatrix}$$

and the vector

$$\mathbf{k} = \begin{pmatrix} k(x_1) \\ k(x_2) \\ \cdot \\ \cdot \\ \cdot \\ k(x_{N-1}) \end{pmatrix}$$

we may write

$$\mathbf{f}(\mathbf{y}) = \mathbf{G}\mathbf{y} + \mathbf{k}$$

The system (7-12) now reduces to the system of linear equations,

$$\mathbf{A}\mathbf{y} = \mathbf{b} \tag{7-13}$$

where

$$\mathbf{A} = \mathbf{J} + h^2\mathbf{B}\mathbf{G}, \qquad \mathbf{b} = \mathbf{a} - h^2\mathbf{B}\mathbf{k} \tag{7-14}$$

Although the order of the linear system (7-13) may be large (indeed, the order tends to infinity as $h \to 0$), its numerical solution is relatively easy owing to the fact that the matrix $\mathbf{A}$ shares with the matrices $\mathbf{J}$ and $\mathbf{B}$ the property of having nonzero elements only on the main diagonal and the two diagonals adjacent to it. Matrices of this kind are called *tridiagonal*. In fact, if $\mathbf{A} = (a_{nm})$, we have by virtue of (7-14)

$$\begin{aligned} a_{n,n} &= 2 + h^2\beta_1 g_n, \quad a_{n,n-1} = -1 + h^2\beta_0 g_{n-1}, \\ a_{n,n+1} &= -1 + h^2\beta_2 g_{n+1}, \end{aligned} \qquad n = 1, 2, \cdots, N-1$$

and $a_{nm} = 0$ for $|n - m| > 1$.

Linear systems involving nonsingular tridiagonal matrices are most easily solved by an adaptation of the Gaussian algorithm which may conveniently be summarized as follows. Assume we have succeeded in determining two nonsingular matrices $\mathbf{L} = (l_{mn})$ and $\mathbf{U} = (u_{mn})$ of the special form

$$\mathbf{L} = \begin{pmatrix} 1 & & & & 0 \\ l_{21} & 1 & & & \\ & l_{32} & 1 & & \\ & & \cdot & \cdot & \\ & & & \cdot & \cdot \\ & & & \cdot & \cdot \\ 0 & & & l_{N-1,N-2} & 1 \end{pmatrix} \qquad \begin{array}{l} (l_{mm} = 1;\ l_{mn} = 0 \text{ for } n > m \\ \text{or } n < m - 1) \end{array}$$

$$\mathbf{U} = \begin{pmatrix} u_{11} & u_{12} & & & 0 \\ & u_{22} & u_{23} & & \\ & \cdot & \cdot & & \\ & & \cdot & \cdot & \\ & & \cdot & \cdot & \\ & & & u_{N-2,N-2} & u_{N-2,N-1} \\ 0 & & & & u_{N-1,N-1} \end{pmatrix} \qquad \begin{array}{l} (u_{mn} = 0 \text{ for } n < m \\ \text{or } n > m + 1) \end{array}$$

having the property that

$$\mathbf{LU} = \mathbf{A} \tag{7-15}$$

In order to solve (7-13), we first determine the vector $\mathbf{z}$ such that

$$\mathbf{Lz} = \mathbf{b} \tag{7-16}$$

and then $\mathbf{y}$ such that

$$\mathbf{Uy} = \mathbf{z} \tag{7-17}$$

Since $\mathbf{y} = \mathbf{U}^{-1}\mathbf{z} = \mathbf{U}^{-1}\mathbf{L}^{-1}\mathbf{b} = \mathbf{A}^{-1}\mathbf{b}$, the vector $\mathbf{y}$ thus determined satisfies (7-13).

In order to carry out this program, we note that (7-15) is equivalent to the relations

(7-18) $$u_{11} = a_{11}$$

(7-19a) $$l_{n,n-1}u_{n-1,n-1} = a_{n,n-1}, \qquad n = 2, 3, \cdots, N-1$$

(7-19b) $$l_{n,n-1}u_{n-1,n} + u_{n,n} = a_{nn}, \qquad n = 2, 3, \cdots, N-1$$

(7-19c) $$u_{n,n+1} = a_{n,n+1}, \qquad n = 1, 2, \cdots, N-2$$

Relation (7-19*c*) immediately yields $u_{n,n+1}$; relations (7-19*a*) and (7-19*b*) may be rearranged to yield $l_{n,n-1}$ and u_{nn} recursively, as follows;

(7-20) $$u_{11} = a_{11}$$

(7-21a) $$l_{n,n-1} = \frac{a_{n,n-1}}{u_{n-1,n-1}}, \qquad n = 2, \cdots, N-1$$

(7-21b) $$u_{nn} = a_{nn} - l_{n,n-1}a_{n-1,n}$$

The algorithm (7-21) breaks down when $u_{n-1,n-1} = 0$. If this is the case, then, denoting by det **A** the determinant of the matrix **A**, we have det **A** = det **U** det **L** $= u_{11}u_{22}\cdots u_{N-1,N-1} = 0$, and **A** is singular.

The vector **z** can be determined simultaneously with **L** from the relations

$$z_1 = b_1$$

$$l_{n,n-1}z_{n-1} + z_n = b_n, \qquad n = 2, \cdots, N-1$$

which may be recursively arranged in the form

(7-22a) $$z_1 = b_1$$

(7-22b) $$z_n = b_n - l_{n,n-1}z_{n-1}, \qquad n = 2, \cdots, N-1$$

If (7-17) is written out in components we find

$$u_{N-1,N-1}y_{N-1} = z_{N-1}$$

$$u_{nn}y_n + u_{n,n+1}y_{n+1} = z_n, \qquad n = 1, 2, \cdots, N-2$$

These relations can be rearranged so as to furnish the components of **y**, *starting with the last component*. One finds

(7-23a) $$y_{N-1} = \frac{z_{N-1}}{u_{N-1,N-1}}$$

(7-23b) $$y_n = \frac{z_n - a_{n,n+1}y_{n+1}}{u_{nn}}, \qquad n = N-2, \cdots, 1$$

The complete procedure may be summarized as follows:

(I) Calculate u_{nn} and z_n, starting with $n = 1$, from the relations:

$$u_{11} = a_{11}, \qquad z_1 = b_1$$

$$l_{n,n-1} = \frac{a_{n,n-1}}{u_{n-1,n-1}}$$

$$\left.\begin{aligned} u_{nn} &= a_{nn} - l_{n,n-1}a_{n-1,n}, \\ z_n &= b_n - l_{n,n-1}z_{n-1}, \end{aligned}\right\} \quad n = 2, 3, \cdots, N-1$$

The quantities $l_{n,n-1}$ need not be saved, unless the same system is solved repeatedly for different nonhomogeneous terms. The quantities a_{nn} and b_n are not used after u_{nn} and z_n have been computed.

(II) Calculate y_n, starting with $n = N - 1$, from the relations:

$$y_{N-1} = \frac{z_{N-1}}{u_{N-1,N-1}}$$

$$y_n = \frac{z_n - a_{n,n+1}y_{n+1}}{u_{nn}}, \qquad n = N-2, \cdots, 1$$

The whole process requires approximately $3N$ additions, $3N$ multiplications, and $2N$ divisions. This compares very favorably to the $\frac{1}{3}N^3$ multiplications alone that have to be performed in the solution of a system with a matrix of order N that has no zero elements.

Numerical example. To solve the tridiagonal system arising from the boundary problem $y'' = 0$, $y(0) = 0$, $y(1) = 1$, using the operator (7-9) with $h = 0.2$. The arrows in Table 7.1 indicate the order in which the auxiliary quantities are calculated.

Table 7.1

Solution of Simple Boundary Value Problem by Gaussian Algorithm

n	$a_{n,n-1}$	a_{nn}	$a_{n,n+1}$	b_n	$l_{n,n-1}$	u_{nn}	z_n	y_n
1		2	−1	0		2 →	0	$\frac{1}{5}$
2	−1	2	−1	0	$-\frac{1}{2}$	$\frac{3}{2}$ →	0	$\frac{2}{5}$ ↑
3	−1	2	−1	0	$-\frac{2}{3}$	$\frac{4}{3}$ →	0	$\frac{3}{5}$ ↑
4	−1	2		1	$-\frac{3}{4}$	$\frac{5}{4}$ →	1 →	$\frac{4}{5}$ ↑

The above algorithm merely serves the purpose of finding the solution $\mathbf{y}$ of (7-13), provided that this solution exists. The existence of a solution (or, what amounts to the same thing, the nonsingularity of $\mathbf{A}$) for problems of class M is proved in §7.2-1.

7.1-4. Solution of the difference scheme: nonlinear case. If the function $f(x, y)$ is not linear in y, one cannot hope to solve the system of equations (7-12) by algebraic methods. Some iterative procedure must be resorted to. The method which we recommend for this purpose is a generalization of the Newton-Raphson method to systems of transcendental equations.

In the case of a single equation, the Newton-Raphson method consists in linearizing the given equation $f(x) = 0$ by replacing $f(x) - f(x^{(0)})$ by its differential at a point $x^{(0)}$ believed to be close to the actual solution, and solving the linearized equation $f(x^{(0)}) + f'(x^{(0)})\Delta x = 0$. The value $x^{(1)} = x^{(0)} + \Delta x$ is then accepted as a better approximation, and the process is continued if necessary. Quite analogously, if $\mathbf{y}^{(0)}$ is a vector believed to be close to the actual solution of the equation

$$\mathbf{J}\mathbf{y} + h^2\mathbf{B}\mathbf{f}(\mathbf{y}) - \mathbf{a} = \mathbf{0} \tag{7-24}$$

so that the *residual vector*

$$\mathbf{r}(\mathbf{y}^{(0)}) = \mathbf{J}\mathbf{y}^{(0)} + h^2\mathbf{B}\mathbf{f}(\mathbf{y}^{(0)}) - \mathbf{a} \tag{7-25}$$

is small, we replace the increments of the functions $\mathbf{r}(\mathbf{y})$ by their differentials at the point $\mathbf{y} = \mathbf{y}^{(0)}$ and solve the resulting linear system of equations for the increment of the vector $\mathbf{y}$, which we shall call $\Delta\mathbf{y}$. Since the expression $\mathbf{J}\mathbf{y}$ is already linear in $\mathbf{y}$, the differential of the vector $\mathbf{r}(\mathbf{y})$ at $\mathbf{y} = \mathbf{y}_0$ is found to be $\mathbf{F}(\mathbf{y}^{(0)})\Delta\mathbf{y}$, where $\mathbf{F}(\mathbf{y})$ denotes the diagonal matrix

$$\mathbf{F}(\mathbf{y}) = \begin{pmatrix} f_y(x_1, y_1) & & & & 0 \\ & f_y(x_2, y_2) & & & \\ & & \cdot & & \\ & & & \cdot & \\ & & & & \cdot \\ 0 & & & & f_y(x_{N-1}, y_{N-1}) \end{pmatrix} \tag{7-26}$$

The linearized system (7-24) thus reads

$$\mathbf{r}(\mathbf{y}^{(0)}) + (\mathbf{J} + h^2\mathbf{B}\mathbf{F}(\mathbf{y}^{(0)}))\Delta\mathbf{y} = \mathbf{0} \tag{7-27}$$

and its solution is given by

$$\Delta\mathbf{y} = \Delta\mathbf{y}^{(0)} = -\mathbf{A}(\mathbf{y}^{(0)})^{-1}\mathbf{r}(\mathbf{y}^{(0)})$$

provided that the inverse of the matrix

$$\mathbf{A}(\mathbf{y}) = \mathbf{J} + h^2\mathbf{B}\mathbf{F}(\mathbf{y}) \tag{7-28}$$

exists for $\mathbf{y} = \mathbf{y}^{(0)}$. If all goes well, the vector $\mathbf{y}^{(1)} = \mathbf{y}^{(0)} + \Delta\mathbf{y}^{(0)}$ will be a better approximation to the exact solution, the residual $\mathbf{r}(\mathbf{y}^{(1)})$ will be smaller, and the process can be repeated with $\mathbf{y}^{(1)}$ taking the place of $\mathbf{y}^{(0)}$, etc., until convergence is achieved.

It is proved in §7.2-4 that for problems of class M the system (7-24) has a unique solution $\mathbf{y}$ for sufficiently small values of h, and that Newton's method produces a sequence of vectors $\mathbf{y}^{(n)}$, $n = 1, 2, \cdots$, which converges rapidly to $\mathbf{y}$ provided that the initial approximation $\mathbf{y}^{(0)}$ is not too bad. In the present section we are mainly concerned with questions of computational technique. In this respect it should be noted that for the solution of (7-27) it is not necessary to calculate the inverse of the matrix $\mathbf{A}(\mathbf{y}^{(0)})$. All that is required is the solution of the system of linear equations

$$\mathbf{A}(y^{(0)})\Delta\mathbf{y} = -\mathbf{r}(\mathbf{y}^{(0)})$$

for the components of $\Delta\mathbf{y}$. This solution is greatly facilitated by the fact that the matrix $\mathbf{A}(\mathbf{y}^{(0)})$ is again *tridiagonal*. In fact, if $\mathbf{A}(\mathbf{y}^{(0)}) = (a_{mn})$, we have

$$\begin{aligned} a_{n,n-1} &= -1 + h^2\beta_0 f_y(x_{n-1}, y_{n-1}^{(0)}), & n &= 2, \cdots, N-1 \\ a_{n,n} &= 2 + h^2\beta_1 f_y(x_n, y_n^{(0)}), & n &= 1, \cdots, N-1 \\ a_{n,n+1} &= -1 + h^2\beta_2 f_y(x_{n+1}, y_{n+1}^{(0)}), & n &= 1, \cdots, N-2 \end{aligned}$$

and all other elements are zero. The method described in §7.1-3 thus is immediately applicable. The only work that is required for one step of Newton's method in addition to the work involved in the solution of a linear system is the evaluation of the residual vector $\mathbf{r}(\mathbf{y}^{(0)})$ and of the partial derivative $f_y(x_n, y_n)$ $(n = 1, 2, \cdots, N-1)$.

Numerical example. To solve the difference equations arising from the nonlinear boundary value problem

$$y'' = -2 + \sinh y, \qquad y(0) = 0, \qquad y(1) = 0 \tag{7-29}$$

using the operator (7-9) with $h = 0.1$. As stated, the problem involves the 9 unknown values $y_1, y_2, \cdots, y_9$. However, the numerical work can be cut almost in half by using the fact that the exact solution $y(x)$ satisfies $y(x) = y(1 - x)$. [This follows readily from the fact that the solution of (7-29) is unique by Theorem 6.1.] We thus may assume that $y_n = y_{9-n}$

and have to compute only the values $y_1, \cdots, y_5$. Using $y_6 = y_4$, the difference equations arising from (7-9) read

$$\begin{aligned}
2y_1 - y_2 + 0.01(-2 + \sinh y_1) &= 0 \\
-y_1 + 2y_2 - y_3 + 0.01(-2 + \sinh y_2) &= 0 \\
\cdots\cdots\cdots\cdots\cdots\cdots\cdots\cdots& \\
-y_3 + 2y_4 - y_5 + 0.01(-2 + \sinh y_4) &= 0 \\
-2y_4 + 2y_5 + 0.01(-2 + \sinh y_5) &= 0
\end{aligned}$$

As a first approximation we pick the values of the function $y(x) = x(1 - x)$ which is the solution of (7-29) if the nonlinear term $\sinh y$ is neglected. There results the residual vector

$$\mathbf{r}(\mathbf{y}^{(0)}) = (0.00090122, 0.00160684, 0.00211547, 0.00242311, -0.00252612)$$

and the matrix

$$\mathbf{A}(\mathbf{y}^{(0)}) = \begin{pmatrix} 2.01004053 & -1 & & & \\ -1 & 2.01012827 & -1 & & \\ & -1 & 2.01022131 & -1 & \\ & & -1 & 2.01028939 & -1 \\ & & & -1 & 2.01031413 \end{pmatrix}$$

(Elements not shown are zero.) The correction $\Delta\mathbf{y}^{(0)}$ is found to be

$$\Delta\mathbf{y}^{(0)} = (-0.0075325, \quad -0.0142394, \quad -0.0194837, \quad -0.0228116, \quad -0.0239511)$$

The new approximation $\mathbf{y}^{(1)} = \mathbf{y}^{(0)} + \Delta\mathbf{y}^{(0)}$ produces the residual vector

$$\mathbf{r}(\mathbf{y}^{(1)}) = (0.0000000, \quad 0.0000002, \quad 0.0000003, \quad 0.0000006, \quad 0.0000008)$$

A further application of the process produces the vector

$$\mathbf{y}^{(2)} = (0.0824662, \quad 0.1457580, \quad 0.1905125, \quad 0.2171837, \quad 0.2260438)$$

for which $\mathbf{r}(\mathbf{y}^{(2)}) = 0$ to the number of digits retained.

In problems such as the present one where the elements of the matrix $\mathbf{A}(\mathbf{y})$ do not depend very sensitively on $\mathbf{y}$, it is frequently permissible to evaluate $\mathbf{A}(\mathbf{y})$ only for $\mathbf{y} = \mathbf{y}^{(0)}$ and to determine the further approximations from the relations

$$\mathbf{A}(\mathbf{y}^{(0)})\Delta\mathbf{y}^{(\nu)} = -\mathbf{r}(\mathbf{y}^{(\nu)}), \qquad \nu = 1, 2, \cdots$$

7.2. Existence of a Solution of the Difference Scheme

In §§7.2-1 and 7.2-2 we shall discuss some notions about matrices which are relevant to the numerical solution of boundary value problems. As a corollary we shall obtain the result that the linear systems (7-12) and (7-27) always have a unique solution. In §§7.2-3 and 7.2-4 we shall discuss Newton's method for a general system of n nonlinear equations with n unknowns. As an application of this discussion we shall be able to establish a theorem guaranteeing the convergence of Newton's method in the form discussed in §7.1-4 for boundary value problems of class M.

7.2-1. Irreducible matrices. Let W be the set of the first n integers, $W = \{1, 2, \cdots, n\}$. A matrix $\mathbf{A} = (a_{ij})$ is called *reducible* if it is possible to decompose W into two nonempty, disjoint subsets S and T, such that $a_{ij} = 0$ for $i \in S$ and $j \in T$. Denoting, as is customary in set theory, by $\cup$ the union and by $\cap$ the intersection of two sets and by $\emptyset$ the empty set, the conditions imposed on S and T may be formally written as $S \cup T = W$, $S \cap T = \emptyset$, $S \neq \emptyset$, $T \neq \emptyset$.

By permuting the sequence $\{1, \cdots, n\}$ in such a way that the first r $(1 \leq r < n)$ elements belong to S and the last $r - n$ elements belong to T, it is seen that a reducible matrix is similar to a matrix of the form

$$\begin{pmatrix} \mathbf{A}_{11} & \mathbf{0} \\ \mathbf{A}_{21} & \mathbf{A}_{22} \end{pmatrix}$$

where $\mathbf{A}_{11}$ and $\mathbf{A}_{22}$ are square matrices of orders r and $n - r$, respectively, $\mathbf{A}_{21}$ is a rectangular $(n - r) \times r$ matrix, and $\mathbf{0}$ is the $r \times (n - r)$ zero matrix. A system of linear equations with a reducible, nonsingular matrix thus has the property that some equations of the system can be solved without paying attention to the remaining equations.

A matrix which is not reducible is called *irreducible*. It will be important for us to establish the irreducibility of the matrix (7-14). In this respect the following theorem is helpful:

THEOREM 7.2. *A matrix $\mathbf{A} = (a_{ij})$ of order $n \geq 2$ is irreducible if and only if for any two integers i and j, $i \in W$, $j \in W$, there exists a sequence of nonzero elements of $\mathbf{A}$ of the form*

$$\{a_{i,i_1}, a_{i_1,i_2}, a_{i_2,i_3}, \cdots, a_{i_{m-1},j}\} \tag{7-30}$$

A sequence of nonzero elements of $\mathbf{A}$ of the form (7-30) is called a *chain*. If $a_{ij} \neq 0$, the chain (7-30) may be assumed to consist of the one element a_{ij}.

Proof. (i) Assume $\mathbf{A}$ is reducible, and let S and T be the sets introduced above. Since all elements in a chain $\{a_{i_{\nu-1},i_\nu}\}$ are different from 0, there are only the three possibilities

$$\begin{array}{llll} \text{(a)} & i_{\nu-1} \in T, & i_\nu \in T & \\ \text{(b)} & i_{\nu-1} \in T, & i_\nu \in S & \\ \text{(c)} & i_{\nu-1} \in S, & i_\nu \in S, & \nu = 1, \cdots, m \end{array}$$

In other words, the indices $(i_0, i_1, \cdots, i_m)$ defining a chain can change their affiliation to one of the sets S and T only once, and if they do change it, they can change it only from T to S. It follows that no chain can exist for $i = i_0 \in S, j = i_n \in T$.

(ii) Assume now that no chain exists for $i = i^*, j = j^*$. Let S be the set of all indices j such that a chain exists for the pair (i^*, j). If $S = \emptyset$, this means that $a_{i^*,j} = 0$ for $j = 1, \cdots, n$, and $\mathbf{A}$ is reducible. If $S \neq \emptyset$, let $T = W - S$. Since $j^* \in T$, $T \neq \emptyset$. Obviously $T \cup S = W$ and $T \cap S = \emptyset$. We assert that $a_{ij} = 0$ for $i \in S, j \in T$, which implies that $\mathbf{A}$ is reducible. If $a_{ij} \neq 0$ for some $i \in S, j \in T$, then, since there exists a chain for (i^*, i), this chain could be extended by attaching a_{ij} at the right end to a chain for (i, j). This would imply that j is in S and not in T, which contradicts the definition of S. This completes the proof of Theorem 7.2.

COROLLARY OF THEOREM 7.2. *A tridiagonal matrix* $\mathbf{A} = (a_{ij})$ *is irreducible if and only if*

$$a_{i,i-1} \neq 0 \ (i = 2, 3, \cdots, n) \text{ and } a_{i,i+1} \neq 0 \ (i = 1, 2, \cdots, n-1) \tag{7-31}$$

Proof. (i) Assume that (7-31) holds, and let (i, j) be a pair of indices with $i \leq j - 2$. A chain is then given by the sequence $\{a_{i,i+1}, a_{i+1,i+2}, \cdots, a_{j-1,j}\}$. A similar argument holds for $i \geq j + 2$. If $a_{ii} = 0$, a suitable chain is $\{a_{i,i+1}, a_{i+1,i}\}$. It follows that $\mathbf{A}$ is irreducible.

(ii) If $a_{m,m-1} = 0$ for some $m \in W$, $m > 1$, let $S = \{m, m+1, \cdots, n\}$, $T = \{1, 2, \cdots, m-1\}$. We then have $a_{ij} = 0$ for $i \in S$ and $j \in T$, and $\mathbf{A}$ is reducible. A similar argument holds if $a_{m,m+1} = 0$ for some m.

As a consequence of this corollary we have that the matrices $\mathbf{A}$ defined by (7-14) and $\mathbf{A}(\mathbf{y})$ defined by (7-28) are irreducible if

$$1 - h^2\beta_\mu g_n \neq 0 \quad \text{and} \quad 1 - h^2\beta_\mu f_y(x_n, y_n) \neq 0$$

respectively, for $n = 1, \cdots, N - 1$ and $\mu = 0, 2$. If L denotes a Lipschitz constant for the function $f(x, y)$, these conditions are satisfied whenever

$$h^2\beta_\mu L < 1, \qquad \mu = 0, 2 \tag{7-32}$$

This is the case if $\beta_0 = \beta_2 = 0$, or if $\beta_\mu \neq 0$, for sufficiently small values of h.

7.2-2. Monotone matrices. We recall that by the notation $\mathbf{z} \geq \mathbf{0}$ we mean that all components z_i of the vector $\mathbf{z}$ satisfy $z_i \geq 0$. Similarly, by $\mathbf{z} > \mathbf{0}$ is meant that all components satisfy $z_i > 0$.

A matrix $\mathbf{A}$ with real elements is called *monotone* if $\mathbf{Az} \geq \mathbf{0}$ implies $\mathbf{z} \geq \mathbf{0}$. If $\mathbf{A}$ is monotone, and $\mathbf{Az} \leq 0$, then, since $-\mathbf{Az} = \mathbf{A}(-\mathbf{z}) \geq \mathbf{0}$, we have $-\mathbf{z} \geq \mathbf{0}$ or $\mathbf{z} \leq \mathbf{0}$. Thus, if $\mathbf{A}$ is monotone and $\mathbf{Az} = \mathbf{0}$, $\mathbf{z}$ must satisfy both $\mathbf{z} \geq \mathbf{0}$ and $\mathbf{z} \leq \mathbf{0}$, hence $\mathbf{z} = \mathbf{0}$. Hence $\det \mathbf{A} \neq \mathbf{0}$, and we have proved: *A monotone matrix is nonsingular.* The statement of the following theorem thus makes sense.

THEOREM 7.3. *A matrix $\mathbf{A}$ is monotone if and only if the elements of the inverse matrix $\mathbf{A}^{-1}$ are nonnegative.*

In analogy to the notation for vectors used above, the condition of the theorem may be stated as $\mathbf{A}^{-1} \geq \mathbf{0}$.

Proof. (i) Let $\mathbf{A}^{-1} \geq \mathbf{0}$ and let $\mathbf{Az} \geq \mathbf{0}$. Then $\mathbf{z} = \mathbf{A}^{-1}(\mathbf{Az}) \geq \mathbf{0}$, since a nonnegative matrix operating on a nonnegative vector yields a nonnegative vector. It follows that $\mathbf{A}$ is monotone.

(ii) Assume that $\mathbf{A}^{-1} = (b_{ij})$ has a negative element b_{rs}. Denote by $\mathbf{e}_s$ the sth column of the unit matrix. Then the rth element of the vector $\mathbf{z} = \mathbf{A}^{-1}\mathbf{e}_s$ equals b_{rs} and hence is negative. Thus $\mathbf{Az} = \mathbf{A}(\mathbf{A}^{-1}\mathbf{e}_s) = \mathbf{e}_s$ is nonnegative, but $\mathbf{z}$ is not nonnegative, hence $\mathbf{A}$ fails to be monotone.

In most practical situations the criterion implied in Theorem 7.3 cannot be used in order to decide whether a given matrix $\mathbf{A}$ is monotone, since the inverse $\mathbf{A}^{-1}$ is not explicitly known. The following result contains an easily applicable sufficient condition for $\mathbf{A}$ to be monotone.

THEOREM 7.4. *Let the matrix $\mathbf{A} = (a_{ij})$ be irreducible and satisfy the conditions*

$$\text{(i)} \qquad a_{ij} \leq 0, \qquad i \neq j; \ i, j = 1, \cdots, n$$

$$\text{(ii)} \qquad \sum_{j=1}^{n} a_{ij} \begin{cases} \geq 0, & i = 1, 2, \cdots, n \\ > 0, & \text{for at least one } i \end{cases}$$

Then $\mathbf{A}$ is monotone.

Condition (ii) means that the sums of the elements in any row of $\mathbf{A}$ are nonnegative and positive for at least one row. In the German literature (ii) is called the "weak row sum criterion."

As a result of Theorem 7.4 we shall be able to assert that the matrix $\mathbf{A}(\mathbf{y})$ defined by (7-28) [this matrix contains (7-14) as a special case] is monotone for boundary value problems of class M. For it has already been observed in the preceding section that these matrices are irreducible for sufficiently small h; for the same values of h the off-diagonal elements are nonpositive. The row sums are given by

$$\sum_{j=1}^{N-1} a_{ij} = h^2\{\beta_0 f_y(x_{i-1}, y_{i-1}) + \beta_1 f_y(x_i, y_i) + \beta_2 f_y(x_{i+1}, y_{i+1})\} \quad \text{for } i = 2, 3, \cdots, N-2 \tag{7-33}$$

and

$$\sum_{j=1}^{N-1} a_{1j} = 1 + h^2\{\beta_1 f_y(x_1, y_1) + \beta_2 f_y(x_2, y_2)\} \tag{7-34}$$

$$\sum_{j=1}^{N-1} a_{N-1,j} = 1 + h^2\{\beta_0 f_y(x_{N-1}, y_{N-1}) + \beta_1 f_y(x_N, y_N)\}$$

If condition (7-40) below is satisfied, the terms on the right of (7-33) and (7-34) are ≥ 0 and >0, respectively. Thus condition (ii) is also satisfied, and it follows that $\mathbf{A}(\mathbf{y})$ is monotone. This fact will be crucial for the error estimates in §7.3.

Proof of Theorem 7.4. Assume that there exists a vector $\mathbf{z}$ with a negative component $z_q < 0$, $q \in W$, but such that $\mathbf{Az} \geq \mathbf{0}$. This assumption is equivalent to assuming that $\mathbf{A}$ is not monotone; we shall show that this contradicts the assumption that $\mathbf{A}$ is irreducible. Denote by $\mathbf{e}$ the vector whose components are all 1; since the components of $\mathbf{Ae}$ are just the row sums of $\mathbf{A}$, we have $\mathbf{Ae} \geq \mathbf{0}$, $\mathbf{Ae} \neq \mathbf{0}$ by (ii). Since the sum of two nonnegative vectors is nonnegative, it follows that for $0 \leq \lambda \leq 1$,

$$\lambda\mathbf{Az} + (1 - \lambda)\mathbf{Ae} = \mathbf{A}[\lambda\mathbf{z} + (1 - \lambda)\mathbf{e}] \geq \mathbf{0} \tag{7-35}$$

Consider the vector

$$\mathbf{w}_\lambda = \lambda\mathbf{z} + (1 - \lambda)\mathbf{e} \tag{7-36}$$

as a function of λ. For $\lambda = 0$ all components of $\mathbf{w}_\lambda$ are positive, namely $+1$; for $\lambda = 1$ there is at least one negative component, namely z_q. The components of $\mathbf{w}_\lambda$ are continuous functions of λ; when λ varies from 0 to 1, at least one component of $\mathbf{w}_\lambda$ must pass through the value 0. Let Λ be the smallest value of λ such that $\mathbf{w}_\lambda$ has a zero component. Clearly, $0 < \Lambda < 1$. Now let S be the set of the indices of the zero components of $\mathbf{w}_\Lambda$, and let $T = W - S$. By construction, $S \neq \emptyset$. But also $T \neq \emptyset$, for if all components of $\mathbf{w}_\Lambda$ were zero, then the vectors $\mathbf{z}$ and $\mathbf{e}$ would be proportional,

$$\mathbf{e} = -\frac{\Lambda}{1 - \Lambda}\mathbf{z}$$

and from $\mathbf{Az} \geq \mathbf{0}$ it would follow that

$$\mathbf{Ae} = -\frac{\Lambda}{1 - \Lambda}\mathbf{Az} \leq \mathbf{0}$$

contradicting (ii). By (7-35), $\mathbf{Aw}_\Lambda \geq 0$; thus in particular, if $i \in S$,

$$(\mathbf{Aw}_\Lambda)_i = \sum_{j \in T} a_{ij} w_{\Lambda j} \geq 0 \tag{7-37}$$

By construction, $w_{\Lambda j} > 0$ if $j \in T$. In view of (i), (7-37) is thus only possible if $a_{ij} = 0$, $i \in S$, $j \in T$, i.e., if $\mathbf{A}$ is reducible. This contradiction proves the assertion of the theorem.

The next result will enable us to find bounds for the elements of $\mathbf{A}^{-1}$ when $\mathbf{A}$ satisfies the hypotheses of Theorem 7.4.

THEOREM 7.5. *Let the matrices* $\mathbf{A}$ *and* $\mathbf{B}$ *be monotone, and assume that*

$$\mathbf{A} - \mathbf{B} \geq \mathbf{0} \tag{7-38}$$

Then

$$\mathbf{B}^{-1} - \mathbf{A}^{-1} \geq \mathbf{0} \tag{7-39}$$

Proof. By Theorem 7.3, the matrices $\mathbf{A}^{-1}$ and $\mathbf{B}^{-1}$ exist and are nonnegative. The product of nonnegative matrices is nonnegative. The required result now follows from the identity

$$\mathbf{B}^{-1} - \mathbf{A}^{-1} = \mathbf{B}^{-1}(\mathbf{A} - \mathbf{B})\mathbf{A}^{-1}$$

since the factors on the right are nonnegative.

Returning to our difference equation (7-27), we now stipulate that

$$\beta_\mu \geq 0, \qquad \mu = 0, 1, 2 \tag{7-40}$$

For problems of class M we then have, if the matrix $\mathbf{A}(\mathbf{y})$ is defined by (7-28),

$$\mathbf{A}(\mathbf{y}) \geq \mathbf{J} \tag{7-41}$$

If $h^2L \leq 1$, $\mathbf{A}(\mathbf{y})$ satisfies the hypotheses of Theorem 7.4, and it follows that

$$\mathbf{0} \leq [\mathbf{A}(\mathbf{y})]^{-1} \leq \mathbf{J}^{-1} \tag{7-42}$$

We shall require bounds for the elements of $[\mathbf{A}(\mathbf{y})]^{-1}$ and shall therefore determine $\mathbf{J}^{-1}$ explicitly. Denoting the mth column of $\mathbf{J}^{-1}$ by $\mathbf{j}_m$ ($m = 1, 2, \cdots, N-1$), we have

$$\mathbf{J}\mathbf{j}_m = \mathbf{e}_m \tag{7-43}$$

where $\mathbf{e}_m$ denotes the mth unit vector. Writing out (7-43) in components, we get, defining $j_{0,m} = j_{nm} = 0$,

$$-j_{n-1,m} + 2j_{nm} - j_{n+1,m} = \begin{cases} 0, & n = 1, \cdots, m-1 \\ 1, & n = m \\ 0, & n = m+1, \cdots, N-1 \end{cases} \tag{7-44}$$

It follows that $j_{nm} = p_m n$ $(n \leq m)$ and $j_{nm} = q_m(N - n)$ $(n \geq m)$, where p_m and q_m are two constants which are to be determined such that the two definitions are consistent for $n = m$ and such that the middle condition (7-44) holds. This leads to the conditions

$$p_m m - q_m(N - m) = 0, \qquad p_m + q_m = 1$$

from which we readily find

$$p_m = \frac{N - m}{N}, \qquad q_m = \frac{m}{N}$$

We thus have $\mathbf{J}^{-1} = (j_{mn})$, where

$$j_{mn} = \begin{cases} \dfrac{(N - m)n}{N}, & n \leq m, \\ \dfrac{m(N - n)}{N}, & n \geq m; \ m, n = 1, 2, \cdots, N - 1 \end{cases} \tag{7-45}$$

Remembering that $n = (x_n - a)h^{-1}$ and $N - m = (b - x_m)h^{-1}$, we now may restate result (7-42) in the following way: *If* (7-2) *and* (7-40) *hold, the elements* d_{ij} *of the matrix* $[\mathbf{A}(\mathbf{y})]^{-1}$ *satisfy*

$$0 \leq d_{ij} \leq \frac{(x_m - a)(b - x_n)}{h(b - a)} \tag{7-46}$$

where $m = \min(i, j)$ *and* $n = \max(i, j)$.

7.2-3. A new vector norm. In the discussion of the convergence of Newton's method, and also for the error estimates to be given later, it will be necessary to measure the magnitude of a vector $\mathbf{v}$ with components v_i $(i = 1, \cdots, n)$. In Chapter 3 we introduced for this purpose the expression

$$\|\mathbf{v}\| = |v_1| + |v_2| + \cdots + |v_n| \tag{7-47}$$

For the purpose at hand, however, it is more convenient to use the quantity

$$|\mathbf{v}| = \max_{1 \leq i \leq n} |v_i| \tag{7-48}$$

for measuring the size of $\mathbf{v}$. No confusion with the concept of the absolute value of a number can arise through this notation, since this concept is not defined for vectors. Evidently the relations

$$|\mathbf{v}| \geq 0, \qquad |\mathbf{v}| = 0 \text{ if and only if } \mathbf{v} = \mathbf{0} \tag{7-49a}$$

$$|\mathbf{v} + \mathbf{w}| \leq |\mathbf{v}| + |\mathbf{w}| \tag{7-49b}$$

$$|\lambda \mathbf{v}| = |\lambda| \, |\mathbf{v}| \tag{7-49c}$$

hold, where $\mathbf{v}$ and $\mathbf{w}$ are arbitrary vectors and λ is an arbitrary real or complex number. These relations characterize the number $|\mathbf{v}|$ as a *norm* in the terminology used in the theory of linear vector spaces. If $\mathbf{v}^{(\nu)}$ $(\nu = 1, 2, \cdots)$ is a sequence of vectors with components $v_i^{(\nu)}$ $(i = 1, \cdots, n)$, it is also clear that $|\mathbf{v}^{(\nu)}| \to 0$ as $n \to \infty$ if and only if $v_i^{(\nu)} \to 0$ $(i = 1, \cdots, n)$.

If $\mathbf{A} = (a_{ij})$ is a matrix with real or complex elements, we shall require a number C which has the property that

$$|\mathbf{A}\mathbf{v}| \leq C|\mathbf{v}| \tag{7-50}$$

for all vectors $\mathbf{v}$. We assert that $C = |\mathbf{A}|$ has this property, where

$$|\mathbf{A}| = \max_{1 \leq i \leq n} \sum_{j=1}^{n} |a_{ij}| \tag{7-51}$$

and that this is the best (i.e., smallest) possible value of C. Indeed,

$$\begin{aligned} |\mathbf{A}\mathbf{v}| &= \max_{1 \leq i \leq n} \left| \sum_{j=1}^{n} a_{ij} v_j \right| \leq \max_{1 \leq i \leq n} \sum_{j=1}^{n} |a_{ij}|\, |v_j| \\ &\leq \max_{1 \leq i \leq n} \left(\max_{1 \leq j \leq n} |v_j| \sum_{1 \leq i \leq n} |a_{ij}| \right) \\ &= |\mathbf{v}|\, |\mathbf{A}| \end{aligned}$$

and (7-50) is satisfied. On the other hand, $C = |\mathbf{A}|$ is clearly the best possible value if $\mathbf{A} = \mathbf{0}$; if $\mathbf{A} \neq \mathbf{0}$, let i be such that

$$|\mathbf{A}| = \sum_{j=1}^{n} |a_{ij}|$$

and define $\mathbf{v}$ by

$$v_j = 0, \text{ if } a_{ij} = 0; \qquad v_j = \frac{\bar{a}_{ij}}{|a_{ij}|}, \text{ if } a_{ij} \neq 0$$

Then $|\mathbf{v}| = 1$, and the ith component of $\mathbf{A}\mathbf{v}$ is given by

$$\sum_{\substack{j=1 \\ a_{ij} \neq 0}}^{n} a_{ij} \frac{\bar{a}_{ij}}{|a_{ij}|} = \sum_{j=1}^{n} |a_{ij}| = |\mathbf{A}|$$

Thus $|\mathbf{A}\mathbf{v}| \geq |\mathbf{A}|\, |\mathbf{v}|$ for this particular vector $\mathbf{v}$, and (7-50) could not be true for any number C less than $|\mathbf{A}|$.

In addition to the relations

$$|\mathbf{A}| \geq 0, \qquad |\mathbf{A}| = 0 \text{ if and only if } \mathbf{A} = \mathbf{0} \tag{7-52a}$$

$$|\mathbf{A} + \mathbf{B}| \leq |\mathbf{A}| + |\mathbf{B}| \tag{7-52b}$$

$$|\lambda \mathbf{A}| = |\lambda|\, |\mathbf{A}| \tag{7-52c}$$

which are analogous to (7-49) and which are similarly proved, the number $|\mathbf{A}|$ associated by (7-51) with the matrix $\mathbf{A}$ has also the property

$$|\mathbf{AB}| \leq |\mathbf{A}|\,|\mathbf{B}| \tag{7-53}$$

This may be proved by noting that, by virtue of (7-50) and (7-51),

$$|\mathbf{A}| = \max_{|\mathbf{y}|=1} |\mathbf{Ay}|$$

and hence

$$|\mathbf{AB}| = \max_{|\mathbf{y}|=1} |\mathbf{ABy}| \leq |\mathbf{A}| \max_{|\mathbf{y}|=1} |\mathbf{By}| = |\mathbf{A}|\,|\mathbf{B}|$$

Relations (7-52) and (7-53) state that, in the terminology of linear vector spaces, the number $|\mathbf{A}|$ is a *norm* of the linear operator $\mathbf{A}$. Since no other norms of operators will be used, we shall call $|\mathbf{A}|$ simply the *norm of the matrix* $\mathbf{A}$.

The following lemma (a special case of a very general result due to Banach) will be needed.

LEMMA 7.1. *Let* $\mathbf{A}$ *be a matrix such that* $|\mathbf{A}| = k < 1$, *and let* $\mathbf{I}$ *denote the unit matrix. Then the matrix* $(\mathbf{I} - \mathbf{A})^{-1}$ *exists, and*

$$|(\mathbf{I} - \mathbf{A})^{-1}| \leq \frac{1}{1-k} \tag{7-54}$$

Proof. Let

$$\mathbf{S}_p = \mathbf{I} + \mathbf{A} + \mathbf{A}^2 + \cdots + \mathbf{A}^{p-1}$$

Then

$$\mathbf{S} = \lim_{p\to\infty} \mathbf{S}_p$$

exists, since

$$\begin{aligned} |\mathbf{S}_{p+q} - \mathbf{S}_p| &= |\mathbf{A}^p + \cdots + \mathbf{A}^{p+q-1}| \\ &\leq |\mathbf{A}^p| + \cdots + |\mathbf{A}^{p+q-1}| \\ &\leq \frac{k^p}{1-k} \end{aligned} \tag{7-55}$$

Letting $p \to \infty$ in the identity

$$(\mathbf{I} - \mathbf{A})\mathbf{S}_p = \mathbf{I} - \mathbf{A}^p$$

we get

$$(\mathbf{I} - \mathbf{A})\mathbf{S} = \mathbf{I}$$

The matrix $\mathbf{S}$ thus is a right inverse, and consequently *the* inverse, of the matrix $\mathbf{I} - \mathbf{A}$. Relation (7-54) follows by making $p = 0$ in (7-55) and letting $q \to \infty$.

One further result of a somewhat more special nature will be required. If the elements a_{ij} of a matrix $\mathbf{A}(\mathbf{y})$ are continuously differentiable functions of $\mathbf{y}$ in a region B containing all vectors $t\mathbf{y}_1 + (1 - t)\mathbf{y}_2$, where $\mathbf{y}_1$ and $\mathbf{y}_2$ are two given vectors and $0 \leq t \leq 1$, then

$$|\mathbf{A}(\mathbf{y}_2) - \mathbf{A}(\mathbf{y}_1)| \leq K\,|\mathbf{y}_2 - \mathbf{y}_1| \tag{7-56}$$

where

$$K = \max_{\substack{1 \leq i \leq n \\ \mathbf{y} \in B}} \sum_{j=1}^{n} \sum_{k=1}^{n} \left| \frac{\partial a_{ij}}{\partial y_k} \right| \tag{7-57}$$

The proof is accomplished by applying the mean value theorem to each element of $\mathbf{A}(\mathbf{y}_2) - \mathbf{A}(\mathbf{y}_1)$.

7.2-4. Newton's method for systems of nonlinear equations. Let the n equations

$$\varphi_i(y_1, y_2, \cdots, y_n) = 0, \qquad i = 1, \cdots, n$$

for the n unknowns $y_1, y_2, \cdots, y_n$ be written in vector form as

$$\boldsymbol{\varphi}(\mathbf{y}) = 0 \tag{7-58}$$

Let $\mathbf{A}(\mathbf{y}) = (a_{ij})$ denote the matrix with the elements

$$a_{ij} = \frac{\partial \varphi_i(\mathbf{y})}{\partial y_j}$$

If the vector $\mathbf{y} = \mathbf{y}^{(0)}$ is believed to be an approximation to a solution of the system (7-58), and if the matrix $\mathbf{A}(\mathbf{y}^{(0)})$ is nonsingular, one may hope that the vector

$$\mathbf{y}^{(1)} = \mathbf{y}^{(0)} - \mathbf{A}(\mathbf{y}^{(0)})^{-1}\boldsymbol{\varphi}(\mathbf{y}^{(0)}) \tag{7-59}$$

obtained by linearizing the system (7-58) at $\mathbf{y} = \mathbf{y}^{(0)}$ is a better approximation to the solution. If the matrices $\mathbf{A}(\mathbf{y}^{(\nu)})$ involved continue to be nonsingular, one may hope to obtain a sequence of successively better approximations $\mathbf{y}^{(\nu)}$ $(\nu = 1, 2, \cdots)$ by the algorithm

$$\mathbf{y}^{(\nu+1)} = \mathbf{y}^{(\nu)} - [\mathbf{A}(\mathbf{y}^{(\nu)})]^{-1}\boldsymbol{\varphi}(\mathbf{y}^{(\nu)}), \qquad \nu = 0, 1, \cdots \tag{7-60}$$

which is known as *Newton's method* for the solution of the system of nonlinear equations (7-58). The question of finding simple sufficient conditions for the convergence of Newton's method to a solution of the system (7-58) was considered a difficult problem of numerical analysis until L. V. Kantorovich in 1937 published a theorem (see Kantorovich [1948]) which guarantees the convergence of Newton's method under very general circumstances, *without even assuming the existence of a solution.* We shall state Kantorovich's result in a form applicable to our immediate purpose, which is the discussion of the convergence of Newton's method when $\boldsymbol{\varphi}(\mathbf{y}) = \mathbf{r}(\mathbf{y})$, where $\mathbf{r}(\mathbf{y})$ is defined by (7-25).

THEOREM 7.6. *Assume that the following conditions are satisfied*:

(i) *For* $\mathbf{y} = \mathbf{y}^{(0)}$, *the initial approximation, the matrix* $\mathbf{A}(\mathbf{y}^{(0)})$ *has an inverse* $\boldsymbol{\Gamma}_0 = \mathbf{A}(\mathbf{y}^{(0)})^{-1}$, *and an estimate for its norm is known*:

$$|\boldsymbol{\Gamma}_0| \le B_0 \tag{7-61}$$

(ii) *The vector* $\mathbf{y}^{(0)}$ *approximately satisfies the system of equations* (7-58) *in the sense that*

$$|\boldsymbol{\Gamma}_0 \boldsymbol{\varphi}(\mathbf{y}^{(0)})| \le \eta_0 \tag{7-62}$$

(iii) *In the region defined by inequality* (7-65) *below, the components of the vector* $\boldsymbol{\varphi}(\mathbf{y})$ *are twice continuously differentiable with respect to the components of* $\mathbf{y}$ *and satisfy*

$$\sum_{j,k=1}^{n} \left| \frac{\partial^2 \varphi_i}{\partial y_j \, \partial y_k} \right| \le K, \qquad i = 1, 2, \cdots, n \tag{7-63}$$

(iv) *The constants* B_0, η_0, *and* K *introduced above satisfy the inequality*

$$h_0 \equiv B_0 \eta_0 K \le \tfrac{1}{2} \tag{7-64}$$

Then the system of equations (7-58) *has a solution* $\mathbf{y}^*$ *which is located in the cube*

$$|\mathbf{y} - \mathbf{y}^{(0)}| \le N(h_0)\eta_0 = \frac{1 - \sqrt{1 - 2h_0}}{h_0} \eta_0 \tag{7-65}$$

Moreover, the successive approximations $\mathbf{y}^{(\nu)}$ *defined by* (7-60) *exist and converge to* $\mathbf{y}$, *and the speed of convergence may be estimated by the inequality*

$$|\mathbf{y}^{(\nu)} - \mathbf{y}^*| \le \frac{1}{2^{\nu-1}} (2h_0)^{2^\nu - 1} \eta_0 \tag{7-66}$$

Even in the one-dimensional case ($n = 1$) this theorem is not trivial.

Proof. We shall first show, setting

$$B_1 = \frac{B_0}{1 - h_0}, \qquad \eta_1 = \frac{1}{2} \frac{h_0 \eta_0}{1 - h_0}, \qquad h_1 = \frac{1}{2} \frac{h_0^2}{(1 - h_0)^2} \tag{7-67}$$

that relations (7-61), (7-62), and (7-64) continue to hold if the index 0 is replaced everywhere by 1.

First of all we have, by (7-59),

$$|\mathbf{y}^{(1)} - \mathbf{y}^{(0)}| = |\boldsymbol{\Gamma}_0 \boldsymbol{\varphi}(\mathbf{y}^{(0)})| \le \eta_0 \tag{7-68}$$

Furthermore, using (7-56) with $\mathbf{A}(\mathbf{y})$ in its present meaning,

$$\begin{aligned} |\mathbf{\Gamma}_0[\mathbf{A}(\mathbf{y}^{(0)}) - \mathbf{A}(\mathbf{y}^{(1)})]| &\le |\mathbf{\Gamma}_0|\,|\mathbf{A}(\mathbf{y}^{(0)}) - \mathbf{A}(\mathbf{y}^{(1)})| \\ &\le B_0 K\,|\mathbf{y}^{(0)} - \mathbf{y}^{(1)}| \\ &\le B_0 K \eta_0 = h_0 < 1 \end{aligned}$$

By Banach's Lemma 7.1 (§7.2-3) it follows that the matrix

$$\mathbf{H} = \mathbf{I} - \mathbf{\Gamma}_0[\mathbf{A}(\mathbf{y}^{(0)}) - \mathbf{A}(\mathbf{y}^{(1)})]$$

has an inverse satisfying

$$|\mathbf{H}^{-1}| \le \frac{1}{1 - h_0} \tag{7-69}$$

Setting $\mathbf{\Gamma}_1 = \mathbf{H}^{-1}\mathbf{\Gamma}_0$ we have in view of $(\mathbf{AB})^{-1} = \mathbf{B}^{-1}\mathbf{A}^{-1}$

$$\begin{aligned} \mathbf{\Gamma}_1 &= [\mathbf{I} - \mathbf{\Gamma}_0(\mathbf{A}(\mathbf{y}^{(0)}) - \mathbf{A}(\mathbf{y}^{(1)}))]^{-1}\mathbf{A}(\mathbf{y}^{(0)})^{-1} \\ &= \{\mathbf{A}(\mathbf{y}^{(0)})[\mathbf{I} - \mathbf{\Gamma}_0(\mathbf{A}(\mathbf{y}^{(0)}) - \mathbf{A}(\mathbf{y}^{(1)}))]\}^{-1} \\ &= \{\mathbf{A}(\mathbf{y}^{(0)}) - \mathbf{A}(\mathbf{y}^{(0)}) + \mathbf{A}(\mathbf{y}^{(1)})\}^{-1} \\ &= [\mathbf{A}(\mathbf{y}^{(1)})]^{-1} \end{aligned}$$

Thus the matrix $\mathbf{A}(\mathbf{y}^{(1)})$ has been shown to have the inverse $\mathbf{\Gamma}_1 = \mathbf{H}^{-1}\mathbf{\Gamma}_0$. By (7-69) we find

$$|[\mathbf{A}(\mathbf{y}^{(1)})]^{-1}| = |\mathbf{H}^{-1}\mathbf{\Gamma}_0| \le \frac{B_0}{1 - h_0} = B_1$$

so that (7-61) holds with the index moved up by 1.

Expanding each component of $\boldsymbol{\varphi}(\mathbf{y})$ by Taylor's formula at $\mathbf{y} = \mathbf{y}^{(0)}$, we have

$$\boldsymbol{\varphi}(\mathbf{y}) = \boldsymbol{\varphi}(\mathbf{y}^{(0)}) + \mathbf{A}(\mathbf{y}^{(0)})(\mathbf{y} - \mathbf{y}^{(0)}) + \tfrac{1}{2}\mathbf{r}(\mathbf{y}) \tag{7-70}$$

where the ith component of $\mathbf{r}$ is given by

$$r_i = \sum_{j,k=1}^{n} \frac{\partial^2 \varphi_i(\mathbf{y}^{(0)} + \theta_i(\mathbf{y} - \mathbf{y}^{(0)}))}{\partial y_j\,\partial y_k}\,(y_j - y_j^{(0)})(y_k - y_k^{(0)}), \qquad 0 < \theta_i < 1$$

and hence, if $\mathbf{y}$ satisfies (7-65),

$$|\mathbf{r}| \le K\,|\mathbf{y} - \mathbf{y}^{(0)}|^2$$

By virtue of $\boldsymbol{\varphi}(\mathbf{y}^{(0)}) + \mathbf{A}(\mathbf{y}^{(0)})(\mathbf{y}^{(1)} - \mathbf{y}^{(0)}) = 0$, it follows that

$$\boldsymbol{\varphi}(\mathbf{y}^{(1)}) = \tfrac{1}{2}\mathbf{r}(\mathbf{y}^{(1)})$$

and consequently

$$|\mathbf{\Gamma}_0\boldsymbol{\varphi}(\mathbf{y}^{(1)})| \le \tfrac{1}{2}B_0 K \eta_0^2 = \tfrac{1}{2}h_0\eta_0$$

Thus finally, since $h_0 \le \frac{1}{2}$,

$$|\mathbf{\Gamma}_1\boldsymbol{\varphi}(\mathbf{y}^{(1)})| \le |\mathbf{H}^{-1}\mathbf{\Gamma}_0\boldsymbol{\varphi}(\mathbf{y}^{(1)})| \le |\mathbf{H}^{-1}|\,|\mathbf{\Gamma}_0\boldsymbol{\varphi}(y^{(0)})|$$
$$\le \tfrac{1}{2}\frac{1}{1-h_0}\, h_0\eta_0 = \eta_1 < \eta_0$$

and (7-62) is shown to be satisfied with the index increased by 1.

If $\mathbf{y}$ lies in the region defined by

$$|\mathbf{y} - \mathbf{y}^{(1)}| \le \frac{1-\sqrt{1-2h_1}}{h_1}\,\eta_1$$

where h_1 is defined by (7-67), then it follows by using (7-68) and the definitions of h_1 and η_1 that

$$|\mathbf{y} - \mathbf{y}^{(0)}| \le |\mathbf{y} - \mathbf{y}^{(1)}| + |\mathbf{y}^{(1)} - \mathbf{y}^{(0)}|$$
$$\le \frac{1-\sqrt{1-2h_1}}{h_1}\,\eta_1 + \eta_0$$
$$= \frac{1-\sqrt{1-2h_0}}{h_0}\,\eta_0$$

so that $\mathbf{y}$ still lies in the region defined by (7-65). Thus (7-65) continues to hold if the index is moved up by 1. Finally, we also have

$$h_1 = B_1\eta_1 K = \frac{B_0}{1-h_0}\frac{1}{2}\frac{h_0\eta_0}{1-h_0}\,K$$
$$= \frac{1}{2}\frac{h_0^2}{(1-h_0)^2} \le 2h_0^2 \le \frac{1}{2}$$

and (7-64) is still satisfied, too.

It is now clear that the above argument can be repeated, with $\mathbf{y}^{(1)}, \mathbf{y}^{(2)}$ taking the place of $\mathbf{y}^{(0)}, \mathbf{y}^{(1)}$. If the numbers η_ν and h_ν are recursively defined by

$$\eta_\nu = \frac{1}{2}\frac{h_{\nu-1}\eta_{\nu-1}}{1-h_{\nu-1}}, \qquad h_\nu = \frac{1}{2}\frac{h_{\nu-1}^2}{(1-h_{\nu-1})^2}, \qquad \nu = 1, 2, \cdots$$

we find that

$$|\mathbf{y}^{(\nu+1)} - \mathbf{y}^{(\nu)}| \le \eta_\nu, \qquad \nu = 1, 2, \cdots$$

By induction we may easily show that

$$h_\nu \le \tfrac{1}{2}(2h_0)^{2^\nu}, \qquad \nu = 1, 2, \cdots$$

and hence

$$\frac{1}{2}\frac{1}{1-h_{\nu-1}} \le 1$$

and

$$\eta_\nu \leq h_{\nu-1}\eta_{\nu-1} \leq \cdots \leq h_{\nu-1} \cdots h_0\eta_0$$

$$\leq \frac{1}{2^\nu}(2h_0)^{2^{\nu-1}+2^{\nu-2}+\cdots+1}\eta_0 = \frac{1}{2^\nu}(2h_0)^{2^\nu-1}\eta_0$$

It follows that for any integer $p \geq 1$

$$|\mathbf{y}^{(\nu+p)} - \mathbf{y}^{(\nu)}| \leq |\mathbf{y}^{(\nu+p)} - \mathbf{y}^{(\nu+p-1)}| + \cdots + |\mathbf{y}^{(\nu+1)} - \mathbf{y}^{(\nu)}|$$

$$\leq \eta_{\nu+p-1} + \eta_{\nu+p-2} + \cdots + \eta_\nu$$

Using the algebraic identity

$$\eta_\nu \frac{1 - \sqrt{1 - 2h_\nu}}{h_\nu} - \eta_{\nu+1}\frac{1 - \sqrt{1 - 2h_{\nu+1}}}{h_{\nu+1}} = \eta_\nu, \qquad \nu = 1, 2, \cdots$$

we deduce that

$$|\mathbf{y}^{(\nu+p)} - \mathbf{y}^{(\nu)}| \leq \eta_\nu \frac{1 - \sqrt{1 - 2h_\nu}}{h_\nu} - \eta_{\nu+p}\frac{1 - \sqrt{1 - 2h_{\nu+p}}}{h_{\nu+p}}$$

$$\leq \eta_\nu \frac{1 - \sqrt{1 - 2h_\nu}}{h_\nu}$$

or, using

$$\frac{1 - \sqrt{1 - 2h_\nu}}{h_\nu} \leq 2$$

that

$$|\mathbf{y}^{(\nu+p)} - \mathbf{y}^{(\nu)}| \leq 2\eta_\nu \leq \frac{1}{2^{\nu-1}}(2h_0)^{2^\nu-1}\eta_0 \tag{7-71}$$

Since the term on the right is independent of p and tends to zero as $\nu \to 0$, it follows that the n sequences $\{y_i^{(\nu)}\}$ formed with the components of $\mathbf{y}^{(\nu)}$ are Cauchy sequences and thus have limits y_i^* $(i = 1, 2, \cdots, n)$. Calling $\mathbf{y}^*$ the vector with the components y_i^*, we thus have

$$\lim_{\nu\to\infty} \mathbf{y}^{(\nu)} = \mathbf{y}^*$$

Since all $\mathbf{y}^{(\nu)}$ belong to the compact set defined by (7-65), $\mathbf{y}^*$ also does. The inequality (7-66) follows by letting $p \to \infty$ in (7-71).

It remains to be shown that $\mathbf{y}^*$ is a solution of the given equation (7-58). This is accomplished by considering the identity

$$\boldsymbol{\varphi}(\mathbf{y}^{(\nu)}) = \mathbf{A}(\mathbf{y}^{(\nu)})(\mathbf{y}^{(\nu)} - \mathbf{y}^{(\nu+1)})$$

Using (7-56) we have

$$\begin{aligned}|\mathbf{A}(\mathbf{y}^{(\nu)})| &\le |\mathbf{A}(\mathbf{y}^{(0)})| + |\mathbf{A}(\mathbf{y}^{(\nu)}) - \mathbf{A}(\mathbf{y}^{(0)})| \\ &\le |\mathbf{A}(\mathbf{y}^{(0)})| + K\frac{1-\sqrt{1-2h_0}}{h_0} \\ &= C\end{aligned}$$

say. Hence

$$|\boldsymbol{\varphi}(\mathbf{y}^{(\nu)})| \le C\,|\mathbf{y}^{(\nu)} - \mathbf{y}^{(\nu+1)}|$$

Letting $\nu \to \infty$ and using the fact that $\boldsymbol{\varphi}(\mathbf{y})$ is continuous, we get

$$|\boldsymbol{\varphi}(\mathbf{y}^*)| = 0 \qquad \text{and thus} \qquad \boldsymbol{\varphi}(\mathbf{y}^*) = \mathbf{0}$$

This completes the proof of Theorem 7.6.

If the bound for $|\mathbf{y}^{(\nu)} - \mathbf{y}^*|$ given by (7-66) is denoted by b_ν, we have

$$\log b_\nu = (\nu - 1)\log \tfrac{1}{2} + (2^\nu - 1)\log 2h_0 + \log \eta_c$$

and hence, if $2h_0 < 1$,

$$\frac{\log b_{\nu+1}}{\log b_\nu} \to 2 \qquad (\nu \to \infty) \tag{7-72}$$

Generally, if b_ν is a bound for the error at the νth step of an iterative method for solving equations, the best possible value of the limit (7-72) is called the *order of convergence of the method.* By considering simple examples (see Problem 14) it is easily shown that the value 2 in (7-72) cannot be improved for Newton's method. The order of convergence of Newton's method is thus 2. This fact is sometimes expressed by saying that the method is *quadratically convergent* or that "the number of correct decimal places is doubled at each step."

It can be shown that if in Newton's algorithm (7-60) the matrices $\mathbf{A}(\mathbf{y}^{(\nu)})$ are replaced by $\mathbf{A}(\mathbf{y}^{(0)})$ at each step, the resulting sequence $\mathbf{y}^{(1)}$, $\mathbf{y}^{(2)}, \cdots$ still converges to $\mathbf{y}^*$, provided that the initial approximation $\mathbf{y}^{(0)}$ is sufficiently close. However, the order of convergence of this modified version of Newton's process is only 1 (see Problem 16).

7.2-5. Convergence of Newton's method for the system of difference equations (7-24). We shall verify that, under certain hypotheses, the conditions of Theorem 7.6 are satisfied for the system of (nonlinear) difference equations (7-24). We shall assume throughout that (7-40) holds, so that $\mathbf{A}(\mathbf{y})$ is monotone. In view of (7-42), since

$$\sum_{n=1}^{N-1} j_{mn} = \frac{m(N-m)}{2} \le \frac{N^2}{8} \qquad (m = 1, 2, \cdots, N-1)$$

it follows that (7-61) is satisfied for

$$B_0 = \frac{(b-a)^2}{8h^2} \tag{7-73}$$

Since

$$\varphi_n(\mathbf{y}) = -y_{n-1} + 2y_n - y_{n+1} + h^2\{\beta_0 f(x_{n-1}, y_{n-1}) + \beta_1 f(x_n, y_n) + \beta_2 f(x_{n+1}, y_{n+1})\}$$

condition (7-63) is in view of $\beta_0 + \beta_1 + \beta_2 = 1$ satisfied by

$$K = h^2 L_2 = h^2 \max_{\substack{x\in[a,b]\\ -\infty<y<\infty}} |f_{yy}(x, y)| \tag{7-74}$$

Let the initial approximation $\mathbf{y}^{(0)}$ be defined by

$$y_n^{(0)} = z(x_n), \qquad n = 1, \cdots, N-1 \tag{7-75}$$

where $z(x)$ is a $p + 2$ times differentiable function satisfying $z(a) = A$, $z(b) = B$, p being the order of the difference operator, and let

$$\begin{aligned} Z &= \max_{a\le x\le b} |z^{(p+2)}(x)| \\ R &= \max_{a\le x\le b} |z''(x) - f(x, z(x))| \end{aligned} \tag{7-76}$$

It then follows by the results of §6.1 that for a certain constant G,

$$|-z(x-h) + 2z(x) - z(x+h) + h^2\{\beta_0 z''(x-h) + \beta_1 z''(x) + \beta_2 z''(x+h)\}| \le h^{p+2}GZ$$

and consequently, in view of $\beta_i \ge 0$, $\beta_0 + \beta_1 + \beta_2 = 1$,

$$\begin{aligned} |r_n(y^{(0)})| = |&-z(x_{n-1}) + 2z(x_n) - z(x_{n+1}) \\ &+ h^2\{\beta_0 f(x_{n-1}, z(x_{n-1})) + \beta_1 f(x_n, z(x_n)) + \beta_2 f(x_{n+1}, z(x_{n+1}))\}| \\ &\le h^2 R + h^{p+2}GZ, \qquad n = 1, 2, \cdots, N-1 \end{aligned}$$

Using (7-73) it follows that (7-62) is satisfied for

$$\eta_0 = \frac{(b-a)^2}{8}(R + h^p GZ) \tag{7-77}$$

The main condition (7-62) of Kantorovich's theorem, which guarantees convergence of the Newton process, thus turns out to be satisfied if

$$\frac{(b-a)^4 L_2(R + h^p GZ)}{64} \le \frac{1}{2} \tag{7-78}$$

Various conclusions may be drawn from this inequality. If as an initial approximation $z(x)$ a polynomial of degree $<(p + 2)$ is chosen, then

$Z = 0$, and the method converges at least if the quantity R (which may be interpreted as the maximum error by which the differential equation fails to hold) satisfies

$$R \le \frac{32}{(b-a)^4 L_2} \tag{7-79}$$

For theoretical purposes, we may assume that the initial approximation $z(x)$ is the *exact* solution $y(x)$. We then have $R = 0$, and it follows from (7-78) that Newton's method will converge to a solution of the system of difference equations if

$$h^p \le \frac{32}{(b-a)^4 L_2 G Z} \tag{7-80}$$

i.e., for all sufficiently small values of h.

Having thus demonstrated the *existence* of a solution of the finite difference scheme, we now shall prove the *uniqueness* of this solution for problems of class M. Indeed, if $\mathbf{y}$ and $\mathbf{z}$ are any two solutions, then we may put

$$f(x_n, z_n) - f(x_n, y_n) = \eta_n(z_n - y_n)$$

where η_n is a value of f_y. By (7-2) and by the Lipschitz condition, $0 \le \eta_n \le L$. The vector $\mathbf{d} = \mathbf{z} - \mathbf{y}$ then is easily seen to satisfy the system

$$(\mathbf{J} + h^2\mathbf{BH})\mathbf{d} = \mathbf{0}$$

where $\mathbf{J}$ and $\mathbf{B}$ are defined in §7.1-3, and where $\mathbf{H}$ is the diagonal matrix with diagonal elements η_n. By Theorem 7.4 the matrix $\mathbf{J} + h^2\mathbf{BH}$ is monotone for $h^2 \le L^{-1}$ and hence nonsingular. It follows that $\mathbf{d} = \mathbf{0}$.

We summarize these statements in the following theorem, which is the principal result of this section.

THEOREM 7.7. *Let the system of finite difference equations* (7-24) *have arisen from a boundary problem of class M. If p denotes the order of the finite difference operator* (7-8), *assume that the exact solution $y(x)$ has a continuous* $(p + 2)$nd *derivative in $[a, b]$, and let*

$$Z = \max_{a \le x \le b} |y^{(p+2)}(x)| \tag{7-81}$$

Assume that G is defined as in §6.1-3, *and that*

$$\beta_i \ge 0, \qquad i = 0, 1, 2, \qquad \text{and} \qquad Lh^2 < 1 \tag{7-82}$$

where L denotes the Lipschitz constant of $f(x, y)$. If L_2 is an upper bound for $f_{yy}(x, y)$ in $a \le x \le b$, $-\infty < y < \infty$, then (7-24) *possesses a unique solution for all values of h satisfying* (7-80). *Provided that the initial approximation is determined by* (7-75), *where $z(x)$ satisfies* (7-78), *this solution can be found by Newton's method.*

This theorem contains the existence of a unique difference solution for a linear boundary value problem as the special case $L_2 = 0$. Newton's method then converges in one step.

7.3. The Discretization Error in Boundary Problems of Class M

As before, we mean by discretization error the quantity $e_n = y_n - y(x_n)$, where y_n is the exact solution of the finite difference scheme (7-24) (calculated without round-off), and $y(x)$ is the exact solution of the boundary value problem. We shall establish both bounds and asymptotic formulas for the discretization error.

7.3-1. A priori bounds. A first bound for the discretization error can be obtained as a corollary of the convergence theorem for Newton's method (Theorem 7.6). If Newton's method is started with $z(x) = y(x)$, then we have, denoting by $\mathbf{e}$ the vector with components e_n,

$$|\mathbf{y}^{(0)} - \mathbf{y}^*| = |\mathbf{e}|$$

Assuming that (7-80) holds, we find from (7-65), using (7-77),

$$|\mathbf{e}| \le \frac{1 - \sqrt{1 - 2h_0}}{h_0} \frac{(b - a)^2 GZh^p}{8} \tag{7-83}$$

where h_0 is defined in (7-64). Using the fact that

$$1 < \frac{1 - \sqrt{1 - 2h_0}}{h_0} \le 2 \qquad \text{for } 0 < h_0 \le \tfrac{1}{2}$$

we can deduce from (7-83) the simpler bound

$$|\mathbf{e}| \le \frac{(b - a)^2 GZh^p}{4} \tag{7-84}$$

We now shall prove a slightly improved and generalized version of this bound without the detour through Newton's method.

THEOREM 7.8. *Instead of* (7-24), *let the values* y_n *satisfy the equations*

$$\begin{aligned} -y_{n-1} + 2y_n - y_{n+1} + h^2\{\beta_0 f(x_{n-1}, y_{n-1}) + \beta_1 f(x_n, y_n) \\ + \beta_2 f(x_{n+1}, y_{n+1})\} = \theta_n K h^{q+2}, \qquad n = 1, 2, \cdots, N - 1 \end{aligned} \tag{7-85}$$

where the θ_n *are arbitrary numbers satisfying* $|\theta_n| \le 1$, *and where* K *and* q *are arbitrary nonnegative constants. Then, in the notation of Theorem* 7.7, *if* (7-80) *holds, the discretization error satisfies*

$$|e_n| \le \frac{(x_n - a)(b - x_n)}{2} (GZh^p + Kh^q), \qquad n = 1, 2, \cdots, N - 1 \tag{7-86}$$

Proof. The exact solution $y(x)$ satisfies, according to §6.1-3,

$$
\begin{aligned}
(7\text{-}87)\quad -y(x_{n-1}) + 2y(x_n) - y(x_{n+1}) + h^2\{\beta_0 f(x_{n-1}, y(x_{n-1})) \\
+ \beta_1 f(x_n, y(x_n)) + \beta_2 f(x_{n+1}, y(x_{n+1}))\} = \theta_n' GZh^{p+2}
\end{aligned}
$$

where $|\theta_n'| \le 1$. Subtracting this relation from (7-85) we get, applying the mean value theorem to the differences $f(x_n, y_n) - f(x_n, y(x_n))$,

$$
\begin{aligned}
-e_{n-1} + 2e_n + e_{n+1} + h^2\{\beta_0 g_{n-1}e_{n-1} + \beta_1 g_n e_n + \beta_2 g_{n+1}e_{n+1}\} \\
= \theta_n''(h^{q+2}K + h^{p+2}GZ)
\end{aligned}
$$

where $0 \le g_n \le L$. Denoting by $\mathbf{G}$ the diagonal matrix with elements g_n $(n = 1, \cdots, N-1)$ and defining the matrices $\mathbf{J}$ and $\mathbf{B}$ as in §7.1-3, we thus have

$$
(\mathbf{J} + h^2\mathbf{BG})\mathbf{e} = (h^{q+2}K + h^{p+2}GZ)\boldsymbol{\theta}
$$

where $\boldsymbol{\theta}$ is a vector whose components numerically do not exceed 1. For $h^2L \le 1$ the matrix $\mathbf{J} + h^2\mathbf{BG}$ satisfies the conditions of Theorem 7.4, and furthermore we have $\mathbf{J} + h^2\mathbf{BG} \ge \mathbf{J}$. From Theorem 7.5 it follows that

$$
\mathbf{0} \le (\mathbf{J} + h^2\mathbf{BG})^{-1} \le \mathbf{J}^{-1}
$$

If $\mathbf{J}^{-1} = (j_{mn})$, we have

$$
|e_m| \le (h^{q+2}K + h^{p+2}GZ)\sum_{n=1}^{N-1} j_{mn}
$$

The assertion of Theorem 7.8 now follows from the fact that

$$
\begin{aligned}
\sum_{n=1}^{N-1} j_{mn} &= \frac{1}{N}[(N-m)(1 + 2 + \cdots + m) + m(1 + 2 + \cdots + N - m - 1)] \\
&= \frac{m(N-m)}{2} = \frac{(x_m - a)(b - x_m)}{2h^2}
\end{aligned}
$$

For $K = 0$ (difference equations satisfied exactly), small values of h, and $x_m = \frac{1}{2}(a + b)$ the bound (7-86) is virtually equivalent to (7-84). The reader is asked to verify that the bound is sharp for the boundary value problem $y'' = x^{p+2}$, $y(a) = 0$, $y(b) = 0$.

7.3-2. Asymptotic behavior of the discretization error. In addition to the hypotheses of Theorem 7.8 we shall now assume that the order $p \ge 2$ and that $K = 0$, so that the difference equations (7-24) are satisfied exactly. It then follows that $|\mathbf{e}| = O(h^2)$. We shall also assume that the difference operator satisfies

$$
(7\text{-}88)\qquad \beta_0 = \beta_2
$$

and that the exact solution $y(x)$ of the problem is $p + 4$ times continuously differentiable. [Both the hypotheses $p \ge 2$ and (7-88) are satisfied for the most commonly used operators (7-9) and (7-10).] We then have

$$\begin{aligned}&-y(x_{n-1}) + 2y(x_n) - y(x_{n+1})\\&+ h^2\{\beta_0 f(x_{n-1}, y(x_{n-1})) + \beta_1 f(x_n, y(x_n)) + \beta_2 f(x_{n+1}, y(x_{n+1}))\}\\&\qquad = -C_{p+2}y^{(p+2)}(x_n)h^{p+2} + O(h^{p+4})\end{aligned}$$

(In view of $\beta_0 + \beta_1 + \beta_2 = 1$, C_{p+2} is identical with the error constant C in the present situation.) Again we subtract from (7-85). Since $e_n = O(h^2)$,

$$f(x_n, y_n) - f(x_n, y(x_n)) = g(x_n)e_n + O(h^4)$$

Furthermore, by virtue of (7-88),

$$y^{(p+2)}(x_n) = \beta_0 y^{(p+2)}(x_{n-1}) + \beta_1 y^{(p+2)}(x_n) + \beta_2 y^{(p+2)}(x_{n+1}) + O(h^2)$$

Dividing by h^p and defining the magnified error $\bar{e}_n$ by $\bar{e}_n = h^{-p}e_n$, we can write the resulting relation in the form

$$-\bar{e}_{n-1} + 2\bar{e}_n - \bar{e}_{n+1} + h^2\{\beta_0\Phi_{n-1} + \beta_1\Phi_n + \beta_2\Phi_{n+1}\} = O(h^4)$$

where

$$\Phi_n = g(x_n)\bar{e}_n - Cy^{(p+2)}(x_n), \qquad n = 1, 2, \cdots, N-1$$

The same relation [without the $O(h^4)$ term on the right-hand side] would result by solving the boundary value problem

(7-89) $$e''(x) = g(x)e(x) - Cy^{(p+2)}(x), \qquad e(a) = e(b) = 0$$

by the finite difference method on hand, denoting by $\bar{e}_n$ the approximation to $e(x_n)$. Since the problem (7-89) is clearly of class M, we may appeal to Theorem 7.8 to conclude that

$$\bar{e}_n = e(x_n) + O(h^2), \qquad n = 1, 2, \cdots, N-1$$

where $e(x)$ is the solution of (7-89). We thus have proved:

THEOREM 7.9. *Under the hypotheses stated at the beginning of this section, the errors e_n satisfy*

(7-90) $$e_n = h^p e(x_n) + O(h^{p+2}), \qquad n = 1, 2, \cdots, N-1$$

where $e(x)$ denotes the solution of the boundary value problem (7-89).

Of interest here is the fact that the exponent in the remainder term in (7-90) is $p + 2$ and not $p + 1$, as might be expected. The practical significance of this result is that Richardson's deferred approach to the limit, if applied as explained in §2.2-7, will improve the accuracy by two orders of magnitude (provided that there is no round-off and that h is already sufficiently small). See also Problem 16 of Chapter 6 in this connection.

Although the analysis of this section assumes that the boundary value problem under consideration is of class M, it is believed that the result (7-90) also holds for some problems which do not satisfy this assumption.

7.3-3. The difference correction. Let us assume that the given differential equation (7-1) is linear. The function $g(x)$ in (7-89) is then simply the coefficient of y in the given differential equation. Still we cannot use (7-89) to determine $e(x)$ in an actual problem, because $y(x)$ and hence $y^{(p+2)}(x)$ is unknown. Heuristically, however, we may proceed as follows. If the $O(h^{p+2})$ term in (7-90) were zero, we would have

$$y_n = y(x_n) + h^p e(x_n) \tag{7-91}$$

and hence, assuming that $p = 2q$ is even and that $e(x)$ is sufficiently differentiable, using the result of Problem 32 of Chapter 5

$$\nabla^{p+2} y_{n+q+1} = h^{p+2} y^{(p+2)}(x_n) + h^{2p+2} e^{(p+2)}(x_n) + O(h^{p+4})$$

Hence

$$y^{(p+2)}(x_n) = h^{-p-2} \nabla^{p+2} y_{n+q+1} + O(h^2)$$

Thus, if in a finite difference approximation for the boundary problem (7-89) the values $y^{(p+2)}(x_n)$ are replaced by $h^{-p-2}\nabla^{p+2}y_{n+q+1}$ an error of only $O(h^4)$ is committed in the difference approximation, and by Theorem 7.8 the solution e_n^* of the difference scheme differs from $e(x_n)$ by only $O(h^2)$. By (7-91), the values $y_n + h^p e_n^*$ then differ from $y(x_n)$ by only $O(h^{p+2})$.

The quantities $h^p e_n^*$ represent what is called the *difference correction* by L. Fox [1957], where numerous examples and refinements of the technique are described. The method is also applicable to nonlinear equations; in this case the quantities $g(x_n)$ are replaced by $f_y(x_n, y_n)$. The exact justification of the method hinges on the fact that, under suitable smoothness assumptions, (7-90) can be improved to

$$e_n = h^p e_p(x_n) + h^{p+2} e_{p+2}(x_n) + O(h^{p+4}) \tag{7-92}$$

where $e_p(x) = e(x)$, and $e_{p+2}(x)$ is a certain continuous function of x. We relegate the proof of (7-92) and the precise determination of $e_{p+2}(x)$ to the problem section (see Problem 21).

7.4. Influence of Round-off Error

7.4-1. A priori bounds. The values $\tilde{y}_n$ resulting from carrying out the algorithm described in §7.1 on a specific computer in general will not satisfy the relations (7-24) exactly. Instead, they satisfy relations which we write in the form

$$(7\text{-}93) \quad -\tilde{y}_{n-1} + 2\tilde{y}_n - \tilde{y}_{n+1} + h^2\{\beta_0 f(x_{n-1}, \tilde{y}_{n-1}) + \beta_1 f(x_n, \tilde{y}_n) + \beta_2 f(x_{n+1}, \tilde{y}_{n+1})\} = \varepsilon_n, \qquad n = 1, 2, \cdots, N-1$$

We shall again call ε_n the *local round-off error.* It must be realized, however, that the presence of ε_n may be due not only to rounding but also to the fact that Newton's method has been terminated after a finite number of steps, or that the system (7-24) has not been solved accurately for other reasons. In any event, ε_n can be determined, *up to a round-off error in the literal sense,* by substituting the calculated values $\tilde{y}_n$ into the given equations.

If we assume that

$$(7\text{-}94) \qquad |\varepsilon_n| \leq \varepsilon, \qquad n = 1, 2, \cdots, N-1$$

a bound for the round-off error $r_n = \tilde{y}_n - y_n$ is readily established by the method by which we proved Theorem 7.8. The result is

$$(7\text{-}95) \qquad |r_n| \leq \frac{(x_n - a)(b - x_n)}{2} \frac{\varepsilon}{h^2}$$

This bound is of the order of h^{-2} which, as we shall see later, represents the true order of magnitude of r_n if all ε_n have the same sign.

7.4-2. Refined bounds. We now replace the assumption (7-94) by

$$|\varepsilon_n| \leq p(x_n)\varepsilon, \qquad n = 1, 2, \cdots, N-1$$

where $p(x)$ is a nonnegative, continuous function for $a \leq x \leq b$ and

$$(7\text{-}96) \qquad \varepsilon \leq K_1 h^3$$

It then follows from Theorem 7.8 [letting $q = 1$ and $K = K_1 \max p(x)$] that $\tilde{y}_n - y(x_n) = O(h)$. We thus have, letting $g(x) = f_y(x, y(x))$,

$$(7\text{-}97) \qquad f(x_n, \tilde{y}_n) - f(x_n, y(x_n)) = g(x_n)r_n + \theta_n K_2 h^2$$

where $|\theta_n| \leq 1$ and K_2 is a constant, independent of h, which is zero if $f_{yy} = 0$, i.e., for linear differential equations. Using (7-97) when subtracting (7-93) from (7-85), we find, letting $g(x_n) = g_n$,

$$-r_{n-1} + 2r_n - r_{n+1} + h^2\{\beta_0 g_{n-1} r_{n-1} + \beta_1 g_n r_n + \beta_2 g_{n+1} r_{n+1}\} = \varepsilon_n + \theta_n K_2 h^4$$

or, using matrix notation and defining the vectors $\mathbf{r}$, $\boldsymbol{\epsilon}$, and $\boldsymbol{\theta}$ in the obvious way,

$$(\mathbf{J} + h^2\mathbf{BG})\mathbf{r} = \boldsymbol{\epsilon} + h^4 K_4 \boldsymbol{\theta}$$

Assuming that (7-40) holds, so that $\mathbf{J} + h^2\mathbf{BG}$ is monotone and hence nonsingular, we have

$$\mathbf{r} = \mathbf{r}^{(1)} + \mathbf{r}^{(2)},$$

where $\mathbf{r}^{(1)}$, called the *primary round-off error*, is given by

$$\mathbf{r}^{(1)} = (\mathbf{J} + h^2\mathbf{BG})^{-1}\boldsymbol{\epsilon}$$

and the *secondary round off error* $\mathbf{r}^{(2)}$ by

$$\mathbf{r}^{(2)} = h^4K_2(\mathbf{J} + h^2\mathbf{BG})^{-1}\boldsymbol{\theta}$$

Using the fact that $\mathbf{0} \le (\mathbf{J} + h^2\mathbf{BG})^{-1} \le \mathbf{J}^{-1}$ and $|\mathbf{J}^{-1}| = (b-a)^2/8h^2$, we readily find that

$$|\mathbf{r}^{(2)}| \le \tfrac{1}{8}h^2K_2(b-a)^2$$

and thus $\mathbf{r}^{(2)} = O(h^2)$, whereas $\mathbf{r}^{(1)}$ must be expected to be $O(h)$. For small values of h, the major contribution to $\mathbf{r}$ thus arises from the primary round-off error, and the effects of nonlinearity, which are represented by $\mathbf{r}^{(2)}$, may be neglected.

We define the matrix $\mathbf{D}$ by

(7-98) $$\mathbf{D} \equiv h^{-1}(d_{mn}) = (\mathbf{J} + h^2\mathbf{BG})^{-1}$$

Although the elements of $\mathbf{D}$ depend on h, they are known to be nonnegative for sufficiently small h and to be bounded by the corresponding elements of $\mathbf{J}^{-1}$. Postponing a more precise determination of $\mathbf{D}$ to the next section, we have

(7-99) $$|r_m^{(1)}| \le \frac{\varepsilon}{h}\sum_{n=1}^{N-1} p_n d_{nm}$$

where $p_n = p(x_n)$.

7.4-3. The matrix D. In this section we shall study the behavior as $h \to 0$ of the elements d_{nm} of the matrix $h\mathbf{D}$.

Let $g(x) = f_y(x, y(x))$, and let the functions $s(x)$ and $t(x)$ be defined respectively as the solutions of the initial value problem

(7-100) $$s'' = g(x)s, \qquad s(a) = 0, \qquad s'(a) = 1$$

and the "end value" problem

(7-101) $$t'' = g(x)t, \qquad t(b) = 0, \qquad t'(b) = -1$$

We note that the Wronskian determinant of $s(x)$ and $t(x)$,

(7-102) $$W(x) = s'(x)t(x) - t'(x)s(x)$$

is constant, since

$$W'(x) = s''(x)t(x) - t''(x)s(x)$$
$$= g(x)[s(x)t(x) - t(x)s(x)] = 0$$

The constant value W of $W(x)$ is different from zero, because, for instance,

$$W(b) = -t'(b)s(b) = s(b)$$

and $s(b) \geq b - a > 0$ by virtue of (7-100).

We now introduce a function $G(x, \xi)$, defined for $x \in [a, b]$ and $\xi \in [a, b]$ by

$$G(x, \xi) = \begin{cases} W^{-1}s(x)t(\xi), & x \leq \xi \\ W^{-1}s(\xi)t(x), & x \geq \xi \end{cases} \tag{7-103}$$

The function $G(x, \xi)$ is called *Green's function* of the boundary value problem

$$y'' = g(x)y, \qquad y(a) = A, \qquad y(b) = B$$

and plays an important role in many theoretical studies of this linear boundary value problem. It is easily shown from the definition of $G(x, \xi)$ and of the functions $s(x)$ and $t(x)$ that the following relations hold:

$$G(a, \xi) = G(b, \xi) = 0, \qquad a \leq \xi \leq b \tag{7-104a}$$

$$\frac{d^2}{dx^2} G(x, \xi) = g(x)G(x, \xi), \qquad a \leq x < \xi,\ \xi < x \leq b \tag{7-104b}$$

$$\lim_{x \to \xi - 0} \frac{d}{dx} G(x, \xi) - \lim_{x \to \xi + 0} \frac{d}{dx} G(x, \xi) = 1, \qquad a < \xi < b \tag{7-104c}$$

Relation (7-104c) states that the derivative of $G(x, \xi)$ with respect to x has a jump discontinuity at $x = \xi$, and that the difference of the left and the right derivative at $x = \xi$ is 1. It can also be shown that $G(x, \xi)$ is the only function defined for $x, \xi \in [a, b]$ which satisfies the three relations (7-104). As a consequence of either (7-103) or (7-104) it also follows that

$$G(x, \xi) = G(\xi, x), \qquad x, \xi \in [a, b] \tag{7-105}$$

We now shall prove:

THEOREM 7.10. *Let p, the order of the finite difference operator* (7–8), *be at least* 2, *and let $g(x)$ be twice continuously differentiable in* $[a, b]$, $g(x) \geq 0$. *Then there exists a constant K such that the numbers d_{mn} defined by* (7-98) *satisfy*

$$|d_{mn} - G(x_m, x_n)| \leq h^2 K \tag{7-106}$$

for all h satisfying (7-32) *and all* $x_m, x_n \in [a, b]$.

Proof. It follows from $(\mathbf{J} + h^2\mathbf{BG})\mathbf{D} = \mathbf{I}$ that

$$-d_{m-1,n} + 2d_{mn} - d_{m+1,n} + h^2\{\beta_0 g_{m-1} d_{m-1,n} + \beta_1 g_m d_{mn} + \beta_2 g_{m+1} d_{m+1,n}\} = \begin{cases} 0, & m \neq n \\ h, & m = n \end{cases} \tag{7-107}$$

We set, for any function $z(x)$ defined in $[a, b]$ and all x such that $x + h$ and $x - h$ are in $[a, b]$,

$$L[z(x);\ h] = -z(x-h) + 2z(x) - z(x+h) + h^2\{\beta_0 g(x-h)z(x-h) + \beta_1 g(x)z(x) + \beta_2 g(x+h)z(x+h)\}$$

By the method used in §6.1-3 we can easily show that, if $g(x)z(x) = z''(x)$ and if $z(x)$ is four times continuously differentiable,

$$|L[z(x);\ h]| \le h^4 \Lambda Z_4 \tag{7-108}$$

where Λ is a constant which depends on β_0, β_1, β_2 only and where Z_4 is a bound for $|z^{\mathrm{iv}}(x)|$.

Now let $0 < m < n$. Assuming that L acts on the first variable in $G(x, \xi)$, we then have

$$L[G(x_m, x_n);\ h] = t(x_n)L[s(x_m);\ h]$$

and thus, denoting by S_i and T_i upper bounds for the functions $|s^{(i)}(x)|$ and $|t^{(i)}(x)|$ respectively ($i = 0, 1, 2, \cdots$),

$$L[G(x_m, x_n);\ h] = \theta_{mn} \frac{T\Lambda S_4}{W} h^4 \tag{7-109a}$$

where $|\theta_{mn}| \le 1$. Similarly, we find for $n < m < N$

$$L[G(x_m, x_n);\ h] = \theta_{mn} \frac{S\Lambda T_4}{W} h^4 \tag{7-109b}$$

where again $|\theta_{mn}| \le 1$. For $m = n$ we have, still assuming that L acts only on the first argument in G:

$$WL[G(x_n, x_n);\ h] = t(x_n)[s(x_n) - s(x_{n-1})] + s(x_n)[t(x_n) - t(x_{n+1})] + h^2\{\beta_0 t(x_n)g(x_{n-1})s(x_{n-1}) + \beta_1 g(x_n)t(x_n)s(x_n) + \beta_2 s(x_n)g(x_{n+1})t(x_{n+1})\}$$

Expanding the expression on the right in powers of h and making use of $gs = s''$, $gt = t''$, we find in view of (7-104c) that

$$L[G(x_n, x_n);\ h] = h + \theta_n h^3(\tfrac{1}{6} + |\beta_0|)\frac{S_3T + ST_3}{W} \tag{7-110}$$

where $|\theta_n| \le 1$.

We now set

$$e_{mn} = d_{mn} - G(x_m, x_n)$$

The above results imply that

$$-e_{m-1,n} + 2e_{m,n} - e_{m+1,n} + h^2\{\beta_0 g_{m-1}e_{m-1,n} + \beta_1 g_m e_{m,n} + \beta_2 g_{m+1}e_{m+1,n}\} = \begin{cases} \theta_{mn}K_1h^4, & m \ne n \\ \theta_n K_2 h^3, & m = n \end{cases} \tag{7-111}$$

where

$$K_1 = \max\left(\frac{T\Lambda S_4}{W}, \frac{S\Lambda T_4}{W}\right), \qquad K_2 = (\tfrac{1}{6} + |\beta_0|)\,\frac{S_3T + T_3S}{W}$$

Regarding (7-111) as a system of $N - 1$ linear equations for the $N - 1$ unknowns $e_{1,n}, \cdots, e_{N-1,n}$ with the matrix $\mathbf{J} + h^2\mathbf{BG}$, it follows from

$$\mathbf{0} \le (\mathbf{J} + h^2\mathbf{BG})^{-1} \le \mathbf{J}^{-1}$$

using the explicit representation (7-45) of $\mathbf{J}^{-1}$, that

$$|e_{m,n}| \le \left[\frac{(b-a)^2K_1}{8} + \frac{(b-a)K_2}{4}\right]h^2$$

establishing the desired result.

Using (7-99) and approximating the sum by an integral, we now have

$$|r_n^{(1)}| \le \frac{\varepsilon}{h^2}\{m(x_n) + O(h)\} \tag{7-112}$$

where

$$m(x) = \int_a^b G(x, \xi)p(\xi)\, d\xi \tag{7-113}$$

Using the properties (7-108), it can be verified that the function $m(x)$ is the solution of the boundary value problem

$$m'' = g(x)m + p(x), \qquad m(a) = m(b) = 0 \tag{7-114}$$

7.4-4. Statistical theory of round-off. Turning to the statistical model of round-off, we now assume that the local round-off errors are random variables with $E(r_n) = \mu_n$, where $|\mu_n| \le \mu p(x_n)$, $p(x)$ being defined as in §7.4-2. We do not assume that the random variables ε_n are independent. Instead, we let

$$\begin{aligned}\text{covar}\,(\boldsymbol{\epsilon}) &= E[(\boldsymbol{\epsilon} - \boldsymbol{\mu})(\boldsymbol{\epsilon}^T - \boldsymbol{\mu}^T)] \\ &= \sigma^2\mathbf{C}\end{aligned}$$

where the matrix $\mathbf{C} = (c_{mn})$ is assumed to be known. In view of $\mathbf{r}^{(1)} = \mathbf{D}\boldsymbol{\epsilon}$ we then have $E(\mathbf{r}^{(1)}) = \mathbf{D}\boldsymbol{\mu}$ and hence

$$|E(r_n^{(1)})| \le \frac{\mu}{h^2}\{m(x_n) + O(h)\}$$

where $m(x)$ is defined by (7-113) or (7-114).

For the covariance matrix we find, using the rules established in §1.5-3,

$$\begin{aligned}\text{covar}(\mathbf{r}) &= E[(\mathbf{r} - E(\mathbf{r}))(\mathbf{r}^T - E(\mathbf{r}^T))] \\ &= E[\mathbf{D}(\boldsymbol{\epsilon} - \boldsymbol{\mu})(\boldsymbol{\epsilon}^T - \boldsymbol{\mu}^T)\mathbf{D}^T] \\ &= \mathbf{D}E[(\boldsymbol{\epsilon} - \boldsymbol{\mu})(\boldsymbol{\epsilon}^T - \boldsymbol{\mu})]\mathbf{D}^T \\ &= \sigma^2\mathbf{D}\mathbf{C}\mathbf{D}^T\end{aligned}$$

Of particular interest are the diagonal elements of the matrix covar (**r**), which represent the variances of the quantities $r_n^{(1)}$. For these elements we find

$$\text{var}(r_m^{(1)}) = \frac{\sigma^2}{h^2}\sum_{n=1}^{N-1}\sum_{k=1}^{N-1} d_{mn}c_{mk}d_{km} \tag{7-115}$$

If there exists a continuous function $c(\xi, \eta)$, defined for $a \le \xi \le b$ and $a \le \eta \le b$, such that $c_{mn} = c(x_n, x_m)$, then the sum in (7-115) may be approximated by a double integral, yielding

$$\text{var}(r_n^{(1)}) = \frac{\sigma^2}{h^4}\{V(x_n) + O(h)\} \tag{7-116}$$

where

$$V(x) = \int_a^b\int_a^b G(x, \xi)c(\xi, \eta)G(\eta, x)\,d\xi\,d\eta \tag{7-117}$$

If, on the other hand, the random variables are independent, the matrix **C** reduces to a diagonal matrix. If its nonzero elements can be represented in the form $c_{nn} = q(x_n)$, where $q(x)$ is a given nonnegative continuous function of x, then the sum in (7-115) reduces to a simple sum, and we have

$$\text{var}(r_n^{(1)}) = \frac{\sigma^2}{h^3}\{v(x_n) + O(h)\} \tag{7-118}$$

where, in view of (7-106),

$$v(x) = \int_a^b G(x, \xi)^2 q(\xi)\,d\xi \tag{7-119}$$

It does not appear to be possible to define either $V(x)$ or $v(x)$ by a simple differential equation. By virtue of (7-46), the inequality

$$v(x) \le (b - x)^2\int_a^x (x - \xi)^2 q(\xi)\,d\xi + (x - a)^2\int_x^b (b - \xi)^2 q(\xi)\,d\xi \tag{7-120}$$

and a similar inequality for $V(x)$ are readily established.

7.5. Problems and Supplementary Remarks

Section 7.1

1. Solve the boundary value problem

$$y'' = (1 + x^2)y, \qquad -1 < x < 1$$
$$y(-1) = y(1) = 1$$

by the fourth order method (7-10), using the step $h = 0.2$. [By virtue of symmetry, only one-half of the interval needs to be considered.]

2. Solve the boundary value problem

$$y'' = 6x + y^3, \qquad 0 < x < 1$$
$$y(0) = y(1) = 0$$

using the difference operator (7-9) with $h = 0.25$. [Solve the nonlinear difference equation by Newton's method, using as initial approximation the solution of the boundary problem obtained by omitting the term y^3.]

3. Solve, by the Gaussian algorithm as described in §7.1-3, the system of equations $\mathbf{J}\mathbf{y}_n = \mathbf{e}_n$, where $\mathbf{e}_n$ is the nth $(N - 1)$-dimensional unit vector $(n = 1, 2, \cdots, N - 1)$.

4. Solve the system

$$\mathbf{J}\mathbf{y} = \begin{pmatrix} 1 \\ 1 \\ \cdot \\ \cdot \\ \cdot \\ 1 \end{pmatrix}$$

and verify that $\mathbf{y} = \sum_{n=1}^{N-1} \mathbf{y}_n$.

5*. An obvious difference analog for the differential equation

$$y^{\text{iv}} + Ay'' = f(x, y) \qquad (A = \text{const.})$$

is given by the difference equation

$$\nabla^4 y_{n+2} + Ah^2\nabla^2 y_{n+1} - h^4 f_n = 0$$

Improve the accuracy of this approximation by suitably determining the constants α, β, and δ in the difference equation

$$\nabla^4 y_{n+2} + Ah^2\{\nabla^2 y_{n+1} + \alpha\nabla^4 y_{n+2}\} = h^4\{f_n + \beta\nabla^2 f_{n+1} + \delta\nabla^4 f_{n+2}\}$$

6. Extend the algorithm given in §7.1-3 for solving $\mathbf{Ax} = \mathbf{b}$ where $\mathbf{A}$ is tridiagonal to the case where $\mathbf{A}$ is quindiagonal, i.e., $a_{ij} = 0$ for $|i - j| > 2$.

7. Consider the system of boundary value problems

$$\mathbf{y}'' = \mathbf{G}(x)\mathbf{y} + \mathbf{b}(x), \qquad \mathbf{y}(a) = \mathbf{l}, \qquad \mathbf{y}(b) = \mathbf{r}$$

Let every differential equation be replaced by the analog of (7-8). If the $(N-1) \times k$-dimensional compound vector $\mathbf{Y}$ with components $\mathbf{Y}_1, \mathbf{Y}_2, \cdots, \mathbf{Y}_{N-1}$ is introduced, the resulting system of linear equations can be written in the form

$$\mathbf{AY} = \mathbf{B} \tag{7-121}$$

where $\mathbf{B}$ is a certain vector depending on $\mathbf{b}(x)$, $\mathbf{l}$, and $\mathbf{r}$, and where $\mathbf{A}$ denotes the compound matrix

$$\mathbf{A} = \begin{pmatrix} 2\mathbf{I} + h^2\mathbf{G}(x_1) & -\mathbf{I} & \mathbf{O} & \cdots & \mathbf{O} \\ -\mathbf{I} & 2\mathbf{I} + h^2\mathbf{G}(x_2) & -\mathbf{I} & \cdots & \mathbf{O} \\ & \cdot & \cdot & \cdot & \\ & \cdot & \cdot & \cdot & \\ & \cdot & \cdot & \cdot & \\ \mathbf{O} & & & -\mathbf{I} & 2\mathbf{I} + h^2\mathbf{G}(x_{N-1}) \end{pmatrix} \tag{7-122}$$

($\mathbf{I} = k \times k$ unit matrix). Develop an algorithm for the numerical solution of the system (7-121) which works entirely with $k \times k$ block matrices.

8. Prove the analog of Theorem 7.1 for the boundary conditions

$$y'(a) - cy(a) = A, \qquad y'(b) + dy(b) = B \tag{7-123}$$

where $c \geq 0$, $d \geq 0$, $c + d > 0$.

[Use $y(a) = \alpha$ as a parameter and show that $y_\alpha(x, \alpha) \geq 1 + c(x - a)$, $y'_\alpha(x, \alpha) \geq c$.]

Section 7.2

9. (a) Show that the unit matrix is reducible.

(b) Let $\mathbf{A} = (a_{ij})$ and $\mathbf{B} = (b_{ij})$ be two matrices of order n. If $\mathbf{A}$ is irreducible and if $a_{ij} \neq 0$ implies $b_{ij} \neq 0$, then $\mathbf{B}$ is irreducible. If $\mathbf{A}$ is reducible and if $a_{ij} = 0$ implies $b_{ij} = 0$, then $\mathbf{B}$ is reducible.

10. A tridiagonal matrix $\mathbf{A} = (a_{ij})$ is reducible if and only if $a_{i,i-1}a_{i-1,i} = 0$ for some i. Prove this statement both by using the definition and by Theorem 7.2.

11. Let α, β, and $\gamma \neq 0$. Which of the 2×2 matrices

(a) $\begin{pmatrix} \alpha & \beta \\ \gamma & 0 \end{pmatrix}$, (b) $\begin{pmatrix} 0 & \beta \\ \gamma & 0 \end{pmatrix}$, (c) $\begin{pmatrix} 0 & 0 \\ \gamma & 0 \end{pmatrix}$

are reducible?

12. The matrix $\mathbf{A}$ defined by (7-122) is nonsingular provided that all matrices $\mathbf{G}(x_n)$ $(n = 1, 2, \cdots, N-1)$ satisfy the conditions of Theorem 7.4.

13. The matrix of the linear system obtained by discretizing the differential equation $y'' = g(x)y + h(x)$ under the boundary conditions (7-123) for sufficiently small values of h satisfies the hypotheses of Theorem 7.4 and hence is monotone.

14. The equation

$$x^2 - a = 0 \tag{7-124}$$

where $a > 0$, is solved by Newton's method, where $x_0 > 0$.

(a) Show that the hypotheses of Kantorovich's Theorem 7.6 are satisfied.

(b) Verify that the solution $x = \sqrt{a}$ lies on the boundary of the region defined by (7-65).

(c) Verify that the successive approximations satisfy the relation

$$\frac{x_n - \sqrt{a}}{x_n + \sqrt{a}} = \left(\frac{x_0 - \sqrt{a}}{x_0 + \sqrt{a}}\right)^{2^n}$$

and compare the actual error $x_n - \sqrt{a}$ with the estimate given by (7-66).

15. What is the order of convergence of the iteration process for solving $x = f(x)$ described in Theorem 5.4?

16.* *Convergence of the modified Newton's process.* Denote by $\mathbf{z}^{(0)}, \mathbf{z}^{(1)}, \cdots$ the successive approximations obtained in the modified Newton's method for solving $\boldsymbol{\varphi}(\mathbf{y}) = 0$ described by

$$\mathbf{z}^{(\nu+1)} = \mathbf{z}^{(\nu)} - [\mathbf{A}(\mathbf{z}^{(0)})]^{-1}\boldsymbol{\varphi}(\mathbf{z}^{(\nu)}), \quad \nu = 0, 1, 2, \cdots \tag{7-125}$$

[The functional matrix $\mathbf{A}(\mathbf{z})$ is inverted only once.] Assuming that the hypotheses of Theorem 7.6 are satisfied and that $h_0 < \frac{1}{2}$, show that

$$|\mathbf{z}^{(\nu)} - \mathbf{y}^*| \le q^{\nu-1}\,|\mathbf{z}^{(1)} - \mathbf{y}^*| \tag{7-126}$$

where $q = 1 - \sqrt{1 - 2h_0} < 1$, and that consequently $\lim_{\nu\to\infty} \mathbf{z}^{(\nu)} = \mathbf{y}^*$. (See Kantorovich [1948], p. 181.)

17. Verify the estimate (7-126) in the application of the modified Newton's method to (7-124). Can you construct a function $f(x)$ for which the estimate is sharp?

Section 7.3

18. Perform the difference calculation for the boundary value problem $y'' = y - 2, y(0) = 2, y(1) = 0$ analytically, using (7-9), and verify the statement of Theorem 7.9.

19. Perform analytically the difference calculation for the boundary value problem (not of class M) $y'' = y, y(0) = 0, y'(1) = 1$, using the operator (7-9), and determine the order of the error, if $y'(1)$ is approximated by (a) $h^{-1}(y_N - y_{N-1})$; (b) $(2h)^{-1}(y_{N+1} - y_{N-1})$.

20. Improve the numerical values found in Problem 2 by calculating the difference correction. Obtain the required values of y^{vi} by differencing and, if necessary, extrapolating the values of $y'' = (1 + x^2)y$ found previously. Compare the values thus obtained with the sixth derivative of the exact solution.

21*. *Justification of difference correction.* Let $L[y(x);\ h]$ denote the difference operator associated with the difference expression (7-8). If $y(x)$ is sufficiently differentiable and

$$L[y(x);\ h] = C_{p+2}y^{(p+2)}(x)h^{p+2} + C_{p+4}y^{(p+4)}(x)h^{p+4} + O(h^{p+6})$$

show that (7-92) holds with $e_{p+2}(x) = \eta(x)$, when

(a) $p = 2, \beta_0 = \beta_2 = 0$, and $\eta(x)$ is the solution of the boundary value problem

$$\eta'' = g(x)\eta + \tfrac{1}{2}f_{yy}(x, y(x))[e(x)]^2 + C_6 y^{\text{vi}}(x) + C_4 e^{\text{iv}}(x), \qquad \eta(a) = \eta(b) = 0$$

(b) $p = 4$, $\beta_0 = \beta_2$, and $\eta(x)$ is the solution of

$$\eta'' = g(x)\eta - (C_8 + \beta_0 C_6)y^{\text{viii}}(x), \qquad \eta(a) = \eta(b) = 0$$

22. *Inclusion Theorem* (Collatz [1960], p. 200). If for a given boundary value problem of class M two functions $y_1(x)$ and $y_2(x)$ satisfying the boundary conditions can be found such that

$$-y_1''(x) + f(x, y_1(x)) \le 0 \le -y_2''(x) + f(x, y_2(x)), \qquad a \le x \le b$$

then the exact solution $y(x)$ satisfies

$$y_1(x) \le y(x) \le y_2(x), \qquad a \le x \le b$$

Section 7.4

23. Show that for $g(x) = 1$,

$$G(x, \xi) = \frac{\sinh\,[\min\,(x, \xi) - a]\sinh\,[b - \max\,(x, \xi)]}{\sinh\,(b - a)}$$

and, assuming $p(x) = q(x) = 1$,

$$m(x) = 1 - \frac{\cosh\,[\frac{1}{2}(b + a) - x]}{\cosh\,[\frac{1}{2}(b - a)]}$$

$$v(x) = \frac{\sinh\,(x - a)\sinh\,(b - x)}{\sinh\,(b - a)} - \frac{(x - a)\sinh^2(b - x) + (b - x)\sinh^2(x - a)}{2\sinh^2(b - a)}$$

24. Prove relations (7-104).

25. Show that the solution of the boundary value problem $y'' = g(x)y + p(x)$, $y(a) = A$, $y(b) = B$ can be represented in the form

$$y(x) = \frac{A}{t(a)}t(x) + \frac{B}{s(b)}s(x) + \int_a^b G(x, \xi)p(\xi)\,d\xi$$

26. Denote by $G_i(x, \xi)$, $i = 1, 2$, the Green's functions associated with the boundary value problems $y'' = g_i(x)y, y(a) = A, y(b) = B$, where $0 \le g_1(x) \le g_2(x)$, $x \in [a, b]$. Show that $0 \le G_2(x, \xi) \le G_1(x, \xi)$, $x, \xi \in [a, b]$.

Notes

§7.1-2. Other discrete variable methods for solving boundary value problems are proposed by Gavurin [1949], Stüssi [1950], Mikelazde [1953a], Karpilovsaya [1953], Glinskaya and Mysovskih [1954], Adachi [1955], Goodman and Lance [1956], Tihonov and Samarskii [1956], Ridley [1957], Fox [1957]. Sturm-Liouville problems are discussed by Farrington, Gregory, and Taub [1957] and by Wachspress [1960]. Difference equations with nonuniform meshsize are studied by Brown [1960].

§7.1-3. The special form of the Gaussian algorithm used here is discussed (sometimes with minor modifications) by Fox [1957], Anonymous [1957], Douglas [1959], Forsythe and Wasow [1960], p. 103, Wachspress [1960]. The method is easily adapted to the case where the matrix **A** has more than three nonzero diagonals; see Rutishauser [1958], Conte and Dames [1958], and Problem 6.

§7.1-4. Numerical experiments on the use of Newton's method for the solution of difference equations arising from nonlinear boundary value problems are reported by Ehrlich [1960] and Mercer [1960]. For another iterative method see Pope [1960].

§7.2-1. Theorem 7.2 is stated by Varga [1959].

§7.2-2. In the definition of a monotone matrix and some of the discussion we follow the treatment by Collatz [1960], p. 43.

§7.2-4. The proof of Theorem 7.6 is a close adaptation of Kantorovich's original proof (Kantorovich [1948]) to the finite-dimensional case, using a special norm.

§7.3-1. Error bounds for finite difference approximations to solutions of more general boundary value problems may be found in some of the papers listed under §7.1-2 and also in Schröder [1956, 1958]. Inherent pitfalls of finite difference methods for boundary value problems not of class M are pointed out by Mitchell [1959].

§7.4. The propagation of round-off error in the solution of linear systems with tridiagonal matrices is discussed by Douglas [1959], Wachspress [1960], Lowan [1960].

Bibliography

An attempt has been made to give a reasonably complete listing of all relevant titles that appeared in book form or in established periodicals during the last decade. Lecture notes, manuscripts, and unpublished research reports are included if they have come to the author's attention and have been deemed relevant. Papers that appeared prior to 1950 are included if they are, in the author's opinion, of lasting importance. More complete bibliographies for the literature before 1952 are to be found in Milne [1953], and for the literature before 1932, in Bennett, Milne, and Bateman [1956].

Auxiliary references not directly concerned with discrete variable methods in ordinary differential equations are marked with an asterisk.

Adachi, R. [1955]: A method on the numerical solution of some differential equations. *Kumamoto J. Sci. Ser. A*, **2,** 244–252.

*Ahlfors, L. V. [1953]: *Complex Analysis*, McGraw-Hill, New York.

Albrecht, J. [1955]: Beiträge zum Runge-Kutta-Verfahren. *Z. angew. Math. Mech.*, **35,** 100–110.

Alonso, R. [1960]: A starting method for the three-point Adams predictor-corrector method. *J. Assoc. Comp. Mach.*, **7,** 176–180.

Anderson, W. H. [1960]: The solution of simultaneous ordinary differential equations using a general purpose digital computer. *Comm. Assoc. Comp. Mach.*, **3,** 355–360.

Anonymous [1957]: Modern Computing Methods. *Notes Appl. Sci. No.* 16, National Physical Laboratory, London.

Antosiewicz, H. A., and Walter Gautschi [1961]: Numerical methods in ordinary differential equations, to appear in *Survey of Numerical Analysis*, J. Todd (ed.), McGraw-Hill, New York.

Artemov, G. A. [1955]: On a modification of Caplygin's method for systems of ordinary differential equations of the first order. *Dokl. Akad. Nauk SSSR* (N. S.), **101,** 197–200.

Azbelev, N. V. [1952]: On an approximate solution of ordinary differential equations based upon S. A. Caplygin's method. *Dokl. Akad. Nauk SSSR* (N. S.), **83,** 517–519.

——— [1953]: On the limits of applicability of S. A. Caplygin's theorem. *Dokl. Akad. Nauk SSSR* (N. S.), **89,** 589–591.

——— [1955]: On the extension of Caplygin's method beyond the limit of applicability of the theorem on differential inequalities. *Dokl. Akad. Nauk SSSR* (N. S.), **102,** 429–430.

Babkin, B. N. [1948]: Angenäherte Lösung gewöhnlicher Differentialgleichungen beliebiger Ordnung mit der Methode der schrittweisen Annäherungen auf Grund eines Satzes von S. A. Caplygin über Differentialgleichungen. *Dokl. Akad. Nauk SSSR* (2), **59,** 419–422.

——— [1949]: On a modification of a method of S. A. Caplygin for approximate integration. *Dokl. Akad. Nauk. SSSR* (N. S.), **67,** 213–216.

——— [1954]: Approximate integration of systems of ordinary differential equations of the first order by the method of S. A. Caplygin. *Izv. Akad. Nauk SSSR Ser. Mat.*, **18,** 477–484.

Bahvalov, N. S. [1955a]: On the estimation of the error in the numerical integration of differential equations by the Adams extrapolation method. *Dokl. Akad. Nauk SSSR* (N. S.), **104,** 683–686.

——— [1955b]: Some remarks concerning the numerical integration of differential equations by the method of finite differences. *Dokl. Akad. Nauk SSSR* (N. S.), **104,** 805–808.

Balakin, V. B. [1959]: Bilateral approximations to the solution of the equation $y^{(n)} = f(x, y)$. *Ukrain. Mat. Z.*, **11,** 203–207.

Bashforth, F., and J. C. Adams [1883]: *Theories of Capillary Action*, Cambridge Univ. Press.

Bennett, A. A., W. E. Milne, and H. Bateman [1956]: *Numerical Integration of Differential Equations*, Dover, New York.

Bieberbach, L. [1944]: *Theorie der Differentialgleichungen*, Dover, New York.

——— [1951]: On the remainder of the Runge-Kutta formula in the theory of ordinary differential equations. *Z. angew. Math. Physik*, **2,** 233–248.

*Birkhoff, G., and S. MacLane [1953]: *A Survey of Modern Algebra*, rev. ed., Macmillan, New York.

Blanch, G. [1952]: On the numerical solution of equations involving differential operators with constant coefficients. *Math. Tables Aids Comput.*, **6,** 219–223.

——— [1957]: Criteria for the choice of an integration formula. Talk delivered at USE meeting, Dayton.

Blum, E. K. [1957]: A modification of the Runge-Kutta fourth-order method. *Numerical Note N-N 80*, Ramo-Wooldridge Corp., Los Angeles.

Brock, P., and F. J. Murray [1952]: The use of exponential sums in step by step integration. *Math. Tables Aids Comput.*, **6,** 63–78, 138–150.

Brodskii, M. L. [1953]: Asymptotic estimates of the errors in the numerical integration of systems of ordinary differential equations by difference methods. *Dokl. Akad. Nauk SSSR* (N. S.), **93,** 599–602.

Brouwer, D. [1937]: On the accumulation of errors in numerical integration. *Astronomical J.*, **46,** 149–153.

Brown, R. R. [1960]: Solution of boundary value problems using nonuniform grids. Ph.D. dissertation, Univ. California, Los Angeles.

Bückner, H. [1952]: Ueber eine Näherungslösung der gewöhnlichen linearen Differentialgleichung 1. Ordnung. *Z. angew. Math. Mech.*, **22,** 143–152.

Budak, B. M. [1956]: On the method of straight lines for certain boundary problems. *Vestnik Moskov. Univ. Ser. Mat. Meh. Astr. Fiz. Him.*, **11,** 3–12.

——— and A. D. Gorbunov [1959]: Stability of calculation processes in the solution of the Cauchy problem for the equation $dy/dx = f(x, y)$ by multipoint difference methods. *Dokl. Akad. Nauk SSSR*, **124,** 1191–1194.

Bukovics, E. [1950]: Eine Verbesserung und Verallgemeinerung des Verfahrens von Blaess zur numerischen Integration gewöhnlicher Differentialgleichungen. *Oesterreich. Ing.-Archiv*, **4,** 338–349.

——— [1953]: Beiträge zur numerischen Integration, I, II. *Monatshefte der Math.*, **57,** 217–245; **57,** 333–350.

——— [1954]: Beiträge zur numerischen Integration, III. *Monatshefte der Math.*, **58,** 258–265.

Capra, V. [1956]: Nuove formule per l'integrazione delle equazioni differenziali ordinarie del 1° e del 2° ordine. *Univ. e Politec. Torino. Rend. Sem. Mat.*, **16,** 301–359.

——— [1957]: Valutazione degli errori nella integrazione numerica dei systemi di equazioni differenziali ordinarie. *Atti Accad. Sci. Torino Cl. Sci. Fis. Mat. Nat.*, **91,** 188–203.

Carr, J. W. [1957]: Lecture notes on ordinary differential equations. Applications of Advanced Numerical Analysis to Digital Computer Problems, Summer session, Univ. Michigan.

——— [1958]: Error bounds for the Runge-Kutta single-step integration process. *J. Assoc. Comp. Mach.*, **5,** 39–44.

Cauchy, A. [1840]: Mémoire sur l'intégration des équations différentielles. *Oeuvres complètes*, IIe série, tome 11, 399–465.

Cernysenko, E. A. [1958]: On a method of approximate solution of Cauchy's problem for ordinary differential equations. *Ukrain. Mat. Z.*, **10,** 89–100.

Certaine, J. [1960]: The solution of ordinary differential equations with large time constants, in *Mathematical Methods for Digital Computers*, A. Ralston and H. Wilf (eds.), Wiley, New York, pp. 128–132.

Ceschino, F. [1954]: Critère d'utilisation du procédé de Runge-Kutta. *C. R. Acad. Sci. Paris*, **238,** 986–988, 1553–1555.

——— [1956]: L'intégration approchée des équations différentielles. *C. R. Acad. Sci. Paris*, **243,** 1478–1479.

——— and J. Kuntzmann [1958]: Impossibilité d'un certain type de formule d'intégration approchée à pas liés. *Chiffres*, **1,** 95–101.

——— [1960]: Faut-il passer à la forme canonique dans les problèmes différentielles de conditions initiales? *Information Processing*, UNESCO, Paris, pp. 33–36.

*Coddington, E. A., and N. Levinson [1955]: *Theory of Ordinary Differential Equations*, McGraw-Hill, New York.

Collatz, L. [1949]: *Eigenwertaufgaben mit technischen Anwendungen*, Akad. Verlagsgesellschaft, Leipzig.

——— [1953]: Ueber die Instabilität beim Verfahren der zentralen Differenzen für Differentialgleichungen zweiter Ordnung. *Z. angew. Math. Physik*, **4,** 153–154.

——— [1960]: *The Numerical Treatment of Differential Equations*, 3rd ed., Springer, Berlin.

——— and R. Zurmühl [1942]: Zur Genauigkeit verschiedener Integrationsverfahren bei gewöhnlichen Differentialgleichungen. *Ing.-Arch.*, **13,** 34–36.

Conte, S. D. [1958]: Stable operators in the numerical solution of second order differential equations. *Numerical Analysis Note N-N 112*, Space Technology Laboratories, Los Angeles.

——— and R. T. Dames [1958]: An alternating direction method for solving the biharmonic equation. *Math. Tables Aids Comput.*, **12,** 198–204.

Conte, S. D., and R. F. Reeves [1956]: A Kutta third-order procedure for solving differential equations requiring minimum storage. *J. Assoc. Comp. Mach.*, **3,** 22–25.

Cowell, P. H., and A. C. D. Crommelin [1910]: Investigation of the motion of Halley's comet from 1759 to 1910. Appendix to Greenwich Observations for 1909, Edinburgh, p. 84.

*Cramér, H. [1946]: *Mathematical Methods of Statistics*, Princeton Univ. Press.

Dahlquist, G. [1951]: Fehlerabschätzungen bei Differenzenmethoden zur numerischen

Integration gewöhnlicher Differentialgleichungen. *Z. angew. Math. Mech.*, **31**, 239–240.

——— [1956]: Convergence and stability in the numerical integration of ordinary differential equations. *Math. Scand.*, **4**, 33–53.

——— [1959]: Stability and error bounds in the numerical integration of ordinary differential equations. *Trans. Roy. Inst. Technol., Stockholm*, Nr. 130.

Dettmar, H. K., and A. Schlüter [1958]: Praktische Lösung von Eigenwertaufgaben vom Hartree-Fockschen Typus. *Z. Angew. Math. Mech.*, **38**, 220–236.

de Vogelaere, R. [1955]: A method for the numerical integration of differential equations of second order without explicit first derivatives. *J. Res. Nat. Bur. Standards*, **54**, 119–125.

——— [1957]: On a paper of Gaunt concerned with the start of numerical solutions of differential equations. *Z. angew. Math. Physik*, **8**, 151–156.

Douglas, Jim [1956]: On the error in analogue solutions of heat conduction problems. *Quart. J. Appl. Math.*, **14**, 333–335.

——— [1959]: Round-off error in the numerical solution of the heat equation. *J. Assoc. Comput. Mach.*, **6**, 48–58.

Ehrlich, L. [1960]: Experience with numerical methods for a boundary value problem. *Technical Note N-N 141*, Space Technology Laboratories, Los Angeles.

Eltermann, H. [1955]: Fehlerabschätzungen bei näherungsweiser Lösung von Systemen von Differentialgleichungen erster Ordnung. *Math. Z.*, **62**, 469–501.

Euler, L. [1913]: *Opera omnia, series prima*, Vol. 11, Leipzig and Berlin.

——— [1914]: *Opera omnia, series prima*, Vol. 12, Leipzig and Berlin.

Farrington, C. C., R. T. Gregory, and A. H. Taub [1957]: On the numerical solution of Sturm-Liouville differential equations. *Math. Tables Aids Comput.*, **11**, 131–150.

Fehlberg, E. [1958]: Eine Methode zur Fehlerverkleinerung beim Runge-Kutta-Verfahren. *Z. angew. Math. Mech.*, **38**, 421–426.

——— [1960]: Neue genauere Runge-Kutta-Formeln für Differentialgleichungen zweiter Ordnung. *Z. angew. Math. Mech.*, **40**, 252–259.

——— [1961]: Numerisch stabile Interpolationsformeln mit günstiger Fehlerfortpflanzung für Differentialgleichungen erster und zweiter Ordnung. *Z. angew. Math. Mech.*, **41**, 101–110.

*Feller, W. [1957]: *An Introduction to Probability Theory and Its Applications*, 2nd ed., Wiley, New York.

Forsythe, G. E. [1959]: Note on rounding-off errors. *SIAM Rev.*, **1**, 66–67.

——— and W. Wasow [1960]: *Finite-Difference Methods for Partial Differential Equations*, Wiley, New York.

Fort, T. [1948]: *Finite Differences and Difference Equations in the Real Domain*, Oxford, Univ. Press.

Fox, L. [1957]: *The Numerical Solution of Two-Point Boundary Problems in Ordinary Differential Equations*, Oxford Univ. Press.

——— [1960]: Some numerical experiments with eigenvalue problems in ordinary differential equations, in *Boundary Problems in Differential Equations*, R. E. Langer (ed.), Univ. of Wisconsin Press, Madison, pp. 243–255.

——— and A. R. Mitchell [1957]: Boundary-value techniques for the numerical solution of initial value problems in ordinary differential equations. *Quart. J. Mech. Appl. Math.*, **10**, 232–243.

Franklin, J. [1952]: On the method of successive approximations. *Tech. Rept. No. 7*, Appl. Math. and Statist. Lab., Stanford Univ.

——— [1959]: Numerical stability in digital and analog computation for diffusion problems. *J. Math. Phys.*, **37**, 305–315.

Frei, T. [1954]: Anwendung der Momente der Integralkurven zur numerischen Lösung von Differentialgleichungen. *Magyar Tud. Akad. Alkalm. Mat. Int. Közl.*, **2,** 395–414.

Fricke, A. [1949]: Ueber die Fehlerabschätzung des Adamsschen Verfahrens zur Integration gewöhnlicher Differentialgleichungen erster Ordnung. *Z. angew. Math. Mech.*, **29,** 165–178.

Gaier, D. [1956]: Ueber die Konvergenz des Adamsschen Extrapolationsverfahrens. *Z. angew. Main. Mech.*, **36,** 230.

Galler, B. A., and D. P. Rozenberg [1960]: A generalization of a theorem of Carr on error bounds for Runge-Kutta procedures. *J. Assoc. Comp. Mach.*, **7,** 57–60.

Garfinkel, B. [1954]: On the choice of mesh in the integration of ordinary differential equations. *Rept. No. 907*, Ballistics Research Labs., Aberdeen Proving Ground, Md.

Garwick, J. V. [1955]: The solution of boundary-value problems by step-by-step methods. *Arch. Math. Naturw.*, **52,** 1–67.

Gaunt, J. A. [1927]: The deferred approach to the limit, II—Interpenetrating lattices. *Trans. Roy. Soc. Lond.*, **226,** 350–361.

Gautschi, Walter [1955]: Ueber den Fehler des Runge-Kutta-Verfahrens für die numerische Integration gewöhnlicher Differentialgleichungen *n*-ter Ordnung. *Z. angew. Math. Physik.*, **6,** 456–461.

Gavurin, M. K. [1949]: On a method of numerical integration of homogeneous linear differential equations convenient for mechanisation of the computation. *Trudy Mat. Inst. Steklov*, **28,** 152–156.

Gill, S. [1951]: A process for the step-by-step integration of differential equations in an automatic digital computing machine. *Proc. Cambridge Philos. Soc.*, **47,** 96–108.

Glinskaya, N. N., and I. P. Mysovskih [1954]: On the numerical solution of a boundary problem for a non-linear ordinary differential equation. *Vestnik Leningrad Univ.*, **9,** 49–54.

Goldberg, S. [1958]: *Introduction to Difference Equations*, Wiley, New York.

Goodman, T. R., and G. N. Lance [1956]: The numerical integration of two-point boundary value problems. *Math. Tables Aids Comput.*, **10,** 82–86.

Gorn, S., and R. Moore [1953]: Automatic error control—the initial value problem in ordinary differential equations. *Rept. No. 893,* Ballistic Research Labs., Aberdeen Proving Ground, Md.

Gray, H. J., Jr. [1955]: Propagation of truncation errors in the numerical solution of ordinary differential equations by repeated closures. *J. Assoc. Comp. Mach.*, **2,** 5–17.

Hamel, G. [1949]: Zur Fehlerabschätzung bei gewöhnlichen Differentialgleichungen erster Ordnung. *Z. angew. Math. Mech.*, **29,** 337–341.

Hammer, P. C., and T. W. Hollingsworth [1955]: Trapezoidal methods of approximating solutions of differential equations. *Math. Tables Aids Comput.*, **9,** 92–96.

Hamming, R. W. [1959]: Stable predictor-corrector methods for ordinary differential equations. *J. Assoc. Comp. Mach.*, **6,** 37–47.

Hanson, J. W. [1960]: A numerical example of integration in a plane without perturbations using Cowell's method. *Math. Note No.* 3, Computation Center, Univ. of North Carolina.

Hartree, D. R. [1952]: *Numerical Analysis*, Oxford Univ. Press.

Henrici, P. [1960]: Theoretical and experimental studies on the accumulation of error in the numerical solution of initial value problems for systems of ordinary differential equations. *Information Processing*, UNESCO, Paris, pp. 36–44.

——— [1961]: The propagation of round-off error in the numerical solution of initial value problems involving ordinary differential equations of the second order.

Proceedings of a Symposium on the Numerical Treatment of Ordinary Differential Equations, Integral Equations, and Integro-differential Equations, Rome.

Herrick, S. [1951]: Step-by-step integration of $\ddot{x} = f(x, y, z, t)$ without a "corrector." *Math. Tables Aids Comput.*, **5**, 61–67.

Heun, K. [1900]: Neue Methode zur approximativen Integration der Differentialgleichungen einer unabhängigen Veränderlichen. *Z. Math. Physik*, **45**, 23–38.

Hildebrand, F. B. [1956]: *Introduction to Numerical Analysis*, McGraw-Hill, New York.

*Hoel, P. G. [1954]. *Introduction to Mathematical Statistics*, 2nd ed., Wiley, New York.

Hull, T. E., and W. A. J. Luxemburg [1960]: Numerical methods and existence theorems for ordinary differential equations. *Num. Math.*, **2**, 30–41.

Hull, T. E., and A. C. R. Newbery [1959]: Error bounds for a family of three-point integration procedures. *J. Soc. Indust. Appl. Math.*, **7**, 402–412.

——— [1961]: Integration procedures which minimize propagated errors. *J. Soc. Indust. Appl. Math.*, **9**, 31–47.

Huskey, H. D. [1949]: On the precision of a certain procedure of numerical integration. With an appendix by Douglas R. Hartree. *J. Research Nat. Bur. Standards*, **42**, 57–62.

Huta, A. [1956]: Une amélioration de la méthode de Runge-Kutta-Nyström pour la résolution numérique des équations différentielles du premier ordre. *Acta Fac. Nat. Univ. Comenian. Math.*, **1**, 201–224.

——— [1957]: Contribution à la formule de sixième ordre dans la méthode de Runge-Kutta-Nyström. *Acta Fac. Nat. Univ. Comenian. Math.*, **2**, 21–24.

Ionescu, D. V. [1954]: Une généralisation d'une propriété intervenant dans là méthode de Runge-Kutta pour l'intégration numérique des équations différentielles. *Acad. Rep. Populare Române, Bul. Stint. Ser. Mat. Fiz.*, **6**, 229–241.

——— [1956]: Une généralisation d'une propriété intervenant dans la méthode de Runge-Kutta pour l'intégration numérique des équations différentielles. *Acad. Rep. Populare Române, Bul. Stint. Ser. Mat. Fiz.*, **8**, 67–100.

Jackson, J. [1924]: Note on the numerical integration of $d^2x/dt^2 = f(x, t)$. *Roy. Ast. Soc. Monthly Notices*, **84**, 602–606.

Jacobsen, L. S. [1952]: On a general method of solving second-order ordinary differential equations by phase-plane displacements. *J. Appl. Mech.*, **19**, 543–553.

*Kaluza, T. [1928]: Ueber die Koeffizienten reziproker Potenzreihen. *Math. Z.*, **28**, 161–170.

Kamke, E. [1943]: *Differentialgleichungen, Lösungsmethoden und Lösungen*, Akademische Verlagsgesellschaft, Leipzig.

——— [1947]: *Differentialgleichungen reeller Funktionen*, Dover, New York.

Kantorovich, L. V. [1948]: Functional analysis and applied mathematics. *Uspekhi Mat. Nauk*, **3**, 89–185. Translated by C. D. Benster and edited by G. E. Forsythe as Nat. Bur. Standards *Rept. No. 1509*.

Karpilovskaya, E. B. [1953]: On the convergence of an interpolation method for ordinary differential equations. *Uspekhi Mat. Nauk* (N. S.), **8**, 111–118.

Keitel, G. H. [1956]: An extension of Milne's three-point method. *J. Assoc. Comp. Mach.*, **3**, 212–222.

Kopal, Z. [1958]: Operational methods in numerical analysis based on rational approximations. *On Numerical Approximation, Proceedings of a Symposium*, R. E. Langer (ed.), Univ. of Wisconsin Press, Madison, pp. 25–43.

Korganoff, A. [1958]: Sur les formules d'intégration numérique des équations différentielles donnant une approximation d'ordre élevé. *Chiffres*, **1**, 171–180.

Kuntzmann, J. [1953]: Remarques sur la méthode de Runge-Kutta. *C. R. Acad. Sci. Paris*, **242**, 2221–2223.

——— [1959a]: Deux formules optimales du type de Runge-Kutta. *Chiffres*, **2,** 21–26.
——— [1959b]: Evaluation de l'erreur sur un pas dans les méthodes à pas séparés. *Chiffres*, **2,** 97–102.
Kutta, W. [1901]: Beitrag zur näherungsweisen Integration totaler Differentialgleichungen. *Z. Math. Phys.*, **46,** 435–453.
Lax, P. D., and R. D. Richtmyer [1956]: Survey of the stability of linear finite difference equations. *Comm. Pure Appl. Math.*, **9,** 267–293.
Levy, H., and E. A. Baggott [1934]: *Numerical Studies in Differential Equations*, Vol. 1, Watts, London. Reprinted under the title *Numerical Solutions of Differential Equations*, Dover, New York, 1950.
Liniger, W. [1957]: Zur Stabilität der numerischen Integrationsmethoden für Differentialgleichungen. Thesis, Univ. Lausanne.
*Loève, M. [1955]: *Probability Theory. Foundation. Random Sequences.* Van Nostrand, New York.
Lotkin, M. [1951]: On the accuracy of Runge-Kutta's method. *Math. Tables Aids Comput.*, **5,** 128–133.
——— [1952]: A new integration procedure of high accuracy. *J. Math. Physics*, **31,** 29–34.
——— [1954]: The propagation of error in numerical integrations. *Proc. Amer. Math. Soc.*, **5,** 869–887.
——— [1955]: On the improvement of accuracy in integration. *Quart. Appl. Math.*, **13,** 47–54.
——— [1956]: A note on the midpoint rule of integration. *J. Assoc. Comp. Mach.*, **3,** 208–211.
Loud, W. S.[1949]: On the long-run error in the numerical solution of certain differential equations. *J. Math. Phys.*, **28,** 45–49.
Lowan, A. N. [1960]: On the propagation of errors in the inversion of certain tridiagonal matrices. *Math. Comput.*, **14,** 333–338.
Löwdin, P. O. [1952]: On the numerical integration of ordinary differential equations of the first order. *Quart. Appl. Math.*, **10,** 97–111.
Lozinskii, S. M. [1953]: Estimate of the error of an approximate solution of a system of ordinary differential equations. *Dokl. Akad. Nauk SSSR* (N. S.), **92,** 225–228.
Luzin, N. N. [1951]: On the method of approximate integration of academician S. A. Caplygin. *Uspehi Matem. Nauk* (N. S.), **6,** 3–27.
Martin, D. W. [1958]: Runge-Kutta methods for integrating differential equations on high speed digital computers. *The Computer J.*, **1,** 118–123.
Matthieu, P. [1951]: Ueber die Fehlerabschätzung beim Extrapolationsverfahren von Adams. I. Gleichungen 1. Ordnung. *Z. angew. Math. Mech.*, **31,** 356–370.
——— [1953]: Ueber die Fehlerabschätzung beim Extrapolationsverfahren von Adams. II. Gleichungen zweiter und höherer Ordnung. *Z. angew. Math. Mech.*, **33,** 26–41.
Mercer, R. J. [1960]: Boundary value solutions of trajectory problems. *Tech. Note PA*-2399-01/2, Space Technology Laboratories, Los Angeles.
Mikelazde, M. S. [1953a]: Numerical solution of a system of differential equations. Application of the method to the computation of rotating shells. *Akad. Nauk SSSR, Prikl. Mat. Meh.*, **17,** 382–386.
——— [1953b]: Numerical solution of boundary problems for non-linear ordinary differential equations. *Soobsc. Adak. Nauk Gruzin. SSR*, **14,** 133–137.
Mikulaschkova, R. [1957]: Rounding-off error in numerical calculation from the point of view of statistics. *Pokroky Mat. Fys. Astr.*, **2,** 697–707.
*Miller, J. C. P. [1955]: *Tables of Weber Parabolic Cylinder Functions, Introduction.* Her Majesty's Stationery Office, London.

Milne, W. E. [1926]: Numerical integration of ordinary differential equations. *Amer. Math. Monthly*, **33,** 455–460.

——— [1949]: A note on the numerical integration of differential equations. *J. Research Nat. Bur. Standards*, **43,** 537–542.

——— [1950]: Note on the Runge-Kutta method. *J. Research Nat. Bur. Standards*, **44,** 549–550.

——— [1953]: *Numerical Solution of Differential Equations*, Wiley, New York.

——— and R. R. Reynolds [1959]: Stability of a numerical solution of differential equations. *J. Assoc. Comp. Mach.*, **6,** 196–203.

——— [1960]: Stability of a numerical solution of differential equations. II. *J. Assoc. Comp. Mach.*, **7,** 46–56.

Milne-Thomson, L. M. [1933]: *The Calculus of Finite Differences*, Macmillan, London.

Mitchell, A. R. [1959]: The influence of critical boundary conditions on finite difference solutions of two-point boundary value problems. *Math. Tables Aids Comput.*, **13,** 252–260.

——— and T. W. Craggs [1953]: Stability of difference relations in the solution of ordinary differential equations. *Math. Tables Aids Comput.*, **7,** 127–129.

Mohr, E. [1951]: Ueber das Verfahren von Adams zur Integration gewöhnlicher Differentialgleichungen. *Math. Nachr.*, **5,** 209–218.

Moore, R. E. [1960]: Non-linear two point boundary problems with applications to the restricted three-body problem. *Rept. LMSD*-895025, Lockheed Missiles and Space Division, Sunnyvale, Calif.

Morel, H. [1956]: Evaluation de l'erreur sur un pas dans la méthode de Runge-Kutta. *C. R. Acad. Sci., Paris*, **243,** 1999–2002.

Morrison, D., and L. Stoller [1958]: A method for the numerical integration of ordinary differential equations. *Math. Tables Aids Comput.*, **12,** 269–272.

Moulton, F. R. [1926]: *New Methods in Exterior Ballistics*, Univ. Chicago Press.

Muhin, I. S. [1952a]: Application of the Markov-Hermite interpolation polynomials for the numerical integration of ordinary differential equations. *Akad. Nauk SSSR Prikl. Mat. Meh.*, **16,** 231–238.

——— [1952b]: On the accumulation of errors in numerical integration of differential equations. *Akad. Nauk SSSR Prikl. Mat. Meh.*, **16,** 753–755.

Murray, F. J. [1950]: Planning and error considerations for the numerical solution of a system of differential equations on a sequence calculator. *Math. Tables Aids Comput.*, **4,** 133–144.

*National Bureau of Standards [1942]: *Tables of Probability Functions*, New York.

*Nörlund, N. E. [1924]: *Vorlesungen über Differenzenrechnung*, Springer, Berlin.

Noumerov, B. V. [1924]: A method of extrapolation of perturbations. *Roy. Ast. Soc. Monthly Notices*, **84,** 592–601.

Nyström, E. J. [1925]: Ueber die numerische Integration von Differentialgleichungen. *Acta Soc. Sci. Fenn.*, **50,** No. 13, 1–55.

Obrechkoff, N. [1942]: Sur les quadratures méchaniques. *Spisanie Bulgar. Akad. Nauk*, **65,** 191–289.

Oldehoeft, A. E. [1959]: A numerical example in which Euler's method is stable and Milne's method is unstable. *Math. Note No.* 2, Computation Center, Univ. of North Carolina.

——— [1960]: The propagation of error in a predictor-corrector method for the solution of $dy/dx = f(x, y)$. *Mathematical Note No.* 5, Computing Center, Univ. of North Carolina.

Papoulis, A. [1952]: On the accumulation of errors in the numerical solution of differential equations. *J. Appl. Phys.*, **23,** 173–176.

Pope, D. A. [1960]: A method of "alternating corrections" for the numerical solution of two-point boundary value problems. *Math. Comput.*, **14,** 354–361.

Quade, W. [1957]: Numerische Integration von gewöhnlichen Differentialgleichungen durch Interpolation nach Hermite. *Z. angew. Math. Mech.*, **37,** 161–169.

——— [1959]: Ueber die Stabilität numerischer Methoden zur Integration gewöhnlicher Differentialgleichungen erster Ordnung. *Z. angew. Math. Mech.*, **39,** 117–134.

Rademacher, H. [1948]: On the accumulation of errors in processes of integration on high-speed calculating machines. Proceedings of a Symposium on Large-Scale Digital Calculating machinery. *Annals Comput. Labor. Harvard Univ.*, **16,** 176–187.

Ralston, A. [1960]: Numerical integration methods for the solution of ordinary differential equations, in *Mathematical Methods for Digital Computers*, A. Ralston and H. Wilf (eds.), Wiley, New York, pp. 95–109.

Rapoport, I. M. [1952]: A new method of approximate integration of ordinary differential equations. *Ukrain. Mat. Z.*, **4,** 399–413.

Rice, J. R. [1960]: Split Runge-Kutta methods for simultaneous equations. *J. Research Nat. Bur. Standards*, **64B,** 151–170.

Richardson, L. F. [1927]: The deferred approach to the limit, I—Single lattice. *Trans. Roy. Soc. London*, **226,** 299–349.

Richter, W. [1951]: Sur l'erreur commise dans la méthode de Milne. *C. R. Acad. Sci. Paris*, **233,** 1342–1344.

——— [1952]: Estimation de l'erreur commise dans la méthode de M. W. E. Milne pour l'intégration d'un système de *n* équations différentielles du premier ordre. Thesis, Univ. Neuchâtel.

Richtmyer, R. D. [1957]: *Difference Methods for Initial-Value Problems*, Interscience, New York.

Ridley, E. C. [1957]: A numerical method of solving second-order linear differential equations with two-point boundary conditions. *Proc. Cambridge Philos. Soc.*, **53,** 442–447.

Robertson, H. H. [1960]: Some new formulae for the numerical integration of ordinary differential equations. *Information Processing*, UNESCO, Paris, pp. 106–108.

Romanelli, M. J. [1960]: Runge-Kutta methods for the solution of ordinary differential equations, in *Mathematical Methods for Digital Computers*, A. Ralston and H. Wilf (eds.), Wiley, New York, pp. 110–120.

Rothe, E. [1930]: Zweidimensionale parabolische Randwertaufgaben als Grenzfall eindimensionaler Randwertaufgaben. *Math. Ann.*, **102,** 650–670.

Runge, C. [1895]: Ueber die numerische Auflösung von Differentialgleichungen. *Math. Ann.*, **46,** 167–178.

Rutishauser, H. [1952]: Ueber die Instabilität von Methoden zur Integration gewöhnlicher Differentialgleichungen. *Z. angew. Math. Physik*, **3,** 65–74.

——— [1955]: Bemerkungen zur numerischen Integration gewöhnlicher Differentialgleichungen *n*-ter Ordnung. *Z. angew. Math. Physik*, **6,** 497–498.

——— [1958]: Solution of eigenvalue problems with the *LR*-transformation. *Nat. Bur. Standards Appl. Math. Ser.*, **49,** 47–81.

——— [1960]: Bemerkungen zur numerischen Integration gewöhnlicher Differentialgleichungen *n*-ter Ordnung. *Numer. Math.*, **2,** 263–279.

Salzer, H. E. [1956]: Osculatory extrapolation and a new method for numerical integration of differential equations. *J. Franklin Inst.*, **262,** 111–120.

——— [1957]: Numerical integration of $y'' = \varphi(x, y, y')$ using osculatory interpolation. *J. Franklin Inst.*, **263,** 401–409.

Schröder, J. [1956]: Ueber das Differenzenverfahren bei nichtlinearen Randwertaufgaben. I. *Z. angew. Math. Mech.*, **36,** 319–331; II. 443–455.

——— [1958]: Fehlerabschätzungen bei gewöhnlichen und partiellen Differentialgleichungen. *Arch. Rational Mech. Anal.*, **2,** 367–392.

——— [1961a]: Fehlerabschätzungen mit Rechenanlagen bei gewöhnlichen Differentialgleichungen erster Ordnung. *Num. Math.*, **3,** 39–61.

——— [1961b]: Verbesserung einer Fehlerabschätzung für gewöhnliche Differentialgleichungen erster Ordnung. *Numer. Math.*, **3,** 125–130.

Sconzo, P. [1954]: Formule d'estrapolazione per l'integrazione numerica delle equazioni differenziali ordinarie. *Boll. Un. Mat. Ital.* (3), **9,** 391–399.

Serrais, F. [1956]: Sur l'estimation des erreurs dans l'intégration numérique des équations différentielles linéaires du second ordre. *Ann. Soc. Sci. Bruxelles, Sér. I,* **70,** 5–8.

Sheldon, J. W., B. Zondek, and M. Friedman [1957]: On the time-step to be used for the computation of orbits by numerical integration. *Math. Tables Aids Comput.*, **11,** 181–189.

Shura-Bura, M. R. [1952]: Estimates of errors of numerical integration of ordinary differential equations. *Akad. Nauk SSSR Prikl. Mat. Meh.*, **16,** 575–588.

Smith, O. K. [1959]: The inverse matrix of solutions of the variational equations. *Tech. Note PA-1951-16/2*, Space Technology Laboratories, Los Angeles.

Stafford, H. R. [1959]: On linear differential and difference equations. M.A. Thesis, Univ. of Kansas.

Sterne, T. E. [1953]: The accuracy of numerical solutions of ordinary differential equations. *Math. Tables Aids Comput.*, **7,** 159–164.

Stohler, K. [1943]: Eine Vereinfachung bei der numerischen Integration gewöhnlicher Differentialgleichungen. *Z. angew. Math. Mech.*, **23,** 120–122.

Störmer, C. [1907]: Sur les trajectoires des corpuscules électrisés. *Arch. sci. phys. nat., Genève*, **24,** 5–18, 113–158, 221–247.

——— [1921]: Méthodes d'intégration numérique des équations différentielles ordinaires. *C. R. congr. intern. math., Strasbourg*, pp. 243–257.

Stüssi, F. [1950]: Numerische Lösung von Randwertproblemen mit Hilfe der Seilpolygongleichung. *Z. angew. Math. Mech.*, **1,** 53–70.

*Taylor, A. E. [1955]: *Advanced Calculus*, Ginn, New York.

Tihonov, A. N., and A. A. Samarskii [1956]: On finite difference schemes for equations with discontinuous coefficients. *Dokl. Akad. Nauk SSSR* (N. S.), **108,** 393–396.

Todd, J. [1950]: Notes on numerical analysis, I. Solution of differential equations by recurrence relations. *Math. Tables Aids Comput.*, **4,** 39–44.

Tollmien, W. [1938]: Ueber die Fehlerabschätzung beim Adamsschen Verfahren zur Integration gewöhnlicher Differentialgleichungen. *Z. angew. Math. Mech.*, **18,** 83–90.

——— [1953]: Bemerkung zur Fehlerabschätzung beim Adamsschen Interpolationsverfahren. *Z. angew. Math. Mech.*, **33,** 151–155.

Uhlmann, W. [1957a]: Fehlerabschätzungen beim Anfangswertproblem gewöhnlicher Differentialgleichungssysteme 1. Ordnung. *Z. angew. Math. Mech.*, **37,** 88–99.

——— [1957b]: Fehlerabschätzung bei Anfangswertaufgaben einer gewöhnlichen Differentialgleichung höherer Ordnung. *Z. angew. Math. Mech.*, **37,** 99–111.

Urabe, M. [1960]: Theory of errors in numerical integration of ordinary differential equations. *Tech. Summary Rept. No. 183*, U.S. Army Mathematics Research Center, Madison, Wis.

——— [1961]: Theory of errors in numerical integration of ordinary differential equations. *J. Sci. Hiroshima Univ., Ser.* A-1, **25,** 3–62.

——— and S. Mise [1955]: A method of numerical integration of analytic differential equations. *J. Sci. Hiroshima Univ. Ser. A*, **19,** 307–320.

Urabe, M., and T. Tsushima [1953]: On numerical integration of ordinary differential equations. *J. Sci. Hiroshima Univ. Ser. A*, **17,** 193–219.

Urabe, M., and H. Yanagihara [1954]: On numerical integration of the differential equation $y^{(n)} = f(x, y)$. *J. Sci. Hiroshima Univ. Ser. A*, **18,** 55–76.

*van der Waerden, B. [1950]: *Modern Algebra*, Vol. I, translated by T. J. Benac, F. Ungar Pub. Co., New York.

van Wijngaarden, A. [1953]: Erreurs d'arrondiment dans les calculs systématiques. Les machines à calculer et la pensée humaine, *Colloq. intern. centre nat. recherche sci., Paris*, no. 37, pp. 285–293.

Varga, R. S. [1959]: *Iterative Numerical Analysis*, Carnegie Institute of Technology, Pittsburgh.

Vietoris, L. [1953a]: Der Richtungsfehler einer durch das Adamssche Interpolationsverfahren gewonnenen Näherungslösung einer Gleichung $y' = f(x, y)$. *Oesterr. Akad. Wiss., Math.-naturw. Kl., Sitzber., Abt. IIa*, **162,** 157–167.

——— [1953b]: Der Richtungsfehler einer durch das Adamssche Interpolationsverfahren gewonnenen Näherungslösung eines Systems von Gleichungen $y'_k = f_k(x, y_1, \cdots, y_m)$. *Oesterr. Akad. Wiss. Math.-naturw. Kl., Abt. IIa*, **162,** 293–299.

Vlasov, I. O., and I. A. Čarnyi [1950]: On a method of numerical integration of ordinary differential equations. *Akad. Nauk SSSR Inženernyĭ Sbornik*, **8,** 181–186.

von Mises, R. [1930]: Zur numerischen Integration von Differentialgleichungen. *Z. angew. Math. Mech.*, **10,** 81–92.

von Oppolzer, T. R. [1880]: *Lehrbuch zur Bahnbestimmung der Kometen und Planeten*, Vol. 2, W. Engelmann, Leipzig.

von Sanden, H. [1945]: *Praxis der Differentialgleichungen*, 3rd. ed., de Gruyter, Berlin.

Vorobev, L. M. [1956]: The applicability of S. A. Caplygin's method of approximate integration to a certain class of ordinary nonlinear differential equations of the second order. *Uspehi Mat. Nauk* (N. S.), **11,** 181–185.

Voronovskaya, E. V. [1955]: On an alteration of Caplygin's method for differential equations of the first order. *Prikl. Mat. Meh.*, **19,** 121–126.

Wachspress, E. L. [1960]: The numerical solution of boundary value problems, in *Mathematical Methods for Digital Computers*, A. Ralston and H. Wilf (eds.), Wiley, New York, pp. 121–127.

Wall, D. D. [1956]: Note on predictor-corrector formulas. *Math. Tables Aids Comput.*, **10,** 167.

Warga, J. [1953]: On a class of iterative procedures for solving normal systems of ordinary differential equations. *J. Math. Physics*, **31,** 223–243.

Wasow, W. [1955]: Discrete approximations to elliptic differential equations. *Z. angew. Math. Physik*, **6,** 81–97.

Weissinger, J. [1950]: Eine verschärfte Fehlerabschätzung zum Extrapolationsverfahren von Adams. *Z. angew. Math. Mech.*, **30,** 356–363.

——— [1952]: Eine Fehlerabschätzung für die Verfahren von Adams und Störmer. *Z. angew. Math. Mech.*, **32,** 62–67.

——— [1953]: Numerische Integration impliziter Differentialgleichungen. *Z. angew. Math. Mech.*, **33,** 63–65.

Wilf, H. S. [1957]: An open formula for the numerical integration of first order differential equations. *Math. Tables Aids Comput.*, **11,** 201–203.

——— [1959]: A stability criterion for numerical integration. *J. Assoc. Comput. Mach.*, **6,** 363–365.

——— [1960]: Maximally stable numerical integration. *J. Soc. Ind. Appl. Math.*, **8,** 537–540.

Young, D. M. [1955]: Gill's method for solving ordinary differential equations. *Numerical Note N-N 4*, Ramo-Wooldridge Corp., Los Angeles.

——— [1957]: Review of Wall [1956]. *Math. Rev.*, **18**, 336.

Zadiraka, K. V. [1951]: Solution by the method of S. A. Caplygin of linear differential equations of the 2nd order with variable coefficients. *Dopovidi Akad. Nauk Ukrain. RSR*, **1951**, 163–170.

——— [1952]: The approximate integration by S. A. Caplygin's method of linear differential equations of the 2nd order with variable coefficients. *Ukrain. Mat. Z.*, **4**, 299–311.

——— [1955]: The construction of upper and lower bounds for the eigenvalues of one-dimensional self-adjoint boundary value problems of even order. *Doklady Akad. Nauk SSSR*, **102**, 681–684.

——— and I. B. Pogrebiskii [1950]: On the application of S. A. Caplygin's method of approximate integration of ordinary differential equations to a boundary problem. *Dopovidi Akad. Nauk Ukrain. RSR*, **1950**, 95–100.

Zondek, B., and J. W. Sheldon [1959]: On the error propagation in Adams' extrapolation method. *Math. Tables Aids Comput.*, **13**, 52–55.

Zurmühl, R. [1948]: Runge-Kutta Verfahren zur numerischen Integration von Differentialgleichungen *n*-ter Ordnung. *Z. Angew. Math. Mech.*, **28**, 173–182.

——— [1952]: Runge-Kutta Verfahren unter Verwendung höherer Ableitungen. *Z. angew. Math. Mech.*, **32**, 153–154.

Index